Property of
American Plastics Council
Automotive Learning Center

TP 1142 .P56 1998

Plastics additives

Plastics Additives

POLYMER SCIENCE AND TECHNOLOGY SERIES

Series editors

Dr Derek Brewis
Inst. of Surface Science & Technology
Loughborough University of
 Technology
Loughborough, Leicestershire
LE11 3TU

Professor David Briggs
Siacon Consultants Ltd
21 Wood Farm Road
Malvern Wells
Worcestershire
WR14 4PL

Advisory board

Professor A. Bantjes
University of Twente
Faculty of Chemical Technology
Department of Macromolecular
 Chemistry and Materials Science
PO Box 217, 7500 AE Enschede
The Netherlands

Dr John R. Ebdon
The Polymer Centre
School of Physics and Chemistry
Lancaster University
Lancaster LA1 4YA
UK

Professor Richard Pethrick
Department of Pure and Applied
 Chemistry
Strathclyde University
Thomas Graham Building
295 Cathedral Street
Glasgow G1 1XL
UK

Dr Chi-Ming Chan
Department of Chemical Engineering
The Hong Kong University of Science
 and Technology
Room 4558, Academic Building
Clear Water Bay, Kowloon
Hong Kong

Professor Robert G. Gilbert
School of Chemistry
University of Sydney
New South Wales 2006
Australia

Dr John F. Rabolt
Materials Science Program
University of Delaware
Spencer Laboratory #201
Newark, Delaware 19716
USA

JOIN US ON THE INTERNET VIA WWW, GOPHER, FTP OR EMAIL:

WWW: http://www.thomson.com
GOPHER: gopher.thomson.com
FTP: ftp.thomson.com
EMAIL: findit@kiosk.thomson.com

A service of I(T)P®

Plastics Additives

An A–Z reference

Edited by

Geoffrey Pritchard

Consultant
Worcester
UK

and

Emeritus Professor of Polymer Science
Kingston University
Surrey
UK

CHAPMAN & HALL
London · Weinheim · New York · Tokyo · Melbourne · Madras

Published by Chapman & Hall, 2–6 Boundary Row, London SE1 8HN, UK

Chapman & Hall, 2–6 Boundary Row, London SE1 8HN, UK

Chapman & Hall GmbH, Pappelallee 3, 69469 Weinheim, Germany

Chapman & Hall USA, 115 Fifth Avenue, New York, NY 10003, USA

Chapman & Hall Japan, ITP-Japan, Kyowa Building, 3F, 2-2-1 Hirakawacho, Chiyoda-ku, Tokyo 102, Japan

Chapman & Hall Australia, 102 Dodds Street, South Melbourne, Victoria 3205, Australia

Chapman & Hall India, R. Seshadri, 32 Second Main Road, CIT East, Madras 600 035, India

First edition 1998

© 1998 Chapman & Hall

Typeset in 10/12 pt Palatino by Academic & Technical, Bristol
Printed in Great Britain by T J International Ltd, Padstow, Cornwall

ISBN 0 412 72720 X

Apart from any fair dealing for the purposes of research or private study, or criticism or review, as permitted under the UK Copyright Designs and Patents Act, 1988, this publication may not be reproduced, stored, or transmitted, in any form or by any means, without the prior permission in writing of the publishers, or in the case of reprographic reproduction only in accordance with the terms of the licences issued by the Copyright Licensing Agency in the UK, or in acccordance with the terms of licences issued by the appropriate Reproduction Rights Organization outside the UK. Enquiries concerning reproduction outside the terms stated here should be sent to the publishers at the London address printed on this page.
 The publisher makes no representation, express or implied, with regard to the accuracy of the information contained in this book and cannot accept any legal responsibility or liability for any errors or omissions that may be made.

A catalogue record for this book is available from the British Library.

∞ Printed on permanent acid-free text paper, manufactured in accordance with ANSI/NISO Z39.48-1992 and ANSI/NISO Z39.48-1984 (Permanence of Paper).

Contents

The items without reference numbers are not entries. They are cross-referenced to relevant entries.

List of contributors	xiii
Preface	xix

Introduction

Additives are essential	3
Quick reference guide	11
Practical methods of mixing additives with polymers	16
Analytical methods for additives in plastics	26
Biodegradation of plastics: monitoring what happens	32

Alphabetical section

Acid scavengers for polyolefins	43
Acrylic processing aids – see Processing aids for vinyl foam	
Algicides – see Biocides *and* Biocides: some kinetic aspects	
Alumina trihydrate – see Flame retardants: inorganic oxide and hydroxide systems *and* Smoke suppressants *and* Flame retardancy: the approaches available	
Aluminium flakes – see Conducting fillers for plastics	
Amphiphiles – see Surfactants	
Analysis of additives – see Analytical methods for additives in plastics, in Introductory section	
Anti-blocking of polymer films	49
Antifouling agents – see Biocides	
Antimicrobial agents – see Biocides *and* Biocides: some kinetic aspects	
Antimony trioxide – see Flame retardants: inorganic oxide and hydroxide systems *and* Flame retardancy: the approaches available	

Antioxidants: an overview	55
Antioxidants: hindered phenols	73
Antioxidants: their analysis in plastics	80
Antioxidants for poly(ethylene terephthalate)	95
Antistatic agents	108
Aramid fibres – see Reinforcing fibres	
Asbestos – see Reinforcing fibres	
Azo compounds – see Dyes for the mass coloration of plastics *and* Pigments for plastics	
Azobisisobutyronitrile – see Blowing agents	
Bacteriocides – see Biocides *and* Biocides: some kinetic aspects	
Biocides	115
Biocides: some kinetic aspects	121
Biodegradation monitoring – see Biodegradation of plastics: monitoring what happens, in Introductory section	
Biodegradation promoters for plastics	135
Blowing agents	142
Borates – see Flame retardants: borates *and* Flame retardancy: the approaches available	
Boron compounds – see Flame retardants: borates *and* Flame retardancy: the approaches available	
Calcium carbonate	148
Carbon black	153
Carbon fibres – see Reinforcing fibres	
Cellulose – see Flame retardants: intumescent systems	
Charring additives – see Flame retardancy: the approaches available *and* Flame retardants: intumescent systems *and* Flame retardants: iron compounds, their effect on fire and smoke in halogenated polymers *and* Flame retardants: poly(vinyl alcohol) and silicon compounds *and* Flame retardants: tin compounds	
Chlorofluorocarbons – see Blowing agents	
Compatibilizers – see Surfactants: applications in plastics *and* Compatibilizers for recycled polyethylene	
Compatibilizers for recycled polyethylene	162
Conducting fillers for plastics: (1) Flakes and fibres	170
Conducting fillers for plastics: (2) Conducting polymer additives	180
Core-shell modifiers – see Impact modifiers: (1) Mechanisms and applications in thermoplastics *and* Impact modifiers: (2) Modifiers for engineering thermoplastics	
Coupling agents	189
Curing agents	197
Diluents and viscosity modifiers for epoxy resins	211
Dispersing agents – see Surface treatments for particulate fillers in plastics	

Dyes for the mass coloration of plastics	217
EMI shielding additives – see Conducting fillers for plastics (1) and (2) *and* Antistatic agents	
Ester lubricants – see Lubricating systems for rigid PVC	
Extenders – see Diluents and viscosity modifiers for epoxy resins	
Fibres: the effect of short glass fibres on the mechanical properties of thermoplastics	226
Fillers	241
Fillers: their effect on the failure modes of plastics	252
Flame retardancy: the approaches available	260
Flame retardants: borates	268
Flame retardants: halogen-free systems (including phosphorus additives)	277
Flame retardants: inorganic oxide and hydroxide systems	287
Flame retardants: intumescent systems	297
Flame retardants: iron compounds, their effect on fire and smoke in halogenated polymers	307
Flame retardants: poly(vinyl alcohol) and silicon compounds	315
Flame retardants: synergisms involving halogens	327
Flame retardants: tin compounds	339
Fluorescent pigments – see Pigments for plastics	
Foam control agents – see Surfactants: applications in plastics	
Fungicides – see Biocides *and* Biocides: some kinetic aspects	
Glass fibres – see Reinforcing fibres *and* Fibres: the effects of short glass fibres on the mechanical properties of thermoplastics	
Glass beads – see Fillers	
Glass spheres – see Hollow microspheres	
HALS – see Hindered amine light stabilizers: introduction *and* Hindered amine light stabilizers: recent developments	
Heat stabilizers – see Surfactants: applications in plastics	
Hindered amine light stabilizers: introduction	353
Hindered amine light stabilizers: recent developments	360
Hollow microspheres	372
Hydrocarbon waxes – see Release agents *and* Lubricating systems for rigid PVC	
Hydrofluorocarbons – see Blowing agents	
Hydrochlorofluorocarbons – see Blowing agents	
Hydrotalcites – see Acid scavengers for polyolefins	
Impact modifiers: (1) Mechanisms and applications in thermoplastics	375
Impact modifiers: (2) Modifiers for engineering thermoplastics	386
Impact modifiers: (3) Their incorporation in epoxy resins	398
Impact modifiers: (4) Organic toughening agents for epoxy resins	406

viii   *Contents*

Impact modifiers: (5) Modifiers for unsaturated polyester and vinyl ester resins	416
Intumescent additives – see Flame retardants: intumescent systems *and* Flame retardancy: the approaches available *and* Flame retardants: halogen-free systems (including phosphorus additives)	
Iron compounds – see Flame retardants: iron compounds, their effect on fire and smoke in halogenated polymers *and* Flame retardancy: the approaches available	
Kaolin – see Fillers	
Lactates – see Acid scavengers for polyolefins	
Light and UV stabilization of polymers	427
Liquid rubber toughening agents – see Impact modifiers	
Low profile additives in thermoset composites	442
Lubricants – see Release agents *and* Surfactants: applications in plastics *and* Lubricating systems for rigid PVC	
Lubricating systems for rigid PVC	450
Magnesium hydroxide – see Flame retardants: inorganic oxide and hydroxide systems *and* Fillers	
Maleic anhydride – see Fibres: the effects of short glass fibres on the mechanical properties of thermoplastics *and* Coupling agents	
Melamine – see Flame retardants: halogen-free systems	
Metal deactivators – see Antioxidants: an overview *and* Antioxidants for poly(ethylene terephthalate)	
Metal flakes and fibres – see Conducting fillers for plastics: (1) Flakes and fibres *and* Reinforcing fibres *and* Fillers	
Metallic soaps – see Lubricating systems for rigid PVC	
Mica	459
Miscibility – see Polymer additives: the miscibility of blends	
Mixing – see Practical methods of mixing additives with polymers, in Introductory section	
Molybdenum trioxide – see Flame retardants: inorganic oxide and hydroxide systems *and* Smoke suppressants *and* Flame retardancy: the approaches available	
Nickel fibres – see Conducting fillers for plastics (1)	
Nucleating agents for thermoplastics	464
Optical brighteners	472
Organometallic esters – see Diluents and viscosity modifiers for epoxy resins	
Organophosphates – see Nucleating agents for thermoplastics	
Paper for resin bonded paper laminates	474
Pearlescent pigments – see Pigments for plastics	
Phenol compounds – see Antioxidants: hindered phenols *and* Light and UV stabilization of polymers	

Phosphates – see Surface treatments for particulate fillers in plastics	
Phosphite esters – see Antioxidants: an overview *and* Recycled plastics: additives and their effects on properties	
Phosphorus compounds for flame retardancy – see Flame retardants: halogen-free systems *and* Flame retardancy: the approaches available	
Photochromic compounds – see Dyes for the mass coloration of plastic	
Photostabilizers – see Light and UV stabilization of polymers	
Phthalate esters – see Plasticizers	
Phthalocyanines – see Light and UV stabilization of polymers *and* Pigments for plastics	
Pigments for plastics	485
Piperidine compounds – see Light and UV stabilization of polymers	
Plasticizers	499
Plasticizers: health aspects	505
Polyacrylates – see Impact modifiers: (1) Mechanisms and applications in thermoplastics	
Polyester and polyamide fibres – see Reinforcing fibres	
Polyetherimides – see Impact modifiers: (4) Organic toughening agents for epoxy resins	
Polyethersulfones – see Impact modifiers: (4) Organic toughening agents for epoxy resins	
Polyethylene fibres – see Reinforcing fibres	
Polymer additives – the miscibility of blends	513
Polypropylene fibres – see Reinforcing fibres	
Polysiloxanes – see Impact modifiers: (1) Mechanisms and applications in thermoplastics	
Poly(vinyl alcohol) – see Release agents *and* Flame retardants: poly(vinyl alcohol) and silicon compounds	
Precipitating elastomers – see Impact modifiers: (4) Organic toughening agents for epoxy resins	
Processing aids: fluoropolymers to improve the conversion of polyolefins	519
Processing aids for vinyl foam	526
Quartz – see Fillers	
Reactive diluents – see Diluents and viscosity modifiers for epoxy resins	
Recycled plastics: additives and their effect on properties	535
Reinforcing fibres	544
Release agents	559
Rice husk ash	561

Contents

Rubber additives – see Impact modifiers: (1) Mechanisms and applications in thermoplastics *and* Surface-modified rubber particles for polyurethanes

Scorch inhibitors for flexible polyurethanes ... 567

Separation of additives – see Analytical methods for additives in plastics, in Introductory section

Shrinkage control agents – see Low profile additives in thermoset composites

Silanes – see Surface treatments for particulate fillers in plastics *and* Coupling agents

Silica – see Fillers

Silicon compounds, silicates – see Fillers *and* Flame retardants: poly(vinyl alcohol) and silicon compounds *and* Coupling agents

Smoke suppressants ... 576

Stabilizers – see Hindered amine light stabilizers: introduction *and* Hindered amine light stabilizers: recent developments *and* Antioxidants: an overview *and* Light and UV stabilization of polymers

Stainless steel fibres – see Conducting fillers for plastics

Stannates – see Flame retardants: tin compounds *and* Flame retardancy: the approaches available *and* Flame retardants: inorganic oxide and hydroxide systems

Starch – see Biodegradation promoters for plastics

Stearates – see Acid scavengers for polyolefins *and* Release agents *and* Lubricating systems for rigid PVC

Surface-modified rubber particles for polyurethanes ... 584
Surface treatments for particulate fillers in plastics ... 590
Surfactants: applications in plastics ... 604
Surfactants: the principles ... 613

Talc – see Fillers

Thermoplastics toughening modifiers – see Impact modifiers: (3) Their incorporation in epoxy resins

Thickening agents for sheet moulding compounds – see Low profile additives in thermoset composites

Thixotropic agents – see Fillers

Thioesters – see Antioxidants for poly(ethylene terephthalate)

Tin oxide – see Flame retardants: inorganic oxide and hydroxide systems *and* Flame retardants: tin compounds

Titanates – see Surface treatments for particulate fillers in plastics

Titanium dioxide – see Pigments for plastics *and* Fillers

Toughening agents – see Impact modifiers

Trimellitate esters – see Plasticizers

Ultraviolet light stabilizers – see Light and UV stabilization of polymers

Viscosity modifiers – see Diluents and viscosity modifiers for epoxy resins
Vitamin E – see Antioxidants: an overview
Wetting agents – see Surfactants: applications in plastics
Wollastonite – see Fillers
Zinc compounds (e.g. stannates) – see Flame retardants: inorganic oxide and hydroxide systems *and* Flame retardants: tin compounds *and* Smoke suppressants *and* Flame retardancy: the approaches available
Zinc oxide – see Acid scavengers for polyolefins
Zirconates – see Surface treatments for particulate fillers in plastics

Index   625

List of contributors

John Accorsi and Michael Yu
Cabot Corporation, Special Blacks Division, 157 Concord Road, Billerica, MA 01821, USA

M.Y. Ahmad Fuad and Z. Ismail
Plastics Technology Center, SIRIM, PO Box 7035, 40911 Shah Alam, Malaysia

N.S. Allen
Department of Chemistry, Manchester Metropolitan University, Chester Street, Manchester, UK

S. Al-Malaika
Polymer Processing and Performance Group, Aston University, Birmingham B4 7ET, UK

Anthony L. Andrady
Camille Dreyfus Laboratory, Research Triangle Institute, Durham, NC 27709, USA

Kenneth E. Atkins
Union Carbide Corporation, Technical Center, R & D, PO Box 8361, South Charleston, NC 25303, USA

Nadka Avramova
University of Sofia, Faculty of Chemistry, 1126 Sofia, Bulgaria

Asoka J. Bandara
Faculty of Science, Kingston University, Penrhyn Road, Kingston upon Thames, Surrey KT1 2EE, UK

Bernard D. Bauman
Composite Particles, Inc., 2330 26th Street S.W., Allentown, PA 18103, USA

S. Bazhenov
Institute of Chemical Physics, Kosygin Street 4, 117977 Moscow, Russia

Lynn A. Bente
Keystone Aniline Corporation, 121 W 17th Street, Dover, OH 44622, USA

Donald M. Bigg
R.G. Barry Corporation, Columbus, Ohio, USA

Thomas J. Blong
Dyneon L.L.C., St Paul, MN 55144-1000, USA

C.C. Briggs
Microfine Minerals Ltd, Raynesway, Derby DE21 7BE, UK

S.C. Brown
Alcan Chemicals, Alcan Laboratories, Southam Road, Banbury, Oxon OX16 7SP, UK

D.F. Cadogan
European Council for Plasticisers and Intermediates (ECPI), Avenue E. van Nieuwenhuyse 4, bte 2, Auderghem, B-1160 Brussels, Belgium

Giovanni Camino
Dipartimento di Chimica IFM dell'Università, Via P. Giuria, 10125 Torino, Italy

Peter Carty
Department of Chemical and Life Sciences, University of Northumbria at Newcastle, Newcastle upon Tyne NE7 7XA, UK

Robert M. Christie
Dominion Colour Corporation, 199 New Toronto Street, Toronto, Ontario M8V 2E9, Canada

J.H. Clint
School of Chemistry, The University of Hull, Hull HU6 7RX, UK

Roger W. Crecely and Charles E. Day
Brandywine Research Laboratory, Inc., 226 West Park Place, Newark, DE 19711, USA

C.A. Cruz, Jr
Plastics Additives Research Department, Rohm & Haas Company, PO Box 219, Bristol, PA 19007, USA

P.A. Cusack
ITRI Ltd, Kingston Lane, Uxbridge, Middlesex UB8 3PJ, UK

List of contributors

John Davis
Albright and Wilson UK Ltd, International Technical Centre, PO Box 800, Trinity Street, Oldbury, Warley, West Midlands B69 4LN, UK

Ed Feltham
W.R. Grace and Co., PO Box 2117, Baltimore, Maryland 21203-2117, USA

Koen Focquet
Dyneon* N.V., B-2070 Zwijndrecht, Belgium (*A 3-MHoechst enterprise)

Marianne Gilbert
Institute of Polymer Technology and Materials Engineering, Loughborough University, Loughborough, Leicestershire LE11 3TU, UK

Robert L. Gray and Robert E. Lee
Great Lakes Chemical Corporation, PO Box 2200, West Lafayette, Indiana 47906, USA

Roberto Greco
Institute of Research and Technology of Plastic Materials of National Research Council of Italy, Via Toiano 6, Arco Felipe, Naples, Italy

G.J.L. Griffin
Ecological Materials Research Group, Epson Industries Ltd., Units 4–6, Ketton Business Estate, Stamford, Lincs PE9 3SZ, UK

K.Z. Gumargaliva and G.E. Zaikov
Russian Academy of Sciences, Kosygin Street 4, 177977 Moscow, Russia

P.S. Hope
BP Chemicals, Applied Technology, Grangemouth, Scotland FK3 9XH, UK

C.J. Howick
European Vinyls Corporation (UK) Ltd, Technical Services Department, PO Box 8, The Heath, Runcorn, Cheshire WA7 4QD, UK

G.P. Karayannidis, I.D. Sideridou and D.X. Zamboulis
Aristotle University of Thessaloniki, Department of Chemistry, GR-54006 Thessaloniki, Greece

Harutun G. Karian, Hidetomo Imajo and Robert W. Smearing
Thermofil, Inc., 815 N 2nd St., Brighton, Michigan 48116, USA

Francesco Paolo La Mantia
Ingegneria Chimica Processi dei Materiali, Università di Palermo, Viale delle Scienze, 90128 Palermo, Italy

List of contributors

Ján Malik and Gilbert Ligner
Clairant Huningue S.A., Avenue de Bale, BP 149, F-68331 Huningue, France

Ronald L. Markezich
Occidental Chemical Corporation, Technology Center, Grand Island, New York 14072, USA

J.E. McIntyre
Department of Textile Industries, The University of Leeds, Leeds LS2 9JT, UK

Z.A. Mohd Ishak and A.K. Mohd Omar
School of Industrial Technology, Universiti Sains Malaysia, 11800 Penang, Malaysia

Salvatore J. Monte
Kenrich Petrochemicals, Inc., Box 32, Bayonne, NJ 07002-0032, USA

Roderick O'Connor
Borax Europe Ltd, Guildford, Surrey GU2 5RQ, UK

Richard G. Ollila
Transmet Corporation, 4290 Perimeter Drive, Columbus, OH 43228, USA

John Patterson
Rohm & Haas Company, Plastics Additives Applications Laboratory, Bristol, PA 19007, USA

Raymond A. Pearson
Lehigh University, Materials Science and Engineering Department, Bethlehem, PA 18015-3195, USA

R.J. Porter
Devon Valley Industries, Devon Valley Mill, Hele, Exeter, Devon EX5 4PJ, UK

Jan Pospíšil
Institute of Macromolecular Chemistry, Academy of Sciences of the Czech Republic, 16206 Prague, Czech Republic

Geoffrey Pritchard
York House, Moseley Road, Hallow, Worcester WR2 6NH, UK

Robert A. Shanks and Bill E. Tiganis
Applied Chemistry, CRC for Polymer Blends, RMIT University, Melbourne, Australia

Kelvin K. Shen
U.S. Borax, Inc., Valencia, California 91355, USA

G.A. Skinner
School of Applied Chemistry, Kingston University, Penrhyn Road,
Kingston upon Thames, Surrey KT1 2EE, UK

Andreas Thürmer
Clariant Huningue S.A., Avenue de Bale, BP 149, F-68331 Huningue,
France

Richard Sobottka
Grace GmbH, Postfach 449, in der Hollerheckel, D-6520 Worms, Germany

Tony Tikuisis and Van Dang
Nova Chemicals Ltd, Nova Chemicals Technical Center,
3620–32 Street N.E., Calgary, Alberta T1Y 6G7, Canada

J.S. Ullett and R.P. Chartoff
The Center for Basic and Applied Polymer Research,
The University of Dayton, Dayton, Ohio 45469-0130, USA

Gregory G. Warr
School of Chemistry, The University of Sydney, NSW 2006, Australia

Stewart White
Anzon Ltd, Cookson House, Willington Quay, Wallsend,
Tyne and Wear NE28 6UQ, UK

Joseph B. Williams, Julia A. Falter and Kenneth S. Geick
Lonza, Inc., Research and Development, 79 Route 22 East, PO Box 993,
Annandale, New Jersey 08801, USA

E.M. Woo
Department of Chemical Engineering, National Cheng Kung University,
Tainan 701-01, Taiwan

Alan Wood
Manchester Materials Science Centre, University of Manchester and
UMIST, Grosvenor Street, Manchester M1 7HS, UK

Guennadi E. Zaikov and Sergei M. Lomakin
Institute of Biochemical Physics, Russian Academy of Sciences,
Kosygin Street 4, 177977 Moscow, Russia

Preface

Plastics without additives would be commercial failures. They would not be processable, or else they would not be marketable, or else they would not be durable, or else they would be too flammable. The growth in the worldwide plastics industry in recent years owes at least as much to developments in additive technology as it does to improvements in polymer science. Probably the improvements in polymer science have been more widely disseminated and better explained.

There are now far more categories of additives than there were 25 years ago. Changes in technology are driven partly by the desire to produce plastics which are ever more closely specified for particular purposes, and partly by environmental pressures and legislative changes. At the time of writing, three additives experiencing those pressures are halogen-based flame retardants, heavy pigment metals, and plasticizers. This book does not deal specifically with environmental issues as a subject (nor with health and safety topics) but it includes within its authorship representatives of both the manufacturing industry sector and the academic world, thus providing a range of views on controversial issues.

There have been several books written about additives in recent years, but surprisingly little actual duplication. It is a big subject. Previous books have fallen into one of two categories. Some have concentrated on providing commercial information, in the form of data about the multitude of additive grades, or about changes in the market. Others have provided detailed accounts of the scientific principles underlying current practice, but targeted at the already experienced technologist in industry.

This book certainly has more in common with the second category. It gives higher priority to promoting understanding than to conveying factual information, but it differs from some other scientific accounts in putting the emphasis on accessibility or, in computer jargon, on being user-friendly. The book is intended for people on the edge of the subject: the final-year university science student; the postgraduate researcher; the recently appointed industrial technologist; the salesman; the manager and executive.

The ways in which accessibility is promoted are as follows.

1. The introduction is written for those with no previous knowledge of plastics additives whatsoever.
2. The subject matter is presented alphabetically for ease of use. Antioxidants are discussed under the letter A, biocides under B, and all flame retardants, whether boron or phosphorus based, under F. Fillers frequently have more than one function and an exception has been made here, but all articles about fillers are cross-referenced in 'Fillers' under F. A very wide range of additives types is covered, including those for thermoplastics and thermosetting resins.
3. All additives have one or more articles, each of which is short enough to be read in one session. The authors have attempted to explain both the technical principles and the reasons for current industrial practice. Sometimes there is overlap between two articles on related subjects, but it has been allowed where having two different perspectives seems beneficial.

Several contributors would like to have provided very extensive literature references, but the decision was taken to limit the number of sources severely, otherwise with about 65 contributions, there might have been over a hundred pages of literature sources. This is not strictly necessary for an introductory book of this size, which is designed to be easily read.

The book may be of incidental interest as a reference source to those who work with similar substances in areas other than the plastics industry. After all, pigments, surfactants, antistatics, flame retardants, antioxidants and many other plastics additives have applications elsewhere as well. There is considerable overlap between the plastics industry and the textiles, rubber and food technology industries.

The editor wishes to thank all the contributors throughout the world for their cooperation in preparing the manuscript, and to acknowledge the role of their employers in allowing them to take part, and to provide access to the necessary information.

Geoffrey Pritchard

January 1997

Introduction

Additives are essential

For twenty years the world has been absorbed in a computer revolution of such intensity that the progress made in other areas of technology, including much of materials technology, has been neglected by the media. This brief introductory section is designed to highlight the way in which plastics additives now constitute a highly successful and essential sector of the chemical industry. The professional scientists and technologists who use this book for reference purposes will already be very familiar with plastics and the additives used in them. It would be understandable if such people take for granted that progress in industrial chemistry and plastics technology is a positive influence on our quality of life. Other readers, and perhaps even some science students, may be more ambivalent. Many people have been influenced by the widespread public suspicion of chemicals in general (and additives in particular, whether in foods or plastics). The benefits of plastics additives can easily be assumed to be marginal. We need to explain that they are not simply optional extras; they are essential ingredients which can make all the difference between success and failure in plastics technology.

I hope, therefore, that readers who are unfamiliar with additives for plastics will take a few moments to read on, while those who are involved professionally every day with optical brighteners, or hindered amine light stabilizers, or low profile additives, or nucleating agents, will forgive the use of some rather elementary examples to illustrate the central theme: additives are essential.

PLASTICS ARE HARDLY VIABLE WITHOUT ADDITIVES

Early plastics were often unsatisfactory. Complaints about plastics articles were common. This was partly because of design faults, such as slavish imitation of shapes already in use with metals, but the failure to appreciate the need for additives to improve processing and durability was also important, and poor durability was commonplace. Nowadays, car components, household appliances, packaging materials, electronic and telecommunications products and the like are made from polymers, but they are not just polymers, or they would be complete technical failures. They are polymers mixed with a complex blend of materials known collectively as additives. There are many nominally organic plastics articles which actually consist of considerably less than 50% organic polymer, the remainder being largely inorganic additives. Additives cost money in the short term, of course, and even after considering raw materials costs, incorporating them into plastics can be an additional expense, but by reducing overall production costs and making products last longer, they help to save money and conserve raw material reserves. Processing plastics to form useful and saleable articles without additives is virtually impossible.

A few examples should illustrate these points.

PROCESSING AIDS

Many fabrication processes essentially consist of melting polymer powder or granules inside a heated tube. This 'melt' is forced through a shaped orifice or die, as in extrusion, or injected into a mould, as in injection moulding, or rolled into sheets on a calendar, or blown into flat film or into bottle shapes using film blowing or bottle blowing equipment attached to an extruder. The ease with which this is done depends on the physical and chemical properties of each plastics material, in particular on its melt viscosity and its resistance to heat and oxidation during processing. These characteristics can be improved through the use of additives known as process aids.

Process aids become liquid during the moulding process, and form a film around coloured particles so that they mix better. Other additives make the individual polymer particles adhere more to each other inside the tube, so that they 'melt' more quickly. This means that the moulding temperature can be lower, which saves energy and prevents or reduces heat damage to the plastic.

Certain plastics, such as PVC, can be very difficult to process because they become viscous and sticky when they melt. Lubricants help to reduce viscosity by creating a film between the polymer melt and the mould, and by lubricating the polymer particles against each other. More intricate shapes can be moulded, and the moulding temperature can also be lowered.

ANTIOXIDANTS AND HEAT STABILIZERS

Most plastics have to be processed at above 180°C, a temperature which can sometimes spoil the colour and weaken or embrittle the plastic.

However, these effects can be prevented or reduced by antioxidants, i.e. organic compounds which help protect the plastics under hostile conditions. Other additives called heat stabilizers help stop plastics, particularly PVC, from decomposing during processing. They are often compounds based on epoxies, or on calcium, zinc, tin and other metals.

Some plastics are subjected continuously to heat throughout their life. We do not need to reach for exotic examples from the space industry here; the humble automatic coffee vending machine will suffice, operating as it does for 24 hours a day, 365 days a year. Where drinks are concerned, the additives used must be rigorously tested to avoid any tainting of the contents of the vessels.

PIGMENTS: FASHION AND FUNCTION

Marketing people have to consider what it is about a plastics object that catches our attention – shape, colour, surface texture. Plastics are coloured using two main methods. The surface can be painted or printed after moulding, or pigments can be incorporated before or during moulding. With this method, colour pigments can create decorative effects that go right through the object and therefore never wear off. This property, coupled with the range of moulding techniques available, gives designers a tremendous freedom.

By manipulating additives, plastics can be colour matched with parts made of other materials such as metal, wood, paint and fabric. Cars, radios and kitchen appliances use this technique.

Fashion is important commercially, not only for clothes and accessories but when considering tableware, kitchenware and office equipment. In all these areas pigments enable plastics to offer an endlessly variable palette of colours, as vivid as any other medium. However, pigments are not just about fashion, and aesthetics. Colour in plastics also has many non-decorative functions. It can be used to cut down light for the protection of the contents of medicine bottles or increase safety by the colour coding of electrical wiring. Designers often use colour to differentiate the controls on machines, and 'day-glow' pigments prevent road accidents. Runners and cyclists wear reflective fabrics and strips, while road, rail and building site workers can easily be seen in their fluorescent helmets and jackets.

To make an opaque moulding, pigments are chosen that absorb or scatter light very well. The most common, cost-effective way of creating solid colour is to use carbon black or titanium dioxide. Carbon black absorbs light, whereas titanium dioxide, with its high refractive index, scatters light, producing a very high level of whiteness and brightness. It is one of a range of inorganic pigments, and is mixed with other colours to create pastel shades. Organic pigments are also good for making bright colours.

IMPACT MODIFIERS AND FIRE RETARDANTS

The domestic appliance market covers, among other products, many housings for electrical gadgets. It is instructive to consider how the functional effectiveness of such products would be affected by an absence of additives.

Consider a vacuum cleaner. Without an impact modifier, a vacuum cleaner will crack if it is treated to normal rough usage. Without light stable pigments, its colour will fade. If it contains no pigments anyway, it will soon look drab and dirty. More worrying in an electrical appliance may be the lack of fire retardants. Some plastics articles burn in fires, and fatalities are often attributed not to heat but to smoke. The addition of smoke suppressants such as alumina trihydrate, halogen or antimony compounds can be very effective in preventing such incidents. An excellent illustration of lives saved by flame retardants in plastics is the conveyor belt in coal mines. For many years fires occurred regularly when pulleys overheated, causing serious accidents and death. But when belting made from PVC containing high levels of flame retardants was introduced in the mid-1950s, these accidents stopped. Clearly the side-effects of additives on weathering, mechanical properties and chemical resistance have to be taken into account.

COST

The additives that assist the moulding of plastics, such as lubricants, process aids, and heat stabilizers, can cost many times more than the raw material, and although only small amounts are used, they are nevertheless essential and greatly enhance the final performance. Other additives such as mineral fillers like chalk, talc and clay, are naturally occurring substances which tend to be cheaper than the raw polymer, although surface treatments are sometimes applied to the filler particles to prevent their agglomeration or to improve their compatibility with the polymer, or to aid processing, and this together with other filler particle processing operations can mean that the cost reduction is not as great as is sometimes thought. Fillers are not necessarily incorporated with the intention of reducing cost, but in order to secure certain technical benefits: talc and chalk increase rigidity, whereas clay improves electrical properties. Mineral fillers also increase the thermal conductivity of plastics so that they heat up and cool down quickly, meaning shorter mould cycle times and more articles produced at a lower cost. A saving of one US cent per moulding may not sound much, but if it involves producing several injection mouldings every few seconds, this small saving can become very significant.

OUTDOOR DURABILITY

Children's toys and garden furniture, packaging, and flooring are some of the products that form the backdrop to our lives, and it is hard to overestimate the rough treatment they have to endure. In sports stadia, more and more spectator seating is moulded in brightly coloured plastics, and playing surfaces are often made of synthetic fibres. Indoor swimming pools may have plastics roofing materials. All of these are exposed to the weather day and night, summer and winter, but a combination of light stabilizers, ultra-violet absorbers and antioxidants ensures consistent high performance. Natural materials usually have to be finished off after manufacture with paints and lacquers. Plastics enjoy the advantage of already incorporating – before or during the moulding process – the additives that prolong their useful lives for many years. This can greatly reduce maintenance costs. Figure 1 shows the improvement obtained by stabilizing polycarbonate against ultra-violet light, both in hot, wet environments and in hot, dry locations. The criterion here is notched

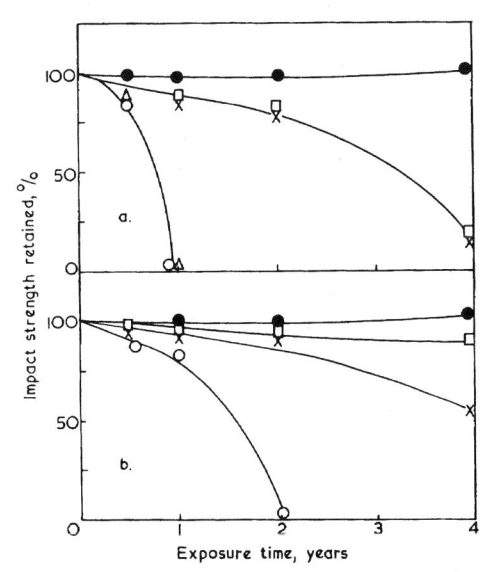

Figure 1 Notched impact strength of polycarbonate in (a) hot, dry and (b) hot, wet climates, with and without ultra-violet stabilizing additives. Solid circles – controls; squares – sheet; triangles - -heat stabilized; crosses – ultra-violet stabilized injection moulding grades. From Davis, A. and Sims, D. (1983) *Weathering of polymers*, Applied Science Publishers, London. Available from Chapman & Hall, London.

impact strength, because of the tendency for the polymer to become brittle in the absence of appropriate additives.

ENERGY SAVING

When certain plastics, notably polyurethanes, are moulded at high temperatures, additives called blowing agents volatilize, or else decompose chemically, to form gases such as nitrogen, carbon dioxide and water vapour. These gases, trapped in the plastics, turn the material into foam, thus increasing the thermal and acoustic insulation and the energy absorption properties, incidentally reducing weight. These foams are so commonplace that their everyday use needs little description – hamburger boxes to keep food hot, cushioning in sports' shoes, buoyancy aids, and automobile parts where lower weight makes large savings in fuel. The kinds of chemicals used as blowing agents have changed dramatically in the past few years, in response to concern about the effects of some of these reactive chemicals on the ozone layer.

FOOD PRODUCTION

Throughout the world, crop yields are boosted by plastics film laid over the soil to trap heat and moisture. Tomato production, for example, has been increased in some areas by 300%. Additives have been developed that allow the sheet to capture the sun's warmth during the growing season, but to break up as soon as the harvest arrives. The sheet disintegrates gradually in sunlight and the fragments can be ploughed into the earth where the soil bacteria quickly break them down into carbon dioxide and water. In areas of predictable climate this process can be timed to an accuracy of within seven days. Where plastics cannot be re-used or recycled, biodegradation could offer a clean, safe method of disposal. In some other applications, biodegradation is an undesirable process from which certain plastics have to be protected by additives known as biocides. The performance of a given plastics material such as flexible PVC in outdoor and underground applications can be revolutionized by appropriate additives.

WASTE DISPOSAL

Plastics waste disposal can cause problems, especially as plastics are usually mixed up with other types of waste such as paper, metals and food. For recycling they really need to be sorted into individual types such as polythene, polystyrene or PVC before being mixed with virgin material. Otherwise they have no strength if remoulded, and may literally fall apart. Sorting can be very difficult.

This is an area in which additives called compatibilizers can help. They are substances which have the right chemical structure and morphology to promote a degree of miscibility between various kinds of polymer, rather as a detergent can promote miscibility between different liquids. Compatibilizers for use with recycled plastics are currently being developed and improved. Mixed plastics waste can be remoulded into fencing, pallets and road markers, thus saving valuable timber. Additives are vital for reprocessing waste plastics into useful products for a second life.

NEW PROBLEMS FOR OLD

The inevitable consequence of any new technology is that there will be new problems which have to be addressed and which were not widely foreseen. The toxicity of certain pigments, both in plastics and in paints, has been a source of concern for many years and it has been a driving force for the development of new, safer pigments which will have their applications in wider areas than those originally envisaged. Other environmental issues have had similar beneficial consequences. The trend towards the incineration of plastics, for example, recovers considerable energy for further use, but thought has to be given to the effects of any additives on the emissions produced.

Other problems involve a simple recognition that some additives are not yet technically completely satisfactory. Flame retardants in exterior building panels have to be colour-stable if unsightly discoluration is not to occur. This goal is not always achieved. Sometimes one additive interferes with, and prevents another from working. Often, biocides are needed only because certain other additives such as plasticizers and organic fillers are susceptible to biological attack. It is the role of the additive technologist and polymer formulator to overcome such problems. Nevertheless the benefits of additives far outweigh the disadvantages.

POLYMER-BOUND ADDITIVE FUNCTIONALITY

It is technically possible, although not necessarily economic, to incorporate additive functional groups within the structure of the polymer itself, thus dispensing with small-molecule additives. There are potential advantages in this approach, which could be applied (for example) to antioxidants, so that they would be stable and would not leach out of the polymer during exposure to rain or other sources of moisture. It must be said that, at present, the main trend is towards having more additives, and using ever more complex formulations to achieve a range of desirable properties. It has becomes more and more difficult to advise how (say) PVC will behave, because there is no unique substance called PVC on the market, only several hundred diverse grades of PVC, all containing specific additives which help to ensure fitness for purpose. The same basic polymer is used for flexible tubing, foam, rigid pipe, outdoor pond lining, clothing, pigmented wire coating, and clear bottles. Anyone examining such a wide range of products from the same base polymer can be left in little doubt about the importance of the additives present.

ACKNOWLEDGEMENT

The author of this chapter acknowledges that it is based in part on an article entitled 'Additives Make Plastics', produced by the Additive Suppliers Group of the British Plastics Federation. Permission to adapt the article in this way has been given by the BPF. The views expressed in the adapted version should not be attributed to the above group, nor to the British Plastics Federation itself.

Quick reference guide

The following list provides a summary of the purposes of many of the common additives used in commercial thermoplastics and thermosetting resins.

Additives	Function
Accelerator	Chemical used to increase the rate at which a process occurs; usually refers to the cure process in thermosetting resins, but in theory the term can be applied much more widely.
Antiblocking agent	These substances prevent plastics films from sticking together, and are used to facilitate handling or for other reasons.
Antifogging agents	These improve packaging film clarity, by preventing any water from the contents of the package from condensing as droplets on the inside surface of the film.
Antioxidant	Substance which protects a polymer against oxidation, whether during processing or in service life.
Antistatic agent	Additive which reduces or eliminates surface electrical charges and hence prevents dust pick-up etc. on polymer surfaces.
Biocide	Additive which protects a plastics article against attack by bacteria, fungi, algae, moulds etc., which in most cases are a problem only where there are additives such as plasticizers present. Biocides come in several types – fungicides, bactericides etc.

Additives	Function
Blowing agent	Substance added to a polymer, so as to generate gas which will have the effect of expanding or foaming the polymer. The gas can be produced chemically or by simple evaporation.
Compatibilizer	Substance, usually polymeric, which when added to a mixture of two rather dissimilar polymers, enables them to become more intimately mixed than before.
Coupling agent	Substance which is used in trace quantities to treat a surface so that bonding occurs between it and another kind of surface, e.g. mineral and polymer.
Curing agent	Reactive chemical which promotes crosslinking in polymers, e.g. peroxides in polyesters, or amines in epoxy formulations.
Diluent	Strictly, a solvent which makes a solution more dilute; but in the context of additives, any substance which reduces resin viscosity and hence makes processing easier. Frequently refers to epoxy resins.
Defoaming agent	Substance which removes trapped air from liquid mixes during compounding.
Exotherm modifier	Substance which reduces the maximum temperature reached during an exothermic crosslinking reaction.
Fibre	Reinforcement for polymers; improves mechanical properties. Length : diameter ratio very high.
Filler	Particulate additive, designed to change polymer physical properties (e.g. fire resistance, modulus, shock resistance) or to lower cost.
Flame retardant	Substance added to reduce or prevent combustion.
Foam catalyst	Substance used in (mainly) polyurethane foam production to control the foaming process and achieve satisfactory foam quality.

Quick reference guide

Additives	Function
Fragrance modifiers	*see* Odour modifiers.
Heat stabilizers	These additives prevent polymers from degrading thermally, even in absence of oxygen, during processing.
Impact modifier	Substance added to use up the energy of crack propagation and hence increase resistance to impact.
Light stabilizer	Chemical added to reduce or eliminate reactions caused by visible or ultra-violet light radiation, which would otherwise cause polymer degradation in outdoor use.
Low profile additive	Substance added to thermosetting moulding compounds, particularly polyesters, to counteract shrinkage during cure, and uneven surface finish.
Lubricant	Two main functions: they prevent a polymer from sticking to the mould or the machinery, and reduce melt viscosity, allowing the molten polymer to pass easily through intricate channels. They also reduce friction between polymer particles before they melt.
Microspheres/ Microballoons	Spherical filler particles, usually hollow, used to reduce weight in a product without much adverse effect on mechanical properties.
Nucleating agent	Substance which promotes or controls the formation of spherulites in crystallizable polymers. Nucleating agents lead to several small spherulites, rather than a few large ones.
Odour modifiers	These substances may be used to mask an undesirable odour, or to add a desirable one. Their value is dependent on the persistence of their action.
Optical brightener	Special fluorescent organic substances used to correct discoloration or enhance whiteness. They absorb ultra-violet radiation and emit it as visible light.

Additives	Function
Peroxide	Source of free radicals, generally for crosslinking thermosetting resins or polyolefins. Also used in the rubber industry and as polymerization initiators.
Plasticizer	Additive designed to space out the polymer molecules, facilitating their movements and leading to enhanced flexibility (lower modulus) and ductility. Widely used to convert PVC from the rigid to the flexible variety. Can sometimes be polymeric.
Processing aid	Related to, but not the same as, lubricants. They are additives which do whatever is necessary to counter processing problems: this varies with the polymer, but often means lowering the melt viscosity, and improving melt homogeneity.
Release agent	Substance designed to ease the parting of a plastics object from its mould; it may be an internal additive or a coating; plastics sheets can also be used.
Slip agent	These are a kind of lubricant which have insufficient compatibility to remain long in the polymer; they migrate to the surface, reducing tack. They can perform secondary functions, e.g. act as antistatic agents.
Smoke suppressant	Substance which changes the nature of the polymer combustion process, if it is not possible to prevent it, so as to reduce smoke formation.
Surfactant	These substances reduce surface tension in liquids. For example they stabilize the cells of polyurethane foams during the foaming process.
Thickening agent	Viscosity increaser. In unsaturated polyester resins, magnesium oxide is able to increase the viscosity so much that a liquid resin becomes a tack-free solid, suitable for making sheet moulding compound (SMC). It achieves this by reacting with carboxyl end groups.

Quick reference guide

Additives	Function
Thixotropic agent	These modify the dependence of viscosity on shear rate, producing low viscosity at high rates, and vice versa.
Ultra-violet stabilizer	As for light stabilizers, but active in the ultra-violet part of the spectrum, and used especially in outdoor applications to preserve polymers against harmful radiation which would otherwise cause degradation.
Wetting agents	These substances wet out solid substrates, e.g. filler particle surfaces, and help their uniform dispersion in a polymer matrix without agglomeration.

Practical methods of mixing additives with polymers

Alan Wood

Industrial mixing operations can usually be broken down into two types, these being batch and continuous.

In the case of batch mixing, the resulting compound is manufactured in discrete quantities. Typical examples of batch mixers are two-roll mills and internal mixers. Continuous mixing is usually carried out using either single or twin-screw extruders, the components of the mix being fed on a continuous basis into the machine.

BATCH MIXING

Two-roll mill

Two-roll mills are generally good dispersive mixers and, with the intervention of the operator, good distributive mixers. The mixing process

1. tends to be slow,
2. is largely manual, and
3. can present problems related to health.

As a result of these limitations, two-roll mills in the plastics industry tend to be used only for laboratory scale activities. The mill consists of a pair of contra-rotating rolls, the typical laboratory mill having rolls around 300 mm wide and 200 mm diameter.

The degree of mixing achieved using a two-roll mill is controlled by the following.

Plastics Additives: An A–Z Reference
Edited by G. Pritchard
Published in 1998 by Chapman & Hall, London. ISBN 0 412 72720 X

1. Controlling the surface speed of the rolls, the speeds often being different. The ratio of the surface speeds of the rolls is referred to as the friction ratio. Increasing the friction ratio increases the level of applied shear in the gap between the two rolls, the nip; this leads to improved dispersive mixing.
2. Controlling the nip gap; the smaller the nip gap, the higher the level of applied shear in the nip and the better the dispersive mixing obtained.
3. Controlling the size of the rolling bank; typically the rolling bank should have a diameter of the order of around a tenth of the mill roll diameter. Thus, this, coupled with the nip gap and the mill roll diameter, limits the batch size that can be mixed on a two-roll mill. Adjustments can be made to the edge plates in order to limit/increase the width of the mill roll being used. Both distributive and dispersive mixing processes occur in the rolling bank.
4. Controlling the roll temperature; in the case of mills used for thermoplastics and thermosets, the rolls are heated. Generally, the lower the roll temperature, the higher the shearing forces generated and thus the better the dispersive mixing. It is often advantageous to have the two rolls at different temperatures.

In order to mix efficiently, the compound must band (see Fig. 1(a)), but not stick, that is it must cling to one of the rolls. The roll to which the compound bands can be influenced by the friction ratio and the roll temperatures. Mixing is difficult if the material bags (see Fig. 1(b)).

The roll temperatures depend heavily on the design of the mill and the material being mixed. The roll temperature should ideally be slightly (approximately 10°C) below that at which the polymer melts in order to prevent the polymer sticking (as opposed to banding) to the rolls. The energy for melting/softening of the polymer comes from two sources:

1. thermal conduction from the hot rolls;
2. dissipation of energy; this arises from the shearing of the material in the nip.

Before mixing any materials on a mill it is important that the safety systems on the mill are checked. The gap between the rolls is usually small and the mill drive powerful, and thus the process is potentially very dangerous. Laboratory mills are usually provided with a cage arrangement that covers the upper section of the mill. This guard is such that it has movable gates, these being linked to electrical limit switches, which, if the gate is opened, stop the mill. Larger mills do not have such guards but usually have Lunn bars. These are hinged bars across the face of the rolls, which, if pushed, will stop the rolls. The techniques used to stop the rolls usually depend on the size of the

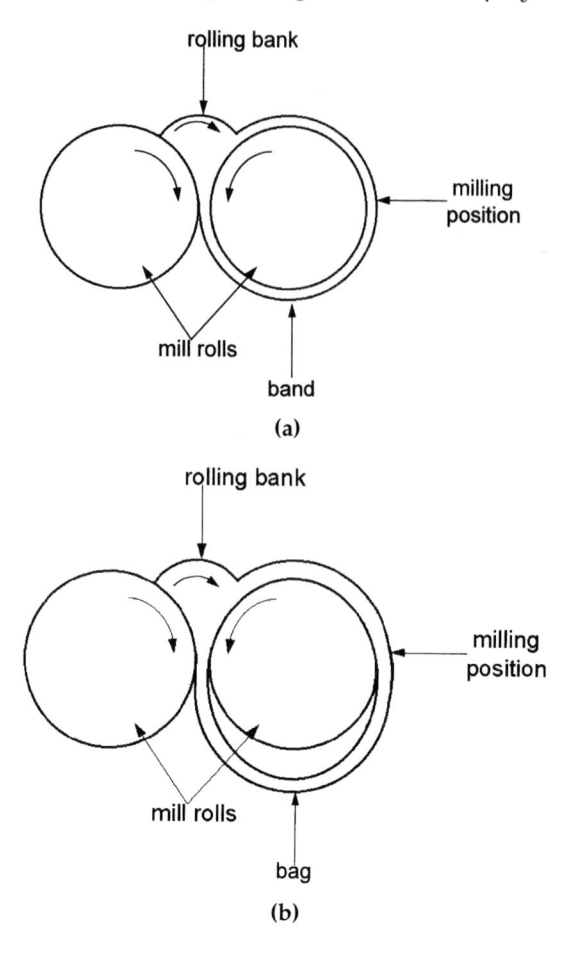

Figure 1 Two-roll mill: (a) the material bands on one roll; (b) the material bags instead.

machine. In the case of laboratory mills it is possible to use disc brake systems in order to stop the mill. On some laboratory mills and on larger commercial machines the technique adopted uses a plugging relay. The plugging relay operates by effectively starting the mill drive motor in reverse, the plugging relay operating for a very short time, just enough to bring the mill to a halt. It is important that the period of operation is controlled as if it is too short the mill would not stop fully and if it is too long, the mill would run in the reverse direction; this also presents a potential hazard. On the edge plates there is normally a set of red arrows. If the operator can reach past the arrows without tripping the safety system then he should not operate the mill. The arrows are positioned to ensure that there is sufficient time, should the safety system be tripped, for the mill to stop before any accident occurs.

The mixing process starts with the banding of the polymer onto one of the rolls, this involving passing the polymer through the nip several times in order to soften and melt it. Once the band is formed, the other components of the mix are slowly added. The rate of addition should be controlled such that no large losses of powder occur and the rolling bank does not becomes too disturbed as this leads to the break up of the band and makes processing difficult. Loose powders can represent a safety hazard, and adequate dust and fume extraction must be used. If the addition rate is too high, powders will often pass through the nip without being incorporated into the band. It is necessary for the operator to cut and fold the compound, material being cut from one side of the roll and fed onto the opposite, as the mixing process occurring has a number of deficiencies, as follows.

1. The distribution of additives only occurs partially through the thickness of the band on the roll. There is a layer of material adjacent to the roll that does not get incorporated into the rolling bank and remains stagnant on the roll.
2. The cross-roll distribution of materials is poor.

Whilst mill mixing is relatively slow and labour intensive, the quality of mixing that can be achieved is often very high. This is in part due to the level of shearing encountered in the process, this being such that the compound temperature can be controlled very well and so prolonged periods of mixing are possible with little risk of degradation.

Internal mixer

An internal mixer essentially consist of a two-roll mill enclosed in a chamber. There are two basic designs of machine:

1. those featuring interlocking rotors; and
2. those featuring non-interlocking rotors.

The capacity of the machines ranges from less than 1 litre up to 500 litres capacity. The rotors in both styles of machine feature surface projections, these often being referred to as nogs or wings. These protrusions generate cross-chamber flows as would normally be produced by the operator on a two-roll mill. In the case of the non-interlocking rotor systems there is usually a friction ratio between the rotors.

The materials to be mixed are loaded, via the feed throat, with the ram in the elevated position. When mixing, the ram is lowered such that it sits on top of the rolling bank formed at the inlet of the nip between the rotors.

The ram is usually operated pneumatically. The load applied to the rolling bank can be varied by altering the air pressure supplied to the pneumatic cylinder that actuates the ram.

There are several regions in the machine where the compound being mixed is sheared as follows.

1. Between the rotors. Owing to the protrusions on the rotors there is a difference in surface speed between adjacent regions of the rotors in the nip region.
2. In the rolling bank. This arises due to the velocity fields set up between the rotors and the ram.
3. Between the rotors and the chamber wall as the rubber returns to the rolling bank after exiting the nip.

Distributive mixing occurs at the exit of the nip as the compound can return to the rolling bank via either of the two rotors.

The level of shear imposed on the compound being processed is controlled by varying the rotor speed, the higher the speed the greater the shear, and the ram pressure, the higher the ram pressure the higher the shear. The most versatile mixers have variable speed drives but often the choice of speed is limited to either one or two. The maximum ram pressure is limited either by the factory supply or by the design of the glands (seals) on the mixer. The glands are used to seal the chamber where the rotor shafts enter the chamber. Sealing is required in order to prevent material from leaking out during the mixing cycle. In certain circumstances, the use of too high a ram pressure can damage the glands and cause excessive leakage.

The mixing process occurring inside an internal mixer requires that the correct volume of material is loaded for each cycle. If the batch size is too large, unmixed material will remain at the base of the feed throat at the end of the mixing cycle. If the batch size is too small then the mixing process will be inefficient and slow. The volume of a batch is typically 66% of the total internal volume of the mixer, the ratio of the batch volume to that of the mixer being known as the fill factor. The fill factor does, however, vary from compound to compound, being higher in the case of highly filled compounds and lower for highly elastic materials. Determination of the correct loading is best done practically by observing the position of the ram at the end of the mixing cycle. Ideally the ram should be bouncing slightly, its position being close to the mechanical limit of its travel.

Initially the batch weight would be calculated based on the volume of the batch being 66% of that of the mixer. This requires a knowledge of the density of the specific components of the compound.

Modern mixing plant is highly sophisticated, often featuring computer controlled weighing systems. Solids are fed via the feed throat. Liquids, such as oil and plasticizer, are fed, via injectors, directly into the mixing chamber. The sequence of additions depends very much on the compound being processed. In the case of a plastics masterbatch, all the ingredients would normally be loaded together at the start of the cycle except, in some cases, for the liquids, these being metered into the chamber during the mixing cycle. Typical cycle times for a plastics masterbatch are of the order of one minute. Generally the ram would be lifted part way through the cycle in order to clear any powders on the upper surface of the ram.

The determination of the end-point in the mixing process can be done using the following methods.

1. Time. The mixing cycle runs to a timed schedule, additions and discharge being carried out as dictated by the schedule;
2. Temperature. This technique is a modification of the time based system, the difference being that the mix is discharged at a specific temperature rather than a specific time;
3. Plateau power. A typical power versus time curve for a mixing cycle shows an initial peak, the power requirements then falling to a steady level. Once this plateau is reached, the useful mixing process is considered to be complete and the mix would be discharged.
4. Total energy. The power versus time data are integrated to yield the total energy consumed in the mixing process. When this reaches a defined level, the mix would be discharged.

The ratio of cooled surface area of a mixer to its volume is important and presents problems when scaling up. These problems are illustrated by the fact that, using a laboratory scale mixer (capacity one litre) with a cold body, it is not possible to generate sufficient work heat to melt a thermoplastic material such as polyethylene. As a consequence, the mixer needs to be heated in order to mix such materials. In the case of an industrial mixer with a capacity of around 40 litres, no external heating is required, the mixer being cooled, the work heat generated being sufficient to melt the polymer.

One problem with internal mixers is contamination. The main source of contamination is from the small clearance between the rotors and the end bodies and from the glands. It is very difficult to clean an internal mixer thoroughly and, as a consequence, mixing processes dealing with black and coloured compounds are often separated from those dealing with white compounds.

CONTINUOUS MIXING

Single-screw extruders

In general, single-screw extruders are poor dispersive and distributive mixers. Generally any additives will have either been:

1. previously compounded into the polymer, the feedstock being a uniform blended material;
2. compounded into a small quantity of the polymer to produce a masterbatch. This is then diluted by mixing it with virgin polymer in the extruder hopper; or
3. added as a liquid system, the system being injected into the extruder part way along the barrel.

The mixing process in a single-screw extruder is laminar. The materials being mixed exist as discrete layers, striations. The process of mixing involves a reduction in the striation thickness, this being related to an increase in interfacial area between the major and minor components of the mixture.

It can be shown that the change in striation thickness on mixing, and hence in the degree of laminar mixing, is a simple function of the total shear strain imposed on the system. However, at the end of the process the components of the mixture still exist as discrete components. The total shear strain exerted on the melt is a function of the residence time of the melt in the process. As a result of the complex velocity profile of the melt in the screw channel, the residence time of the melt in the channel varies as a function of the position of the melt in the screw channel as well as the down-channel velocity of the melt.

Practically, mixing can be affected by such things as the following.

1. Die-head pressure. The higher the pressure, the lower the flow rate and the higher the melt temperature.
2. Screw cooling. Generally the centres of extruder screws are hollow. This allows cooling fluid to be passed along the screw in order to cool the screw at various points. The commonest areas to cool are:
   (a) in the region at the base of the feed hopper; this prevents premature melting of the polymer and the formation of melt rings, but has no effect on mixing;
   (b) the screw tip; this acts to control the degradation effects due to the residence time of material on the screw tip.
   It is not advisable to flood cool the whole length of the screw with cooling water as high melt pressures, in excess of those required to deform permanently the barrel of the extruder, can be developed.
3. Reducing the barrel temperature. This tends to increase the shear exerted in the melt. However, care must be taken if the extruder is stopped, as the melt in the screw channel may solidify.

Further improvements in the quality of mixing obtained can be brought about by modifying the screw or by the use of special mixing sections, these usually being located in the metering section of the screw. Typical examples of such sections are pins, barriers, Barr type mixing sections and cavity transfer mixers.

It is important to remember that it is not possible to control the temperature of the melt produced on a single screw extruder, merely to influence it. Hence, particularly with thermally sensitive materials, compounding is not usually carried out using single-screw extruders.

There are specialized versions of single-screw extruders that have the compounding ability of a twin-screw extruder. These machines feature interlocking screw and barrel arrangements which have kneading elements. The operation of this type of machine is more akin to that used with twin-screw extruders rather than single-screw machines.

Twin-screw extruders

Twin-screw extruders were first developed and patented in the late 1930s, the original development work having been done by Colombo (co-rotating) and Pasquetti (counter-rotating).

Twin-screw extruders can be classified as follows.

1. Non-intermeshing. These are sometimes referred to as tangential twin-screw extruders and are effectively two single-screw extruders but with a single barrel. This type of machine is of little use for mixing.
2. Intermeshing. This type of twin-screw extruder can be further classified as:
   (a) co-rotating or counter-rotating depending on the relative motion of the screws to one another;
   (b) conjugated (highly intermeshing) or non-conjugated (loosely meshing);
   (c) parallel or conical, depending on the shape of the screw.

Continuous compounding processes require continuous feeding, this being achieved by the use of volumetric or gravimetric (loss-in-weight) feeder systems. Gravimetric systems are more accurate but are more expensive. Volumetric feeders have to be calibrated prior to use in order to ensure that if process conditions are altered, the correct feed composition can be maintained. Normally twin-screw extruders are starve fed. The main-screw and feeder-screw speeds are adjusted to obtain optimum quality.

Modern intermeshing machines are often based on a system of segmented screw and barrel components, this allowing the development of specific screw and barrel geometries to suit a wide range of compounds.

The advantages of twin-screw extruders over single-screw machines, in the case of intermeshing screws, include:

1. more positive pumping;
2. better mixing; and
3. less sensitivity to die head pressure.

The principles of operation depend heavily on the particular configuration of machine being used.

It is necessary, as a result of the intermeshing nature of the screws, that one flight must pass freely through the channel of the other.

With counter-rotating screws, in the intermeshing region, the flight of one screw enters the channel of the other and remains in the same position relative to the flanks (sides of the channel) of the other screw throughout the region. The flights, relative to the point of intermeshing, are moving in the same direction and, as a consequence, the shape of the flights is not important in respect of the mechanical operation of the machines.

In the case of rectangular section flights, the clearance between the flanks of the flights is maintained throughout the region of intermeshing. If the screws are perfectly conjugated, this generates a very positive pumping action, the material being retains in the 'C' shaped chamber formed. This does, however, lead to some problems regarding mixing, the generation of a homogeneous mixture being easier if the material in one chamber is allowed to enter another.

There are the following ways around these problems.

1. Increase the clearances, that is reduce the degree of conjugation. This, however, reduces the pumping efficiency.
2. Change the shape of the flights. An example of this is trapezoidal flights. In the plane of the screws, the flights are perfectly conjugated, but once away from this position, due to the angled flanks, the clearances open. This allows flow from one chamber into another. However, the conjugation in the plane of the screws ensures that the melt cannot rotate with the screws and hence good pumping is still achieved.
3. Introduction of mixing sections, these being regions of the screw where the degree of conjugation is reduced.
4. Use split or interrupted flights.
5. Use multiple start sections and changes in pitch.

High shear zones are present between the flanks (sides) of the flights, this possibly giving rise to problems related to excessive temperature and degradation.

The screw geometry in the case of co-rotating machines is more complex than that of counter-rotating machines. This arises from the relative motion of the flights. To conjugate, the flights must have angled flanks, trapezoidal flights being used.

In respect of mixing, owing to the positions of the flights in the intermeshing region, flows do occur between channels, the screw channels having a figure-of-eight geometry. As a result of the ease of flow around the screws the pressure around the screw is fairly constant. No high shear zones exist as they do with the counter-rotating systems.

The screws are still conjugated and hence positive pumping is obtained, along with a self-cleaning action.

Mixing can be achieved by the use of special mixing sections or by the truncation of the flights, the mixing ability in this case being expressed as:

$$m = \frac{e - E}{p - 2E}$$

where p is the screw pitch, E is the width of the top of the flights and e is the width of the base of the screw channel.

BIBLIOGRAPHY

Health and Safety Commission (1991) *Safeguarding of nips in the rubber industry*, UK Health and Safety Executive.
Horten, H.E. (1970) *Plastics and Rubber Machinery*, Elsevier, London.
Rauwendaal, C. (1990) *Polymer Extrusion*, Hanser Publishers, Munich.
Martelli, F.G. (1983) *Twin-Screw Extruders: a Basic Understanding*, Van Nostrand Reinhold, New York.
Hepburn, C. (1982) *Rubber Technology and Manufacture*, 2nd edn, published for the Plastics and Rubber Institute by Butterworth Scientific, London.

Keywords: batch, continuous, mixing, mill, extruder, compounding, masterbatch, co-rotating machines.

Analytical methods for additives in plastics

Roger W. Crecely and Charles E. Day

INTRODUCTION

Additives are responsible for the acceptance of plastics worldwide. The amount of plasticizer in polyvinyl chloride determines whether or not it is rigid (as is pipe) or flexible (as is wallpaper). The analytical chemist is frequently called upon to identify these additives and their concentrations.

The analysis of additives is a challenge for the analytical chemist, as these substances cover the range of chemical materials from organic to totally inorganic. Furthermore they are seldom pure, often being mixtures of isomers or technical grade substances. Additional components can complicate the analysis. Traces of curing agents or low molecular weight oligomers can interfere. The concentration of the additives also plays an important role in the method selected. Additives at high concentration, such as plasticizers, are relatively easy to classify. On the other hand, ultra-violet stabilizers at the low parts per million level can be difficult to isolate. The polymer matrix is an important factor in choosing an analysis method. The physical and chemical properties of the polymer determine how the additives can be separated. Some of these problems can be overcome by a good understanding of the plastics material under study. A few minutes spent gathering related information will save hours in the laboratory.

STRATEGY

The analyst should consider the laboratory support available at his location. Normally several techniques are required for the analysis. Typically

Plastics Additives: An A–Z Reference
Edited by G. Pritchard
Published in 1998 by Chapman & Hall, London. ISBN 0 412 72720 X

the additive must be isolated from the sample in pure form, and identified. An estimate of its concentration can then be obtained. Separation is the most difficult challenge. Techniques such as solvent extraction, high temperature distillation, ashing, and thin layer or column chromatography can easily be done in most laboratories. Gas chromatography (GC) and high pressure liquid chromatography (HPLC) require more complex instrumentation and method development. Identification techniques require sophisticated instruments. They usually involve spectroscopy. Infra-red (IR) spectrophotometric analysis is the most powerful and cost effective method available for general analysis. Much of this discussion will be centered around this technique. Other instrumental techniques such as ultra-violet (UV), nuclear magnetic resonance (NMR), mass spectrometry (MS), emission/absorption methods (ICP, AA) and X-ray (diffraction, XFA) require more complex laboratory support. Preliminary considerations and subsequent analysis methods are presented in the next sections.

GATHER BASIC INFORMATION

Obtain a clear understanding from the person requesting the analysis concerning his requirements. Does he need to verify a specific component or assay for an unknown material? Is a generic identification and semi-quantitative analysis satisfactory? Or is a more precise analysis required involving more time and cost? If he is asking to determine the presence of a particular additive, a minimum detection limit must be established.

Research the type of additives and normal concentrations. Learn about their properties. Do they have a unique element or chemical tag that can be used for quantitative measurements? Is an analysis method available from the supplier or in the literature? Many sources are available: scientific journals, trade journals, technical application sheets from instrument manufacturers and standards compendia. Any method selected must be validated in your own laboratory. Learn as much as possible about the polymer matrix. Thermoplastics can be hot pressed into thin films. Many polymers are solvent resistant, so extraction techniques can be used to isolate the additives.

VERIFY THE SAMPLE

Is the polymer matrix as claimed? Infra-red analysis is a good technique for this determination. IR spectra can be obtained quickly with minimum sample preparation. Liquids and films can be run directly. The dust from filing a solid sample can be ground with potassium bromide (KBr) and a pellet pressed for analysis. Is the additive readily apparent in the spectrum?

Table 1 Direct infra-red measurements

Polymer	Press temp. (°C)	Additive	Peak (μm)[a]	Absorptivity (A/mil/100%)[b]
Ethylene	140	Talc	9.85	10
Ethylene	140	Kaolin clay	9.95	8.5
Ethylene	140	Zinc stearate	6.47	3.4
Ethylene	140	Stearic acid	5.85	5.3
Acrylonitrile/ Butadiene/Styrene	150	Ethylene distearamide	3.03	4.0
Acetal	200	Nylon	6.08	8.5
Propylene	160	Antimony oxide	26	2.1

$$\frac{\text{measured absorbance} \times 100}{\text{absorptivity} \times \text{sample thickness (in mils)}} = \text{weight percent}$$

[a] A given wavelength, x (μm) and wave number, y (cm^{-3}) can be interconverted by the formula $xy = 10\,000$.
[b] 1 mil = 0.001 1 inch = 0.0254 mm.

Often simple preliminary testing can be used as a clue to the best analysis method. Examination under a stereomicroscope will give information about homogeneity and about the form of additives such as chopped fibers. Ashing a sample on the end of a spatula may leave an inorganic residue. If a flame retardant is claimed to be present, the sample should not support combustion.

ANALYSIS BY DIRECT MEASUREMENT WITHOUT SEPARATIONS

Thin films of thermoplastic polymers may be prepared for IR by hot pressing at an appropriate temperature and thickness. Their infra-red spectra will display peaks unique to certain additives. Quantitative data can be obtained from measured absorbance, measured film thickness and the absorptivity given in Table 1. This assumes that the polymer is the solvent. Polyolefins lend themselves best to this technique, because their IR spectra have few interfering bands. Usually the quantitative measure of organic additives can be expected to be at least ±10% relative, except in the case of inorganics where the result is only a reasonable estimate of concentration.

Direct measurement of concentration can also be used for some UV absorbing additives by using thin films and a UV spectrophotometer.

SEPARATION BY DISSOLVING THE POLYMER

Solvent extraction can be effective, not only for extracting the additives, but also to remove the polymer matrix. Table 2 lists some common

Table 2 Solvents to dissolve polymers

Polymers	Solvent
Polyvinyl alcohol, polyacrylic salts	Water
Polyvinylidene fluoride, cellulose acetate or nitrate	Acetone
Polyamides (nylons)	Formic acid
Polyurethanes	Dimethyl formamide
Polyvinyl chloride	Tetrhydrofuran
Polystyrene, acrylics, polycarbonates	Chloroform or methylene chloride
Polyolefins	Toluene
Perfluorocarbons, cured polymers aromatic polyesters	Insoluble

polymer solvents. It is a guideline to selecting suitable solvents to remove the polymer and leave behind the insoluble additives.

Dissolving the polymer is a useful technique to isolate inorganic fillers. Filtering or centrifuging can then be used to collect the insoluble solids.

Dissolving the entire sample and then reprecipitating the polymer is an excellent method for quantitative analysis, because it ensures that all of the additive is released. The process may have to be repeated several times to achieve a clean separation.

SEPARATION BY DISSOLVING ADDITIVES

If the additive can be extracted in pure form from the polymer, both identity and concentration can be determined. After evaporation of the solvent IR, NMR or MS can be used to identify the additive. If quantitative analysis is desired, the sample should be finely divided and multiple extractions performed. A Soxhlet extraction apparatus is useful. If the identification technique indicates that more than one additive is present, further separation will be necessary. The use of another extracting solvent may be effective. For example, a methylene chloride extract which has been dried on a watch glass can be rapidly tested using a small amount of a different solvent. Hexane will remove aliphatic additives and leave behind the more polar materials. Table 3 lists some appropriate solvents.

COLUMN CHROMATOGRAPHY

Simple liquid column chromatography based on solvent elution polarity can be used to separate mixtures. A typical column for the separation is an 8–10 mm glass tube approximately 250 mm long, containing 125 mm of chromatographic alumina supported by glass wool. The polymer is extracted with methylene chloride and then with methanol. The

Table 3 Additive solvents for some polymer matrices

Solvent	Polymer	Additive
Methanol	Ethylene	Amide slip agent, stearate salts
	Ethylene/vinylacetate	Sodium acetate residue
Acetone	Phenylene oxide/styrene	Triphenyl phosphate plasticizer
	Ethylene	Phenolic antioxidants
Hexane	Vinylchloride	Phthalate plasticizers
	Urethanes	Phosphate flame retardants
Chloroform	Ethylene	Phenolic antioxidants, waxes
	Acetals	Phenolic antioxidants

combined extracts are dried, mixed with a small amount of column material and added to the top of the column. Elution should then begin with a nonpolar solvent such as *n*-hexane, followed by a more polar solvent such as methylene chloride or chloroform and finally by isopropanol or methanol/ethanol. The cuts ($\sim$2 cm^3) may be collected in small vials and dried in a hood or cast on salt plates for IR analysis. Table 4 gives the expected component separations. This method is particularly useful, since it provides enough separated material for further identification. GC and HPLC are good techniques for quantitative measurements once the individual components are known.

THERMAL SEPARATIONS

Thermal techniques can be used to separate additives which are volatile at high temperature or which sublime. A temperature gradient tube can easily be constructed in any laboratory. A pyrex glass tube 6–8 mm in diameter and 300 mm long is closed at one end. A small amount of sample is placed in the bottom of the tube. It is heated at the closed end so as to establish a thermal gradient along the tube. A simple method of establishing a gradient is to insert the tube into a sleeve wrapped with heating

Table 4 Liquid column separations

Additives	Solvent
Relatively nonpolar additives: oils, waxes, plasticizers, flame retardants	Hexane
More polar additives: antioxidants, stabilizers	Chlorinated solvents
Polar additives: amide slip agents, surfactants, carboxylates (stearate salts)	Alcohols

wire. A rheostat can be used to control the temperature. This technique may be used to separate as little as 50 µg of additives from polymers.

Any additive that will distil or sublime at less than the degradation temperature of the polymer can be collected and analyzed by IR. For example, antioxidants such as butylated hydroxy toluene in polyethylene can be collected and identified by this method.

Thermogravimetric analysis (TGA) is a more refined technique. It provides quantitative data on volatile components. Not only can the concentration of a volatile component be determined, but the transition temperature provides qualitative information.

Ashing the sample is a good technique for inorganics and also yields quantitative data. A moderate temperature (400–500°C) should be used to avoid decomposition of fillers such as carbonates.

In this chapter we have provided a guide to approaching a very complicated analytical problem. The analysis methods were chosen for their simplicity. In some cases a choice of methods is available: plasticizers can be analyzed by extraction or by thermal analysis. Modern paired instruments, such as GC–MS and TGA–IR, provide very powerful techniques for separation and identification. Much more detail is available in the references listed in the bibliography. *Atlas of Polymer and Plastics Analysis* is a comprehensive source, particularly for IR analysis.

BIBLIOGRAPHY

Scholl, F. (1981) *Atlas of Polymer and Plastics Analysis, Vol 3, Additives and Processing Aids*, Verlag Chemie, Basle.

Crompton, T.R. (1977) *Chemical Analysis of Additives in Plastics*, Pergamon Press, Oxford.

Haslam, J., Willis, H.A. and Squirrell, D.C.M. (1972) *Identification and Analysis of Plastics*, Butterworth and Co. Ltd., London.

Keywords: additives, chemical analysis, plastics, gas chromatography, HPLC, infrared spectroscopy, UV, NMR, ICP, atomic absorption spectrometry, X-ray diffraction, thermogravimetric analysis.

Biodegradation of plastics: monitoring what happens

Anthony L. Andrady

THE EFFECT OF LIVING ORGANISMS ON ORGANIC SUBSTANCES

Biodegradation refers to the process of chemical breakdown of a substance due to the action of living organisms. Generally, it is the action of microorganisms present in soil, water or special environments such as compost heaps that is responsible for biodegradation. These environments support large populations of different varieties of bacteria, fungi and actinomycetes species [1]. The chemistry of the breakdown process varies with the substrate in question, but in the case of polymers usually involves an initial breakdown of the linear chain-like molecules into smaller units. These smaller units may in turn be further biodegraded in one or more steps, invariably yielding small inorganic molecules, mainly carbon dioxide, water and ammonia (where the substrate is nitrogenous). The process of complete biodegradation of a polymer or any organic material can be represented by the following simple chemical process.

$$C_xH_yO_z + \left(x - \frac{z}{2} + \frac{y}{4}\right)O_2 \rightarrow xCO_2 + \frac{y}{2}H_2O \qquad (1)$$

Note that the process requires oxygen and is hence an aerobic biodegradation process. In some environments such as in deep-sea sediments, anaerobic digesters, or marshy soil depleted of oxygen, alternate chemical routes are available for biodegradation. Such anaerobic biodegradation processes yield a different mix of products such as methane. But the original material, the substrate, is nevertheless broken down into

Plastics Additives: An A–Z Reference
Edited by G. Pritchard
Published in 1998 by Chapman & Hall, London. ISBN 0 412 72720 X

simple, mostly gaseous products. This process is sometimes referred to as mineralization. A material such as plastic film is sometimes attacked by insects or rodents and reduced to small shreds. As this process involves little or no chemical change, it is not a biodegradation but only a biodeterioration.

The microorganisms facilitate this sequence of biodegradation reactions using extracellular enzymes they secrete. In the process they utilize the substrate using it to derive energy, to grow in size, and to reproduce. This process is referred to as metabolism. Energy is a product on the right-hand side of the above equation but is used by the microorganism. As might be expected, the microorganism prefers substrates that need only a few enzymes to break down and are hence easily biodegradable. With some complex organic compounds and polymers, several enzyme systems or even several different types of microorganisms might be needed to bring about biodegradation. A particularly complex organic polymer such as humus can be recalcitrant in soil because its biodegradation requires such a large array of enzymes.

All organic materials biodegrade and will invariably be reduced to small molecules if left exposed to a biotic environment, and are strictly speaking biodegradable. However, the term biodegradable polymer is commonly used to mean those polymers that biodegrade at a rapid rate compared with commodity plastics such as polyethylene. The rate at which biodegradation proceeds varies widely with the chemical nature of the substance. With most synthetic organic polymers, including plastics, the process is extremely slow compared with that of other organic materials or even with naturally occurring polymers such as cellulose. This is anticipated to some extent as the complex chemical reactions that constitute biodegradation can only occur at the ends of long chain-like polymer molecules. Therefore, the higher the molecular weight of a plastic, the slower it will generally biodegrade. For instance, in the case of polyethylene, only about 0.1% per year of the carbon that makes up the polymer is transformed into CO_2 by biodegradation even under the best laboratory exposure conditions! The most abundant biopolymer, cellulose, under similar conditions will degrade at a much faster rate of over 10% mineralization per week and will be consumed in less than a year. A water soluble simple organic compound such as glucose is fully mineralized in mere days. Any of these substrate materials that are inherently biodegradable will of course not biodegrade in an abiotic environment. This is why organic materials such as human hair and corn seeds stored in burial chambers in pyramids are known to last for hundreds of years.

There are exceptions to the rule that high polymers, including plastics, are resistant to biodegradation. Synthetic aliphatic polyesters such as polycaprolactone, poly(lactic acid) or its copolymers, poly(vinyl alcohol), and some polyethers do undergo rapid biodegradation under biotic exposure conditions. Certain additives affect biodegradability (see under 'Biocides').

MEASURES OF BIODEGRADATION

Several different measures of the biodegradation process exist [2] as follows.

- The obvious direct measure of biodegradation of a polymer will be the loss of polymer from the system. This measure essentially assesses the early stage of biodegradation when a high polymer material is broken down into a low polymer, an oligomer or a non-polymeric product.

  In the case of a plastic material, the gradual biodegradation of the polymer results in a decrease in its average molecular weight. This is also associated with a decrease in strength and other mechanical properties of the polymer. Monitoring any of these properties provides a reliable measure of the early stages of biodegradation.
- The percent conversion of the carbon in the polymer into carbon dioxide is a good measure of biodegradation of a polymer. In the case of an anaerobic process, methane or some other product might be measured in place of (or in addition to) carbon dioxide.
- As microorganisms metabolize the substrate, their growth and reproduction will increase the amount of living microbial tissue or biomass in the system. Therefore, increase in biomass is a good measure of biodegradation as well. The respiratory processes of the biomass that facilitate aerobic biodegradation will require oxygen. The demand for oxygen by a biotic system therefore provides another measure of the biodegradation process.

Which of these approaches are adopted for monitoring biodegradation depends very much on the reasons for monitoring it in the first place. In some applications such as in plastic netting used in aquaculture, it might be sufficient merely to find out the rate at which the strength of the product decreases with the duration of exposure. Where the environmental fate of a plastic needs to be ascertained, the percent conversion of the polymer to carbon dioxide might be employed. Standardized tests are available from various national and international testing organizations that employ these different approaches to measuring biodegradability. Some of these are shown in Table 1.

Biodegradation tests may use either a single species of a microorganism or a consortium of microorganisms for the purpose. The single-species tests provide some information but are not suitable for assessment of the biodegradation of a polymer under more complex real environments.

Table 1 Standard test method for assessment of polymer biodegradability

Environment	ASTM test method	Microorganisms
Single species	D 5247-92, G21, G22	Species of bacteria and fungi specified
Sewage sludge	D5209-91, D5271-92	Aerobic activated sludge organisms
Marine environment	D5437-93	Marine algae and foulant invertebrates
Compost pile	D5338-93	Thermophilic microbes in compost
Anaerobic environment	D5210-92	Anaerobic microbes from activated sludge

Not only will the diversity of the biotic fraction enhance biodegradation in a mixed culture, but co-metabolism can also contribute to the degradation process. A single species capable of biodegradation of a polymer may not survive as a significant species in a mixed culture in spite of its ability to utilize the polymer as a carbon source. Therefore, the potential of a given single species for biodegradation of a particular polymer cannot always be extrapolated to real-life environments. In general, test methods that rely on mixed cultures of microorganisms tend to be more informative.

A biodegradation test employing a mixed culture of microorganisms might be carried out in a natural biotic exposure site or in a laboratory biotic medium. In either case the biotic potential of the medium must be quantified, in terms of the concentration of biomass at the natural location or in the laboratory media, or the amount of active microbial concentrate (the inoculum) used to create it. Using the natural environment has the advantage that the amount of biodegradation obtained is likely to be similar to that obtained under field conditions. With laboratory methods the advantages are improved reproducibility and some degree of acceleration of the process, allowing the results to be obtained faster. Scientific literature and standardized test methods by the ASTM and OECD refer to procedures for conducting biodegradation tests under several biotic environments [3].

TESTS FOR THE SUSCEPTIBILITY OF POLYMERS TO BIODEGRADATION

Tests designed to assess the biodegradability of polymeric materials consist of the following general steps.

- Preparation of a biotic medium using a suitable inoculum.

- Incubation of plastic material sample, a blank, and reference materials in the medium. A readily biodegradable reference material is commonly used. In addition a bio-inert material might be used as a negative control.
- Periodic measurement of a pertinent property of the plastic material, the biotic system, or both. Typical properties monitored are:
  - respiratory gases from biomass (CO_2 and/or O_2);
  - change in the amount of polymer or in a selected property of the test material such as the tensile strength.

TESTING FOR COMPLETE BIODEGRADATION OF A POLYMER

In studies of the environmental fate of polymers it is often necessary to study the rate and extent to which a polymer is completely biodegraded. This is conveniently assessed in terms of mineralization by employing the modified Sturm test [4]. It is a laboratory respirometric method that uses sewage sludge inocula as the biotic fraction, but the user may use other microbial environments of interest. With certain types of polymers that are readily biodegradable (such as aliphatic polyesters or polyacids) and with biopolymer materials (paper, cotton cloth, ligno-cellulose, cellophane, chitin, chitosan) it is convenient to use a specially designed flask [5] for this purpose (Figure 1).

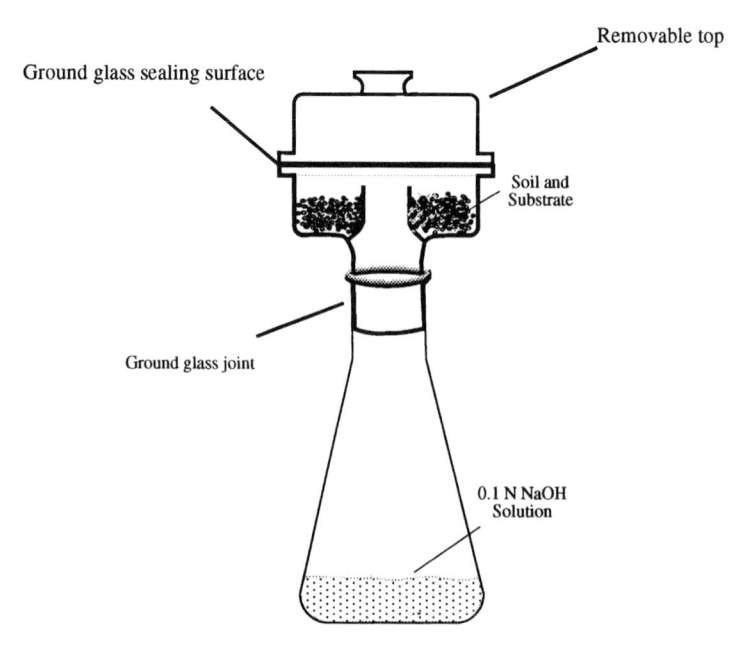

Figure 1 Biodegradation flask [5].

The general procedure for carrying out a modified Sturm test is discussed below. The basic methods of media preparation and incubation are common to other available tests and can be modified using alternate sources of inoculum. Where monitoring the weight of the plastic material or mechanical properties of the sample is desired, a film sample of polymer is used in place of the finely divided form.

Biotic medium

The test might be carried in water or soil, but the latter is discussed here. In both cases a suitable volume of activated sludge to obtain complete mineralization of the sample in about a month is used as an inoculum. About 10 cm^3 per liter is recommended for aqueous systems and we have used 5 cm^3 per 50 g with soil media successfully. The inoculum must be used the same day as collected and kept aerated until used. Also sufficient urea and potassium hydrogen phosphate (0.1 and 0.05 percent of weight of polymer substrate) are added to the medium to fortify it and to promote faster microbial growth.

The soil water content is adjusted to 60% of its maximum water-holding capacity and the system is incubated at 32°C for a day.

Incubation of sample

About 50 g quantities of the soil are transferred into the upper chamber of each biodegradation flask assembly (Figure 1). A small quantity (usually 0.1 g per 50 g soil) of the plastic, preferably in a finely divided form, is added to the soil in test flasks. A blank with no polymer mixed with the soil and a flask with a reference material such as cellulose (cellophane or filter paper) are generally used.

A known volume of 0.1N KOH is placed in the bottom of the flask, and it is closed firmly with the upper part using a very thin film of high vacuum grease to ensure air-tightness. Under incubation biodegradation proceeds yielding CO_2 that flows down from the upper chamber into the bottom of the flask to dissolve in the alkali. The amount of CO_2 reacted with the alkali can be determined by titration against HCl. While the precision of the titrimetry is quite good, it is still desirable to carry out the test in triplicate (making a total of nine flasks for a single substrate test).

Monitoring

At fixed intervals, usually daily, the bottom part of the assembly is removed and titrated against 0.1N HCl using a suitable indicator. A second flask with a fresh quantity of the alkali is substituted in its place

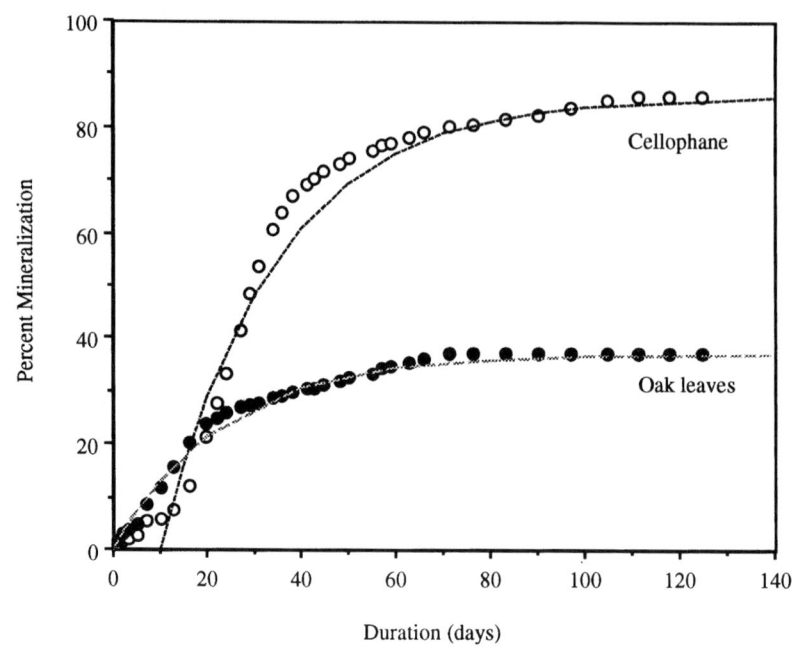

Figure 2 Mineralization curves for cellophane and oak leaves.

and the flask assembly re-incubated for a further 24 hour period. The amount of carbon dioxide liberated in the test flask is corrected for background levels given by the blank determination. The test is continued until 3–4 days of near zero net difference in CO_2 is obtained between the blank and the test flasks.

Typical results [2] obtained from such a mineralization test are shown in Figure 2 in the form of a mineralization curve. The vertical axis shows the percentage of carbon present in the substrate that was converted into carbon dioxide. The test results shown are for oven-dried oak leaves and a cellophane (cellulose) reference material. Note that the biodegradation of the cellophane, which is low molecular weight cellulose, is faster than that of oak leaves which are a composite material made of cellulose and lignin. The latter is more difficult to biodegrade, hence the slower rate. Within about a month the cellophane reference material undergoes extensive carbon conversion (of over 60%). A material that yields at least 60% of its carbon content as CO_2 within 28 days of incubation is regarded as being 'readily biodegradable'. The data, however, are relative as the biotic composition of the inoculum and several other variables remain uncontrolled. The conclusions must compare the result for the sample with that for the reference material.

The shape of the curve in Figure 2 suggests that it can be fitted with the empirical equation

$$y = a\{1 - \exp[-k(t - c)]\}$$
$$= 0 \quad \text{for } t < c \tag{2}$$

where y = the cumulative amount of carbon dioxide evolved as a percentage of the theoretically expected value (the theoretically expected value is the amount of CO_2 obtained when all the carbon in the polymer sample (0.1 g) is fully mineralized), t = time, c = lag time, k = rate constant and a = asymptote of the curve. The value of k might be used as a relative empirical measure of the rate of biodegradability. When comparing the ease of biodegradability of several polymers or a polymer and a reference material, k provides a useful means of quantifying the data.

Precautions and limitations

Several experimental and theoretical limitations must be appreciated in the interpretation of test results.

- The air volume in the flask assembly must be large enough to avoid depletion of oxygen during the monitoring period (one day). The amount of alkali in the flask must be sufficient to absorb all the CO_2 generated in the same period.
- A negligible amount of CO_2 from ambient air might react with the alkali while removing the top chamber and replacing the bottom part of the flask assembly, every monitoring period. With readily biodegradable materials this error is negligible. But with slowly biodegradable material it might be necessary to carry out this operation (and even incubate the flasks) in a glove box with air free of CO_2.
- A 100% conversion of the carbon in polymer to CO_2 should not be expected as some of the carbon is used to increase the microbial biomass. This is why only a 60% conversion is adequate to regard the substrate as being biodegradable. However, the ratio of carbon used to increase biomass to that converted to carbon dioxide varies with the substrate as well as the microbial species.
- The microbial inoculum used in the test might turn out to be a consortium that is specially adapted to particular types of substrates. For instance, soil microbes from a process waste water stream of a starch manufacturing operation are likely to be rich in organisms particularly good at utilizing starch. Such nutritional bias in the inoculum must be taken into consideration in interpreting the test data.

REFERENCES

E.A. Paul and F.E. Clark (1989) *Soil Microbiology and Biochemistry*, Academic Press, New York.
A.L. Andrady (1994) *J. Macromolecular Science, Reviews in Macromolecular Chemistry and Physics*, C34(1), 25.
R. Narayan (1994) In *Biodegradable Plastics and Polymers* (Eds Y. Doi and K. Fukuda) Elsevier Science, New York, p. 261.
R.N. Sturm (1973) *J. American Oil Chemists Association*, 50, 159.
A.L. Andrady, J.E. Pegram and S. Nakatsuka (1993) *J. Environmental Polymer Degradation*, 1(1), 31.

Keywords: biodegradation, bacteria, fungi, actinomycetes, anaerobic processes, metabolism, biodeterioration, cellulose, biomass, algae, mineralization, incubation, inoculum.

See also: Biocides;
Biocides: some kinetic aspects;
Biodegradation promoters for plastics.

Alphabetical section

Acid scavengers for polyolefins

Andreas Thürmer

INTRODUCTION

The transformation of a polyolefin material on its way from the reactor fluff to the finished article usually comprises several steps including processing operations at elevated temperatures. During these steps the polymer material suffers from various degradation processes caused by the combined influence of heat, shear and oxygen, which can significantly alter the processing characteristics of the polymer melt and/or the mechanical and aesthetic properties of the final article. The negative effects of these degradation processes can be restrained by the use of suitable stabilizers, which are usually employed together with co-additives as (for example) acid scavengers. All the additives and co-additives used for the stabilization of a given polymer grade constitute an 'additive system' or 'additive package'.

The base additive packages required for the stabilization of polyolefins usually comprise combinations of phenolic antioxidants (primary antioxidants, radical scavengers), phosphites or phosphonites (secondary antioxidants) and acid scavengers. The co-operative performance of such an additive combination is certainly influenced by the proper choice and concentration of all individual components. Even acid scavengers can play an important role in melt viscosity retention during processing as well as in the long term stability of the final polymer article.

PROPERTIES OF ACID SCAVENGERS APPLIED TO POLYOLEFINS

Acid scavengers (or antacids) are commonly salts of weak organic or inorganic acids, i.e. the conjugated bases are able to neutralize acidity

Plastics Additives: An A–Z Reference
Edited by G. Pritchard
Published in 1998 by Chapman & Hall, London. ISBN 0 412 72720 X

effectively. The efficiency of the acid scavengers is determined by the reactivity of the salt achieved in the apolar polyolefin matrix, and on the other hand, by the acidity of the impurities which expel the weaker acid from its salt. Besides the traditional salts of fatty acids, some inorganic compounds, such as synthetic hydrotalcites or zinc oxide, ZnO, are frequently used.

Antacids serve several important purposes in the additive formulations for polyolefins. The following are examples.

1. The presence of catalyst residues (e.g. from Ziegler–Natta catalyst systems) in a polymer matrix generates free acidity after the catalyst deactivation by steam stripping or solvent treatment. Antacids neutralize this acidity and prevent thereby many undesired side effects such as corrosion of processing equipment, chloride stress-cracking etc. [1].
2. The stearate-type acid scavengers especially act like internal and external slipping agents, which is important especially for the processing of HMW polyolefins and film production, in general, in order to reduce shear forces during the extrusion step.
3. Certain Na and Ca salts of higher fatty acids (C28–C33), besides their acid scavenging property, influence the crystallization behaviour of polyolefins as well as of some engineering plastics such as PET (polyethylene terephthalate) and PA (polyamides). They exhibit certain nucleating effects, i.e. acceleration of crystalline kinetics and enhancement of mechanical properties of the finished article.
4. Improvement in the performance of hindered amine light stabilizers [2] (see the two articles on these stabilizers later in this book) as well as their resistance to pesticides (important for greenhouse films) can be achieved by acid scavengers as well.

Requirements of acid scavengers in polyolefins

Acid scavengers applied in polyolefins have to fulfil further requirements, given by the properties of polyolefins as well as by the transformation processes applied.

1. Compatibility with the polymer matrix, which is for example given by many metallic stearates or other fatty acid derivatives. These are miscible under processing conditions when molten together with (or prior to) the addition of the polymer. For the non-melting inorganic salts (e.g. synthetic hydrotalcites), a coating is recommended for better compatibilization.
2. Optimal particle size in order to achieve even distribution and high homogeneity (i.e. pre-mixes with polymer fluff, prior to extrusion).

Concerning the inorganic salts mentioned, micronized products are used in order to enhance surface area and the homogeneity of dispersion in the extrusion step.
3. Low volatility and high thermal stability referring to the processing conditions of polyolefins, avoiding blooming out, charring or discoloration.
4. High purity of products, so that other components of the additive package are not influenced by undesirable side reactions (e.g., from impurities traced in by antacids). Polyolefin insoluble by-products should be minimized, partly in order to avoid pressure rise during melt filtration and partly to achieve melt clarity.

TYPES OF ACID SCAVENGERS USED IN POLYOLEFINS

Referring to the polyolefin industry, metal stearates (especially calcium, sodium and zinc) and synthetic dihydrotalcites are of paramount importance as acid scavengers applied in additive formulations. A minor part is occupied by the calcium lactates/lactylates (salts of lactic acid), ZnO and other oxides.

Metal stearates

The majority of the commercial metal stearates (besides special grades) are produced by a reaction of certain metal hydroxides or oxides with hydrogenated tallow fatty acids (HTFA) or mixtures of stearic and palmitic acids. Dependent on the production process applied (direct fusion in molten fatty acid or precipitation in an excess of water) a broad range of qualities (from neutral to high base excess) is available in various commercial forms. Concerning polymer applications, a nearly neutral grade of high purity and heat stability, i.e. with low acid content and low amount of water soluble salts (basicity), should be preferred.

Calcium lactates/lactylates

Like calcium stearates, calcium lactates are produced by the reaction of $Ca(OH)_2$ with lactic acid and/or stearic acid to form the calcium stearoyl-2-lactylate, respectively. Besides the conventional acid scavenging performance which is analogous to the stearates, the lactic acid derivatives are able to form chelate complexes with residual amounts of aluminium and titanium via the free OH group present.

By scavenging the metal traces from the catalyst system with high Lewis acidity, discoloration in combination with phenols may be reduced.

However lactates are thermally not as stable as pure stearate derivatives. A certain compromise is achieved with the lactylates.

Inorganic acid scavengers DHT and ZnO

Synthetic hydrotalcites are layered AlMg hydroxycarbonates containing exchangeable anions, which perform as efficient halogen (Cl) scavengers. The abbreviation DHT is widely used and it corresponds to the trade name of the first commercially available synthetic hydrotalcite, from Kyowa Chemical, Inc. Resulting from synthesis, a finely divided free flowing, amorphous powder is yielded, which shows a reasonably high neutralizing capacity in polyolefins compared with conventional Ca stearates, probably supported by absorption effects of the layered structure. The DHT products are characterized by their molar ratio of Al and Mg. Owing to the basic character of the DHT (pH 9.5), higher concentrations can easily lead to an over-basing of the entire additive system, eventually resulting in discolorations by side reactions of phenolic antioxidants present. As observed with all mineral antacids, the proper dispersion in the polymer matrix of under-the-processing-conditions non-melting salts is of vital importance. In the case of DHT, dispersion can be enhanced by a coating of, e.g., sodium stearate, which also gives an additional acid scavenging effect.

The amphoteric ZnO, produced in micronized form by the French process (burning of Zn vapour in an oxygen environment) entered the field of polyolefin stabilization by its UV-'absorbing' property, analogously to TiO_2.

The acid scavenging performance of ZnO may be related to the formation of basic hydroxy-chlorides, $ZnCl_2 \cdot 4Zn(OH)_2$ hydrates, which are known to be formed in the presence of HCl. Absorption effects on the particle surfaces can contribute to the acid scavenging performance of ZnO, as well.

Being a bright white, pigment like product, the ZnO contributes to cover certain discoloration effects at an application concentration of e.g. 500–1000 ppm, without generating a reasonable opacity (micronized grades only).

Mixtures of DHT and ZnO with conventional metal stearates can improve the overall performance of additive packages.

INFLUENCE OF ANTACIDS ON STABILIZATION PACKAGES FOR POLYOLEFINS

Acid scavengers primarily control the Lewis acidity and Cl acidity, introduced into the polymer from the catalyst residues, and also contribute to

Table 1 Influence of antacids on the processing stabilization of PP

Antacid	Ca stearate	Na stearate	Zn stearate	DHT
MFI (g/10 min)	4.46	2.96	3.44	2.64
Yellowness index	1.0	2.2	1.2	3.5

keeping a suitable, more or less neutral environment in the polymer matrix by balancing and also buffering the additive system.

Colour/melt stabilization

The kind and extent of the contribution of acid scavengers to the melt and colour stabilization of polypropylene during multiple extrusion testing is illustrated in the example given in Table 1. All the samples contained 500 ppm phenolic antioxidant, 700 ppm secondary phosphorus based processing stabilizer, and 1000 ppm acid scavenger. The variations in the melt stabilization efficiency (MFI) as well as in the corresponding discoloration (yellowness) are clearly seen for the different acid scavengers used.

The results in Table 1 illustrate that the proper choice of the antacid can improve the overall performance of antioxidant systems. However, the performance ranking of the individual acid scavengers (as shown in Table 1) is influenced, furthermore, by the type of phosphorus processing stabilizer, and therefore, the optimal polymer stabilization should preferably be achieved by adjusting the whole 'additive package' and not only by the one component (e.g. acid scavenger) variation.

Studies [3] on processing stabilization of polyethylenes reported no significant influence of the acid scavengers – Ca stearate and Zn stearate – on the melt stabilization, independent of the polyethylene type (HDPE Cr- or Ziegler-catalysed, LLDPE gas phase- and solution-process). On the other hand, the colour stability was significantly influenced by proper choice of the metal stearate or its combinations with DHT. However, based on the published results, no general recommendation on the type of acid scavenger for polyethylenes can be made; an additive package including acid scavenger should be adjusted and tested for each polymer type separately.

Long-term stabilization

The proper choice of the acid scavenger system is particularly important for the additive packages containing hindered amine light stabilizers (HALS). The amine functionality of HALS is easily protonated in the presence of acidic species, yielding an inactive ammonium salt which does not act as a light stabilizer. It has been shown [2] that the light stability of PP films containing an oligomeric HALS can be enhanced when sodium stearate is used instead of calcium stearate. The improvement was also observed in accelerated weathering of PP fibers which were light stabilized by a HALS containing package [2]. The stabilization efficiency was increased when a less suited acid scavenger, Ca stearate, was replaced by DHT or even better by DHT coated with Na stearate.

CONCLUSION

This chapter is dedicated to the frequently underestimated role of acid scavengers in polyolefin stabilization systems. Acid scavengers significantly contribute to the over-all performance of the whole additive package, required for polyolefin stabilization. Although they are comparatively cheap components, due attention to acid scavengers can give a substantial enhancement to the performance of an additive package. The proper selection and dosing of the acid scavenger is then reflected in improved processing characteristics (e.g. melt stability and colour retention) of polyolefins, better mechanical and physical properties, and a higher long-term stability of the polyolefin articles.

REFERENCES

1. Drake, W.O., Pauquet, J-R. and Todesco, R.V. (1991) Stabilization of New Generation Polypropylene, *Polyolefins VII International Conference*, February 24–27, Houston, Texas.
2. Thürmer, A., Bechtold, K. and Malík, J. (1994) Stabilization of Polypropylene – Aspects of an Enhanced HALS Application, *5th European Polymer Symposium on Polymeric Materials*, October 9–12, Basel, Switzerland.
3. Todesco, R.V. (1992) Recent Advances in Stabilization of Polyethylene: Co-Additives, *MAACK 'Polyethylene World Congress'*, December 7–9, Zürich, Switzerland.

Keywords: polyolefin, stabilizer, catalyst, fatty acid, nucleating, particle size, stearates, aluminium, magnesium, antioxidant, Lewis acidity, chlorine acidity, MFI, yellowness, DHT, HALS.

See also: Antioxidants;
            Hindered amine light stabilizers: an introduction;
            Hindered amine light stabilizers: recent developments.

Anti-blocking of polymer films

Richard Sobottka and Ed Feltham

INTRODUCTION

To some degree, all polymer films tend to stick to themselves. This adhesion is termed 'blocking'. The primary causes of blocking are pressures or temperatures that the film may encounter during extrusion, use, or storage. The degree or severity of the blocking is a function of the film's characteristics as well as the effects of outside forces acting on the film. Blocking can affect the film's entire life cycle, from processing during manufacture of the film to its performance during end use.

Film blocking is experienced most frequently on polyolefin films such as polyethylene and polypropylene. Film hardness and the presence of additives such as plasticizers have a substantial influence on blocking. A relatively soft PVC will block to a greater degree than a relatively hard film such as PET or HDPE. Blocking also occurs on films made from linear polyesters. For any type of film, increasing temperature, pressure, or processing time under varying temperatures and pressures will increase the degree of blocking.

In order to prevent film surfaces from blocking, small amounts of powdered materials, called anti-blocking agents, are incorporated into the polymers. These anti-blocking agents create imperfections on the film's surface and prevent the total contact of the two film layers, reducing blocking.

MEASUREMENT OF BLOCKING

Blocking is measured by calculating the force required to separate two layers of film, and may be expressed in Newtons per unit area or grams

Plastics Additives: An A–Z Reference
Edited by G. Pritchard
Published in 1998 by Chapman & Hall, London. ISBN 0 412 72720 X

required. Several methods exist to determine film blocking, from the simple finger snap test to more standard, scientific methods.

The most simple test of blocking is to rub two layers of film between the thumb and forefinger and note the ease of separation. A slightly more sophisticated but still very subjective test involves inserting one Deutsche Mark or other heavy coin between two layers of film. The film is then shaken until the coin passes completely through the two layers. The harder the film must be shaken, the higher the blocking forces present.

The American Society for Testing and Materials defines a scientific test and apparatus in Method D-3354-89, which measures the force required to separate two layers of film attached to a balance beam device similar to an analytical balance (Figure 1). One sheet of blocked film is attached to an aluminum block secured to the device's base. The other sheet of film is attached to a second aluminum block which hangs

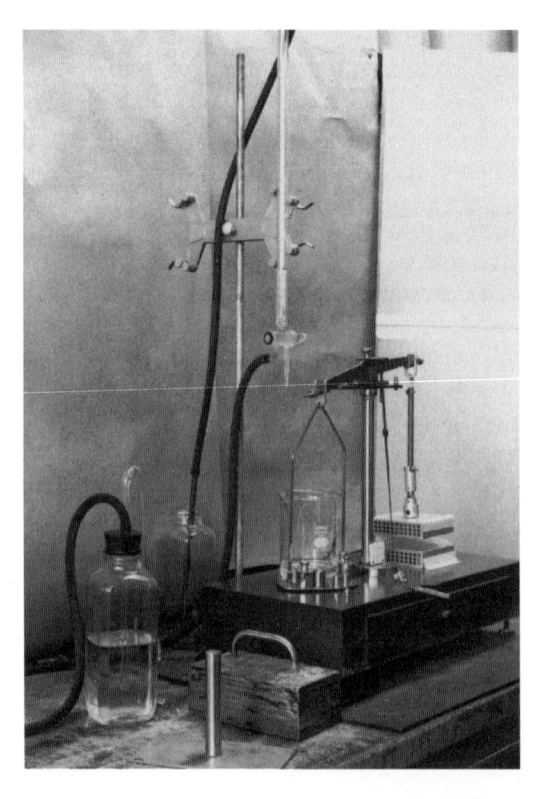

Figure 1 American Society for Testing and Materials (ASTM) Method D-3354-89, which measures the force required to separate two layers of film. Reprinted, with permission, from the Annual Book of ASTM Standards, copyright American Society for Testing and Materials, 100 Barr Harbor Drive, West Conshohocken, PA 19428-2959.

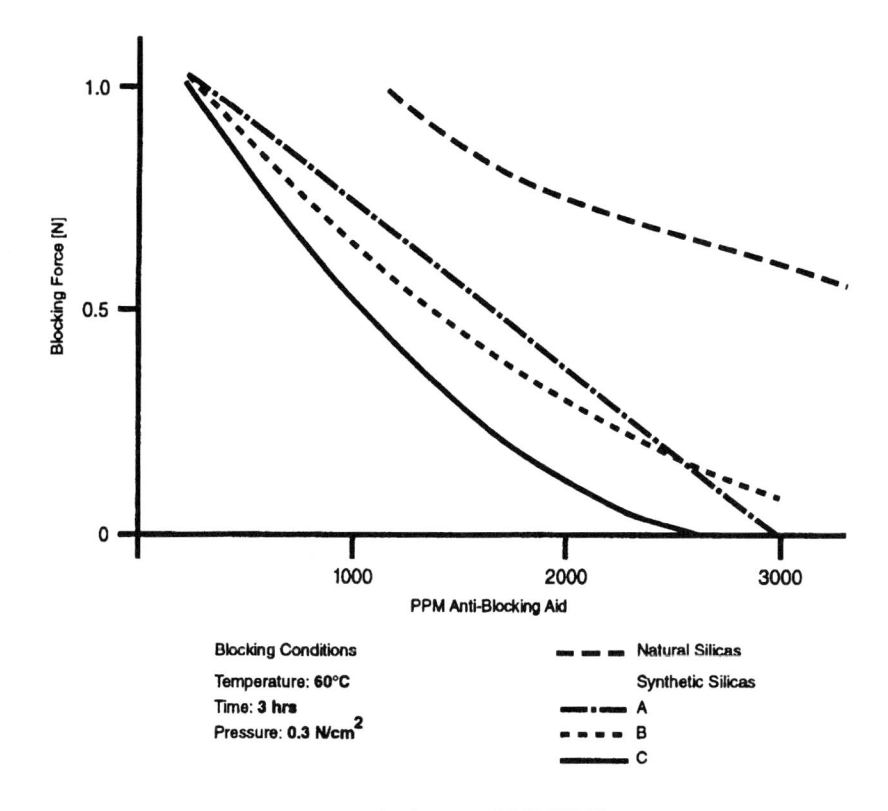

Figure 2 Blocking on LLD-PE films.

from one end of the beam. Weight is added to the other end of the beam in units of $90 \pm 10\,\text{g/m}$, until the two layers either totally separate or reach 1.905 cm separation.

In any test, the films being compared should be conditioned at identical times, temperatures and pressures (Figure 2: anti-blocking curve). These conditioning parameters should be chosen so that the blocking forces are within the scale of the test equipment. Conditioning parameters that are too harsh or too light make it impossible to obtain blocking force readings on the test equipment.

ANTI-BLOCKING AGENTS

Blocking can have detrimental effects on the processing and performance of films, sometimes to the extent that the films are unusable. To lessen the direct contact between layers of film and to counteract blocking forces, anti-blocking and slip-aids may be incorporated into polymers, creating a proper processing environment for the films. Anti-blocking additives must be highly efficient and exhibit constant and dependable quality, with little or no effect over other film properties. Additives should meet the requirements of various countries for products in contact with food. These additives should have no harmful consequences for end-user health and safety.

Anti-blocking can be accomplished either by introducing a thin coating of a non-sticking compound of micron-sized solid particles (slip aids) between the layers, or by creating microscopic roughness on the film's surface to separate layers. For high-quality packaging films, synthetic silicas such as fumed silica, silica gel or zeolites predominate as anti-blocking agents. Naturally occurring silicas and minerals such as clay, diatomaceous earth (DE), quartz or talc, are mainly used in lower-quality and pigmented film grades. The synthetic products have an added benefit in that they do not contain a crystalline phase that is present in many naturally occurring materials. Materials containing a crystalline phase require special handling to minimize dust, and also require special labeling.

Among synthetic silicas (precipitated, fumed, gel), silica gel in micronized form has the greatest potential in anti-blocking applications. In the micronizing process, the silica particle size and particle distribution is closely controlled. The particle sizes of the silica can be adjusted with respect to film thickness – for example, a two micro particle size for very thin films and a ten micron particle size for thicker films. The high porosity of the silicas is an additional benefit, with a large quantity of available particles. According to Moh's scale, silica's hardness factor is low, two to three, preserving equipment and ensuring extended operating life.

The loading of anti-blocking agents in any film is dependent on the film characteristics, the end use, and the efficiency of the anti-block. Loadings for synthetic silica gel may vary from 1000 ppm in an LLDPE film in North America or Europe, to 3000 ppm in a PP film in South East Asia. Talcs are generally used at loadings of 5000 ppm in LLDPE in Europe or North America. Loadings may range higher in certain special films or end uses.

As many film formulations contain both anti-blocking agents and slip aids, combination products are sold by some silica manufacturers. The combination products consist of synthetic silica and a slip aid such as erucamide or oleamide. The combination products show additional benefits through more rapid development of slip and greater anti-blocking efficiency due to better dispersion of both additives.

FACTORS AFFECTING ANTI-BLOCK PERFORMANCE

As anti-blocking agents function by creating surface roughness, their performance is affected by size and shape. Properties such as average particle size, particle size distribution, particle shape and particle strength all influence the efficiency of anti-blocking agents.

In films such as LDPE or PP, the preferred average particle size is generally in the range of 6–20% of the film thickness. Three-dimensional particles, for example DE, quartz, or silica gel, are more efficient than two-dimensional platelet structures, such as clay or talc. The most efficient anti-blocking agents are the synthetic silicas which have efficient average particle sizes and also have narrow, controlled particle size distributions. The narrow distributions ensure that the largest number of particles are of the most efficient size. The high efficiency of the silica gel anti-blocks ensures their usage in high clarity cast polypropylene and PVC films where low loadings have a minimal effect on film properties.

The strength of the anti-blocking agent particles is important in order to retain size and shape during the high shear conditions encountered when the anti-blocking agent is incorporated in the polymer resin.

INCORPORATION OF ANTI-BLOCKING AGENTS

Anti-blocking agents may be incorporated into the polymer resin either by direct addition at the final concentration, or via a concentrate or masterbatch that is added to barefoot resin to achieve the final concentration.

As anti-blocks must be well-dispersed in order to have maximum efficiency, mixing equipment with good dispersive characteristics rather than good distributive characteristics are necessary. The small size of the anti-blocks also necessitates the use of highly dispersive mixing.

Generally, the anti-blocks are dry blended with other additives and stabilizers in a carrier resin. The concentration of anti-block may range from 5 to 10% for synthetic silica to 30–50% for DE or talc. The other additives will be in concentrations that yield the proper final additive levels when let down. To achieve the most thorough mixing, the blended materials are then fed into a mixer/compounder which may be batch equipment, such as a Banbury or Henschel; or into a twin-screw compounding extruder and melt compounded. The pelletized concentrate will be blended with barefoot pellets and let down to the desired final concentration.

In some processes, particularly LDPE, the dry blended additives may be fed through a sidearm extruder into the main resin body to achieve the desired final concentration.

CONCLUSION

The incorporation of micronized silica particles into thermoplastic polymers improves the processing and use of the polymers in thin films.

Keywords: anti-blocking agents, blocking, silica, synthetic silica, clay, diatomaceous earth, talc, coefficient of friction, quartz, masterbatches, slip, slip aids.

ANTIBLOCKING AGENT SUPPLIERS, AND SOURCES OF FURTHER INFORMATION

Grace Davison (Silica gel)
P.O. Box 2117, Baltimore, MD 21202-2117, USA
Tel: (410) 659-9000, Fax: (410) 659-9213

Grace GmbH (Silica gel)
Postfach 449, In der Hollerhecke I, D 6520 Worms, Germany
Tel: 6241-4030

Crosfield Group (Silica gel)
Warrington, WA5 1AB, England
Tel: 925 416100

Specialty Minerals Inc. (Talc)
640 North 13th Street, Easton, PA 18042, USA
Tel: (610) 250-3348

Celite Corporation (Diatomaceous earth)
P.O. Box 519, Lompoc, CA 93438-0519, USA

Antioxidants: an overview

S. Al-Malaika

POLYMER OXIDATION AND ANTIOXIDANTS

The performance of polymer artifacts is adversely affected if degradation occurs during the various stages of polymer manufacture, fabrication, and subsequent exposure to the environment. Molecular oxygen is the major cause of polymer degradation and is responsible for the ultimate mechanical failure of polymer artifacts. The deleterious effect of molecular oxygen is accelerated by many other factors: sunlight; heat; ozone; atmospheric pollutants; water; mechanical stress; adventitious metal and metal ion contaminants.

Polymer degradation during both thermal processing and weathering proceeds essentially through an autoxidative free radical chain reaction process, Scheme 1. This process involves (a) the generation of free radicals (Scheme 1a), (b) propagation reactions, which lead to the formation of hydroperoxides (Scheme 1b and 1c), and (c) termination reactions in which free radicals are eliminated from the autoxidizing system (Scheme 1g, h, and i). Hydroperoxides are inherently unstable to heat, light and metal ions and readily decompose to yield further radicals (Scheme 1d) which would continue to initiate the chain reaction. The role of hydroperoxides in the autoxidation process is to maintain the kinetic chain reaction, which would otherwise rapidly self terminate through radical coupling reactions of the main propagating species (Scheme 1g, 1h, 1i). This autoxidation process normally starts slowly but autoaccelerates leading, in most cases, to catastrophic failure of the polymer artifact.

Plastics Additives: An A–Z Reference
Edited by G. Pritchard
Published in 1998 by Chapman & Hall, London. ISBN 0 412 72720 X

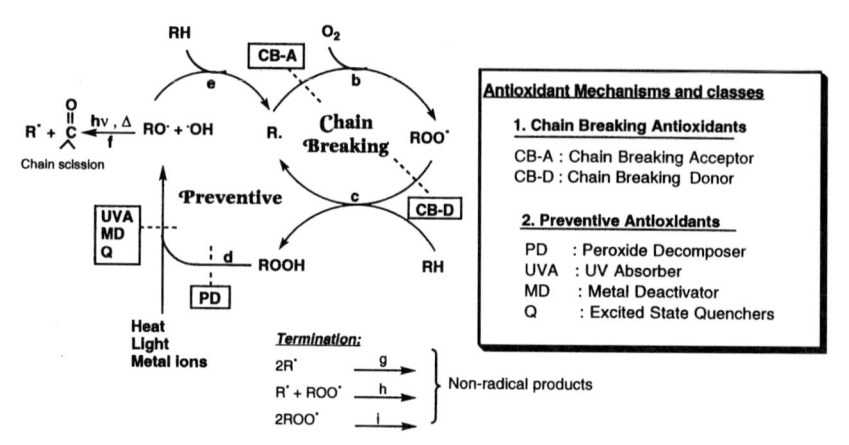

Scheme 1 Oxidative degradation processes and antioxidant mechanisms and classification.

Hydroperoxides, the primary products of autoxidation, are therefore the main initiators in both thermal- and photo-oxidation. Hydroperoxides, and their decomposition products, are ultimately responsible for the changes in molecular structure and overall molar mass of the polymer which are manifested in practice by the loss of mechanical properties (e.g. impact, flexural, tensile, elongation) and by changes in the physical properties of the polymer surface (e.g. loss of gloss, reduced transparency, cracking, chalking, yellowing).

The prior thermal-oxidative history of polymers therefore determines, to a large extent, their photo-oxidative behaviour in service. Processing and fabrication of polymers normally involves the application of mechanical stress at high temperatures in the presence of small quantities of oxygen (Scheme 1). Oxygen and sunlight are the principal degrading agents for hydrocarbon polymers during outdoor weathering.

Inhibition of this oxidation process is, therefore, very important and can be achieved by the use of low levels (0.02–1% but usually 0.05–0.25%) of antioxidants and stabilizers, incorporated during the fabrication process. Almost all synthetic polymers require stabilization against the adverse effects of processing, fabrication, storage and end-use environment. Antioxidants and stabilizers, therefore, occupy a key position in the market of compounding ingredients for polymers, in particular the high volume commodity polymers, e.g. polypropylene, polyethylene and polyvinyl chloride. The term antioxidant is used in this chapter to describe comprehensively all chemical agents which inhibit the oxidation of a polymer matrix arising from the adverse effects of mechanical, thermal, photochemical and environmental factors, whether this oxidation occurs during the processing of the polymeric material, or storage, or throughout the first and subsequent lives of the end-use product.

MECHANISMS OF ACTION AND CLASSIFICATION OF ANTIOXIDANTS

Antioxidants cover different classes of compound which can interfere with the oxidative cycles to inhibit or retard the oxidative degradation of polymers. Scheme 1 shows an outline of the two major antioxidant mechanisms. Two main classes of antioxidant are therefore identified according to the way they interrupt the overall oxidation process: the chain breaking and the preventive antioxidants. Table 1 shows some typical examples of different classes of commercial antioxidants.

Chain breaking (CB) antioxidants (sometimes referred to as primary antioxidants) interrupt the primary oxidation cycle by removing the propagating radicals, $ROO^{\bullet}$ and $R^{\bullet}$. Chain breaking donor antioxidants (CB-D) are electron or hydrogen atom donors which are capable of reducing $ROO^{\bullet}$ to $ROOH$, see reaction 1a. To perform their function, CB-D antioxidants must however be able to compete effectively with the chain propagating step (Scheme 1, reaction c) and the antioxidant radical ($A^{\bullet}$) produced from reaction 1a must lead to stable molecular products, i.e. $A^{\bullet}$ does not continue the kinetic chain either by hydrogen abstraction (reaction 2a) or by reaction with oxygen (reaction 2b). Hindered phenols and aromatic amines (e.g. see Table 1, AOs 1–11) are important examples of commercial CB-D antioxidants. Chain breaking acceptor antioxidants (CB-A) act by oxidizing alkyl radicals in a stoichiometric reaction ($R^{\bullet}$ are removed from the autoxidizing system in competition with the chain propagating reaction, Scheme 1, reaction b, and hence are only effective under oxygen deficient conditions); see reaction 3. Quinones and stable free radicals which can act as alkyl radical trapping agents are good examples of CB-A antioxidants.

Stabilization reaction

$$AH + ROO^{\bullet} \xrightarrow{CB\text{-}D} ROOH + A^{\bullet} \rightarrow \text{Non-radical products} \quad (1a)$$

Propagation reactions

$$A^{\bullet} \begin{cases} \xrightarrow{+RH} AH + R^{\bullet} \longrightarrow ROO^{\bullet} & (2a) \\ \xrightarrow{+O_2/RH} AOOH + R^{\bullet} \longrightarrow ROO^{\bullet} & (2b) \end{cases}$$

Stabilization reaction

$$Q + R^{\bullet} \xrightarrow[O_2\text{-deficient}]{CB\text{-}A} \text{Non-radical products} \quad (3)$$

where A is a CB-D antioxidant and $A^{\bullet}$ is the antioxidant radical. Q is a CB-A antioxidant.

Table 1 Some commercial antioxidants classified according to the two different antioxidant mechanisms

Antioxidant (AO)	Commercial or common name

1. CHAIN-BREAKING (CB) ANTIOXIDANTS

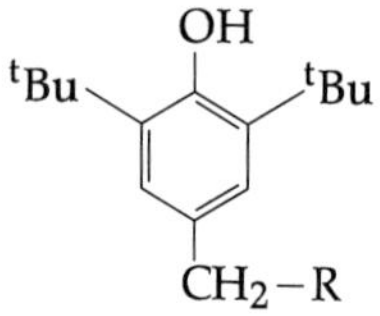

2,6-di-tert-butyl-4-(CH₂–R)phenol (AH)

1 BHT

R = –H — 2 Topanol O
R = –CH₂CO₂C₁₈H₁₇ — 3 Irganox 1076
R = –(CH₂CO₂CH₂)₄C — 4 Irganox 1010

R = (2,4,6-trimethylphenyl substituent) — 5 Ethanox 330, Irganox 1330

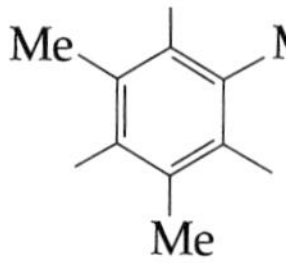

R = (triazine-2,4,6-trione / isocyanurate) — 6 Irganox 3114, Goodrite 3114

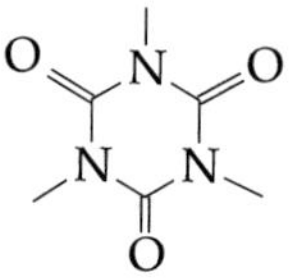

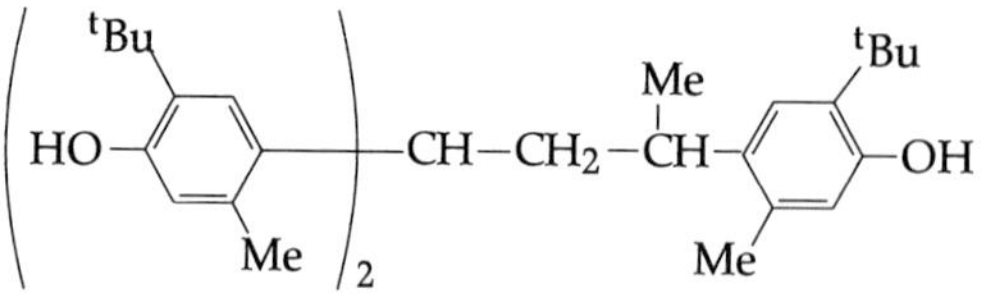

(bis(tBu,Me-hydroxyphenyl)–CH–CH₂–CH(Me)–(tBu,Me-hydroxyphenyl)) — 7 Topanol C

bis(tBu,CH₃-hydroxyphenyl)methane — 8 Cyanox 2246

chromanol with CH₃ groups and R side chain — 9 α-Tocopherol

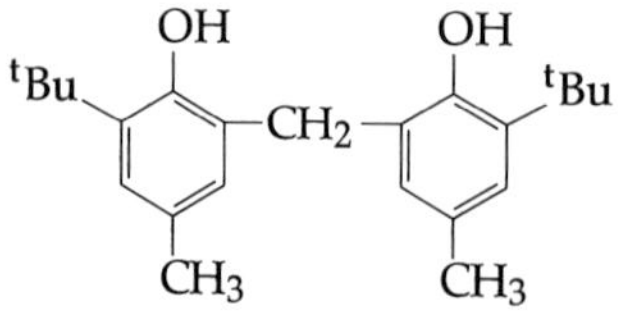

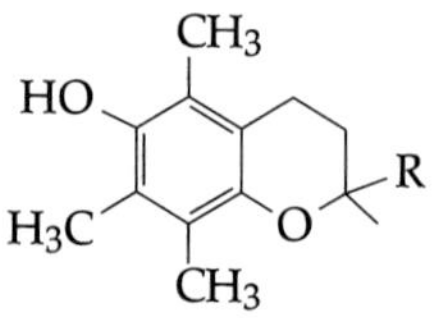

R = (isoprenoid side chain)

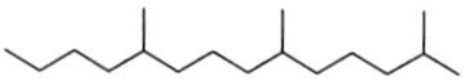

Table 1 Continued

Antioxidant (AO)	Commercial or common name

1. CHAIN-BREAKING (CB) ANTIOXIDANTS – Continued

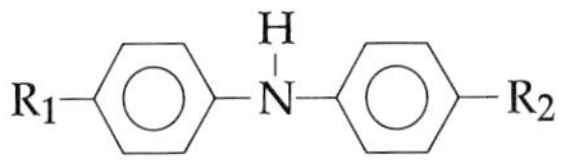

R₁–⟨○⟩–N(H)–⟨○⟩–R₂

10 $R_1 = R_2 = {}^tOct$, Nonox OD

11 $R_1 = H$, $R_2 = HN$–⟨○⟩, Nonox DPPD

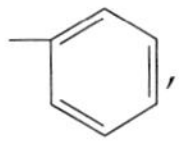

2. PREVENTIVE ANTIOXIDANTS

2.1. Peroxide decomposers (PD)

$(C_{12}H_{25}O)_3P$

12 Phosclere P312, Ultranox TLP

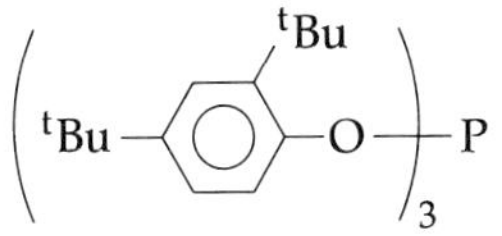

$({}^tBu\text{–⟨○⟩(}{}^tBu\text{)–O–})_3P$

13 Irgafos 168

${}^tBu$–⟨○⟩(${}^tBu$)–O–P(O–)(O–)–⊗–P(O–)(O–)–O–⟨○⟩(${}^tBu$)–${}^tBu$

14 Ultranox 626

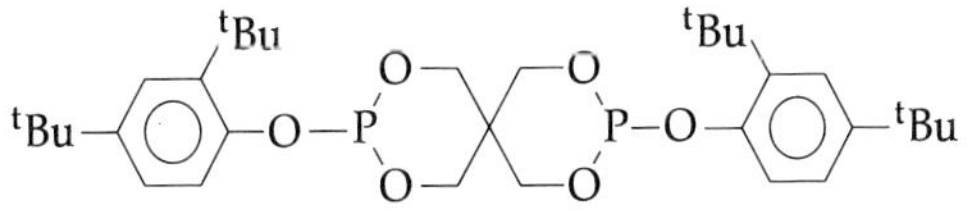

$[RO–\overset{O}{\underset{\|}{C}}–CH_2CH_2]_2S$

$R = C_{18}H_{37}$ 15 Irganox PS802
$R = C_{12}H_{25}$ 16 Irganox PS800

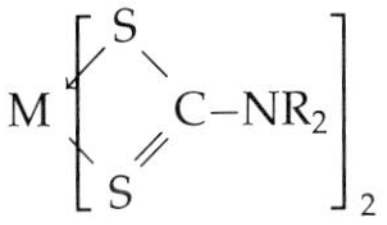

$M[S\text{–}C(=S)\text{–}NR_2]_2$

$M = Zn$, $R = C_4H_9$ 17 Robec Z bud
$M = Fe$ 18 Iron dithiocarbamate

2.2. Metal deactivators (MD)

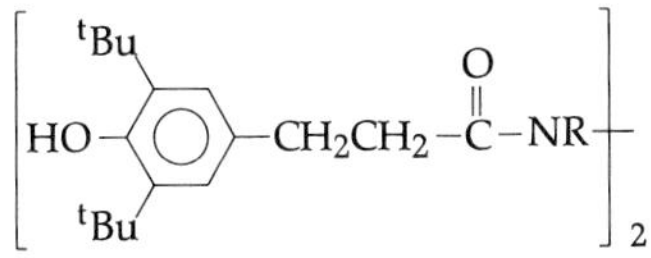

$[HO–⟨○⟩({}^tBu)({}^tBu)–CH_2CH_2–\overset{O}{\underset{\|}{C}}–NR–]_2$

$R = H$ 19 Irganox MD-1024
$R = H–(CH_2)_3$ 20 Irganox 1098

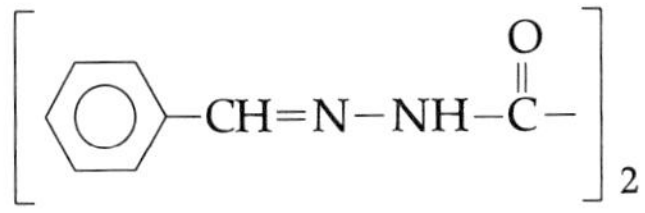

$[⟨○⟩–CH=N–NH–\overset{O}{\underset{\|}{C}}–]_2$

21 Eastman OABH

2.3. Hydrogen chloride scavengers

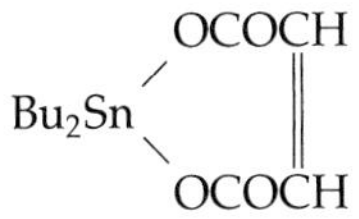

$Bu_2Sn(OCOCH=)(OCOCH=)$

22 Dibutyl tin maleate (DBTM)

60   Antioxidants: an overview

Table 1 Continued

Antioxidant (AO)	Commercial or common name

2.3. Hydrogen chloride scavengers – Continued

Oct$_2$Sn(SCH$_2$$\overset{\overset{\displaystyle O}{\|}}{C}$O–Oct)$_2$ 23 Dioctyl thioglycollate (DOTG)

2.4. Photo-antioxidants

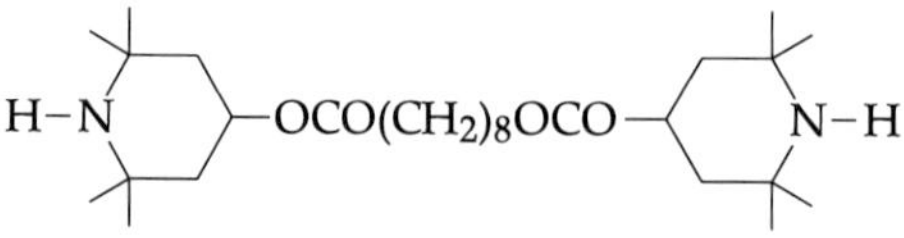

H–N⟨ ⟩–OCO(CH$_2$)$_8$OCO–⟨ ⟩N–H   24 Tinuvin 770

[structure with triazine, piperidine rings, –N–(CH$_2$)$_6$–N–, HN–C(CH$_3$)$_2$CH$_2$C(CH$_3$)$_3$]$_n$ 25 Chimasorb 944

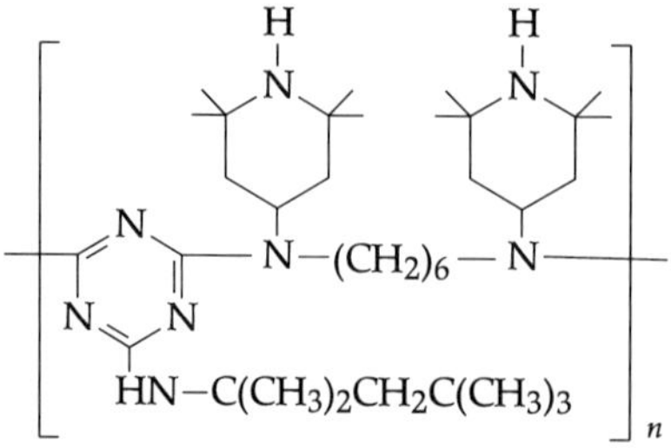

2.5. UV absorbers (UVA)

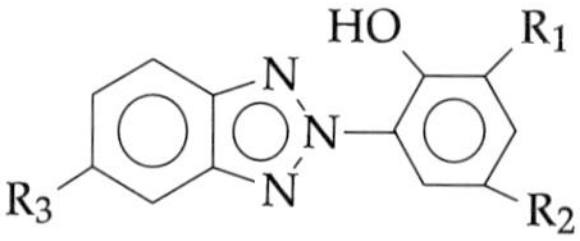

[benzotriazole structure with R$_1$, R$_2$, R$_3$ substituents]

26 Tinuvin 326
   R$_1$ = ^{t}Bu, R$_2$ = CH$_3$, R$_3$ = Cl

27 Tinuvin P
   R$_1$ = H, R$_2$ = CH$_3$, R$_3$ = H

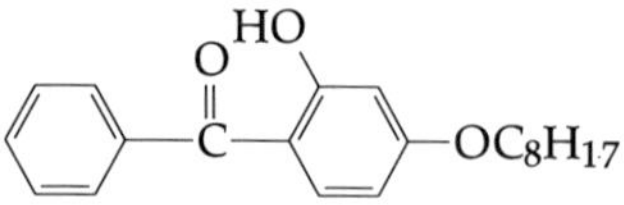

[benzophenone structure with HO and –OC$_8$H$_{17}$]

28 Cyasorb UV531,
   Chimasorb 81

2.6. Nickel complexes

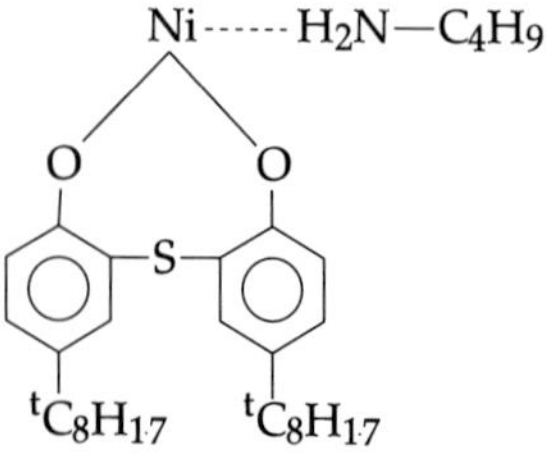

[Ni complex with H$_2$N–C$_4$H$_9$, S-bridged bisphenolate, ^{t}C$_8$H$_{17}$ substituents]

29 Cyasorb UV1084,
   Chimasorb N-705

Table 1 Continued

Antioxidant (AO)		Commercial or common name

2.6. Nickel complexes – Continued

$$\left[\text{HO}-\underset{^{t}Bu}{\overset{^{t}Bu}{\bigcirc}}-CH_2-\overset{O}{\underset{O}{P}}-O-C_2H_5 \right]_2 Ni \qquad 30 \text{ Irgastab 2002}$$

$$\left[(RO)_2-P\underset{S}{\overset{S}{\diagup\!\!\!\diagdown}} Ni \right]_2 \qquad 31 \text{ Nickel dialkyl dithiophosphate (NiDRP)}$$

$$\left[R_2N-C\underset{S}{\overset{S}{\diagup\!\!\!\diagdown}} Ni \right]_2 \qquad 32 \text{ Nickel dialkyl dithiocarbamate (NiDRC)}$$

Preventive antioxidants (sometimes referred to as secondary antioxidants), on the other hand, interrupt the second oxidative cycle by preventing or inhibiting the generation of free radicals. The most important preventive mechanism is the non-radical hydroperoxide decomposition, PD. Phosphite esters and sulphur-containing compounds, e.g. AO 12–18 in Table 1, are the most important classes of peroxide decomposers. The simple trialkyl phosphites (e.g. Table 1, AO 12) decompose hydroperoxides stoichiometrically (PD-S) to yield phosphates and alcohols, see reaction 4. Sulphur compounds, e.g. thioethers and esters of thiodipropionic acid and metal dithiolates (Table 1, AO 15–18, 31, 32), decompose hydroperoxides catalytically (PD-C) whereby one antioxidant molecule destroys several hydroperoxides through the intermediacy of sulphur acids, see reaction 5. References 1 and 2 give detailed discussion on antioxidant mechanisms.

$$P(OR)_3 + ROOH \xrightarrow{\text{PD-S}} O=P(OR)_3 + ROH \qquad (4)$$

$$R-S-R' + ROOH \xrightarrow{\text{PD-C}} O=S\underset{R'}{\overset{R}{\diagdown}} + ROH \qquad (5)$$

$$\downarrow [O]$$

Sulphur acids
(ionic catalysts for HP decomposition)

THE ROLE OF ANTIOXIDANTS DURING PROCESSING AND UNDER CONDITIONS OF SERVICE

Processing antioxidants

Polymer degradation during processing inside a screw extruder or a shearing mixer involves a high rate of radical formation by mechanical scission of the polymer backbone. Because the oxygen concentration under these conditions is only small, alkyl radicals play an important role and become involved in the termination of the autoxidation process. Under normal processing conditions, both alkyl and alkylperoxyl radicals are formed. Antioxidants which can remove macroradicals (in a chain breaking mechanism, CB-D and CB-A) and destroy hydroperoxides (in a preventive mechanism, PD) should, therefore, provide effective stabilization for polymer melts.

The choice of antioxidants for melt stabilization varies depending on the oxidizability of the base polymer, the extrusion temperature, and the performance targets of the end-use applications. Polypropylene (PP), for example, is very sensitive towards oxidation and cannot be processed without adequate stabilization (processing temperatures are in the range 200 to 280°C, and even up to 300°C in certain cases). High density polyethylene (HDPE) is less sensitive to oxidation than PP and lower antioxidant concentrations are normally needed, whereas LDPE (processed around 200°C) is more stable still and may even be used (for certain applications) without the addition of an antioxidant. Unlike PS, rubber modified styrene blends and copolymers, e.g. HIPS, ABS, SAN, are very sensitive to oxidation during processing and require substantial levels of stabilizers during melt processing.

The effectiveness of melt processing antioxidants is normally assessed by subjecting the antioxidant-containing polymer to processing conditions similar to those used in practice, e.g. multiple extrusion, and measuring the changes in melt flow index, MFI, of the polymer. These values are compared with control samples in the absence of the antioxidant. Chain breaking antioxidants are generally used to stabilize the melt in most hydrocarbon polymers. Hindered phenols (CB-D, e.g. AO 1–4, Table 1) are very effective processing (melt) antioxidants for most hydrocarbon polymers. Aromatic amines (also CB-D), on the other hand, have limited use in thermoplastic polymers because of the formation of highly coloured conjugated quinonoid structures during their antioxidant function (their application is therefore limited to stabilizing rubbers). Hindered phenols do not suffer as much from the problem of colour during melt processing with polymers, but yellowing can still occur due to the formation of coloured oxidation products of phenols. Polymer discoloration can be minimized by incorporating peroxidolytic antioxidants, e.g. phosphites (see AO 12–14, Table 1) with hindered phenols during processing.

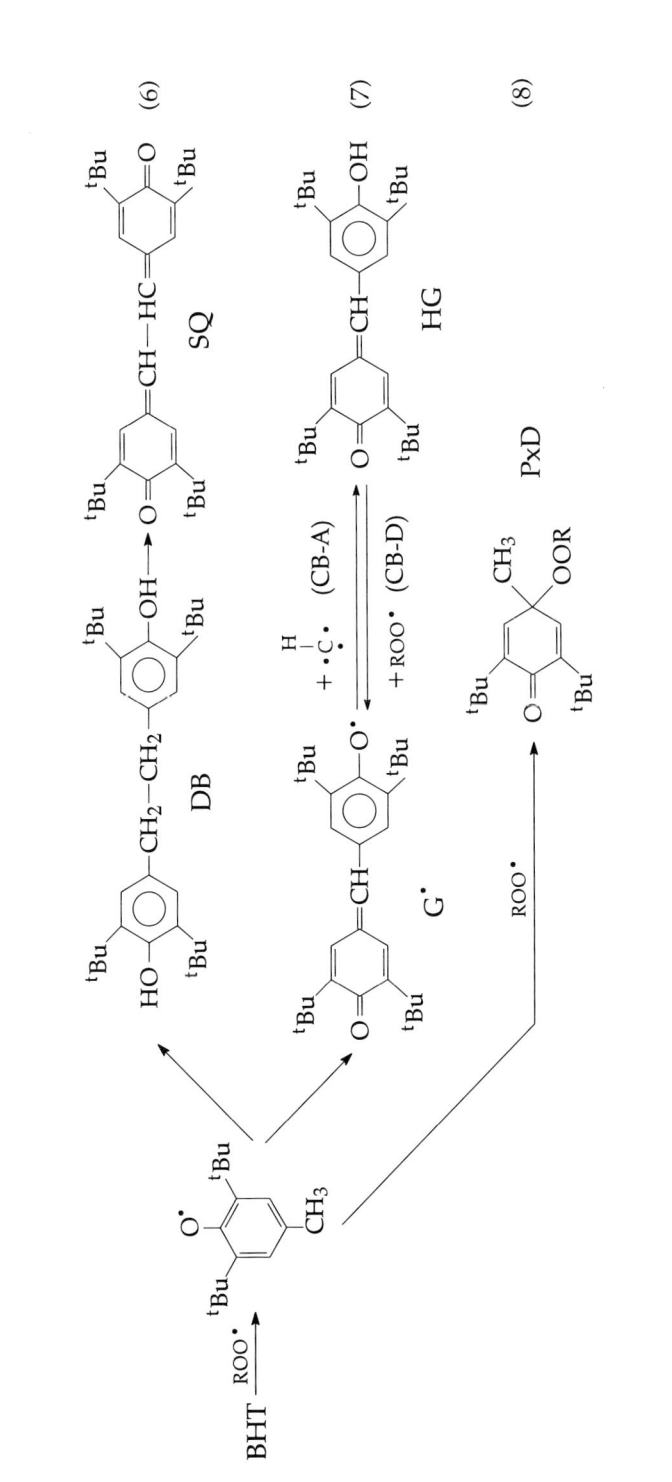

Transformation (oxidation) products of antioxidants formed during melt processing may exert either anti- or pro-oxidant effects; the extent of their contribution determines the overall effectiveness of the antioxidant. For example, in BHT, peroxydienones, PxD (see reactions 8) lead to pro-oxidant effects, due to the presence of the labile peroxide bonds, whereas quinonoid oxidation products, SQ and G$^\bullet$ (reactions 6 and 7), are antioxidants and are more effective than BHT as melt stabilizers for PP. The quinones are effective CB-A antioxidants and those which are stable in their oxidized and reduced forms (e.g. galvinoxyl, G$^\bullet$, and its reduced form, hydrogalvinoxyl, HG) may deactivate both alkyl (via CB-A mechanism) and alkylperoxyl (via CB-D mechanism) radicals in a redox reaction (reaction 7).

Melt stabilization of PVC must be aimed at reducing the formation of HCl and of hydroperoxides in the polymer, and at removing the developing unsaturation which is the source of both colour and further instability. In general, plasticized PVC is processed at lower temperatures than rigid PVC, and less damage is expected during melt processing of the former. Stabilization of PVC has been extensively researched and reviewed. The elimination of labile chlorine atoms from the polymer backbone is the most important stabilization mechanism for PVC. Metal carboxylates (e.g. lead carbonate, calcium, barium and zinc soaps) and tetravalent derivatives of tin (e.g. dibutyl tin maleate, DBTM, AO 22, Table 1) are frequently used for PVC stabilization. Both dialkyl tin maleates and thioglycollates (e.g. DOTG, AO 23, Table 1) function by eliminating HCl (e.g. reaction 9). In addition, the maleates function by removing the unsaturation and limiting colour development, whereas the thioglycollates have an additional peroxidolytic function; see [3] for a review on processing antioxidants.

$$\begin{array}{c} R' \\ \diagdown \\ \diagup \\ R' \end{array} Sn(XR)_2 \xrightarrow{HCl} \begin{array}{c} R' \\ \diagdown \\ \diagup \\ R' \end{array} SnCl_2 + 2RXH \qquad (9)$$

e.g., DOTG$^\bullet$ where XR = SCH$_2$CO$_2$R and R$'$ = Octyl.

Photo-antioxidants

The second most important stage of the polymer lifecycle where oxidative degradation is a major cause for concern occurs during outdoor exposure. The chemical impurities, e.g. hydroperoxides, carbonyl compounds and unsaturation, which are generated during polymer manufacture and fabrication can further sensitize and accelerate polymer degradation under service conditions. Photoxidation of polymers is initiated (at least in the early stages) by hydroperoxides formed during processing. It results in discoloration, gradual loss of mechanical properties and embrittlement of the polymer, with adverse economic and health consequences.

Many polymers which are intrinsically transparent to the shorter wavelengths of the sun's spectrum (<290 nm), become sensitized by the presence of light absorbing impurities and trace level of metals and adventitious species produced during manufacture and fabrication. Hence the susceptibility of commercial polymers (e.g. PE and PP) to outdoor weathering. The outdoor performance of polymers can, however, be markedly improved by photoantioxidants and UV-stabilizers.

The basic requirement for an effective photoantioxidant is that it should not be lost by physical means (because of its high solubility, diffusion, volatility and/or extractability) from the polymer and that the parent antioxidant and its transformation products (formed during melt processing and thermo- and photo-oxidation during in-service) are photostable under continuous exposure to UV light; they must not be lost or transformed into sensitizing products. Other factors which can affect the ultimate photostability of polymers are: sample thickness; polymer crystallinity; and the presence of other additives, e.g., pigments and fillers. For example, chain breaking donor (CB-D) antioxidants such as the hindered phenols are relatively ineffective under photo-oxidative conditions as they are generally unstable to UV light and some of their oxidative transformation products (e.g. (PxD) are powerful photoinitiators). Similarly, many sulphur containing antioxidants which are very effective peroxide decomposers (PD) and thermal antioxidants are not effective photoantioxidants; for example in thiodipropionate esters AO 15 and 16, Table 1, the intermediate sulphoxide which is formed during their antioxidant action photodissociates readily to give free radicals which dramatically reduce their photoantioxidant activity. However, both the hindered phenols and these sulphide antioxidants can synergize with UV stabilizers and become much more effective photoantioxidants.

In contrast to simple sulphides, many metal thiolates (e.g., MDRC, MDRP) are very effective photoantioxidants due to their much higher UV stability. For example, nickel complexes of dithioic acids (see AO 31, 32, Table 1) are excellent photo- and peroxidolytic (PD-C) antioxidants. The metal ion in these compounds plays a crucial part in their overall effectiveness as UV stabilizers. Transition metal complexes containing Ni, Co and Cu are more photostable than group II metal complexes, e.g. Zn, and hence have better overall effectiveness. Other nickel-containing complexes inhibit the photoxidation of polymers by absorbing UV light strongly, e.g. AO 29, 30, Table 1 and, like most other antioxidants, also function by other mechanisms: in this case as excited state quenchers (Q).

The UV absorber (UVA) class (e.g. 2-hydroxybenzophenones and benzotriazoles, see AO 26–28, Table 1) operate primarily by absorbing harmful UV light and dissipating it harmlessly as thermal energy (e.g. via excited state keto-enol tautomerism). Most known UV stabilizers act by more than one mechanism. For example, UV531 (AO 28, Table 1) functions as a UV screen and as a sacrificial antioxidant removing chain initiating radicals (e.g. alkoxyl radicals) via a weak CB-D antioxidant activity. The main limitation of all hydroxybenzophenones is their instability to hydroperoxides and carbonyl compounds during photoxidation. UV531 is often used, therefore, in combination with peroxide decomposers.

Significant commercial development in UV stabilization has been based on hindered piperidines (known as HALS, Hindered Amine Light Stabilizers), see Table 1, AO 24 and 25, which have been used successfully as photoantioxidants in a large number of commercial polymers. See also articles entitled 'Hindered amine light stabilizers: introduction' and 'Hindered amine light stabilizers: recent developments' in this book. HALS is a unique class of photostabilizers which do not subscribe to the general mechanisms of photostabilization: they are not UV screens, do not quench single oxygen or triplet carbonyls, and do not catalyse hydroperoxide decomposition. Their effectiveness, however, is due to their transformation product, the corresponding nitroxyl radicals, which is the real stabilizing species (reaction 10a). Hindered nitroxyl radicals are effective CB-A antioxidants which act by trapping alkyl radicals to give alkyl hydroxylamines (reaction 10b) or/and hydroxylamines (reaction 10c); the former regenerates nitroxyl via a CB-D process (reaction 10d). The overall high efficiency of HALS as UV stabilizers in polymers is attributed to the regeneration of nitroxyl radicals and the complementarity of the CB-A/CB-D antioxidant mechanisms involved. See [4] for a review of photostabilization.

The efficiency of photoantioxidants is tested by first subjecting the stabilized polymer to accelerated weathering and/or outdoor exposure conditions. Accelerated weathering devices have different types of light sources and configuration and include, for example, combinations of fluorescent lamps, xenon arc or carbon arc, and operate in the presence or absence of a combination of other factors, e.g. humidity, temperature, dark and light cycles. The 'weathered' polymers are tested for physical and chemical changes in the presence and absence of antioxidants. These include changes in physical properties (e.g. molecular weight, mechanical properties), chemical microstructure (e.g. build up of carbonyl functions during photoxidative degradation), as well as visual characteristics (e.g. gloss, colour, chalking).

The largest market for photoantioxidants and light stabilizers is in polyolefins. PP is very sensitive to UV light and photostabilization for outdoor and indoor end-use applications is essential. Although PE is less photosensitive than PP, it is still necessary, in general, to stabilize it against UV light for outdoor applications. In the case of LDPE, which is used mainly in films, e.g. in agriculture (mulching films, greenhouse) and for the construction industry (e.g. film covers), the choice of UV stabilizers is more restricted due to limited solubility and blooming of many UV stabilizers. See also the article entitled 'Light and UV stabilization of polymers' in this book.

Thermoxidative antioxidants for stabilization during service

It is important to point out here that stabilizing a polymer for service under elevated temperature conditions is very different from stabilizing it against the high temperatures experienced during processing and fabrication of the polymer. Processing temperatures are very much higher (e.g. for PP, 200–300°C) than those experienced by the polymer under service ageing conditions, which are generally below the polymer melting point (e.g. for PP, rarely above 100°C). The efficiency of thermoxidative antioxidants is assessed following accelerated ageing of the polymer, containing the antioxidant, in a circulating air oven at elevated temperatures by methods similar to those discussed above for testing photoantioxidants (e.g. monitoring carbonyl concentration, changes in tensile strength, elongation-to-break and time to embrittlement of the polymer).

Under normal conditions of thermal oxidation, the ratio of alkylperoxyl radicals to alkyl radicals is very high. The most effective antioxidants for the above conditions of high temperature ageing are CB-D antioxidants, e.g. hindered phenols and aromatic amines. Effective processing antioxidants with high chemical intrinsic activity, however, may not necessarily serve as good antioxidants under the lower temperature service thermoxidative ageing conditions unless they possess suitable physical characteristics, e.g. low volatility. Therefore, stabilizers with higher molecular weight and lower volatility, e.g. AO 3 and 4, Table 1, are potentially more effective than those which have the same antioxidant function but lower molecular weight, such as BHT (AO 1, Table 1).

Peroxide decomposers, e.g. sulphur-containing compounds, enhance the performance of high molecular weight phenols under high temperature service conditions. For example, in polyolefins, dialkyl sulphides (see antioxidants 15 and 16, Table 1) are often used as the peroxide decomposer synergists, while for PVC, organotin stabilizers containing a sulphur moiety (e.g. DRTG, AO 23) are more effective as thermal antioxidants than the traditional organotin compounds without the sulphur (e.g. DRTM, AO 22).

Metal deactivators

The main function of metal deactivators (MD) is to retard efficiently the metal-catalysed oxidation of polymers. Polymer contact with metals occurs widely, for example, when certain fillers, reinforcements and pigments are added to polymers, and, more importantly, when polymers such as polyolefins and PVC are used as insulation materials for copper wires and power cables (copper is a pro-oxidant because it accelerates the decomposition of hydroperoxides to free radicals which initiate polymer oxidation). The deactivators are normally polyfunctional chelating compounds with ligands containing atoms such as N, O, S and P (e.g. see Table 1, AO 19–21) that can chelate with metals and decrease their catalytic activity. Depending on their chemical structures, many metal deactivators function also by other antioxidant mechanisms, e.g. AO 19 and 20 contain the hindered phenol moiety and would function also as CB-D antioxidants.

FACTORS CONTROLLING THE EFFICIENCY AND ACCEPTABILITY OF ANTIOXIDANTS

The commercial success of antioxidants depends not only on their intrinsic antioxidant efficiency but also on their chemical, physical and toxicological characteristics. Chemically, antioxidants should be stable to high temperatures throughout the conversion and fabrication processes of the polymer (it is quite normal for a portion of the antioxidant to be chemically consumed, via oxidative transformations, as a consequence of its antioxidant action). Antioxidants should be hydrolytically stable: this is particularly important for phosphite antioxidants, which suffer from hydrolytic instability giving rise to inorganic acids, hence increasing the risk of corrosion of metal surfaces and the formation of dark coloured spots in the fabricated polymer. Colour instability of antioxidants is also very important, and is mainly due to the formation of coloured oxidation products of the antioxidant (e.g. quinonoid structures from hindered phenol antioxidants); the extent of polymer discoloration (described as 'yellowing'), depends on the chemical structure (e.g. extent of conjugation) of these products.

The physical behaviour of antioxidants is a major factor affecting their efficiency and acceptability. In order to inhibit the oxidation of polymers, the antioxidants have to be present in sufficient concentration at the various oxidation sites. Antioxidants must, therefore, be physically retained in polymers during the high temperature processing/fabrication operation and under aggressive service conditions. Physical factors which control antioxidant retention include diffusivity, solubility, volatility, nature (gas, liquid, solid) of the surrounding medium, polymer morphology and surface-to-volume ratio of the polymer. Antioxidants are usually highly soluble at the elevated processing temperatures but often pass through the equilibrium solubility on cooling and the solid polymer becomes supersaturated with the stabilizer: this can result in 'blooming' or exudation of the antioxidant to the polymer surface. As a result, a concentration gradient near the surface is created, giving rise to further antioxidant migration from the bulk to the surface to restore equilibrium concentrations. Consequently, an antioxidant with low solubility and high diffusion rate is prone to blooming and loss to the surrounding medium, leaving behind an unprotected polymer surface [5].

Migration and ultimate physical loss of antioxidants into the contact media can have severe toxicological consequences in addition to the risks associated with premature failure of the polymer product. Safety of antioxidants is a major issue in polymer stabilization and stringent regulations are already in place, in most countries, on the use of antioxidants in applications involving the human environment, e.g. food packaging and medical implants.

Overall, all aspects of environmental effects, antioxidant safety, optimum concentration levels and antioxidant permanence under the specified service conditions have to be considered and balanced with the initial cost of the antioxidant; the cost effectiveness of antioxidant packages is a decisive factor in all industrial applications.

REACTIVE AND BIOLOGICAL ANTIOXIDANTS

It is imperative to address both problems of the technical capabilities of an antioxidant system and the potential toxicity of the final product during fabrication, storage and use. Research to address these problems has led to various approaches to deal with enhancing the efficiency, substantivity and safety of antioxidants.

One approach to overcoming the problem of antioxidant physical loss is to use reactive antioxidants which attach chemically to the polymer melt during processing (reactive processing). The grafted antioxidant cannot be detached from the polymer matrix except through severance of chemical bonds and should, therefore, lead to a highly substantive and effective antioxidant system, especially in an aggressive leaching environment, with the added advantage of using standard processing machinery to achieve the antioxidant grafting reactions on the polymer backbone. Monomeric antioxidants containing a polymerizable group (e.g. vinyl, acryloyl, methacryloyl) are used normally in the presence of a very small concentration of a free radical initiator to give polymer bound antioxidant functions. A modified approach involves the use of an additional reactive co-monomer which enables very high levels of attachment of the antioxidant, resulting in greater polymer performance and durability, especially under extractive conditions [6].

Another approach to addressing safety aspects of polymer stabilization involves the use of a biological antioxidant such as vitamin E (α-tocopherol is the active form of vitamin E, AO 9, Table 1a). α-tocopherol is essentially a hindered phenol which acts as an effective chain breaking donor (CB-D) antioxidant. α-tocopherol has been shown to be a very effective melt stabilizer for polyolefins, that offers high protection to the polymers at very low concentration. It is highly effective at about a quarter of the 'normal' concentrations used in the case of commercial synthetic hindered phenol antioxidants [7].

SYNERGISTIC AND ANTAGONISTIC EFFECTS OF ANTIOXIDANTS

A co-operative interaction between antioxidants (or antioxidant function) which leads to a greater overall antioxidant effect than the sum of the effects of the separate individual antioxidants is referred to as synergism. Synergism can arise from the combined action of two chemically similar antioxidants, e.g. two hindered phenols (homosynergism), or when two different antioxidant functions are present in the same molecule (auto-synergism, e.g. Irgastab 2002, AO 30, Table 1, having CB and UVA activity), or from the co-operative effects between mechanistically different classes of antioxidants, e.g. the combined effects of chain breaking antioxidants and peroxide decomposers (heterosynergism).

Synergistic mixtures of hindered phenols (mainly chain breaking antioxidants) and phosphites (mainly peroxide decomposers) offer excellent melt and thermal antioxidancy. The synergistic effect is due to a co-operative effect between the high radical scavenging ability of the hindered phenols and the peroxide decomposing activity of the phosphites (hydroperoxides are formed during the first antioxidant step of hindered phenols, see reaction 1a). The phosphites also tend to control the discoloration caused by phenols through their reaction with the coloured phenol oxidation products to give non-coloured products.

Highly effective UV stabilizing systems can also be achieved by using synergistic mixtures of photoantioxidants/stabilizers with complementary antioxidant mechanisms, e.g. the use of combinations of nickel complexes of dithioic acids, NiDRC, NiDRP, with UV absorbers such as UV531, or hindered piperidine light stabilizers with UV absorbers, e.g. the benzotriazoles.

Antisynergistic effects (antagonism) can also occur when two antioxidants interact antagonistically causing a decrease in the sum of their individual effects. For example, antagonism occurs during photoxidation of PP when phenolic antioxidants (CB) are used in combination with metal dithiolates (PD): this is attributed to the sensitized photoxidation of dithiolates by the oxidation products of phenols, particularly, stilbenequinones. Hindered piperidines exhibit a complex behaviour when present in combination with other antioxidants and stabilizers. As discussed earlier they have to be oxidized to the corresponding nitroxyl radical before becoming effective. Consequently, both CB-D and PD antioxidants which remove alkyl peroxyl radicals and hydroperoxides, respectively, antagonize the UV stabilizing action of this class of compounds. However, since the hindered piperidines themselves are neither melt- nor heat-stabilizers for polymers, they are normally incorporated with other conventional antioxidants and stabilizers. See [8] for a review.

REFERENCES

1. Scott, G. (1993) *Atmospheric Oxidation and Antioxidants*, Vol. 1 (ed. G. Scott), Elsevier Applied Science, London and New York, pp. 121–160.
2. Al-Malaika, S. (1993) *Atmospheric Oxidation and Antioxidants*, Vol. 1 (ed. G. Scott), Elsevier Applied Science, London and New York, pp. 161–224.
3. Scott, G. (1993) *Atmospheric Oxidation and Antioxidants*, Vol. 2 (ed. G. Scott), Elsevier Applied Science, London and New York, pp. 141–218.
4. Gugumus, F. (1980) *Oxidation Inhibition of Organic Materials*, Vol. 2 (eds P. Klemchuk and J. Pospisil), CRC Press, Boca Raton, pp. 29–162.
5. Billingham, N.S. (1993) *Atmospheric Oxidation and Antioxidants*, Vol. 2 (ed. G. Scott), Elsevier Applied Science, London and New York, pp. 219–278.
6. Al-Malaika, S. (1993) *Macromolecules – 1992* (ed. J. Kahovec), VSP, the Netherlands, pp. 501–515.
7. Al-Malaika, S. and Issenhuth, S. (1996) *Advances in Chemistry Series – 249* (eds R.L. Clough, K.T. Gillen and N.C. Billingham), ACS, Washington, pp. 425–440.
8. Scott, G. (1993) *Atmospheric Oxidation and Antioxidants*, Vol. 2 (ed. G. Scott), Elsevier Applied Science, London and New York, pp. 431–460.

Keywords: sunlight, heat, hydroperoxides, autoxidation, weathering, photochemical degradation, processing, phosphites, sulphur compounds, metal thiolates, nickel complexes, UV absorbers, HALS, metal deactivators, hindered phenols.

See also: Antioxidants: hindered phenols; Antioxidants for poly(ethylene terephthalate);
Antioxidants: their analysis in plastics;
Hindered amine light stabilizers: introduction;
Hindered amine light stabilizers: recent developments;
Light and UV stabilization of polymers;
Scorch inhibitors for flexible polyurethanes.

Antioxidants: hindered phenols

Jan Pospíšil

INTRODUCTION

Monohydric phenols differing in steric hindrance of the phenolic moiety and in their character of substitution at position 4 are typical chain-breaking antioxidants for commodity and engineering plastics. They have been used as processing stabilizers at temperatures up to 300°C, e.g. in polyolefins. In processing stabilization, they have been mostly combined with hydrolysis-resistant phosphites (secondary antioxidants). Long-term stabilization of plastics is the other principal application area. In this case, phenols are combined with photoantioxidants (hindered amine stabilizers, HAS) and UV absorbers (light stabilizers). The concentration level of phenols added to plastics is generally between 250 and 3000 ppm, depending on the character and expected lifetime of the polymer substrate [1]. Similar concentration levels of phenolics are used for the restabilization of recycled plastics. Permitted levels are administratively regulated for plastics in contact with food. Some suppliers of phenolic antioxidants are listed in Table 1.

MECHANISM OF ACTION

The stabilizing activity of phenols is based on scavenging of the alkyl-peroxy radicals (POO$^\bullet$) generated in oxidizing polymers. Participation of POO$^\bullet$ in the oxidation chain transfer is thus reduced. Reactivity of phenolics with POO$^\bullet$ results, however, in chemical transformation of the original phenolic structure (antioxidant consumption). Hence, the protection of the polymer matrix is diminished in stepwise fashion. Crossconjugated dienoide compounds, e.g. quinone methides, account

Plastics Additives: An A–Z Reference
Edited by G. Pritchard
Published in 1998 by Chapman & Hall, London. ISBN 0 412 72720 X

Antioxidants: hindered phenols

Table 1 Suppliers of persistent hindered phenolic antioxidants

Asahi Denka Kogyo K. K., Furukawa Bldg 3-14, Nihonbashi-Muromachi 2-chome, Chuo-ku, Tokyo, 103 Japan.
BF Goodrich Company, Specialty Polymers and Chemicals Division, 6100 Oak Tree Boulevard, Cleveland, OH 44131, USA.
Chemische Werke Lowi, GmbH, Teplitzer Strasse 14–16, 84469 Waldkreiburg, Germay.
Ciba-Geigy, AG, Head Office, CH-4002 Basel, Switzerland.
Cytec Industries Inc., 5 Garrett Mountain Plaza, West Paterson, NJ 07424, USA.
Ethyl Corporation, Chemicals Group, 451 Florida Blvd, Baton Rouge, LA 70801, USA.
Goodyear Chemicals Co., 1485 Archwood Ave, Akron, OH 44316, USA.
Great Lakes Chemical Italia, S.r.I, Via Maritano 26, S. Donato Milanese, 20097 Milano, Italy.
Hoechst AG, Waxes and Plastics Additives, Gersthofen, 8900 Augsburg 1, Germany.
Imperial Chemical Industries, Specialty Chemicals, Cleeve Road, Leatherhead, Surrey KT22 7SW, UK.
Monsanto Chemical Company, 800 N. Lindbergh Blvd, St. Louis, MO 63166, USA.
Seiko Chemical Co., Ltd., Hirahuma Bldg, 6-2-chome, Kanda Tsukasa-cho, Chiyoda-ku, Tokyo, Japan.
Sumitomo Chemical Company, Ltd., Sumitomo Bldg, 5-33-Kitahama, 4-chome, Chuo-ku, Osaka 541, Japan.
Uniroyal Chemical Company, Inc., Specialty Chemicals, Middlebury, CT 06749, USA.

for some polymer discoloration, and are formed from phenols via phenoxy radicals [2, 3]. The consumption of phenols in the stabilization mechanism has been classified as a sacrificial process.

Twenty-two numbered antioxidant structures are given in this article and some of the more important examples are mentioned in the next few paragraphs.

STRUCTURES OF HINDERED PHENOLS

High inherent chemical efficiency of phenolics is a prerequisite for the successful protection of plastics against degradation. It is controlled by the architecture of the stabilizer molecule. Steric hindrance of the phenolic moiety is one of the factors governing antioxidant efficiency. Most hindered phenols contain one tertiary butyl group combined with a methyl group or two tertiary butyls in positions 2 and 6 [3]. Compounds I and II are typical representatives of broad-spectrum antioxidants for plastics. Some phenolic antioxidants have one of the ortho positions (relative to the HO group) unsubstituted. Compounds III–V are examples.

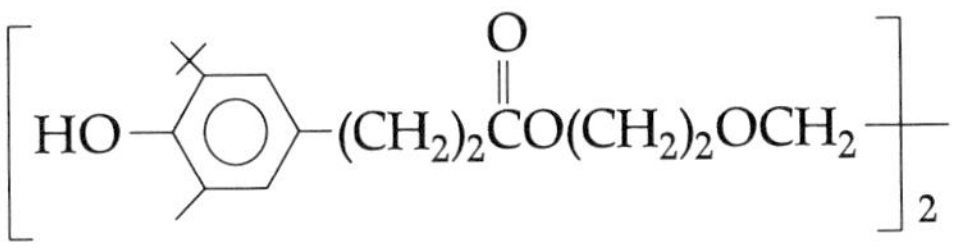

Structures of hindered phenols

[Structures I–V shown: various hindered phenol compounds]

I

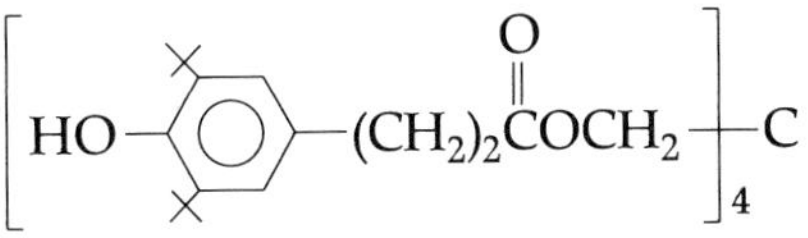

II

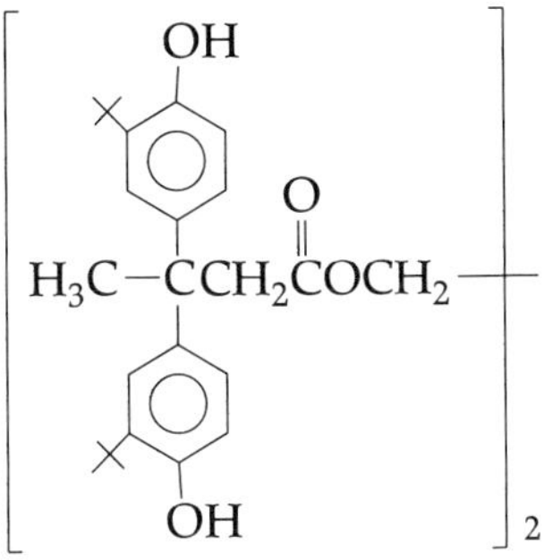

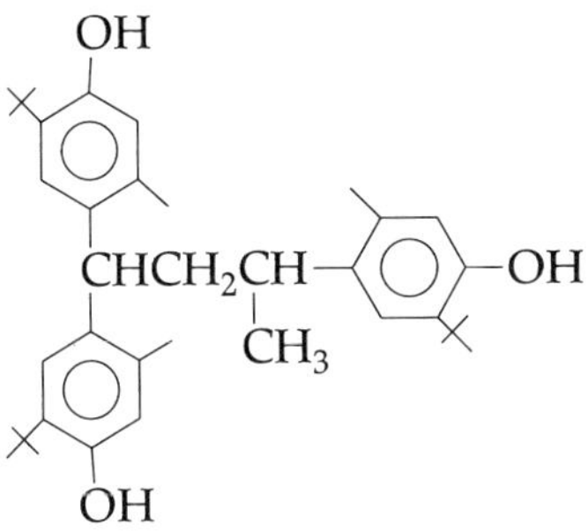

III

IV

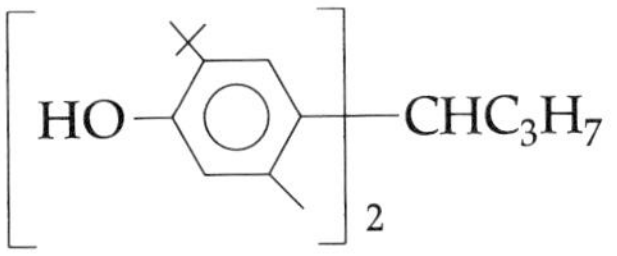

V

Mechanistic elucidations, and analysis of in-polymer tests, has revealed [2, 4] the importance of the structure of the moiety bound to position 4 of the phenolic nucleus. Three main structural types have been used. The first consists of compounds bearing a methyl or a mono- or a disubstituted methyl group, as in IV to VII. After their sacrificial consumption, these phenols transform into relatively stable quinone methides, generally VIII (A, B = residues of the molecule).

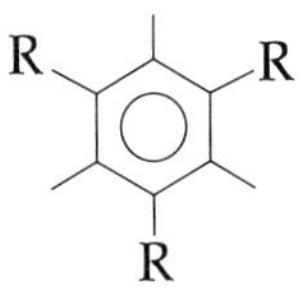

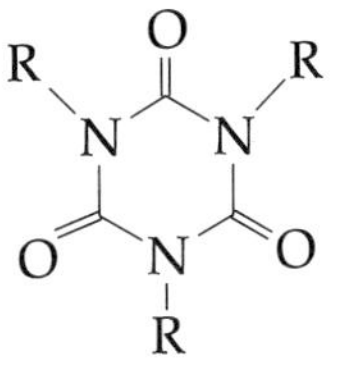

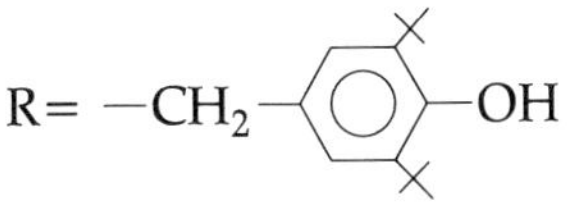

VI

VII

$R = -CH_2--OH$

Antioxidants: hindered phenols

VIII (structure with A, B substituents on quinone methide)
IX CHCH₂COR
X CH=CHCOR

The contribution of this type of quinone methide to the integral stability of plastics is rather low. The second type is substituted in position 4, with a residue of propionic acid derivatives (generally −CH₂CH₂COX), as in I, II, XI–XIV and XVI). Derived quinone methides, e.g. IX, are reactive and prone to isomerize to derivatives of 4-hydroxycinnamic acid (X). The intramolecular rearrangement results in regeneration of the hindered phenolic function, scavenging another POO˙ and, therefore, effectively stabilizing plastics after the 'primary' consumption of the phenol. This

$$[HO-\text{Ar}-(CH_2)_2CNH(CH_2)_3-]_2 \quad \textbf{XI}$$

$$[HO-\text{Ar}-(CH_2)_2CO(CH_2)_3-]_2 \quad \textbf{XII}$$

$$[HO-\text{Ar}-(CH_2)_2CO(CH_2)_2NHC-] \quad \textbf{XIII}$$

$$[HO-\text{Ar}-(CH_2)_2CNH-]_2 \quad \textbf{XIV}$$

XV (bis-phenol with CH₂ bridge and OCCH=CH₂ group)

Structures of hindered phenols

makes the propionate-type of phenols a very attractive group of antioxidants [4]. The third group of phenolics, like III, does not form quinone methides, due to the absence of a hydrogen atom on the α-carbon of the para substituent.

Structural modification of the propionate type of phenolics optimizes or enhances the activity. Esters I and II are antioxidants for polyolefins and styrenics, amide XI is suitable for aliphatic polyamides, ester XII for polyesters. Incorporation of oxalamide or hydrazide moieties in bridges connecting phenolic nuclei in XIII and XIV, respectively, results in antioxidants with metal chelating activity.

Antioxidant bifunctionality is also imparted by the hindered semi-acrylate XV, prone to scavenging P˙ and POO˙ radicals, and by phenolics containing sulfidic (i.e. hydroperoxide decomposing) functions. An efficient antioxidant for crosslinked polyethylene contains the sulfidic function in the propionate bridge (XVI), and antioxidants for styrene copolymers in the bridge directly connecting phenolic nuclei (XVII) or in substituents of the molecule (XVIII).

The high chemical efficiency of phenolics does not itself guarantee successful long-term protection of plastics in aggressive environments. Physical loss of stabilizers by volatility or leaching reduces the final effect. This problem was solved by using phenolics with higher molecular weights which were readily soluble in and compatible with the host polymer matrix. Bi- and polynuclear phenols, exemplified here, fulfil most requirements for physical persistence in demanding applications. Another approach resulting in increased persistence consists of using salts, e.g. XIX. Better dispersion of polymer-insoluble salts is secured by simultaneous application of polyethylene waxes, delivered as one-pack blends with the antioxidant.

XVI

XVII

XVIII

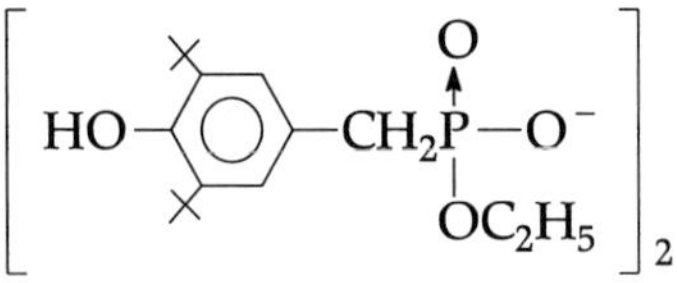

Antioxidants: hindered phenols

XIX 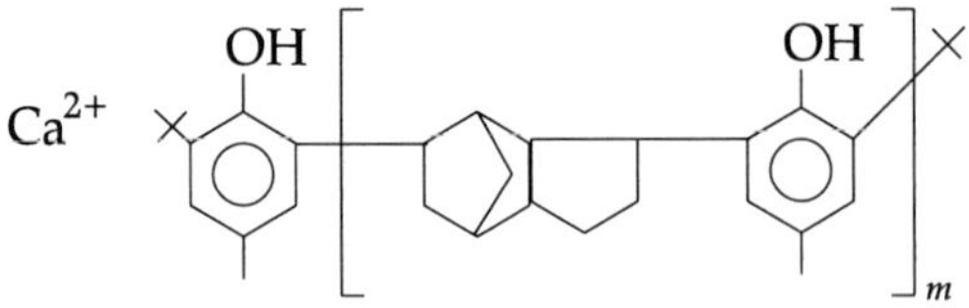 **XX**

For very aggressive conditions, e.g. for the stabilization of plastics for use in the automotive industry, or for materials in contact with hot extracting media, oligomeric or polymeric stabilizers are an option [5]. These systems have to be tailor-made for a particular host polymer, because of potential problems with reduced compatibility. Two examples are the oligomeric phenol XX, and the antioxidant XXI prepared by free-radical copolymerization or antioxidant XXII obtained by functionalization of a conventional polymer. In these persistent phenolics, the strict structural requirements assuring high chemical efficiency are fulfilled as well.

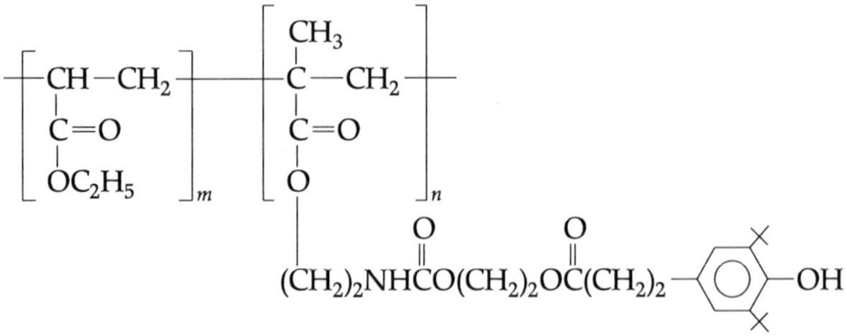

XXI

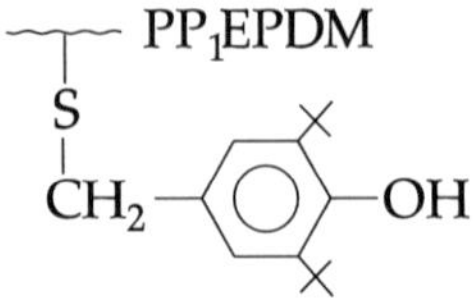

XXII

Hindered phenolic antioxidants are delivered as low-dusting granulates or flakes and are safe additives when properly used according to the rules of industrial hygiene. For specific applications, especially for the stabilization of packaging materials for food contact, the safety sheets issued by producers should be consulted.

REFERENCES

1. Gugumus, F. (1990) Stabilization of plastics against thermal oxidation, in *Oxidation Inhibition in Organic Materials* (eds J. Pospíšil and P. Klemchuk), CRC Press, Boca Raton, Vol. I, pp. 61–172.
2. Pospíšil, J. (1993) Chemical and photochemical behaviour of phenolic antioxidants in polymer stabilization – a state of the art report, Parts I, II, *Polymer Degradation and Stability*, **39**, 103–115; **40**, 217–232.
3. Pospíšil, J. (1990) Antioxidants and related stabilizers, in *Oxidation Inhibition in Organic Materials* (eds J. Pospíšil and P. Klemchuk), CRC Press, Boca Raton, Vol. I, pp. 33–59.
4. Zweifel, H. (1996) Effect of stabilization of polypropylene during processing and its influence on long-term behavior under thermal stress, in *Polymer Durability: Degradation, Stabilization, and Lifetime Prediction* (eds R.L. Clough, N.C. Billingham and K.T. Gillen), Advances in Chemistry Series 249, American Chemical Society, Washington, pp. 375–396.
5. Pospíšil, J. (1991) Functionalized oligomers and polymers as stabilizers for conventional polymers, *Advances in Polymer Science*, **101**, 65–167.

Keywords: light stabilizer, long-term stabilization, photoantioxidant, physical persistence, polynuclear phenols, processing, quinone methide, stabilizers, sterically hindered phenol.

See also: Antioxidants: an overview;
Antioxidants for poly(ethylene terephthalate);
Antioxidants: their analysis in plastics;
Hindered amine light stabilizers: introduction;
Hindered amine light stabilizers: recent developments;
Light and UV stabilization of polymers

Antioxidants: their analysis in plastics

Tony Tikuisis and Van Dang

ANTIOXIDANTS

Most types of commercial plastic materials (including both thermoplastic and thermoset resins) are susceptible to degradation. Plastic or polymer degradation is initiated by a combination of radical generation and/or oxidation reactions, and can occur at any time in the life cycle of the material. The free radical degradation process can be activated by one or more of the following conditions: heat, ultra-violet (UV) radiation and mechanical shear. Reactive impurities such as catalyst residues and the molecular architecture of the polymer (e.g. degree of unsaturation) can also promote and even accelerate the degradation process.

Polymer degradation can result in discoloration and other associated visual defects (such as gels and black specks), inconsistent flow characteristics and a general deterioration in physical properties. Molecular weight can increase as a result of crosslinking reactions or decrease as a result of chain scission reactions.

Antioxidants function by preferentially reacting with the radical intermediates, thereby protecting the polymer and extending its usage life. There are two types of antioxidants that are typically used in commercial polymer stabilization: primary and secondary. The majority of primary antioxidants are either hindered phenolics or secondary aryl-amines. Both hindered phenolics and aryl-amines have one or more reactive OH and NH groups. The hydrogen atoms which are liberated from the alcohol or amine groups readily react with free radicals to form stable species. Secondary antioxidants are usually phosphites or thioesters.

Plastics Additives: An A–Z Reference
Edited by G. Pritchard
Published in 1998 by Chapman & Hall, London. ISBN 0 412 72720 X

They tend to react preferentially, preventing the regeneration of new free radicals from the decomposition of the hydroperoxides. Primary and secondary antioxidants are often used together because they exhibit a synergism which provides an effective mechanism for polymer stabilization.

Characterization of chemical additives such as antioxidants in plastics is often described as a daunting and challenging task. The need for complete compositional analysis of additive packages in industrial plastics for both research and quality control applications has led to the development of numerous analyte-specific test procedures in recent years. The methodology employed in these analyses must overcome many obstacles: the relative instability and high reactivity of many types of additives, the residence of the additives in what is essentially an insoluble polymer material, and the relatively low concentration of these additives in the polymer matrix [1]. In addition, the analysis technique used must be specific and should not be susceptible to interference from other additives which may also be present in the polymer.

Analytical techniques for the quantitative determination of antioxidants in plastics generally fall into two classes: indirect (or destructive) and direct (or non-destructive). Indirect or destructive methods, as the name implies, require a significant alteration to the sample so that the antioxidant can be removed from the plastic material for subsequent detection. Direct or non-destructive methods, on the other hand, involve the direct examination of the plastic sample with minimal sample preparation.

INDIRECT METHODS

Extraction procedures

The majority of the indirect methods used for the analysis of antioxidants in plastic materials utilize some type of extraction process in order to separate the additive from the polymer matrix. Once the desired analyte has been extracted, its concentration level can be readily determined by a variety of techniques. The goal of any extraction technique is to obtain an extraction efficiency for the analyte which meets the analytical requirements in the shortest time possible. The extraction process should be selective so that only the analytes of interest are removed from the polymer. In addition, the actual extraction step should not degrade, modify or adulterate the additives in any way.

Traditional extraction methods have been based on the use of hydrocarbon solvents (in either a direct polymer/solvent reflux or Soxhlet extraction mode). In most cases, preparation of the sample involves modifying its morphology prior to the extraction process by a grinding step in order to maximize the surface-to-volume ratio of the material, thereby shortening the extraction time as well as improving the efficiency of the extraction. Common laboratory grinders such as a Wiley mill can granulate most plastic materials into a powder-like form with a particle size ranging anywhere from less than 100 up to 1000 µm. In some instances, caution should be exercised during the sample preparation step to prevent alteration or loss of the desired analyte. For example, the common polymer antioxidant butylated hydroxy toluene (BHT) is very volatile (mol. wt. 220 g/mol, m.p. 69°C) and can be easily lost during the grinding step, which generates significant frictional heat. Freezing the polymer first with a cryogen such as liquid nitrogen, or using a grinder with coolant circulation, effectively mitigates this problem. The efficacy of the extraction process is influenced by several factors, including: the molecular weight of the additive, the solubility of the additive in the extraction solvent, the density of the polymer, and the amorphous content and degree of crystallinity in the polymer.

One of the drawbacks of conventional solvent extraction is the relatively long extraction times required to obtain quantitative recovery of the desired additive.

Extraction times can vary from as little as 2–3 hours to 24–48 hours for some Soxhlet extractions. The relatively long extraction times usually prohibit the use of these methods for quality control analysis applications in a plastics manufacturing plant. More recently, developments such as the use of microwave oven heating or ultrasonic bath agitation have significantly shortened the extraction time, down to 20–60 min [2].

In addition to extracting the additives of interest, a portion of the polymer matrix may also be solvated during the extraction step. Polyolefins such as polyethylene and polypropylene contain a mixture of polymer chains of varying molecular weights. The low molecular weight or oligomeric (wax) fraction of these polymers is soluble in several of the commonly used extraction solvents, such as cyclohexane, hexane, ethyl acetate, methylene chloride and chloroform. In most cases the soluble polymer fraction interferes with the subsequent detection and quantitation of the extracted additives. Filtration of the solvent extract using a fine filter medium (e.g. Millipore with an average pore size of 0.5 µm) is an effective method of removing the wax.

In some cases, direct solvent extraction does not yield a sufficient additive recovery level. This can occur for a variety of reasons. For example, the molecular weight of the antioxidant can be too high or the polymer matrix could have a relatively high degree of crystallinity, which prohibits the extraction solvent from effectively permeating the polymer and solvating the antioxidant (e.g. the extraction of Irganox 1010 in HDPE). When this occurs, complete dissolution of the sample is an effective way of liberating the additives from the host polymer. A common technique used for polyethylene and polypropylene samples is to dissolve the sample using solvents such as xylene, toluene and di- or trichlorobenzene heated to temperatures as high as 130–150°C. After the plastic sample has been solvated, the polymeric component is precipitated by cooling and/or adding a cold non-solvent such as acetone, methanol or isopropanol. The solvated additives are then separated from the solvent/polymer mixture by filtration. In some instances the polymer does not need to be precipitated out of the solution. For example, polyurethane samples can be readily dissolved using hot dimethylformamide (DMF) and the solution can then by directly analyzed for antioxidant content by high temperature gel chromatography, employing an in-line guard column to filter out the polymer.

Techniques for analysis

After the extraction step has been completed, the antioxidants and other additives present in the solvent can be readily analyzed. The majority of the analytical methods used for measuring the antioxidant content of polymer extract solutions are based on chromatography, such as high performance or high pressure liquid (HPLC), gas (GC), and gel permeation (GPC) or size exclusion (SEC). As most polymer antioxidants contain a phenyl ring or other chromophoric moiety in their structure, UV detection (usually between 200 and 280 nm) is ideally suited for their measurement. The chromatogram shown in Figure 1 aptly demonstrates the utility of

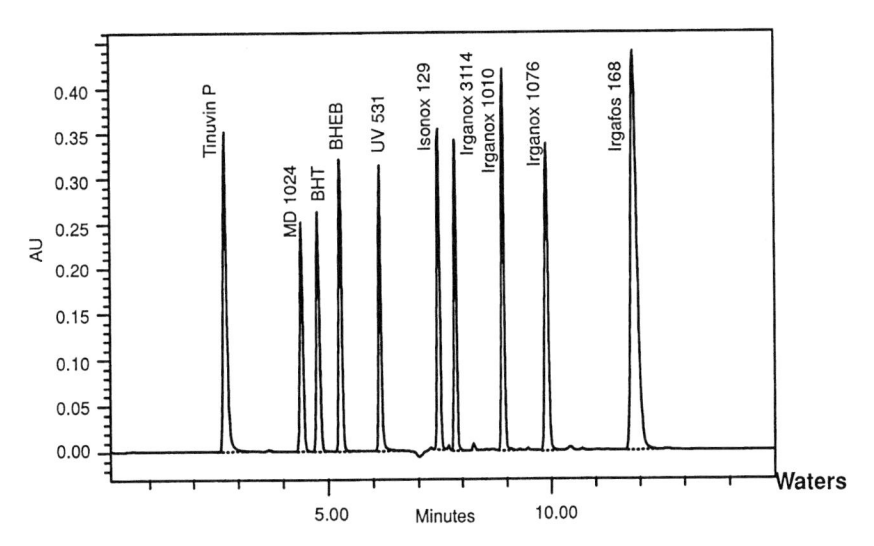

Figure 1 HPLC antioxidant separation with UV detection (courtesy of Waters Corp.^{TM}).

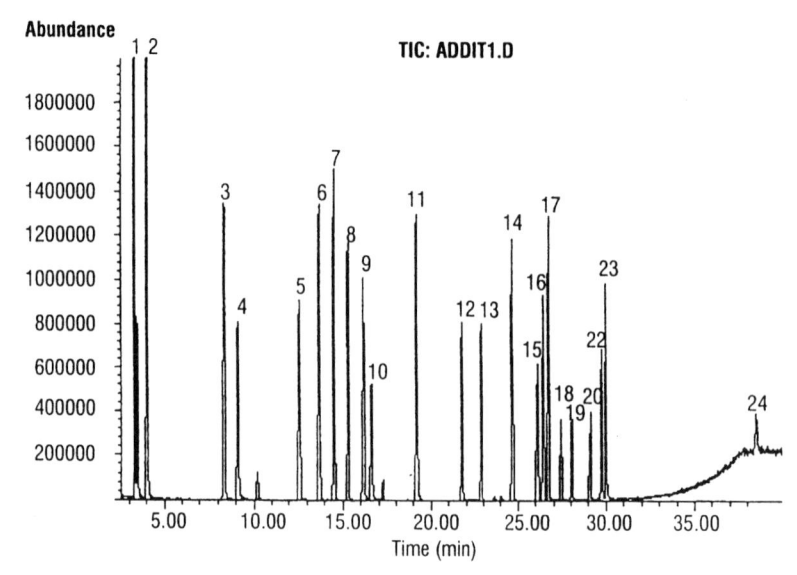

Figure 2 GC-MS analysis of a sample mixture of polymer antioxidants and other additives: (1) BHA; (5) triphenylphosphate; (12) Irgafos 168; (13) Irganox 1076; (15) Irganox 245; (16) Irganox 259; (17) Irganox 565; (20) Irganox 1098; (21) Irganox 3114; (26) Irganox 1330; (23) Irganox PS 802; (24) Irganox 1010 (courtesy of Hewlett Packard[TM]).

using HPLC for the separation and quantitation of common polymer antioxidants [3]. Gas chromatography with mass spectrometric detection (GC/MS) is also a good technique for this application (refer to Figure 2). A simple UV spectrophotometer can also be used. However, in most cases the presence of other additives can interfere with the additive measurement and a chromatographic separation is required prior to detection. The various chromatographic techniques used each have their own advantages and disadvantages. For example, additives which are thermally labile and/or have a high molecular weight can be difficult to analyze by GC, whereas HPLC lacks chromatographic separation efficiency (i.e. resolution) as compared with GC.

SUPERCRITICAL FLUIDS

A relatively new and emerging technique for the extraction of polymer additives is the use of supercritical fluids. The unique properties of supercritical fluids (such as their high diffusion coefficients, approx. 10^{-3}–10^{-4} cm^2/s) make them ideally suited to function as both an extraction medium and mobile phase in chromatographic analysis. Alteration of the density of a supercritical fluid, achieved through temperature and

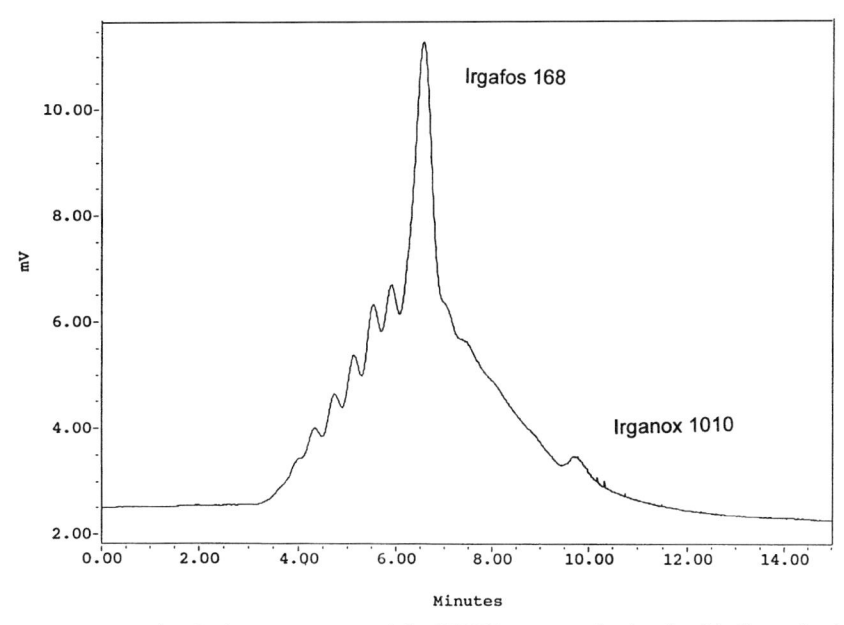

Figure 3 SFE/SFC chromatogram of the HDPE extract obtained with flame ionization detection.

pressure programmed variation, allows the solvation strength of the fluid to be modified and controlled. This is analogous to temperature or solvent gradient programming performed in conventional chromatography (GC or HPLC) to achieve the desired separation of components.

An example of an on-line SFE/SFC (supercritical fluid extraction/chromatography) analysis conducted on a sample of HDPE (high density polyethylene) is shown in Figures 3 and 4. The extraction was performed with a sample size of 0.5 mg using carbon dioxide at 450 atm., at a temperature of 150°C for 30 minutes.

The polyethylene oligomers which are also extracted with the antioxidants are clearly visible with flame ionization detection (FID) and directly interfere with the peak area/height measurement of the two antioxidant peaks (Figure 3). The oligomeric fraction is manifested by a series of regularly spaced peaks in a somewhat normal distribution (representing a mixture of hydrocarbons ranging from C_{10} to C_{30}). When the UV detector is used (Figure 4) the oligomeric species are not detected as they contain few chromophoric functional groups. The two antioxidant peaks (representing Irgafos 168 and Irganox 1010) are clearly visible and can be easily integrated. This example demonstrates the importance of selecting the appropriate detection configuration for the analysis.

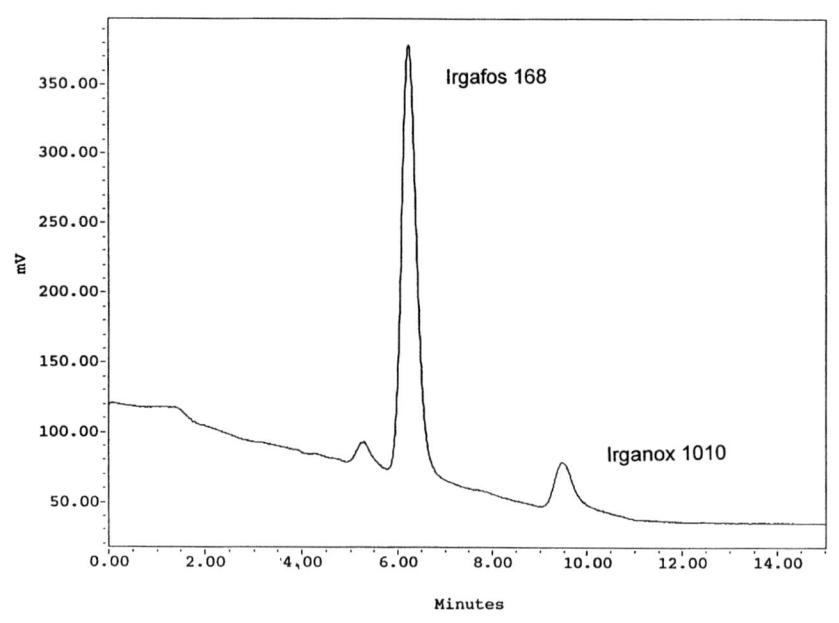

Figure 4 SFE/SFC chromatogram of the HDPE extract obtained with UV detection.

Other destructive techniques used involve the digestion or decomposition of the sample. Wet ashing (i.e. digestion of the sample with a heated mineral acid) or destruction of the sample using a combustion bomb pressurized with oxygen can be used to oxidize the hydrocarbon matrix of the polymer, leaving behind various ionic species in the acidic aqueous solution. For example, the sulfur or phosphorus ions generated from the thioester and/or phosphite type antioxidants originally present in a plastic sample can be quantitated using specific ion electrodes or ion chromatography (IC). Nitrogenous based antioxidants (e.g. aryl amines) present in a plastic sample can be measured by traditional Kjeldahl analysis or pyrolysis/chemiluminescence using an automated nitrogen analyzer.

Thermal analysis

Thermal analysis is another technique which can be used to determine the antioxidant concentration in a polymer sample. The measurement of the oxidative induction time (OIT) of a sample determined with a differential scanning calorimeter (DSC) is a popular method. The OIT of a plastic material is determined by the thermo-analytical measurement of the time interval to the onset of exothermic oxidation of the sample at the specified temperature in an aerobic atmosphere (either air or

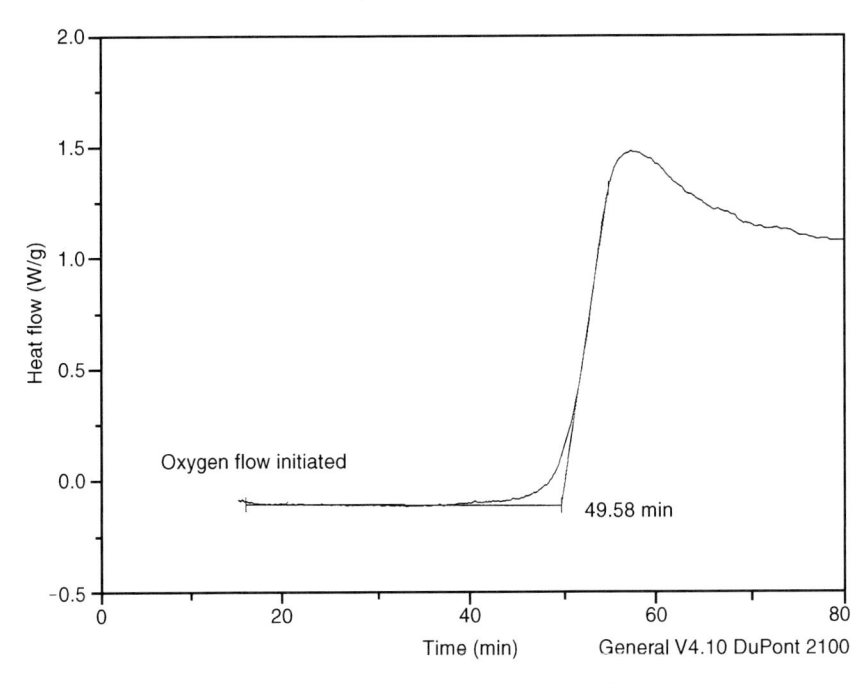

Figure 5 OIT thermal curve of a polyethylene film sample (OIT – 49.6 minutes).

100% oxygen). The onset of oxidation is signaled by an abrupt increase in the sample's evolved heat or temperature (Figure 5).

The OIT test is typically used as a quality control measure to monitor the stabilization level in formulated resins as received from a supplier, prior to extrusion. It is important to note that the OIT test does not provide an identification of which antioxidants are present in the sample. However, quantitative relationships between OIT test data and the level of stabilization can be established for some resin formulations (Figure 6). It is critical that the sample matrix remain relatively constant for the correlation to be maintained. The OIT test is commonly used because of its relative ease and speed; OIT measurements usually require less than one hour at appropriately chosen test conditions.

Not all antioxidants respond to the OIT test; in most cases the phenolic antioxidant concentration can be monitored by measuring the OIT, whereas phosphite antioxidants and hindered amine light stabilizers (HALS) in particular display little response. Because of this, the OIT test is a poor tool for polymer service life predictions as there is no real correlation between the OIT test results and the product's performance in the field. Additional test data such as heat oven aging are often required to supplement the OIT data in order to improve the accuracy of predicting long-term product performance.

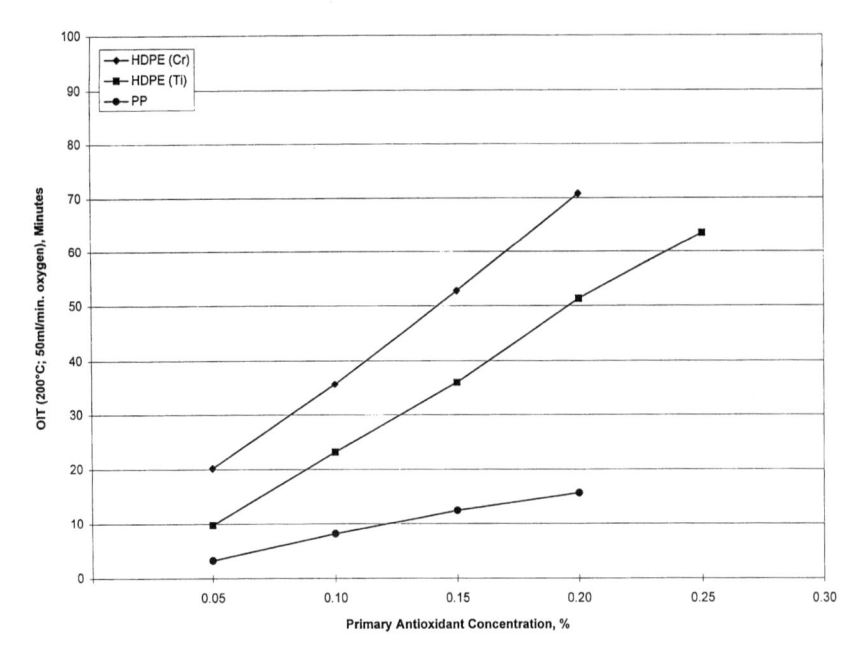

Figure 6 OIT of different polyolefins as a function of the concentration of primary antioxidant (courtesy of Ciba-Geigy).

DIRECT METHODS

Spectroscopic methods

One of the most commonly used direct methods for additive analysis in polymers is based on infrared spectroscopy [4]. With the advent of Fourier transform infrared spectrometers (FTIR) and the development of sophisticated quantitative software programs, direct *in-situ* analysis of antioxidants is a well established technique.

Usually, the only sample treatment required prior to infrared analysis is a modification to the morphology of the sample (e.g. from resin pellets to a film or plaque). This step can be achieved by compression molding the material or by cutting out a test sample coupon from an extruded or processed article (such as an injection molded part). Because the usage level of antioxidants in plastics is generally low (250–2000 ppm) a relatively thick sample pathlength is required (0.25–0.75 mm) for the absorption bands of the antioxidants to be visible. The actual sample thickness required for analysis depends on factors such as the concentration and extinction coefficient of the antioxidants in the sample and the opacity of the sample (pigmented or non-pigmented).

One of the key requirements for infrared analysis is to select an absorption band of the antioxidant which does not interfere with or directly overlap any of the absorption peaks of the host polymer. Once an appropriate absorption band is identified, its absorbance can be readily determined by peak area/height integration. The concentration of the antioxidant in the polymer is then determined by relating the integrated absorbance to a previously established calibration curve. The mid IR domain (i.e. 2.5–25 µm or 4000–400 cm^{-1}) has been used extensively for additive analysis in polymers, as most organic functional groups are infrared active in this spectral region. For example, the carbonyl groups of phenolic antioxidants and the oxygen–phosphorus–phenyl groups of phosphite antioxidants have strong absorbances at approximately 1740 and 890 cm^{-1} respectively.

The infrared spectrum shown in Figure 7 displays a polyethylene sample containing two common polymer antioxidants, Irganox 1010 and Irgafos 168. Note that at the sample pathlength used (approx. 500 µm) the major absorption peaks of the polyethylene are off-scale, while the magnitude of the two antioxidant absorption bands are of sufficient size to be measured. Infrared spectra obtained from a set of polyethylene calibration standards formulated with varying levels of Irganox 1010 and Irgafos 168 are shown superimposed in Figure 8. The expanded spectral region displayed reveals the two antioxidant peaks as well as some polyethylene bands. As expected, the relative peak sizes of the two antioxidant bands increase as a function of concentration while the size of the polyethylene bands remains relatively constant.

One of the main advantages of direct methods of analysis such as infrared spectroscopy is the convenience and speed of the technique. No more than a few minutes are required to generate the infrared spectrum and to calculate the concentration of the antioxidants present. However, infrared methods tend to be very matrix dependent. Any change to the composition of a plastic sample (such as the addition of another stabilizer) can affect the quantitative relationship, and a new calibration curve may be required. Because of this, infrared methods are generally suited only for QC (quality control) applications where they can be used routinely to monitor the addition of antioxidants in the manufacturing of commercial polymers. Development of an accurate calibration curve for the quantitative determination of an antioxidant in a plastic material by infrared analysis can be difficult. Unlike chromatographic analysis, where calibration only requires the gravimetric dissolution of the antioxidant(s) in the appropriate solvent, infrared analysis requires the antioxidant(s) to be homogeneously dispersed throughout the polymer matrix at the correct concentration.

A related technique also very useful for antioxidant analysis in plastic materials is near infrared spectroscopy (NIR) [5], which utilizes the

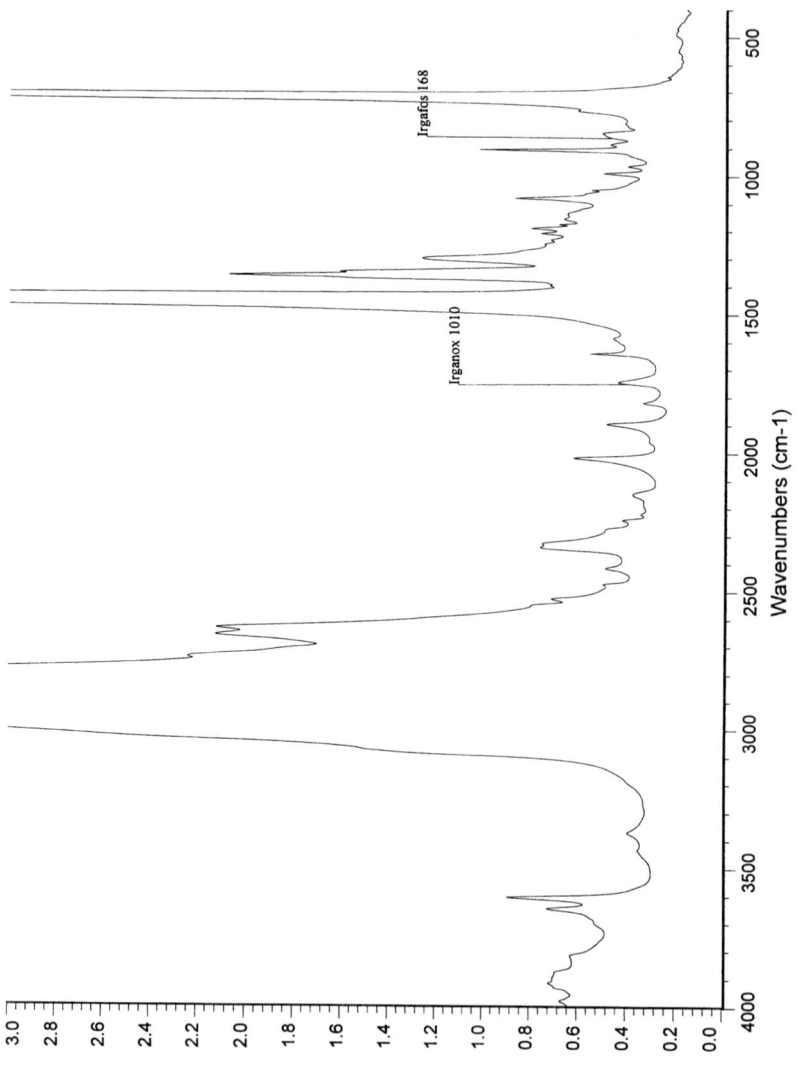

Figure 7 FTIR spectrum of a polyethylene sample formulated with Irganox 1010 and Irgafos 168.

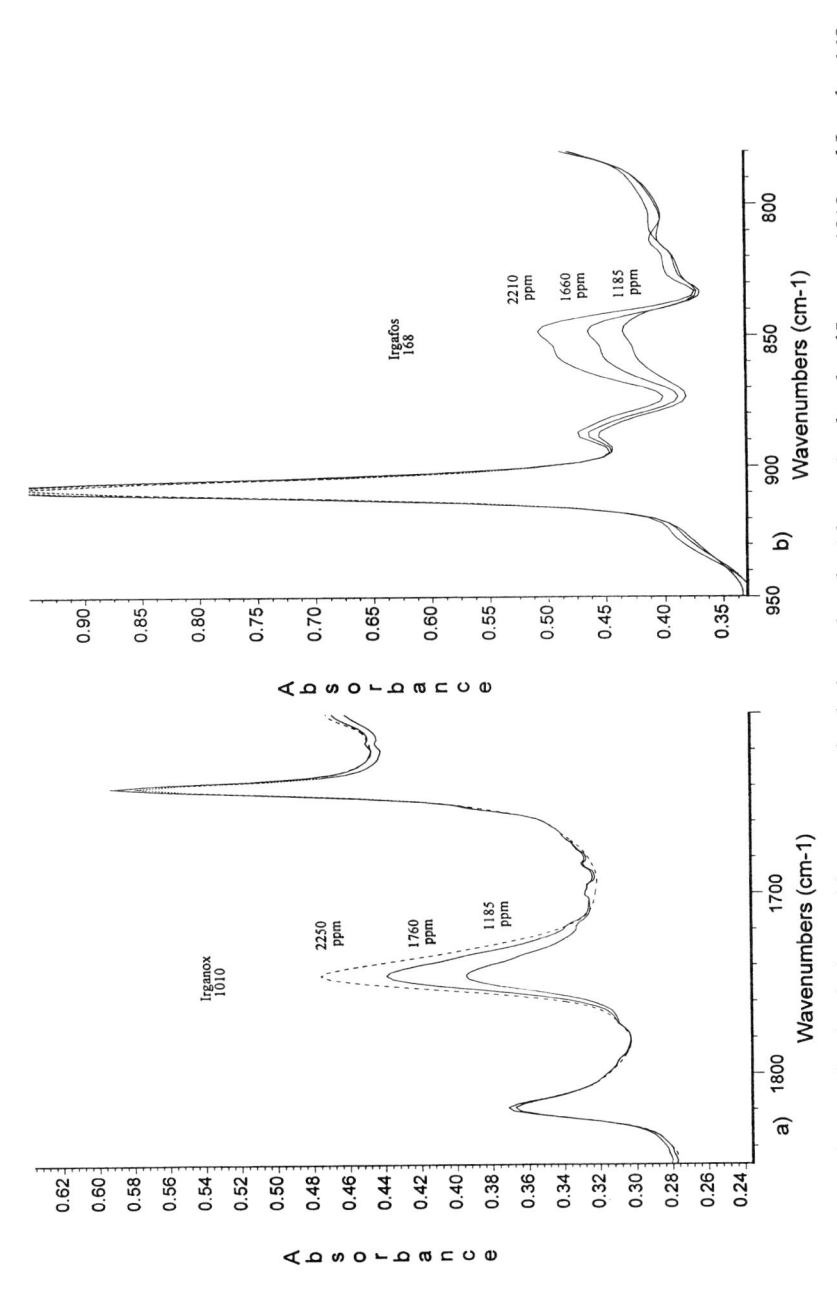

Figure 8 FTIR spectra of a set of polyethylene calibration standards formulated with varying levels of Irganox 1010 and Irgafos 168.

near IR region of the electromagnetic spectrum (i.e., 2.5–0.7 µm or 4000–14 300 cm^{-1}). Hydrogen-terminated groups such as OH, CH and NH groups which are strong mid IR absorbers also give rise to weaker bands in the near IR domain. These near IR absorption bands are caused by overtones/combinations of the fundamental absorptions occurring in the mid IR region.

The lower absorptivity of near IR energy enables it to penetrate more deeply into a sample (up to several centimeters is possible). Because of this, samples subjected to NIR analysis can be almost any convenient size or shape (for applications such as the direct analysis of plastic pellets). As NIR analysis can be conducted on relatively large sample sizes (e.g. 10–100 g of plastic pellets), any minor variation in sample content homogeneity, such as the degree of dispersion of the antioxidant(s), does not impose any significant precision problems on the analytical result obtained. Care must be exercised, however, to ensure that the particle shape and size distribution do not vary significantly from sample to sample because NIR spectra are easily influenced by different sample morphologies, which can affect the absorption band intensity. Observed intensity differences are caused by what is commonly referred to as multiplicative scatter effects. Particle size variations can cause the NIR radiation to scatter differently between samples, which in turn alters the effective pathlength of the NIR radiation into the sample.

Similar to FTIR, NIR spectroscopy is suited primarily for QC type applications where the sample matrix is relatively constant. As NIR radiation is not absorbed by ordinary glass, it can be transported with a fiber optics link. Through the use of a sampling probe the measurement can be brought to the sample rather than bringing the sample to the instrument. With this type of sampling flexibility NIR measurements can be used to provide real-time data (e.g. monitoring the addition level of one or more antioxidants to a polymer in a pelletizing extrusion process, using a sampling probe inserted directly into the polymer melt stream).

X-ray fluorescence

X-Ray fluorescence spectroscopy (XRF) is another commonly used method for the analysis of additives in polymers. As the technique is not suitable for use with elements of lower atomic number than fluorine, primary antioxidants which contain either oxygen or nitrogen as the active element cannot be analyzed by XRF. Secondary antioxidants, however, which contain either phosphorus or sulfur as the active element, can be readily analyzed by this method.

As with infrared analysis, the sample format used for XRF measurement is typically a molded film or plaque. The XRF spectra shown in Figure 9 display the superimposed X-ray emission profiles of four

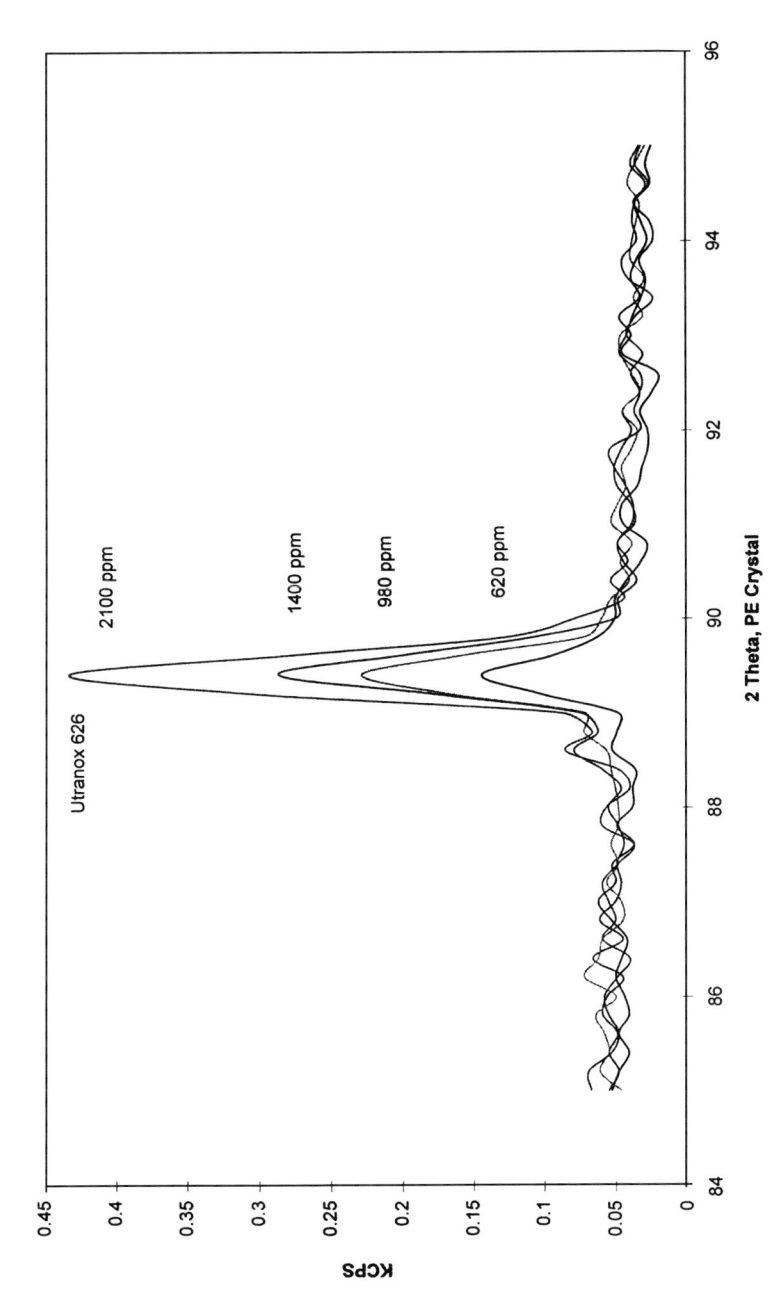

Figure 9 XRF spectra of a set of polyethylene calibration standards formulated with varying levels of Ultranox 626.

polyethylene calibration standards formulated with varying concentrations of Ultranox 626, a phosphite antioxidant. The subsequent calibration curve which is generated can then be used to determine the phosphorus content of unknown samples.

As the XRF technique only provides elemental concentration data, it is a non-specific method (i.e. the same phosphorus calibration curve can be employed for a variety of phosphite antioxidants). Once the phosphorus content of an unknown sample has been obtained, the value is converted to the antioxidant concentration using the appropriate stoichiometric factor (e.g. Ultranox 626, a phosphite antioxidant, has a phosphorus content of 10.2%). Knowledge of which type of antioxidant is formulated into the sample is therefore required. As a result, XRF analysis is more suited for QC type applications, although it can be employed as an effective screening tool to determine the presence of secondary antioxidants in unknown samples.

Unlike some of the other common test methods, the XRF technique does not provide any structural information about the antioxidants detected. In addition to the total concentration, chromatographic and infrared based methods can also provide information on the degree of antioxidant degradation products (such as the phosphate content, which is the oxidized form of phosphite antioxidants, normally created during polymer processing). The XRF technique also has limited utility when more than one phosphite or thioester antioxidant is present in the sample, because the total phosphorus or sulfur concentration does not provide the degree of distribution from each antioxidant.

REFERENCES

1. Crompton, T.R. (1989) *Analysis of Polymers: An Introduction*, Pergamon Press, Ontario.
2. Nielson, Richard C. (1991) Extraction and Quantitation of Polyolefin Additives, *Journal of Liquid Chromatography*, Marcel Dekker, Inc. **14**(3), 503–519.
3. Munteanu, Dan, Isfan, A., Isfan, C. and Tincul, I. (1987) Analysis of Antioxidants and Light Stabilisers in Polymers by Modern Liquid Chromatography, *Chromatographia*, **23**(1), 7–14, pp. 211–315.
4. Haslam, J., Willis, H.A. and Squirrel, D.C.M. (1981) *Identification and Analysis of Plastics*, Heyden and Son Ltd.
5. Spatafore, R. and McDermott, L. (1991) Near-IR reflectance analysis quantifies polyolefin additives, *Plastics Compounding*, September/October 1991, pp. 68–71.

Keywords: degradation, hindered phenolics, secondary aryl-amines, extraction, solvents, HPLC, GC/MS, supercritical fluids, thermal analysis, OIT test, FT-IR, near infrared spectroscopy (NIR), XRF.

See also: Analytical methods for additives in plastics; Antioxidants: an overview.

Antioxidants for poly(ethylene terephthalate)

G.P. Karayannidis, I.D. Sideridou and D.X. Zamboulis

INTRODUCTION

Polymers undergo oxidation reactions when heated in air. Heating of the polymer takes place at any stage of the life of the polymer, e.g. during polymerization, processing or end use of the ready product. Oxidation results in chain scission, branching, crosslinking or formation of oxygen-containing functional groups and leads to deterioration of physical properties and discoloration of the polymer. Antioxidants are added at low concentrations during production or processing of polymers to minimize the adverse effects of thermo-oxidation. Now that recycling of post-consumer polymers is of increasing importance, the use of antioxidants is a good way to avoid deterioration of physical properties and discoloration of the polymer during reprocessing.

Oxidation reactions are chain reactions which follow a free radical mechanism. In chain reactions three distinct steps are present: initiation, propagation and termination. During initiation a free macroradical (P˙) is generated in the polymer by heating, radiation or stress. This reacts readily with oxygen to yield a peroxy radical (POO˙). the peroxy free-radical abstracts hydrogen from another polymer molecule (PH) creating a new macroradical and a hydroperoxide (POOH). Then the hydroperoxide decomposes to two new free radicals, which are also initiators of the chain reaction. These chain reactions have severe consequences for the polymer, and cause extensive localized oxidation.

Plastics Additives: An A–Z Reference
Edited by G. Pritchard
Published in 1998 by Chapman & Hall, London. ISBN 0 412 72720 X

Antioxidants act either by radical trapping, interfering in the radical propagation, or by decomposition of the hydroperoxide formed. In this way they inhibit the formation of new radicals. The former are known as chain breaking or chain terminator antioxidants (primary antioxidants). The latter, the hydroperoxide decomposing agents, are also known as preventive antioxidants (secondary antioxidants).

The primary antioxidants are compounds with an active hydrogen (AH) atom such as hindered phenols or secondary aromatic amines (see Table 1) and they react with the peroxy radicals much faster than the polymer:

$$POO^{\bullet} + AH \rightarrow POOH + A^{\bullet} \qquad (1)$$

The radical $A^{\bullet}$ is a stable unreactive free radical which is not capable of propagating the chain reaction. However, a disadvantage for hindered phenols and aromatic amines is that they do not trap 'carbon' free radicals. Carbon free radical trapping reduces hydroperoxide formation to a great extent and is preferable to peroxy radical trapping. Hindered amines, which are well known as photostabilizers, can trap alkyl radicals and constitute a new class of radical trapping antioxidants.

$$P^{\bullet} + AH \rightarrow PH + A^{\bullet} \qquad (2)$$

The secondary antioxidants are essentially phosphites (see Table 1) or thioesters. These react by a heterolytic mechanism with hydroperoxides to yield non-radical products:

$$POOH + P(OR)_3 \rightarrow POH + PO(OR)_3 \qquad (3)$$

$$POOH + R_1-S-R_2 \rightarrow POH + R_1(SO)R_2 \qquad (4)$$

Primary and secondary antioxidants work in a different way during degradation of the polymer. Therefore, their combination may exhibit a synergistic effect. In the presence of a primary antioxidant the oxidative chain length – the average number of oxygen molecules consumed in a chain reaction – and the concentration of hydroperoxides are reduced. As a result the secondary antioxidant, which is a hydroperoxide decomposer is, in effect, present at a higher concentration relative to the hydroperoxide formed. On the other hand, the decomposition of hydroperoxides to stable products caused by the secondary antioxidant increases the effective concentration of the primary antioxidant, because there are fewer oxidative chains to be interrupted. Thus, in the synergistic mixture of antioxidants each constituent functions so as to increase the effective concentration of the other. Thioesters were found to be very good synergistic companions for hindered phenols.

The oxidation of polymers is also catalyzed by certain metal ions, which can exist in two oxidation states, e.g. copper, iron, cobalt. These ions exhibit a pre-oxidant effect by stimulating the decomposition of hydroperoxides. Metal deactivators are employed to form complex compounds with them. The metal ions are partially or fully deactivated and the polymer is stabilized. Alkyl and aryl phosphates are used for polymer stabilization (see Table 1).

The selection of the appropriate stabilizing system depends on the structure of the polymer, conditions of processing, end use, and on the physical properties of the antioxidant and/or metal deactivator (its compatibility with or solubility in the polymer, volatility, thermal stability, toxicity, odour etc.). There is no ideal system that will work in all circumstances.

Some polymers, such as the highly unsaturated polyolefins, are much more susceptible to oxidation while others, for example the thermoplastic polyesters (poly(ethylene terephthalate) (PET) and poly(butylene terephthalate) (PBT)), show little tendency for oxidation. However, even the latter need protection against oxidation because of the high temperatures and shear rates encountered during manufacture and processing. For example, in PET processes the temperature can be as high as 250 to 350°C and shear rates of 10^4 to $10^5 \, N/m^2$ can be developed. Under these severe conditions, in parallel to the thermo-oxidative degradation, thermal (pyrolytic) and hydrolytic degradation occur.

The thermo-oxidative, thermal and hydrolytic stabilization of PET, PBT and relative polymers are further discussed below.

POLY(ETHYLENE TEREPHTHALATE)

PET is an important commercial material, widely known as one of the major synthetic fibres. It is also used in the manufacturing of soft drink bottles, microwaveable oven ware, photographic films, audio and video recording tapes, films for food packaging or even as electrical insulating material for capacitors. When PET is melted for processing (spinning or extrusion) degradation occurs. Recently, post-consumer collection programmes have made PET bottles readily available for recycling into fibers. With proper stabilization of the original PET and the correct additives in the melt/recycle processing, PET could be recycled several times.

In the first stage of PET production, the precondensate bis(2-hydroxyethyl) terephthalate (BHET) is formed, together with oligomers from the reaction of dimethyl terephthalate (DMT) or terephthalic acid (TPA)

Table 1 Chemicals to stabilize thermoplastic polyesters

Chemical formula and name	Commercial name and supplier	M.W.	m.p. (°C)	CAS No.	Appearance
STERICALLY HINDERED PHENOL ANTIOXIDANTS					
Stearyl 3,5-di-tert-butyl-4-hydroxyhydrocinnamate	AO-50 (Asahi) Anox PP 18 (Great Lakes) Irganox 1076 (Ciba-Geigy) Naugard 76 (Uniroyal) Oxi-chek 116 (Ferro) Ultranox 276 (GE)	531	55	2082-79-3	White crystalline powder
Tetrakis [methylene(3,5-di-tert-butyl-4-hydroxyhydrocinnamate)] methane	Anox 20 (Great Lakes) AO-60 (Asahi) Irganox 1010 (Ciba-Geigy) Ralox 630 (Rashig) Ultranox 210 (GE)	1178	115	6683-19-8	White powder
	AO-80 (Asahi)	741	125	–	White powder
1,3,5-Trimethyl-2,4,6-tris(3,5-di-tert-butyl-hydroxy benzyl) benzene	AO-330 (Asahi) Ethanox 330 (Ethyl) Ionox 330 (Shell) Irganox 1330 (Ciba-Geigy)	774		1709-70-2	

Structure	Name	MW	CAS	Appearance
Tris(3,5-di-tert-butyl-4-hydroxybenzyl) isocyanurate	Anox IC-14 (Great Lakes) AO-20 (Asahi) Irganox 3116 (Ciba-Geigy) Vanox GT (Vanderbilt)	784	221	27676-62-6 White powder
Tris(4-tert-butyl-3-hydroxy-2,6-dimethyl benzyl) isocyanurate	Cyanox 1790 (Cytec/American Cyanamide)	700	–	40601-76-1 White powder

PHOSPHITE ANTIOXIDANTS

Structure	Name	MW	CAS	Appearance
Triisodecyl phosphite	Weston TDP (GE)	502		25448-25-3 Clear liquid

Table 1 Continued

Chemical formula and name	Commercial name and supplier	M.W.	m.p. (°C)	CAS No.	Appearance
PHOSPHITE ANTIOXIDANTS – continued					
Triphenyl phosphite	Weston TPP (GE) Weston (Akzo)	310		101-02-0	Clear liquid
Tsi(p-nonylphenyl) phosphite	Interstab CH-55 (Akzo) Lowinox TNPP (Great Lakes) Naugard P (Uniroyal) Weston TNPP, 399 (GE)	688		26523-78-4	Clear liquid
Diisodecyl pentaerythritol diphosphite	Weston 600 (GE)	508	–	26544-27-4	Clear liquid
Distearyl pentaerythritol diphosphite	Mark 5060 (Argus) PEP-8 (Asahi) Weston 618F (GE)	732	37–46	3806-34-6	White waxy flakes
Bis(2,4-di-tert-butyl phenyl) pentaerythritol diphosphite	PEP-24 (Asahi) Ultranox 626 (GE)	604	160–175	26741-53-7	White powder

PHOSPHITE ANTIOXIDANTS – continued

[Structure: Bis(2,6-di-tert-butyl-4-methyl phenyl) pentaerythritol diphosphite]

PEP-36 (Asahi) | 633 | 237 | – | White powder

Bis(2,6-di-tert-butyl-4-methyl phenyl) diphosphite

PHOSPHATE STABILIZERS

$(CH_3O)_3P=O$
Trimethyl phosphate | 140 | −46 | 512-56-1

$(C_2H_5O)_3P=O$
Triethyl phosphate | 182 | – | 78-40-0

$(C_4H_9O)_3P=O$
Tributyl phosphate | 266 | −79 | 126-73-8

[Structure: Triphenyl phosphate] | 326 | 52 | 115-86-6

Triphenyl phosphate

STERICALLY HINDERED CARBODIIMIDE – DEHYDRATING AGENT

[Structure: Bis(2,6-di-tert-butyl-phenyl) carbodiimide with CH(CH₃)₂, (H₃C)₂HC groups and N=C=N linkage]

Stabilizer 7000 (Rashig) | 362.6 | 48 | – | White to light yellow crystals

Bis(2,6-di-tert-butyl-phenyl) carbodiimide

with ethylene glycol (EG). The reaction is carried out at temperatures ranging from 200 to 260°C and is catalysed by metal salts, usually acetates of Zn, Mn, Co, Li or alkyl titanates

$$CH_3O-\underset{O}{\overset{\parallel}{C}}-\underset{DMT}{\bigcirc}-\underset{O}{\overset{\parallel}{C}}-OCH_3 + HOCH_2CH_2OH \quad EG$$

$$\downarrow -2CH_3OH$$

$$HOCH_2CH_2O-\underset{O}{\overset{\parallel}{C}}-\bigcirc-\underset{O}{\overset{\parallel}{C}}-OCH_2CH_2OH \quad BHET \qquad (5)$$

$$\uparrow -2H_2O$$

$$HO-\underset{O}{\overset{\parallel}{C}}-\underset{TPA}{\bigcirc}-\underset{O}{\overset{\parallel}{C}}-OH + HOCH_2CH_2OH \quad EG$$

The second stage, the polycondensation, is carried out under high vacuum (<1 kPa) and temperatures around 280°C in the presence of Sb_2O_3 as catalyst.

$$nBHET \xrightarrow[280°C]{Sb_2O_3} H\left[OCH_2CH_2O-\underset{O}{\overset{\parallel}{C}}-\bigcirc-\underset{O}{\overset{\parallel}{C}}-OCH_2CH_2OH\right]_n PET$$

$$+(n-1)HOCH_2CH_2OH \quad EG \qquad (6)$$

THERMO-OXIDATIVE DEGRADATION OF PET

PET is much more stable towards auto-oxidation than other polymers, e.g. polyolefins. However, during contact of molten PET with atmospheric oxygen rapid degradation occurs. This is noticeable by colour formation, increase in carboxyl end group content and decrease of molecular weight. The high molecular weight and the physical structure of PET make the direct study of the mechanism of oxidation difficult. Comparison of thermo-oxidation products of PET with those obtained from simplified model compounds has led to the following mechanism [1]:

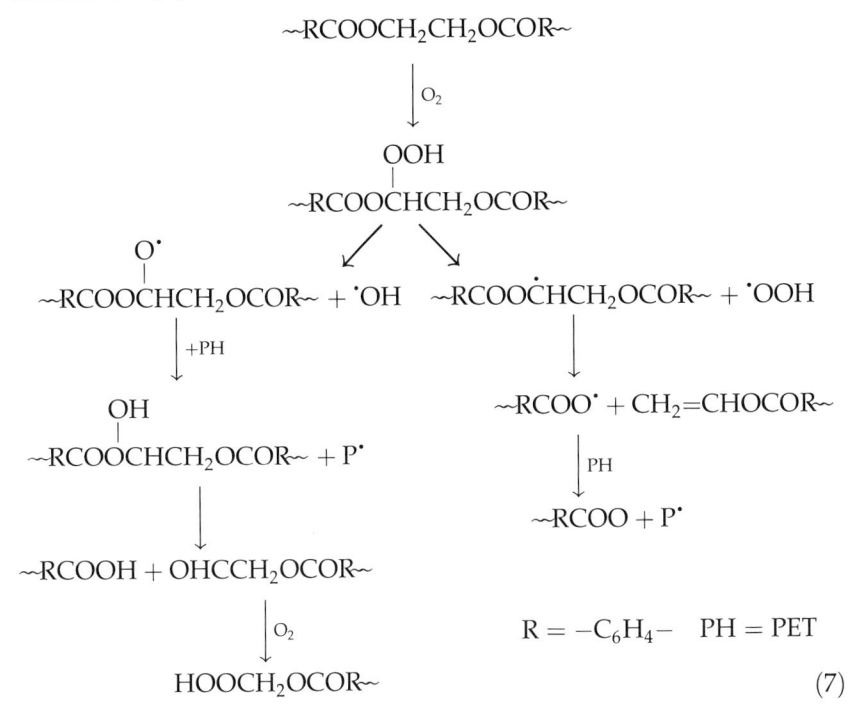

$$\sim\text{RCOOCH}_2\text{CH}_2\text{OCOR}\sim$$
$$\downarrow \text{O}_2$$
$$\sim\text{RCOOCH(OOH)CH}_2\text{OCOR}\sim$$

$$\swarrow \qquad \searrow$$

$\sim\text{RCOOCH(O}^\bullet\text{)CH}_2\text{OCOR}\sim + {}^\bullet\text{OH}$ $\sim\text{RCOOC}^\bullet\text{HCH}_2\text{OCOR}\sim + {}^\bullet\text{OOH}$

$$\downarrow +\text{PH} \qquad\qquad\qquad\qquad \downarrow$$

$\sim\text{RCOOCH(OH)CH}_2\text{OCOR}\sim + \text{P}^\bullet$ $\sim\text{RCOO}^\bullet + \text{CH}_2=\text{CHOCOR}\sim$

$$\downarrow \qquad\qquad\qquad\qquad \downarrow \text{PH}$$

$\sim\text{RCOOH} + \text{OHCCH}_2\text{OCOR}\sim$ $\sim\text{RCOO} + \text{P}^\bullet$

$$\downarrow \text{O}_2 \qquad\qquad R = -\text{C}_6\text{H}_4- \quad \text{PH} = \text{PET}$$

$$\text{HOOCH}_2\text{OCOR}\sim \qquad\qquad\qquad\qquad (7)$$

The main weak point in the polyester chain is the β-methylene group. When diethylene glycol links are formed, the stability of polyesters in oxidation is lowered even more. Phosphites and hindered phenols (see Table 1) are currently used in PET to improve thermo-oxidative stability. It was shown that after five melt extrusions the stabilized PET retained its polymer qualities (colour, clarity, viscosity and carboxyl end group content) and that a blend of a hindered aryl phosphite and a hindered phenolic antioxidant (4:1) was more effective than each one of the components [2].

Thermo-oxidation of PET is catalysed by the metal ions (Zn^{2+}, Mn^{2+} etc.) used as catalysts in transesterification of dimethyl terephthalate. Compounds which form complexes with these metal ions inhibit their catalytic action in general and improve the stability of PET towards oxidation. These compounds are mainly esters of phosphoric acid, e.g. triphenyl phosphate (see Table 1).

THERMAL DEGRADATION

If PET is maintained molten (280–300°C) in a nitrogen atmosphere it degrades slowly to gaseous and solid low molecular weight products.

The gaseous products consist of CO, CO_2, H_2O, CH_3CHO, C_2H_4, CH_4 and C_6H_6 in proportions depending appreciably on temperature. However, acetaldehyde is always the major product (~80%). Solid products consist primarily of terephthalic acid and acidic or cyclic oligomers. The white sublimate, formed during melt spinning of PET, is due to the solid products of thermal degradation.

Thermal degradation of PET occurs by a molecular mechanism via a cyclic transition state with random chain scission at the ester linkages [1]:

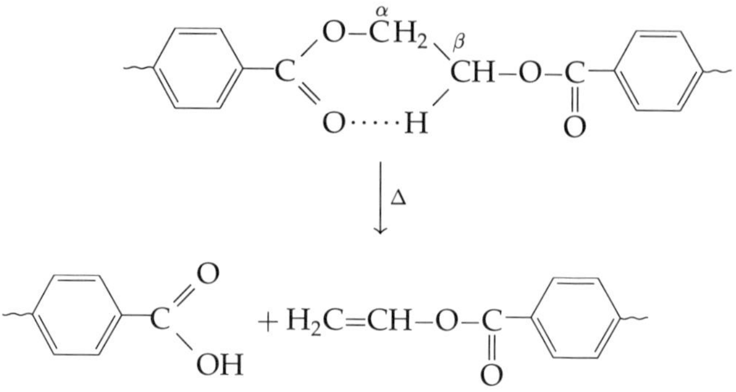

$$\text{Ar}-C\begin{subarray}{c}O-\overset{\alpha}{C}H_2\\ \| \\ O\cdots H\end{subarray}\overset{\beta}{C}H-O-\underset{\|}{C}-\text{Ar} \tag{8}$$

$$\downarrow \Delta$$

$$\text{Ar}-C\begin{subarray}{c}O\\ \| \\ OH\end{subarray} + H_2C=CH-O-\underset{\|}{\overset{O}{C}}-\text{Ar}$$

This is the mechanism generally accepted for the pyrolysis of esters containing at least one β-hydrogen atom. In the transition state the C–H and alkoxy C–O bonds are partially broken while the C=C and O–H bonds are partially formed. The α-carbon atom shows some carbonium ion character and the C–O bond breakage is heterolytic. The thermal degradation of esters and PET cannot be inhibited by free radical trapping agents. It has been shown that the chain scission of the ester linkages in PET is catalysed by the transesterification catalysts used in its production [3]. The coordination of a metal ion with the carbonyl oxygen of the ester group is polarized. Thus, the scission of the proton is favoured and a vinyl group is formed. Polycondensation catalysts which are usually antimony derivatives do not affect the thermal degradation of PET.

Scission of ester linkages causes a reduction in the molecular weight of PET. However, if hydroxyl end groups are present in the PET they react with vinyl end groups formed during oxidation to build molecular weight:

$$\sim R-COOCH_2CH_2OH + CH_2=CHOCO-R\sim$$
$$\downarrow \tag{9}$$
$$\sim R-COOCH_2CH_2OCO-R\sim + CH_3CHO$$

When hydroxyl groups have been consumed the molecular weight will begin to fall. Unfortunately, acetaldehyde is also formed from the above reaction and it strongly affects the quality of the final product. Its presence in concentrations even up to a few ppm gives a flavour to soft drinks bottled in PET. If acetaldehyde remains in the melt it yields polyene, aldehydes and water:

$$nCH_3CHO \longrightarrow CH_3\text{\textcent}CH=CH\text{\textcent}_{n-1}CHO + (n-1)H_2O \qquad (10)$$

Polyene aldehydes discolour PET from colourless to yellow, while the water can cause hydrolytic chain scission of ester linkages. The rate of thermal degradation in closed systems such as in the continuous preparation and processing of PET was found to be about three times higher than that in an open stirred system under flowing nitrogen where acetaldehyde can escape.

The development of colour in PET is also attributed to polyenes formed from those products of thermal degradation of PET which contain vinyl end groups. These products can also lead to the formation of gels and cause serious problems to processing machines and end product (film, fibre).

Protection of PET from thermal degradation is obtained by removal or deactivation of transesterification catalysts present in PET. Esters of phosphoric acid such as trialkyl or triaryl phosphates can be used for this purpose. Organic phosphites, which are well known to decompose hydroperoxides as secondary antioxidants, are also reported to deactivate metal ions (traces and/or catalyst residues) in polyolefins [4]. In PET, it is suspected that they also deactivate the metal species that catalyse the transesterification reaction.

HYDROLYTIC DEGRADATION

PET also undergoes hydrolytic degradation, especially at the high temperatures of the melt. So, it is important for it to be dried to a moisture level of less than 0.005% prior to melt processing. Hydrolytic degradation is catalysed by terminal carboxyl groups of the polyester. Dehydrating agents such as hindered carbodiimides (see Table 1) can be employed to protect PET from hydrolysis during processing and resulting end applications.

EVALUATION OF STABILIZER ACTION

The choice of stabilizers for PET (antioxidant and/or metal deactivator) is based mainly on the method of additive incorporation into the polymer and the end use/application of the polymer. For example, if it is preferable for the additive to be added in the polymerization reactor, then the volatility and the thermal stability of the stabilizer need consideration because of the high temperatures and vacuum prevailing in the reactor. PET thin films require maximum colour improvement and clarity but PET fibres for tyre cords and geotextiles have other requirements.

Usually, to evaluate the performance of the stabilizer, the polymer is treated as if under real conditions of processing in miniature processing machines (extruder, spinner). Alternatively, in a simple melt mixing apparatus, such as the Brabender mixing chamber, stabilized and unstabilized PET resins can be exposed to various temperatures and shear conditions. If degradation occurs then it is accompanied by an increase in the carboxyl end group content, acetaldehyde and colour and by a decrease in viscosity. Therefore, all these indices and others such as torque and yellowness indices are useful to evaluate the extent of degradation of PET. Differential scanning calorimetry and thermogravimetry are used for evaluating the induction time under isothermal conditions and the stabilization coefficient [5]. Measuring more than one index can lead to contradictory results concerning the effectiveness of the stabilizer. The method to be adopted for evaluation of the stabilizer depends on the end product needed.

The thermoplastic polyesters PET and PBT are semicrystalline polymers showing little tendency towards thermo-oxidative degradation under normal conditions. However, under processing conditions degradation can be extensive. It can lead to loss of molecular weight and the development of undesirable characteristics such as colour and odour in the final product. The main purpose of the primary and secondary antioxidants or metal deactivators is to protect the polymer during manufacturing or processing of virgin and recycled resins. For the stabilization of PET and related polymers an antioxidant can be used with a metal deactivator on its own or in a blend in quantities varying from 0.05 to 1.0% in single or multiple doses during different processing steps. The choice of the stabilizer depends on its own performance and for special purposes on its approval for use in resins destined for contact with food.

REFERENCES

1. Buxbaum, L.H. (1968) The Degradation of Poly(ethylene terephthalate), *Angew. Chem. Intern. Edit.*, **7**(3), 182–290.
2. Solera, P. (1993) Stabilization and Color Enhancement of Virgin and Reprocessed PET, *SPE RETEC Proceedings*, October 11–13, 1993, Orlando, FL, 161–180.
3. Zimmerman, H. and Kim, N.T. (1980) Investigations of Thermal and Hydrolytic Degradation of Poly(ethylene terephthalate). *Polymer Engineering and Science*, **20**(10), 680–683.
4. Pobedimskii, D.G., Kurashov, V.I. and Kirpichnikov, P.A. (1983) Metal Complexes with Ligands – Organic Phosphites as Polyolefin Antioxidants. *J. Polym. Sci., Polym. Chem. Ed.*, **21**, 55–66.
5. Karayannidis, G., Sideridou, I., Zamboulis, D., Stalidis, G., Bikiaris, D. and Wilmes, A. (1994) Effect of Some Current Antioxidants on the Thermo-oxidative Stability of Poly(ethylene terephthalate). *Polym. Degr. and Stab.*, **44**, 9–15.

Keywords: Antioxidants, dehydrating agents, PBT, PET, phosphate stabilizers, phosphite antioxidants, stabilizers, sterically hindered carbodiimides, sterically hindered phenols, thermal degradation, thermo-oxidative degradation, thermo-oxidative stability.

See also: Antioxidants: an overview;
Antioxidants: their analysis in plastics;
Hindered amine light stabilizers: introduction;
Hindered amine light stabilizers: recent developments;
Light and UV stabilization of polymers.

Antistatic agents

Geoffrey Pritchard

STATIC ELECTRICITY

Static electricity is a considerable problem in plastics, because with very few exceptions plastics are highly insulating. Antistatic agents are used mainly to eliminate spark discharges and to minimize or prevent the accumulation of surface dust and dirt. A similar problem exists with polymer based textile fibres where it is necessary to reduce fabric cling. Even picking up a plastics bag can generate several thousand volts if the humidity is low enough.

The static is caused by the transfer of electrons across surfaces brought into frictional contact with each other. As long as the two surfaces are in contact, there is no problem. But when they are separated, the distribution of electrons between the two is often unequal. Very few electrons need be involved (say one electron every $100\,nm^2$). Consequently, both surfaces acquire electrical charges, with one having a surfeit of electrons, and the other a deficit. There is no change within the body of the material, it is purely a surface effect. The practical consequences are seen in too much dust attraction, reducing clarity, and damaging aesthetic appeal. Static charge is also reflected in the tendency for films to cling. Processing is sometimes affected, including reduced ease of removal from the mould, and there is the possibility of damage to processing machinery. Static charge can damage electronic components in computers. It can generate explosions by producing sparks that are capable of igniting flammable gases.

APPLICATIONS

The automotive industry requires static elimination in plastics used for fuel systems, electrical and electronic parts, and engine components.

Plastics Additives: An A–Z Reference
Edited by G. Pritchard
Published in 1998 by Chapman & Hall, London. ISBN 0 412 72720 X

The electronics industry needs static protection in relation to packaging of components, flooring and furniture. The housings for electronic goods such as computers, printers, photocopiers and other office equipment should have low static formulations. Video cassettes are also vulnerable. Packaging of food, drugs, medical supplies and hazardous materials constitutes another market for antistatic agents, as does materials handling equipment such as conveyor belts. The chemical process equipment industry needs to eliminate static discharge because of the widespread use of flammable liquids and gases, and since the industry uses plastics or reinforced plastics extensively, this is a significant area, although internal antistatic agents are not necessarily the most appropriate countermeasures here.

ABS is very widely used in the business machine, automotive parts and domestic appliance industry. Polyamides, in addition to their use in the critical area of carpets, also feature in business machine internal parts and to some extent in packaging. The polyolefins are widely used in packaging, pipes, storage bins and foam. PVC is used in conveyor belts, packaging and floors. Polystyrene and polycarbonate are significant materials in some of these areas.

It will be apparent from the above examples that some applications have strict requirements for very good static dissipation performance, i.e. low surface resistivities, on safety grounds, whereas others have more modest requirements because the effect of static is mainly a nuisance.

COUNTER MEASURES

In order to dissipate the electrical charge on the surface of the plastic, we must render the surface electrically conducting in some way. This enables the electrons to be dissipated. There is no problem if a plastics material is already conducting, because of the additives it contains, or because the base polymer is intrinsically conducting – the latter situation is uncommon.

One approach is to assume that static charges will unavoidably be accumulated, and therefore the plastic must either contain an internal additive or possess a surface coating which renders the surface sufficiently conductive to dissipate the static electricity. Other methods of dealing with the problem are available, but this article is confined to the internal additive method, which is widely used.

INTERNAL ANTISTATIC AGENTS

Internal antistatic agents have the advantages that at the levels normally used, they do not significantly alter the physical and mechanical properties, and their cost is reasonably low compared with many other additives. They do not prevent a plastics component from being transparent or affect its colour. Some antistatic agents have useful anti-blocking characteristics. Unfortunately, many of the currently available internal antistatic agents, other than carbon black, aluminium powder and polypyrrole, do not survive temperatures much higher than 180°C, which leaves high temperature plastics with a reduced choice of antistatic protection. Some chemical antistatic agents such as tertiary amine derivatives are thought to accentuate the corrosion of metals in electronic parts. As internal antistatic agents take some time to migrate to the surface, they may not reach full effectiveness immediately.

One reason why internal antistatics are often preferred is that even if they are removed from the surface by various means, such as erosion, there is more additive available within the polymer to migrate and replace what was lost. External, i.e. surface, antistatic agents can be applied in the form of a liquid solution in water or perhaps an alcohol, but to survive they must resist being removed by the processing or handling operations to which the plastics articles are subjected. They may interfere with, or prevent, other coating operations and sometimes they can encourage microorganisms. There are several factors to consider which could eliminate a given antistatic agent. Food contact considerations provide one example.

Internal antistatic agents are compounded into the plastics during or before the shaping stages. Such a process would normally result in the more or less uniform distribution of the antistat in the interior of the finished product, but the additive is deliberately selected so that it has insufficient compatibility with the polymer, to encourage it to migrate to the surfaces. The rate of migration is important as, if it is too slow, it is not effective quickly enough, and static builds up, and if it is too rapid, the surface characteristics of the article may be adversely affected. The rate of migration can be reduced by other formulation constituents such as fillers.

When it has reached the surface, the hydrophilic groups of the additive attract ambient moisture, and this causes the electrical charge on the surface to be dissipated. For this reason, the effectiveness of internal additives declines along with a decline in the ambient relative humidity. They can also act by increasing ionic conductivity.

Polymeric antistatic agents have been employed in textile fibres. Ionically conductive copolyethers such as ethylene oxide/propylene oxide copolymers and other similar polymers are examples. Polyoxyethylene derivatives with high electrical conductivities caused by proton migration are not highly dependent on relative humidity and have found some use in thermoplastics.

SURFACTANTS

Cationic surfactants have hydrophilic groups with positive charges; anionic ones have negative charges, nonionic surfactants have hydrophilic groups with no charge, and amphoteric surfactants have hydrophilic groups with both positive and negative charges, but they are not much used in plastics. Lubricants, which are added to reduce frictional forces during processing, also incidentally reduce static build-up. Any liquid of high dielectric constant should be an antistatic agent – water is only the most obvious example – but such liquids tend to be volatile, and therefore their benefit is temporary.

Surfactants lower the contact angle between water and the plastics substrate, enabling the water to spread more evenly over the surface.

Cationic and anionic antistatic agents perform best in polar polymers, but their interaction with PVC heat stabilizers must be considered. Fatty acid based amines can react with the chlorine in PVC. Nonionic antistats are deployed in polyolefins, ABS and styrene polymers.

HOW MUCH PROTECTION IS REQUIRED?

This depends on the application. Some products require very low surface resistivity. The most common examples are associated with EM or RF shielding. Sources of electromagnetic interference include natural phenomena (cosmic radiation, lightning, corona and static discharge), and man-made devices (electrical and electronic equipment including TV and radio, power tools, ultrasonic cleaners, computers, arc welders, car ignition systems and cash registers). Many appliances are adversely affected by interference, especially communications equipment, hi-fi, and industrial instrumentation. Potential sources of interference are subject to strict regulation.

We should distinguish between the applications requiring materials with intermediate resistivity values, and those requiring electrically conducting, or at least semi-conducting, behaviour. Most plastics, before any antistatic treatment, have very high volume resistivities, of the order of 10^{14} ohm cm, and surface resistivities of the order of 10^{14} ohm/square, at ambient temperatures. Completely effective EMI shielding requires volume resistivities lower than 10 ohm/cm, which is rather beyond the values typically achievable with carbon black filler. Metal fillers can enable these values to be reached, although there may be side-effects such as the abrasiveness of the metal particles. Surface resistivities need to be less than 10^5 ohm/square for EM shielding.

Static dissipative materials are an intermediate category, used where there is a need for surface resistivities in the range 10^5–10^9 ohm/square. These values can be achieved using carbon black or similar fillers. They are relevant to products where static build-up causes arcing or short circuits.

Ordinary antistatic materials are in a still higher surface resistivity range, between 10^9 and 10^{14} ohm/square. Conventional antistatic agents can normally meet most of the requirements in this category, although there may be some associated disadvantages, as mentioned elsewhere in this article.

There are other articles in this book, entitled 'Carbon black' and 'Conducting fillers for plastics', which relate to materials with higher conductivity. It should be remembered that carbon black is sometimes used as a pigment or as a UV stabilizer and will act incidentally as an antistatic agent.

PLASTICS REQUIRING ANTISTATIC AGENTS

Examples of plastics which require conventional antistatic agents are the polyolefins, polystyrene, polyesters, polyurethanes, PVC and acrylics. The polyolefins can be rendered fairly static-free with only low levels (typically less than 1%) of a glyceride or an alkoxylated amine, such as the ethoxylated glyceryl ester of a fatty acid. The agents commonly used are liquids and can be added by a masterbatch method. Polyolefins for electronic packaging need antistatic additives at much higher levels, e.g. up to 10%. PVC requires different additives, e.g. 1–2% quaternary ammonium compounds in rigid, unplasticized PVC and 2–6% ethoxylated alcohol or alkoxylated fatty acid esters in plasticized PVC. High levels of epichlorhydrin copolymers have been recommended for use as antistatic agents in both thermoplastics and thermosets.

ABS, cellulose acetate, and polycarbonate can also be modified by the addition of antistatic agents. Styrene based polymers are too readily compatible with some antistatic agents and care must be taken to ensure that the additive is chosen so that migration to the surface occurs at a sufficient rate. Glycerol derivatives and ethoxylated amines can be used.

TYPICAL COMPOUNDS USED

The internal additives most commonly used in plastics are the non-ionic surfactants, although some cationic and anionic surfactants are also employed (see above). Examples of the non-ionic variety are the ethoxylated fatty alkylamines, the fatty diethanolamines, and the mono- and di-glycerides. The ethoxylated fatty alkylamines are used in hydrocarbon-rich plastics such as the polyolefins and in some of the styrenated polymers. They work well even at low relative humidity levels; their effectiveness persists, they can be used at low concentrations, and many are FDA approved for use in indirect food contact. Certain non-ionic antistatic agents such as glycerol monostearate do not have the persistence associated with the ethoxylated fatty alkylamines. They are therefore suitable only for protection against static charges during processing.

Effectiveness

Examples of cationic antistatic agents include the long chain quaternary ammonium and sulfonium alkyl salts such as chlorides, nitrates and methosulfates. They are used in high concentrations in polar polymers such as PVC, and also in various styrene based polymers. Anionic antistats are usually alkali salts of alkyl phosphonic, sulfonic or dithiocarbamic acids. Examples of their application are the styrene polymers, and the saturated (thermoplastic) polyesters.

Two specific examples of antistatic compounds which can be used internally or applied as a solution are: (3-lauramidopropyl)trimethylammonium methyl sulfate, for use in PVC, polystyrene, polythene and ABS; and stearamidopropyldimethyl-2-hydroxyethyl ammonium nitrate, as a 50% solution in an ispropanol–water mixture, for use in PVC.

EFFECTIVENESS

The effectiveness of antistatic agents can be measured. This is done by observing the rate of decay of surface charge on a plastics article. In the case of high density polyethylene, for example, most of the measurable surface charge on the sample immediately after moulding should have dissipated under ambient conditions within 7 days, but the rate of decline will be dependent on a large number of factors.

Migration rate can depend on the molecular weight of the additive as well as on its compatibility with the polymer. If migration rate is too high, higher molecular weight substances of a similar type can be used.

It is difficult to find effective antistatic agents for engineering thermoplastics, because of the high processing temperatures.

It is possible that an antistatic agent will prove technically effective, but interfere unacceptably with other additives in the polymer, such as stabilizers. On the other hand, some additives can successfully combine an antistatic role with one or more other important functions.

Some suppliers are listed in Table 1.

Table 1 Some suppliers of antistatic agents

Akzo Chemical International BV	ICI Surfactants
Cabot Corporation Special Blacks Division	Kenrich Petrochemicals
Cookson Plastics	Myoshi Oil & Fats Co.
Croxton & Garry Ltd	PPG Industries
Dow	Rhone Poulenc
DuPont	Chemical Ltd
Eastman Chemical	Rohm & Haas
Himont	RTP Co.
Hoechst Aktiengesellschaft	Sandoz Chemical
Henkel KgaA Cok Plastics and Coatings Technology	Sumitomo Seika Chemicals Co.

BIBLIOGRAPHY

Johnson, K. (1972) *Antistatic Agents, Technology and Applications*, Noyes Data Corp., New York.

Johnson, K. (1976) *Antistatic Compositions for Textiles and Plastics*, Noyes Data Corp., New York.

Trost, T. (1995) Electrostatic discharge (esd) – facts and faults – a review, *Packaging Technol. Sci.*, **8**, 231–247 and **6**, 303–313.

Walp, L.E. (1992) Antistatic agents, in *Kirk-Othmer Encyclopedia of Chemical Technology* (ed. Mary Howe-Grant) 4th edn, Wiley, New York.

Keywords: electrical conductivity, resistivity, packaging, dust, migration, ethoxylated amines, quaternary ammonium salts, carbon black, glycerol monostearate, relative humidity, non-ionic surfactant.

See also: Carbon black;
Conducting fillers for plastics.

Biocides

Geoffrey Pritchard

INTRODUCTION

This entry is about the preservatives used to protect plastics against attack by mildew, bacteria, fungi, algae, insects, rodents and marine fouling organisms. The preferred method for most but not all of these problems is to mix an additive with the plastics material, chosen for its antimicrobial action. The main considerations are the efficacy, environmental safety, ease of dispersion, and cost of these additives. Under the heading of efficacy we should consider not only the capacity to kill microorganisms but also the duration of that capacity. Many biocides are too transient. (There is a related entry in this book, entitled: 'Biocides: some kinetic aspects'.)

MICROORGANISMS

Microorganisms are common almost everywhere. They can be roughly classified for present purposes as:

1. bacteria, i.e. simple single-celled organisms, usually less than $3\,\mu$m. Examples include *Pseudomonas aeruginosa, Staphylococcus aereus* and *Escherichia coli*;
2. fungi, which are more complex and larger, multicellular organisms, with not all cells identical; important examples include *Aspergillus niger, Penicillium funiculosum* and *Chaetomium globosum*; and
3. yeasts, similar to fungi but single celled (e.g. *Candida albicans*).

A fourth category, *actinomycetes*, exists as soil microorganisms and can be held responsible for the pink stains sometimes observed on PVC. They can be grouped under the first heading, with bacteria.

Plastics Additives: An A–Z Reference
Edited by G. Pritchard
Published in 1998 by Chapman & Hall, London. ISBN 0 412 72720 X

The more complex the organism, the more likely it is that the biocides required for effective treatment could be harmful to higher forms of life.

SUSCEPTIBILITY OF PLASTICS TO MICROBIAL ACTION

Most plastics are commendably resistant to microbial attack. Examples of the more resistant polymers are the polyolefins, most vinyl polymers and the polyesters (saturated and unsaturated), which altogether constitute a very large part of the total plastics market. Other resistant polymers include polycarbonate, polyformaldehyde, ABS, PTFE, epoxies, silicones, the phenolic and urea-formaldehyde thermosetting resins. Polyvinyl acetate is rather exceptional among vinyl polymers in being attacked by bacteria. Certain cellulosic polymers are also liable to attack; for instance cellulose nitrate, and the expanded form of cellulose acetate, but not acetate butyrate, and not propionate. (Starch is biodegradable and is deliberately used to promote biodegradation.) Polyester based polyurethanes are also susceptible because most ester groups can undergo enzyme-induced chain scission.

The microorganisms causing degradation are not necessarily attached to the polymer itself, but may be anchored in a material contacting the polymer, such as a surface coating, or an adhesive layer.

ADDITIVES IN PLASTICS – THE VULNERABLE CONSTITUENTS

It is often the additives contained in plastics, rather than the synthetic polymers themselves, which are most susceptible to microbial action. The vulnerability of the ester groups in many plasticizers makes flexible PVC a prime area for biocides, especially as this material is widely used in moist environments. The affected plasticizers are mainly the aliphatic ones, e.g. butyrates, sebacates, laurates and adipates, rather than the phthalates, although the latter are not totally immune (ester groups are enzymatically cleaved at very different rates). Some epoxy based plasticizers promote resistance to fungi. The 4,5 epoxytetrahydrophthalates are said to be more resistant than the average phthalates. Some common fillers, such as woodflour and starch, together with oils and lubricants, and certain pigments, are notable supporters of attack by bacteria, fungi or algae.

There are several additives such as organometallic stabilizers, which are incorporated in plastics for reasons unconnected with microbial action, but which nevertheless actually help to promote resistance. Some of the additives which provide incidental resistance to biological action are used in resins which do not require any protection. Others are inadequate because of lack of stability during processing and/or

EFFECTS OF MICROBIAL ATTACK

The visible effects of microorganisms include discoloration, notably a pink stain, or black pitting. There can sometimes be an odour, and there can be changes in the physical properties of the plastics material. Pink stains on PVC and some of its copolymers, in the presence of plasticizers or other suitable additives, are attributed to certain species of *Penicillium, Streptomyces* and *Brevibacteria*. Pink stains have been caused in vinyl polymers by the action of *Actinomycetes*. Fungi can grow on polymer surfaces, with no further effect, or they can produce enzymes which break down macromolecular structures. Serious attack of this kind can result in microcracking and failure of the material. There can be a change in the electrical characteristics of an insulating plastic, following an increase in microcracking, porosity and moisture absorption capacity.

Damage to plastics packaging by rodents is usually mechanical and is not very amenable to countering by chemical additives in the plastics formulation. Several insects are harmful to cellulosic polymers in an essentially similar way, notably boring beetles and termites.

BIOCIDES USED

The agents used will be selective in their action and consideration has to be given to the desired effect, as using more powerful biocides than necessary could cause problems in respect of health, safety or ecology. Biocides are very strictly regulated and the costs of meeting requirements for approval have reduced the number of substances commercially available.

The effectiveness of biocides has to be established by controlled laboratory screening tests against well characterized biological agents. First there are preliminary tests and, if promising, these are followed by long duration field trials. The development phase also establishes what is the appropriate dose level and how sustained the action is likely to be. Common U.S. test methods include ASTM G21 for fungicides, G22 for *Pseudomonas aeruginosa*, G29 for algae, and D3083 for soil burial situations.

There are some circumstances in which all living microorganisms must be destroyed; if so, the agent may be called a sterilant. A sporicide kills spores, while an 'x-cide' selectively and specifically kills 'x-type' organisms and an 'x-static' agent prevents their growth.

So we have bacteriostats, algaestats, fungistats etc. The amount used typically ranges from 1 to 5 parts per hundred by weight.

We can classify biocides into the following categories, of which many are not relevant to the plastics industry.

Structure I 10,10' oxybisphenoxyarsine

1. Quaternary ammonium compounds, which interfere with the bacterial cell membrane; these are very little used in plastics although a few are used as fungicides.
2. Phenolics and chlorinated phenolics; again these are very little used with plastics. Chlorinated hydroxyphenyl ethers have been used as bactericides.
3. Metal-containing organic substances. A very widely used metal-organic biocide indeed is 10,10-oxybisphenoxyarsine (see Structure I). One of its common uses is in plasticized PVC, which may be used for garden pool linings, upholstery, cable insulation or shower curtains. The biocide is available in liquid form. Another example is copper-bis-(8-hydroxyquinoline) (see Structure II). Metal-organic substances have been widely successful as constituents of antifouling paints (see later) and tin compounds are used as fungicides and rodenticides.
4. Organosulfur compounds are mainly slimicides with little application in plastics. The trichloromethyl thiophthalimides are used as fungicides in PVC.
5. Heterocyclic compounds are used in coatings rather than moulded plastics. There are water-insoluble polymeric biocides, such as the polystyrene hydantoins (see Structure III). Some of these latter substances can be rendered inactive by reducing agents or even by excessive doses of bacteria, but it is claimed that they can be reactivated by sodium hypochlorite.

Other biocides, among many which have been tried, include cresol derivatives, naphthenates, metal salts such as copper oleate and barium metaborate, and dithiocarbamates. Zinc pyrithione based biocides are effective in preventing mildew and mould growth.

Structure II Copper bis(8-hydroxyquinoline)

$$-(\underset{\underset{\text{R'}}{|}}{\overset{\underset{\text{R'}}{|}}{C}}-CH_2)_n-$$

[Structure with phenyl group attached to hydantoin ring bearing N-Cl, N-Cl substituents and two C=O groups]

Structure III Polystyrene hydantoin. From Worley, S.D. and Sun, G. (1996) Biocidal polystyrene hydantoins, in *Polymeric Materials Encyclopedia* (ed. Joseph C. Salamons), CRC, Boca Raton, FL.

The selection of a biocide depends, like antistatic agents, on the rate of migration, which in turn depends on the overall formulation (fillers can retard migration considerably). The need is to maintain a steady flow of biocide to the surface throughout the life of the article being protected.

Some biocides may be unacceptable on account of their strong colour (for example, certain copper salts, such as copper 8-quinolinolate) or because of their effects on aquatic life. Toxicity is a problem because the additives are by definition biologically potent, and they require appropriate handling precautions.

Biocides for PVC can affect thermal stability, compatibility with other formulation constituents, or weathering performance.

ANTIFOULING

Conventional biocidal additives are effective against marine organisms, such as barnacles, protozoans, algae, bacteria and fungi. The plastics which are exposed to such organisms include boat hulls, epoxy coatings on offshore structures and bridge supports, and any other ocean-immersed structures. The adverse effects of barnacle growth and slime on fuel consumption in ships is well known. A major concern is the need for ecologically acceptable biocides which are nevertheless functionally effective. For the most part, antifouling paints have traditionally been based on copper or organotin chemistry and are properly discussed elsewhere, as they are not genuinely plastics additives. Early formulations, based on cuprous oxide, were effective enough, but they had a short lifetime. Tributyl tin compounds had a typical lifetime of about 6 years, with an effective action on barnacles at a release rate of less than 1 µg/cm^2/day. Similar substances (tributyl tin oxide, Structure IV,

$$(C_4H_9)_3Sn-O-Sn(C_4H_9)_3$$

Structure IV Tributyl tin oxide

tributyl tin benzoate and salicylate for instance) are used as industrial biocides in plastics and paints.

Recycled plastics generally need re-stabilization against microbial action.

BIBLIOGRAPHY

Rossmore, H.W. (ed.) (1995) *Handbook of Biocides and Preservative Use*, Blackie.
Smith, A. and Springle, R. (1995) *World Guide to Industrial Biocides*, Paint Research Association, Teddington, UK.
Paulus, W. (1993) *Microbicides for the Protection of Materials, A Handbook*, Chapman & Hall, London.
Thorp, K.E.G., Crasto, A.S., Gu, J.-D. and Mitchell, R. (1994) Biodegradation of composite materials, in: *Proceedings of the Tri-Service Conference on Corrosion* (ed. T. Naguy), Orlando, Florida.
Flick, E.W. (1987) *Fungicides, Biocides and Preservatives for Industrial and Agricultural Applications*, Noyes, New York.

Keywords: microbial, fungicides, *Penicillium, Streptomyces, Actinomycetes,* moulds, algae, stain, discoloration, pitting, plasticizer, PVC, starch, antifouling, preservatives, cellulose.

See also: Biocides: some kinetic aspects;
Biodegradation of plastics: monitoring what happens.

Biocides: some kinetic aspects

K.Z. Gumargaliva and G.E. Zaikov

INTRODUCTION

Biological influences can be very important in the degradation of polymers, and of many other materials such as wood. It may be useful to consider biocides as a whole, regardless of the material to which they are applied, so that their role in the protection of plastics can be seen in its wider context.

Microorganisms such as bacteria, microscopic fungi and similar species can have mechanical, chemical and biological effects. Adverse effects of these kinds are referred to as biodegradation processes. Chemical compositions which provide protection from biodegradation processes are known as biocides or, in the case of fungi, fungicides.

The first serious research into substances capable of providing protection against biodegradation was carried out by materials technologists, biologists and chemists in the 1920s and 1930s. At that time the research was concerned with textiles, natural leather, paper, glues etc. A decline in the availability of natural materials and resources together with increased demand for manufactured products made it all the more important to conserve and protect such products for as long as practicable. Protection from biodegradation formed a crucial part of the strategy.

The chief protecting agents are chemical substances, namely biocides, which can either be directly incorporated into the material to be protected, or applied as surface coatings. They can alternatively be added to lubricating compositions, or to the atmosphere, or to the water, or otherwise introduced indirectly into the environment. Several thousand compounds with biocidal efficacy are known. Examples include alcohols, compounds or compositions containing heavy metals, and certain surfactants. Table 1

Plastics Additives: An A–Z Reference
Edited by G. Pritchard
Published in 1998 by Chapman & Hall, London. ISBN 0 412 72720 X

Table 1 Properties of some well-known volatile biocides (fumigants)

Name	Molecular mass	Temperature of boiling (°C)	Solubility in water	Inhibiting concentration, (mg/l)	Penetrability to material	Biocidic activity
Methylbromide	95	4.6	Weak	3500	Excellent	Weak
Propylene oxide	58	34	Good	800–2000	Excellent	Excellent
Formaldehyde	30	−21	Good	3–10	Excellent	Excellent
Ethylene oxide	44	10.4	Complete	400–1000	Moderate	Moderate
β-propyolactone	72	162	Moderate	2–5	Does not penetrate, sterilizes from surface	Moderate

lists some important properties of a selection of well-known volatile biocides, used as fumigants. Tables 2 and 3 show examples of the suppression of microorganism growth by various substances.

The effectiveness of biocides is explained by their ability to penetrate into the cells of microorganisms or to accumulate at the cell surface, affecting vitally important processes within the microorganism. For example, halogen and sulfur-containing biocides have adverse effects on microorganism respiration. Iodides and fluorides decrease the activity of certain enzymes. The toxicity of metal–organic compounds containing heavy metal atoms is also related to enzyme action, for example their effect on sulfohydryl groups in enzymes.

REQUIREMENTS FOR, AND TOXICOLOGICAL CONTROL OF, BIOCIDES

The requirement for a good modern biocide is high activity and effectiveness in combating microorganisms, combined with reliability of application and, increasingly important, the absence of any harmful effects on the environment. There should also be sufficient versatility in its effects on fungi and bacteria. Certain applications also require stability at high temperatures, and in the presence of water. The effectiveness of biocides should ideally be maintained for a long time. Obviously, these additives should not adversely affect the properties of the materials being protected, whether physical, chemical or mechanical. The application of a biocide should not result in accelerated ageing compared with the normal rate of deterioration. As with any other additive discussed in this book, biocides should not adversely affect the processing characteristics of the material, or any other production aspect.

Besides these general demands, biocides must meet certain special requirements, associated with the material's production technology and service conditions. One illustration is that biocides designed to protect polymer and paint or varnish should be readily soluble in organic solvents, and should be readily miscible with other constituents of the material, or at least should form stable emulsions or suspensions.

Biocides are widely used to protect wood. They should easily penetrate the wood and chemically interact with the cellulose component, but not affect its adhesive properties. The fact that wood requires protection against microorganisms leads us to the observation that cellulose based plastics are among the synthetic materials most vulnerable to biodegradation.

By definition, biocides are toxicologically active. It is therefore important to ensure that they are not harmful, in the doses used, to humans and other warm blooded animals. In particular, no cumulative properties

Table 2 Suppression of microorganism growth by using various additives

Petrol additive	Concentration (%)	Mixture of fungi	Glagosporium resinae	Pseudomonas pyncyaneum	Mycobacterium lacticolum
Mixture of aliphatic amines	0.1	100	100	100	100
Dimethyl aminomethyl-parachlorphenol	0.5	60	80	100	100
Trialkylphenyl ester of cyclohexyl phosphine	0.5	40	40	80	80
T-1* petrol with no additives	–		Plentiful growth		

* Russian trade mark

Table 3 Fungicidity of heterocyclic arsenic derivatives [2]

Substance	Concentration (cm^3/l)				
	1	10	50	75	150
10-Phenylphenoxarsin	−	−	−	−	−
10-Chlorphenoxarsin	+	+	+	+	−
10-Ethylphenoxarsin	+	+	−	−	−
10,10-bis(Phenoxarsin) oxide	+	−	−	−	−
10-fluor-5,10-dihydrophenoxarsin	+	+	−	−	−

should be displayed. Specific concerns are possible carcinogenesis, mutagenity, embryotoxicity, and promotion of allergies.

A CLASSIFICATION OF BIOCIDES

Chemical biodegradation protectors are classified according to their biological action, chemical composition, and purposes. The following subdivision is based on biological action:

1. fungicides, used for protecting materials and appliances from degradation by fungi;
2. bactericides, used for protecting from putrid, acid-forming or other bacteria;
3. algicides, used for protecting ships and technical buildings from algae and molluscs;
4. insecticides;
5. zoocides (for protection from rodents).

Biocides can also be classified rather differently, according to their technical purpose, i.e. for the protection of:

1. wood, paper and cardboard;
2. synthetic polymers, notably plastics and rubber;
3. textiles;
4. leather;
5. petroleum products;
6. lubricants;
7. paints and varnishes

Biocides can be subdivided according to their chemical structure into seven main types: inorganic compounds; hydrocarbons, halogenated hydrocarbons and nitrocompounds; alcohols, phenols and their derivatives; aldehydes, ketones, organic acids and their derivatives; amines, amine salts and quaternary ammonium compositions; metal–organic compositions; heterocyclic compounds. The characteristics of the most useful biocides are shown in Table 4.

Table 4 Fungicides protecting nonmetal materials from mold fungi

Class	Purpose	Substance	Solubility in water (%)	Toxicity (D_{50})
1. Inorganic compounds	Wood, paper, cotton and flax fibers	Sodium silicofluoride Na_2SiF_6	0.66% at 20°C	High, 50 mg/kg
	Wood antiseptic	Sodium fluoride NaF	4% at 20°C	High, 100–200 mg/kg
	Wood antiseptic	Sodium dichromate $Na_2Cr_2O_7$	200% at 0°C	Very high, 7–8 mg/kg
2. Hydrocarbons	Paper	Diphenyl	Insoluble	Low, 3000 mg/kg (rats)
halogen hydrocarbons alcohols and phenols	Natural (real) leather Cotton thread	2-hydroxydiphenyl	~0.07% at 25°C	Low, 2700 mg/kg
3. Aldehydes, ketones, carbonic and carboammonium acids	Paint and varnish coverings, PVC, polymer materials etc.	Phthalan N-(trichlormethyl thio)phthalimine	Insoluble	Very low, ~10^4 mg/kg
	Film materials (leather, PVC)	Cyclohexilimide of dichlormaleic acid (cimide)	Insoluble	Moderate 5200–7500 mg/kg

	Rubber composites	Tetramethyl-thiuramdisulfide $(CH_3)_2NCSSCN(CH_3)_2$ with $\overset{\parallel}{S}\quad\overset{\parallel}{S}$	Insoluble	High, 780 mg/kg
4. Amines, amine salts, quaternary ammonium compositions	Volatile fungicide for optical appliances protection	Cyclohexylamine chromate $[\text{C}_6\text{H}_{11}-\text{NH}_2]\; H_2CrO_4$	4%	High, 16–20 mg/kg
5. Elementorganic compounds	Paper, foamy plastics, optical surfaces	Ethylmercurothiosalicilic acid, Na-salt (benzene ring with COONa and SHgC$_4$H$_5$ substituents)	100%	High, 44 mg/kg
6. Heterocyclic compositions	Petroleum lubricating materials	8-hydroxyquinoline (quinoline with OH at 8-position)	Low soluble	Moderate, 1000 mg/kg

TECHNIQUES FOR INVESTIGATING BIOCIDES

Traditional methods are based on analysis of the inhibiting capability of the biocide with respect to microorganism development, e.g. biomass, the radius of colonies during their incubation on a solid agarized surface. Fungicide activity is determined by the following equation:

$$R = (d_0 - d_1) \times 100/d_0 \qquad (1)$$

where d_0 is the diameter of the colonies in mm and d_1 is their diameter in the presence of the fungicide. It is determined in a similar way by estimating microorganism biomass in a fluid culture medium.

One new approach to estimating biocide efficiency is to combine qualitative evaluation with the use of kinetic parameters chosen to characterize the properties of microorganisms and materials. A quantitative estimate of effectiveness is based on studies of biomass growth deceleration, with the help of rate constants for biomass growth.

Traditional methods determine final biomass, and the threshold radius of colony growth on surfaces in the presence of inhibitors after the growth cycle has ended. Or the biocides can simply be subject to a growth/no-growth verdict. Initial rate of microorganism development should be measured; this involves some experimental difficulties.

A radioisotope technique is available for determining the initial mass. Polymer materials, in suspension in water or in a Chapek–Doj culture medium of microscopic fungi of concentration one million cells per cubic centimeter, are incubated in tritiated water vapour. The accumulation of tritium in the biomass is proportional to the microorganism biomass growth; there is typically about 85% water in the biomass. The quantity of biomass is determined by the irradiation intensity of the samples, using a fluid oscillation counter. The experimentally determined kinetic curves are exponential, and the equilibrium biomass value m_∞ can be found along with the initial growth rate V_0 and the effective rate constant. Table 5 shows these parameters for *Aspergillus niger* microscopic fungus, at a cell concentration 1 million per cubic centimeter, for the specific polymers shown.

Table 5 Initial rates of biomass accumulation V_0 and threshold m_∞ values on the surface of different polymer materials

Support	m_∞ ($\mu g/cm^2$)	V_0 ($\mu g/cm^2/day$)	$k_{eff} \times 10^6$ (s^{-1})
Cellulose	10.5 ± 1	0.60 ± 0.05	1.0
Polyethylene terephthalate	2.4 ± 0.15	0.16 ± 0.02	1.2
Polyethylene	1.5 ± 0.20	0.27 ± 0.02	0.9
Polytetrafluorethylene	1.1 ± 0.1	0.04 ± 0.01	0.7

Kinetics of biomass growth in a medium containing biocides 129

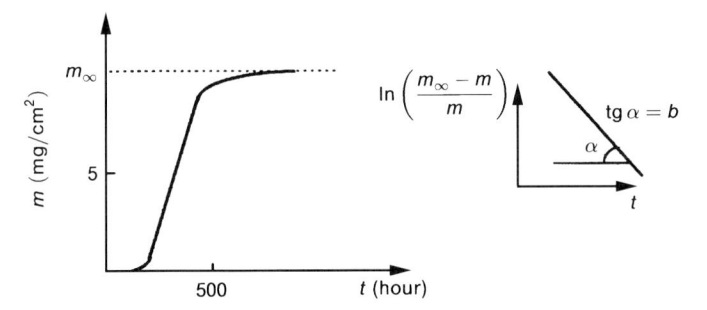

Figure 1 *Aspergillus niger* biomass dependence on cultivation time in Chapek–Doj medium.

KINETICS OF BIOMASS GROWTH IN A MEDIUM CONTAINING BIOCIDES

Figure 1 shows the typical shape of the growth curve for *Aspergillus niger* in liquid Chapek–Doj culture medium. The curve is described by the exponential function:

$$m = m_\infty(1 + a\exp[-b(t - L)]) \qquad (2)$$

where a and b are kinetic parameters, m_∞ is the threshold biomass value, and L is the time for the microorganism to adapt to the culture medium, i.e. the lag period. The physical constants a and b possess physical meaning: a characterizes the ability of a spore to form biomass under the prevailing conditions and b represents the specific microorganism growth rate in the culture medium. Values of a and b are determined graphically, as shown in Figure 1, by converting equation 2 to the linear form below:

$$\ln(m_\infty/m - 1) = \ln a - b(t - L) \qquad (3)$$

The parameters a, b and L are concentration dependent.

Injection of water soluble biocides into the culture medium causes no change in the kinetic dependence of biomass growth (Figure 2).

The use of agarized culture media for the qualitative estimation of biocide properties of water insoluble substances, including many fungicides, is not appropriate because of the irregular distribution of the biocide in the medium. Hydrogel supports made of hydroxy ethyl methacrylate polymers should be used.

Increasing the biocide concentration increases L (Figure 3) according to the following equation:

$$L = L_0 \exp K_L C \qquad (4)$$

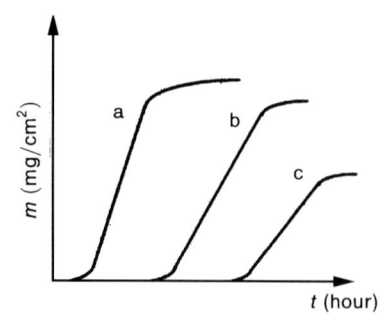

Figure 2 Dependence of biomass growth on cultivation time in liquid culture medium containing different concentration of biocides: (a) 0.1 mg/l; (b) 0.3 mg/l; (c) 0.5 mg/l of merthiolate (sodium salt of ethylmercursalicylic acid).

where L_0 is the lag time in a biocide-free culture medium, C is the biocide concentration and K_L is a kinetic constant obtained from the linear form of the above equation, as indicated in the figure.

The specific growth rate of a microorganism, b, decreases after biocide injection, all other factors being equal, and the inhibited specific growth rate is described as follows:

$$b_i - b_0 K_C/(K_C + C) \tag{4}$$

Here, b_0 is the specific growth rate in absence of the biocide, and K_C is determined from experimental data, as shown in Figure 4, by linearizing equation (4).

The constant K_C for this biocide does not depend on concentration and may be used for quantitative estimations of the biocide activity of substances. The higher the K_C value is, the lower the efficiency observed.

Biocide activity constants, K_C and K_L, are clearly related to each other and to biocide effectiveness. The values obtained for these substances

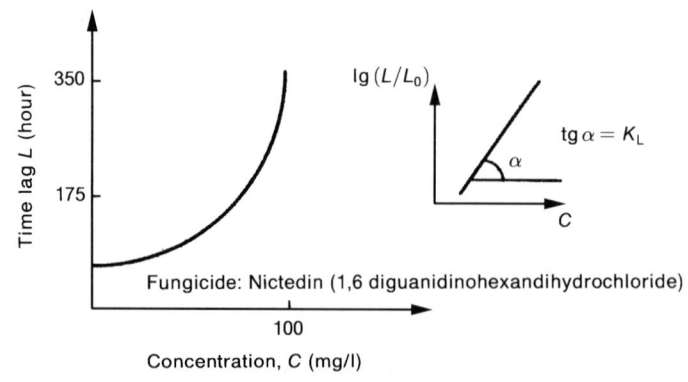

Figure 3 Concentration dependence of the lag time for *Aspergillus niger*.

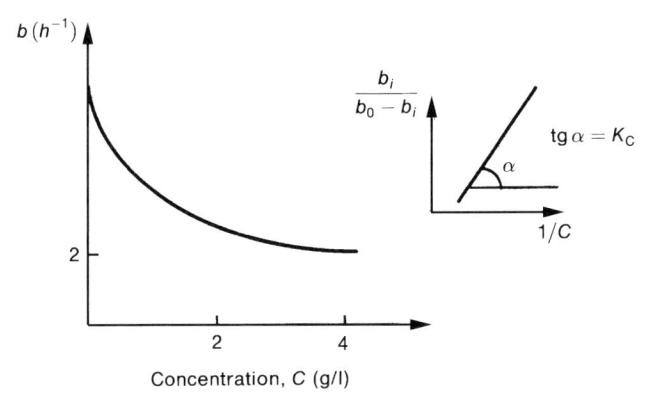

Figure 4 The dependence of parameter b on biocide concentration (*Aspergillus niger* development in presence of $CuSO_4$).

characterize their effectiveness in $\ln(1/K_C) - \ln K_L$ coordinates. They fit the straight line (Figure 5).

Such regularity is used for comparative estimation of biocides and preliminary determination of their concentrations in a material required for reliable protection from biodegradation. General characteristics of biocides which inhibit microorganism development based on kinetic estimation are shown in Tables 6 and 7.

Quantitative testing of known and newly synthesized biocides gives the specificity of each class of microorganism for a particular substrate-support. Such an approach should promote research into synergisms based on the increase in biocidic activity.

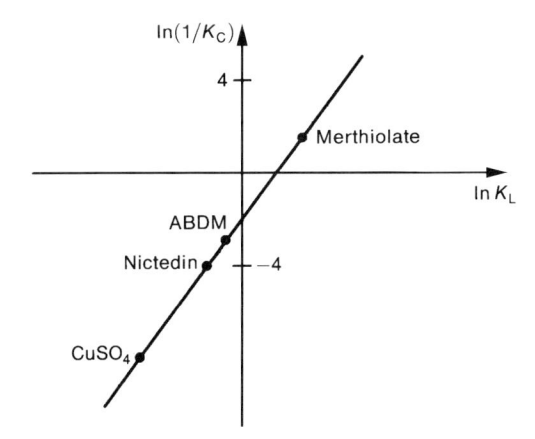

Figure 5 Kinetic constants of the biocide activity of some substances. ABDM is alkylbenzylmethylammonium chloride.

Table 6 Constants of biocide action effectiveness of chemical substances

Biocides	Water soluble biocides K_L (l/mg)	Water soluble biocides K_C (mg/l)	Water insoluble biocides K_L (cm^2/mg)	Water insoluble biocides K_C (mg/cm^2)	b_0 (day^{-1})	L_0 (day)
Sodium merthiolate (ethylmercursalicilic acid salt)	6.7	0.76	–	–	0.0015	1.67
ABDM (alkylbenzene dimethylammonium chloride)	0.17	8.9	–	–	0.0015	1.67
1,6-diguanidinohexadihydrochloride (nictedin)	0.028	80.5	–	–	0.0015	1.67
Copper(II) sulfate (CuSO$_4$)	0.0005	1950	–	–	0.0015	1.67
Oxydiphenyl (ODP)	250	0.004	–	–	0.0015	1.67
N-paratolylmonoimide (PTMI)	62.8	0.05	–	–	0.0015	1.67
Bis(0,0-1-chlor-tri-brom-isopropyl)-3-chlor-2-brompropylphosphanate (phlamal)	–	–	1.9	0.7	0.0015	1.67
2,6-ditertbutyl-4-methyl-phenol (ionol)	–	–	0.2	17.5	0.0015	1.67
Pentachlorphenol	–	–	125.2	0.012	0.002	2.8
Salicilanilide	–	–	44.5	3.79	0.0016	2.2
Trilan	–	–	21.2	0.02	0.001	1.67
Biocine	–	–	20700	9.95	0.0025	

Table 7 Kinetic parameters of *Aspergillus niger* biomass growth. The fungicide is copper(II) sulfate ($CuSO_4$)

Conditions of fungus growth	Fungicide concentration (mg/L)							
	C = 0				C = 2000			
	Kinetic parameters							
	$b\,(h^{-1})$	$L\,(h)$	a	$m_\infty\,(\mu g/cm^2)$	$b\,(h^{-1})$	$L\,(h)$	a	$m_\infty\,(\mu g/cm^2)$
Liquid Chapek–Doj culture medium	0.037	40	230	8.5	0.019	120	200	7.8
Hydrogel	0.035	40	113	4.0	0.020	120	90	3.5

FURTHER READING

Collection (1983) *Actual questions of biodegradation*, Nauka, Moscow.
Gumargalieva, K.Z. and Zaikov, G.E. (1995) Biodegradation of polymer materials. Generalized kinetic data, *Chemical Physics*, **14**(10), 29–38.

Keywords: Biocide, bacteria, fungi, fungicide, microorganism, wood, cellulose, biomass, kinetics, toxicity, cultivation medium, *Aspergillus niger*, biodegradation.

See also: Biocides;
       Biodegradation of plastics: monitoring what happens
          (Introductory section)

Biodegradation promoters for plastics

G.J.L. Griffin

THE PERCEIVED ENVIRONMENTAL PROBLEM

The politically favoured concept of sustainability has focused attention on the need to provide materials with low manufacturing energy demands and preferably made from renewable resources. That such materials should also offer biodegradability as an option in waste disposal, not conflicting with recycling or incineration activities, has served to motivate the development of degradable polymer compositions whose applications have already achieved a modest position in world markets. Claims that goods are overpacked are countered by drawing attention to the need for food hygiene and safety. A call for the recycling of all plastics disposables appears admirable until one learns of the extreme economic and technical problems which beset the recovery of the mixed plastics from MSW (municipal solid waste). Although European politicians have set a 50% target for plastics recycling, recent studies indicate that much more modest levels are realistic. Furthermore the progressive thermal degradation associated with each polymer melt processing operation in the recycling loop would surely eliminate the possibility of continuously cycling a 'national' pool of polymer material. An understandable 'green' urge to return to natural products brings some unwelcome reminders of the cost of paper and card, especially when food contact quality is demanded, as well as recollections of the unsuitability of paper for packaging wet goods. We are also reminded that paper cannot really be considered as a 'natural product' and is separated from its botanical source materials by massive industrial chemical technology such as

Plastics Additives: An A–Z Reference
Edited by G. Pritchard
Published in 1998 by Chapman & Hall, London. ISBN 0 412 72720 X

pulp making, bleaching, coating, glazing and printing techniques. The necessity to conserve non-renewable resources must remain a dominant consideration. We must also face the surprising fact that natural fibre resources would not be adequate to meet the needs of present population levels and expectations.

TECHNICAL OBJECTIVES

The general and product specification targets for 'ideal' biodegradable plastics films, as recognized in laboratories concerned with the development of materials for one-trip packaging applications, can be listed as follows. They should:

1. have a controlled lifetime at ambient temperatures terminated by abrupt mechanical collapse and further slow biological breakdown by aerobic or anaerobic microbiological metabolization into soil humus, and eventual return to the carbon cycle in accelerated MSW (municipal solid waste) composting;
2. be able to be recycled in-plant or as part of post-consumer waste schemes;
3. have no greater calorific value than conventional film, and therefore be no more difficult to incinerate where this is the practice;
4. preferably replace a significant amount of a non-renewable resource material, such as polyolefins, by a renewable resource material such as starch or cellulose;
5. run on existing processing plant without major machine modification;
6. not increase the price of plastics film by more than about 15%;
7. not show extraction figures above the level specified for food contact applications to ensure eventual full approval to national status;
8. allow the degradation property of products to be readily checked by short term QC procedures eventually to become national standards;
9. allow the technology to be transferable without difficulty to other polymer processing methods such as injection moulding, extrusion coating and film casting;
10. have degradation products which are essentially the same as for the decomposition of familiar natural products and should not, therefore, create undesirable secondary pollutants.

THE 'SMART' MATERIAL CONCEPT

We have to accept that the polyolefin group of plastics are remarkably successful as one-trip packaging materials in the form of films, blown bottles, and injection moulded containers. They are, however, astonishingly durable materials provided only that they are not exposed to sunlight. The technical challenge has been to devise some formulation modification which either imposes a predetermined lifetime on the plastic or, alternatively, exploits some means for enabling the plastics to recognize the difference between their status as functioning packages and their disposed status, and to react in a self-destructive manner as a result of this recognition. If we consider the fate of waste plastics materials, a small proportion find their way into our streets, drains, countryside, rivers and oceans as litter, the major proportion into the urban solid-waste disposal stream, and currently about 20% goes into recycling activities. The litter stream is distinguished by getting exposure to sunlight whilst all three routes involve contact with active microbiological systems. Some rather complex solutions have been proposed, involving multilayer systems where an extremely thin outer skin is expected to fracture in the mechanical handling processes of the waste industry, thus exposing a biodegradable or water soluble inner layer to the biologically active waste environment. These ingenious schemes suffer from the penalty of considerable added cost and have not been generally adopted. A considerable range of intrinsically biodegradable polymers have been designed and synthesized with structures planned to be susceptible to the known enzyme systems associated with the microbiological fauna of waste streams. Other similar products have been isolated from the metabolisms of certain bacteria, the polyhydroxybutyrates and butyrate/valerate co-polymers. These polymers all suffer from the penalty of costing at least four times as much as the familiar polyolefins and have, therefore, only found application in very specialized packaging applications where the price is of less consequence. Also they cannot be used where the packaged products themselves may be biologically active.

PHOTODEGRADABLE POLYMERS

Historically the sunlight exposure aspect was the first to enable a true 'recognition of disposal' situation to be exploited and various chemical additives have been developed which can catalyse the photodegradation process. These photodegradable plastics have had some commercial success, particularly in the making of the famous six-pack drink-can rings to reduce the problem of trapping sea birds. The photochemical destruction of polyolefins has been extensively studied and the mode of chain breaking recognized. This process causes a loss of physical properties usually manifested as a severe drop in tensile strength and a dramatic increase in brittleness. Laboratory studies have shown that a certain degree of biodegradability is developed in intensively irradiated samples but these conditions cannot be assumed to apply in the real environment.

AUTOXIDANT SYSTEMS

Development

Early studies of the fate of normal polyolefin films in municipal composting plants showed a rate of strength loss far greater than would be observed if the same materials were buried in garden or agricultural soil. Searching for the relevant difference in the exposure conditions resulted in the observation that fatty residues from kitchen waste were selectively absorbed by the hydrocarbon polymer. These fatty materials were largely unsaturated vegetable fats which, in these contaminated conditions, were rapidly oxidizing through classical rancidity routes which generated peroxides. Typical fat contents were 3–4% and the peroxide content was at a level that could readily be assessed by iodometric tiration. It was hardly surprising that the polymer itself underwent oxidation under these circumstances and the resulting strength loss could be monitored. Attempts to exploit this situation were troubled by the unpredictability of the fat pick-up process and the uncertain proximity of the essential transition metal contamination to provide the oxidation catalyst. The logical step was to produce a polyethylene film which contained the necessary elements of an autoxidation system and this would ensure that the breakdown of the polymer would occur. The next important question was controlling the time that the process would take, and here it was found possible to exploit the induction period phenomenon, well known in chemical reaction dynamics. Because polyolefins are subject to oxidation at melt processing temperatures it is normal for minor amounts of anti-oxidants to be added to them at an early stage in their manufacture. These anti-oxidants are usually of the sacrificial type which scavenge oxygen and free radicals for a time dependent on the added amount and their reactivity. When both anti-oxidant and an autoxidation system of transition metal catalyst plus unsaturated organic compound are present it is possible to produce polyethylene formulations which have lifetimes, free from degradation at ambient temperatures, measured in weeks or years according to amounts of the various additives. Monitoring the molecular weight distribution of these formulations shows virtually no change until the induction period has finished, whereupon the molecular weights fall dramatically to the point where substantial amounts of molecular fragments with $\overline{M}w$ values measured in 100s rather than the usual 10 000s. These fragments, terminated with oxidized groupings such as carboxylic acids, are therefore hydrophilic, in contrast to their hydrophobic polymer precursors, and are readily metabolized by microbial action. It has been found advantageous to facilitate the entry of fungi and

Autoxidant systems 139

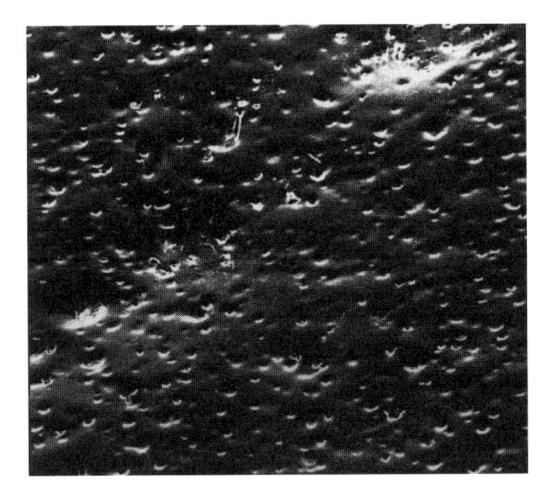

Figure 1 Scanning electron micrograph of the surface of extrusion-blown LDPE film containing 10% of maize starch. As the extruded tube is inflated and stretched, characteristic mounds are raised by the solid starch particles, average diameter 15 µm. Magnification circa ×350.

bacteria into the disintegrating polymer by incorporating a few percent of natural starch particles which, being themselves highly biodegradable, act as stepping stones for the invading organisms (Figures 1 and 2).

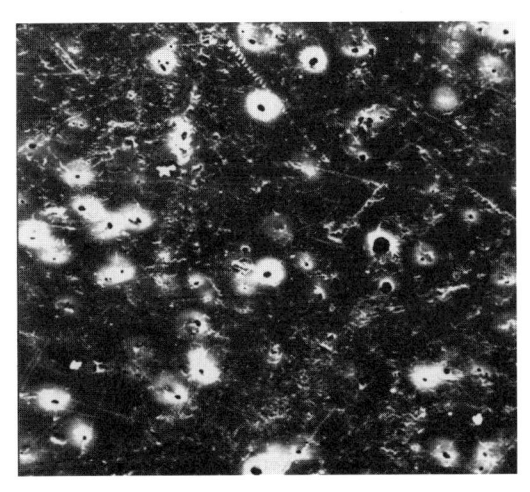

Figure 2 Scanning electron micrograph of film shown in Figure 1 after ageing and immersion in starch dissolving enzyme. Each particle has been converted to sugar which has attracted water osmotically, thus swelling and rupturing the oxidized and weakened polythene and hugely increasing the surface area accessible to bacteria and fungi. Magnification circa ×350.

Choice of catalyst

The use of transition metals as electron movers in oxidation processes is well known to chemists and has been widely exploited commercially in the formulation of drying oil paint finishes. The preferred metals in the paint industry have been cobalt, iron, and copper, usually in the form of organic salts chosen for oil solubility. The requirements for plastics formulations called for low toxicity, low odour, and very pale colour. Cerium and manganese soaps were favoured and manganese was the preferred choice because of its high activity and its inclusion in the EPA list of GRAS (generally recognized as safe) substances.

Manufacturing processes

The five key ingredients (polymer, biodegradation promoter antioxidant, autoxidation promoter and catalyst) have to be brought together using conventional melt processing equipment but in such a manner as to ensure that a conventional granular product can be presented to polymer converters ready for use on their film lines or injection moulding machines with no possibility of premature degradation. This is accomplished by compounding the transition metal soap catalysts with a minor portion of the polymer, containing only its normal antioxidant and no autoxidation promoter. This catalyst granulate has been safely warehoused for several years. The bulk of the polymer with its regular antioxidant content is separately compounded with the autoxidation promoter, usually an SBR elastomer, and the biodegradation promoter, usually pure dry starch derived from maize or rice. It has been found possible to formulate these two compositions at high concentrations of active ingredients so that the cold granules can be blended to provide a masterbatch which, used typically at 15% addition level to normal polymer granules, gives the optimum concentration of the active elements in the finished product film or moulding. The masterbatch blend of two active types of granulate only begins its internal chemistry when the metal catalyst meets the autoxidation promoter upon the mingling of the melts in the extruder barrel.

BIBLIOGRAPHY

Wallhauser, K H. (1972) Mull ü Abfall, **1**, p. 10.
Scott, G. (1995) Degradable Polymers, eds. G. Scott and D. Gilead, Chapman & Hall.
Griffin, G.J.L. (Ed.) (1994) The Chemistry and Technology of Biodegradable Polymers, Blackie Academic.

Vert, M., Feijen, J., Albertsson, A-C., Scott, G. and Chiellinin, E. (Eds) (1992) Biodegradable Polymers and Plastics, Royal Society of Chemistry.
Guillet, J. (Ed.) (1973) Polymers and Ecological Problems, Plenum.

See also: Biodegradation of plastics: monitoring what happens (Introductory section).

Blowing agents

Geoffrey Pritchard

POLYMER FOAMS

The production of a foam normally but not inevitably involves using a blowing agent to generate gas within the polymer, while the latter is temporarily in a viscous liquid state. Foam formation has the advantage of greatly decreasing the weight of the finished product. It also reduces the modulus, while improving other properties such as energy absorption and thermal and acoustic insulation performance. The insulation characteristics of a foam can depend markedly on the gas contained in the cells, and therefore the choice of blowing agent is not simply a matter for consideration during the processing stage. It also affects the ultimate properties. These properties depend on the detailed microstructure of the foam and on the volume fraction of cavities. This in turn depends on the surface tension of the liquid, the rheology of the mix at the time of foam formation, the prevailing temperature, the number of gas bubbles per unit volume, the average wall thickness of the cells etc. The formation of a foam has to be synchronized with the formation of the finished part.

PVC and polyurethane constitute together the majority of expanded plastics, but expanded polystyrene packaging is also important because it has excellent protective qualities for the transport of delicate articles, and is widely used in the catering industry. Crosslinked polyethylene foams are also important and there are several other commercially successful varieties.

PRINCIPLES OF BLOWING AGENTS

Blowing agents are additives which generate a gas during the period when the polymer is beginning to solidify. The polymer solidifies when

Plastics Additives: An A–Z Reference
Edited by G. Pritchard
Published in 1998 by Chapman & Hall, London. ISBN 0 412 72720 X

its viscosity rises sufficiently, for example because of an increase in the molecular weight, or in the case of a crosslinkable polymer, when its network structure reaches a stage of development known as gelation. At gelation, the network is described as an infinite network, and the polymer becomes infusible. The timing of the gas generation process can be crucial.

The use of blowing agents is not the only way to make a cellular or foamed polymer. Other methods include the mechanical generation of a froth, or the direct incorporation of large quantities of thin-walled, hollow microspheres (see the entry 'Hollow microspheres', in this book).

There are two broad categories of blowing agent – physical ones, which change from liquid to gas when the temperature rises or the pressure falls, and chemical ones, which decompose during the processing operation, liberating a gaseous product such as carbon dioxide or nitrogen by means of a chemical reaction. Commercial processes sometimes employ both types together in the same system; carbon dioxide may be generated chemically at an early stage, and a physical blowing agent used to maintain the cell structure. The heat evolved or absorbed during the action of a blowing agent can be used to help control the other reactions taking place. It is possible to use a mixture of endothermic and exothermic chemical blowing agents together if this gives the right thermal characteristics.

The selection of a blowing agent depends on the polymer being foamed, and on the processes involved. The aim is not just to produce a foam of a given density, but to optimize its performance in all respects. Specific blowing agents are recommended for specific applications, e.g. one for business machine housings, another for packaging trays. Many chemicals have been evaluated, but only a small number are commercially successful.

PHYSICAL BLOWING AGENTS

Physical blowing agents are relatively low cost, and they leave no solid residue, but using them requires special equipment. They are non-reactive in the chemical sense, and may consist either of volatile liquids (organic, or inorganic, e.g. water) or of compressed gases, such as carbon dioxide or nitrogen, which are dissolved under pressure in the polymer so as to form a foam when the pressure is released. Physical blowing agents generally lower the viscosity of the reaction mix if present in appreciable quantities.

The solubility of a gas in a polymer is highly pressure and temperature dependent. When the pressure falls, the gas becomes much less soluble and a phase separation results, with the gas forming cells. The process of evaporation counteracts the heat produced by the exothermic polymerization process. The most important application of compressed gas technology is probably in the formation of foamed products during the injection moulding of thermoplastics.

Where the blowing agent is a solid or a liquid, the amount of blowing agent used ranges from less than 1% up to about 15%, depending on circumstances.

LIQUID PHYSICAL BLOWING AGENTS

A range of chlorofluorocarbons (CFCs) used to be successfully employed in rigid polyurethane foams; they not only generated a foamed structure, but also improved the final properties. In recent years, however, there has been a rapid movement away from CFCs because of the problem of the ozone layer. Sweden banned CFCs in 1991, followed by Switzerland, the USA etc. and CFC use is now prohibited.

Hydrochlorofluorocarbons (HCFCs) have been widely adopted, although they still have a finite degree of ozone depletion potential (ODP), typically in the range zero to 0.05, compared with CFCs typically in the range 0.2 to 1. Hydrofluorocarbons (HFCs) such as hexafluorobutane are also being used at the time of writing, but further restrictions on the available choice of substances are being applied, resulting in the elimination of many other chemicals over the next few years. On the other hand the list of banned substances can be pruned occasionally; acetone was withdrawn from a list of blowing agents regulated by the U.S. Environmental Protection Agency, having been shown to have little or no impact on smog and ozone problems. When a new blowing agent is substituted for one that has been withdrawn, the physical properties of the finished product will be altered, and additives such as surfactants and softening agents may be invoked to modify properties such as insulation characteristics.

Methylene chloride is widely used as a physical blowing agent. As with many other agents, there have been problems in achieving the correct rate of volatilization, but it has the great advantage that it is not flammable, unlike the aliphatic hydrocarbon blowing agents. It has one of the lowest thermal conductivities, at $0.0063 \, W/m/K$, compared with $0.0259 \, W/m/K$ for air and $0.0168 \, W/m/K$ for carbon dioxide, thus imparting better themal insulation qualities to the foam. (When the cells of a foam are of the closed variety, they retain much of the original blowing agent, and this contributes significantly to the insulation characteristics of the foam.)

Normal pentane, cyclopentane and isopentane have proved useful, low cost, readily available physical blowing agents, giving foams with good thermal properties. However, the pentanes form potentially explosive mixtures with air. This means that manufacturing processes have to be modified to deal with the risks – a factor which has limited their appeal.

CHEMICAL BLOWING AGENTS

Chemical blowing agents produce gases by a chemical reaction which occurs at the same time and in the same temperature range as the processing operation. The majority of chemical blowing agents are liquids and the decomposition reactions are usually but not invariably exothermic. Chemical blowing agents are favoured for PVC foam production.

It is essential that whatever the gas producing reaction, it occurs within the temperature range of the plastics processing operation. In some circumstances this means adding a second chemical as a 'kicker' to lower the temperature of decomposition of the main blowing agent. Kickers include polyols, urea and amines. Some additives used in plastics formulations, such as basic lead, pigments, fillers, zinc oxide etc. can also act incidentally as kickers. The end product of a blowing agent decomposition reaction is significantly altered by the inclusion of kickers, and they may overpromote the blowing agent decomposition, leading to a need to damp down the kicker reaction with suitable modifiers, such as dicarboxylic acids or anhydrides.

The most important chemical blowing agents are the azo compounds, such as azodicarbonamide (AZDC) and azoisobutyronitrile. Azoisobutyronitrile decomposes on heating, to produce free radicals and nitrogen gas:

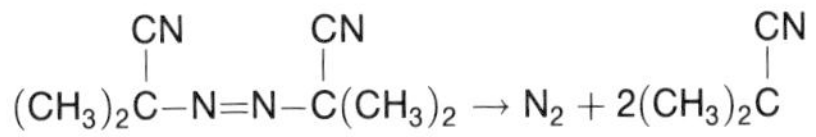

$$(CH_3)_2\overset{\underset{|}{CN}}{C}-N=N-\overset{\underset{|}{CN}}{C}(CH_3)_2 \rightarrow N_2 + 2(CH_3)_2\overset{\underset{|}{CN}}{C}$$

Figure 1 Decomposition of azoisobisbutyronitrile to give nitrogen (chemical blowing agent).

The decomposition of azodicarbonamide follows the reaction:

$$H_2NCO \cdot N=N \cdot CO \cdot NH_2 \rightarrow N_2 + CO + H_2N-CO-NH_2 \rightarrow NH_3 + HNCO$$

with a competing reaction forming hydrazodicarbonamide:

$$H_2NCO \cdot N=N \cdot CO \cdot NH_2 \rightarrow H_2N-CO-NHNHCONH_2$$

In the presence of moisture, AZDC forms hydrazodicarbonamide together with nitrogen, carbon dioxide and ammonia.

Other chemical blowing agents include hydrazine derivatives, semicarbazides, tetrazoles, benzoxazines, and inorganic substances such as sodium bicarbonate in conjunction with citric acid.

Mixed inorganic and organic systems, e.g. those involving 1,1-azobisformamide and sodium bicarbonate, have also been used. Generally, organic chemical blowing agents offer better control than inorganic ones. But the number of factors which have to be taken into account is large: the toxicity; the effect on colour and on mould durability; the volatility of any residual blowing agent, and the possibility of adverse effects on the polymer's fire performance or its thermal or photo-oxidative stability.

The ease of addition to the polymer mix is a practical issue. Chemical blowing agents may be added directly as powder, as solid pellets, or as concentrated liquids, but need accurate metering.

One of the most important foam materials is flexible polyurethane, which is customarily blown by using the carbon dioxide formed as a result of the chemical reaction between di-isocyanates and water. The process is accompanied by the evolution of carbon dioxide, as a product of the reaction between water or acids and isocyanate groups:

$$ONCRCNO + R'COOH = ONCRNHCOR' + CO_2$$

$$ONCRCNO + H_2O = ONCRNHCONHR + CO_2$$

$$ONCRNCO + R'OH = ONCRNHCOOR'$$

These equations are simplifications of the true situation; both isocyanate groups react in practice, and while the R, R' groups in the above equations can be envisaged as simple alkyl or aryl groups (as in toluene di-isocyanate) some may be more complex. The carbon dioxide can be augmented by another, physical blowing agent such as methylene chloride or trichloromonofluoromethane. The importance of the water–isocyanate reaction is not simply in acting as part of the blowing process, but also in providing exothermic heat to help complete the polymerization and remove any unwanted volatiles. The same reaction also produces the hard segments of the polyurethane structure. (The same considerations do not apply with rigid polyurethane foams, where there is already enough exothermic heat from the main reaction.) As well as polyurethane foams, isocyanates are involved in the production of other foamed polymers, such as polyamides and polyimides.

Thermosetting resins, such as unsaturated polyesters and their fibre reinforced analogues, can be produced in foam form and have considerable potential in weight saving applications in transportation. It is possible to foresee more applications in vehicles and transportation generally, where energy saving will again become a strong motivation. In the medium term, there is likely to be a sustained emphasis on achieving existing performance levels (i.e. good quality cellular materials) with environmentally more acceptable blowing agents. There is also a trend towards using more endothermic blowing agents.

Examples of blowing agents are listed in Table 1, and some suppliers are cited in Table 2.

Table 1 Examples of substances used (or formerly used) as blowing agents

Substance	Category
Argon; nitrogen; carbon dioxide	Inert gases
Trichloro fluoromethane	Physical; CFC
Difluorodichloromethane	Physical; CFC
Dichlorofluoromethane	Physical; HCFC
1-chloro,2-2' difluoroethane	Physical; HCFC
1,1,1-trifluoro,1-fluoroethane	Physical; HFC
1,1,1-trifluoro,2-fluoroethane	Physical; HFC
Hexafluorobutane	Physical; HFC
n-pentane	Physical; hydrocarbon
Cyclopentane	Physical; hydrocarbon
Methylene chloride	Physical; chlorinated hydrocarbon
Water	Physical
Sodium bicarbonate/citric acid	Chemical; evolves CO_2, H_2O, 150–230°C
Azoisobutyronitrile	Chemical; evolves N_2 at 70°C
Azodicarbonamide	Chemical; evolves N_2, CO, NH_3, CO_2 at 205–215°C or 150°C and above if activated
2,4,6 trihydrazino-1,3,5 triazine	Chemical; evolves N_2, NH_3 at 245–285°C
p-toluene sulfohydrazide	Chemical; evolves N_2, H_2O, 105–110°C
The following are more commonly used with rubbers than plastics:	
4,4' oxybis(benzenesulfohydrazide)	Chemical; evolves N_2, H_2O at 155–165°C
Dinitrosopentamethylene tetramine	Chemical; evolves N_2, NH_3, HCHO at 195°C

Table 2 Some suppliers of blowing agents

Allied Signal	J M Huber
Bayer	High Polymer Labs.
Boehringer-Ingelheim	Kum Yung Co.
Dong Jin Chemical Industry Co.	Otsuka Chemical Co.
Eiwo Chemical Industry Co.	Sankyo Kasei Co.
Elf Atochem	Solvay Chemicals
Hebron	Toyo Hydrazine Industry
Hoechst A G	Uniroyal
ICI	

BIBLIOGRAPHY

D. Klempner and Kurt C. Frisch (1991) *Handbook of Polymeric Foams and Foam Technology*, Hanser, Munich.
G. Woods (1987) *The ICI Polyurethanes Book*, ICI/Wiley.
Lorna J. Gibson and Michael F. Ashby (1988) *Cellular Solids – Structure and Properties*, Pergamon.

Keywords: foam, cell, polyurethane, CFC, HCFC, carbon dioxide, pentane, inert gas, azo-compounds, isocyanate, thermal insulation.

Calcium carbonate

C.C. Briggs

INTRODUCTION

Calcium carbonate accounts for about 65% of total filler consumption in plastics with an annual world-wide usage of about six million tonnes. It is also the most widely used filler in terms of number of applications. The cheapest grades are low in price and are used primarily to reduce costs. By contrast, the finest grades are an order of magnitude higher in price and are used to modify various properties, both during processing and of the final compounds.

The main classification is between natural minerals and synthetically precipitated grades.

NATURAL MINERAL CALCIUM CARBONATES

All commercial grades of natural calcium carbonate are based upon deposits of calcite, which vary in purity, colour and crystallite size. These factors depend upon genesis and subsequent temperature–pressure history. Chalk is usually off-white in colour and is composed of small crystals which are easily disintegrated. Marble, by metamorphosis, has large crystals and very white deposits can be found. Limestone is one of a number of intermediate forms and varies from white to grey or buff in colour, due mainly to iron impurities.

PROCESSING OF NATURAL CALCIUM CARBONATES

Dry processing involves drying, milling, air classification and sometimes surface treatment. Roller mills, ball mills, peg mills and pin mills are used

Plastics Additives: An A–Z Reference
Edited by G. Pritchard
Published in 1998 by Chapman & Hall, London. ISBN 0 412 72720 X

in conjunction with a wide range of air classifiers. The classifier controls particle size 'top cut' (97% by mass finer). This top size can vary from 100 μm to less than 5 μm.

The median particle size (50% finer) is influenced not only by the classifier but also by the type of mill, the original crystallite size and resistance to comminution. It is important to distinguish between particles and crystals. The former can be whole single crystals, fragments of crystals or strong agglomerates of crystals.

SURFACE TREATMENT OF NATURAL CALCIUM CARBONATES

Silanes are not capable of reacting with a carbonate surface and titanates find only limited use, mainly for viscosity reduction in thermosetting resins. By contrast, very large quantities of calcium carbonate are surface treated with stearic acid. This is normally carried out at elevated temperature under high shear mixing and the stearic acid is largely converted into calcium stearate.

While such stearate coated products are widely used in thermoplastics and thermosets, the major volume usage is in PVC compounds of all types (i.e. uPVC, plasticized PVC and PVC plastisols). The reasons for using a stearate coating vary from application to application. These include improved processing characteristics, mechanical properties and electrical properties (due to better moisture resistance), and reduced absorption of minor additives.

SYNTHETICALLY PRECIPITATED CALCIUM CARBONATES (PCC)

PCC is normally manufactured from natural calcium carbonate by the following route:

(a) calcination to give CaO (+CO_2)
(b) hydration to give $Ca(OH)_2$ slurry
(c) dissolution, filtration and precipitation of $CaCO_3$ using carbon dioxide
(d) filtration, drying and de-agglomeration

Careful control of the precipitation is the key step since this controls:

(i) whether calcite or aragonite is formed
(ii) the crystallite size (often less than 0.1 μm)
(iii) the particle size (agglomerated crystals are often larger than 1 μm)

Surface treatment, most frequently with stearic acid, is carried out at the slurry stage, since this is an efficient coating route but also because the hydrophobic stearate coating gives more economical dewatering. Conversion to calcium stearate is normally 100%.

PCC is also surface treated with materials such as carboxylated polybutadiene. In this case the carboxyl groups react with the surface in a similar manner to stearic acid but the double bonds are available for reaction with polymers. The filler can be pre-coated, or the carboxylated polybutadiene can be added separately during compounding.

This type of coating is generally used for rubber or thermoplastic elastomers where the double bonds can be used to give coupling and reinforcement. In general, PCC is used where fine particle size is the key factor. This primarily influences mechanical properties and surface finish. Good colour and high purity are also beneficial properties in certain applications.

In one specific area, the chemical properties are critical. This is in PVC cable compounds specifically designed to limit hydrogen chloride evolution in a fire. PCC, having a much higher specific surface area than natural calcium carbonates, reacts much faster with the evolved HCl, reducing emissions to the atmosphere.

APPLICATIONS

PVC

PVC compounds account for about 65% of calcium carbonate usage in plastics but this covers a very wide range of applications:

(i) Traditional PVC floor tiles normally employ coarse, cheap, uncoated $CaCO_3$, primarily to reduce cost, but also to add weight. More sophisticated grades such as coated PCC can be used to give better stain and wear resistance but, more importantly, to give rheology suitable to control the colour pattern effects which can be achieved by mixing differently coloured compounds during processing.

(ii) PVC cable compounds, for sheathing, insulation and bedding applications, traditionally employ natural calcium carbonate as the primary filler. The key characteristics required are a top size of about 20 μm, a good colour (though colour consistency is often more important than having an excellent white colour), a low content of iron and other transition metals (important for heat and light stability) and high electrical resistivity.

Stearate coated grades are widely used in cable compounds. They have a beneficial effect on processing, surface finish and electrical properties.

PVC cable compounds remain a major market for $CaCO_3$ fillers but market share in cables as a whole is falling. This is because halogen-free cable compounds now have a substantial market share and normally incorporate only flame retardant fillers.

(iii) There is a wide variety of other plasticized PVC applications including hose, footwear and film. The more expensive finer grades, including PCC, are used only where properties such as surface finish, scratch resistance and crease-whitening are important.
(iv) Unplasticized PVC, sometimes known as rigid PVC, is used primarily in pipes, ducting and profiles. This is a large and still growing sector for both natural calcium carbonate and PCC.

The quality and loading level of fillers are important with regard to processing and final compound properties. One of the primary reasons for the addition of $CaCO_3$ is to accelerate and increase the level of PVC fusion.

Particle size, loading level and coating level all have an influence on fusion characteristics, the final level of gelation and mechanical properties, particularly impact strength.

Coarse particles can act as crack initiators and reduce impact strength. Very fine grades (PCC or natural) increase impact strength and are essential in order to meet certain specifications (e.g. water pressure pipe). Additions of fine $CaCO_3$ can lead to a doubling of impact strength at about 15 phr loading. Higher loadings are usually detrimental.

Window profile is an example where consistent high whiteness is an additional requirement.
(v) PVC plastisols, for applications such as wallpaper, are required to have high filler loadings on grounds of cost, but, at the same time, to retain acceptably low viscosity. Grades of calcium carbonate with a low resin demand, often with superfines removed, are therefore preferred.

By contrast, PVC plastisols for sealant applications require high viscosity with pseudo-plasticity (strongly shear thinning) rheology. This is generally achieved by using a blend of fine particle size PCC with fine particle size natural $CaCO_3$.

Polyester

For calcium carbonate, polyester (GRP, SMC, DMC) is by far the most important of the thermosetting resins and accounts for about 20% of total $CaCO_3$ usage in plastics. Low viscosity is often important, especially for sheet moulding compounds, and surface treated grades are beneficial in this respect. Fine particle size grades are used where good surface finish is required, and colour is often critical.

Polyolefins

Polypropylene and polyethylene each account for about 5% of $CaCO_3$ usage in plastics.

The importance of calcium carbonate fillers in PP homopolymer mirrors the situation for uPVC in that the correct choice of filler and loading level can give marked improvements in impact strength, especially for unnotched samples. Coarse particles (+10 µm) or poorly dispersed agglomerates of fine particles are detrimental to impact strength. However, finer grades with a median-size in the region of 0.5 µm to 1 µm, incorporated at a level of about 40 phr, can give multifold improvements in impact strength. It is believed that this effect is primarily related to changes in crystallinity of the polypropylene, but crack pinning may also be a major factor. A further doubling of impact strength is achieved by stearate coating the $CaCO_3$ at about 1% to 2% by weight. This improves dispersion of filler during compounding. Calcium carbonate is widely used in polyethylene masterbatch, primarily for film applications. The main objective is cost reduction, but a useful side effect is improved anti-blocking characteristics.

CONCLUSIONS

The predominance of calcium carbonate in filled plastics is primarily related to its widespread occurrence as white and pure mineral deposits, combined with the low cost of processing. However, in many large volume applications, it gives important functional benefits and would still be the filler of choice, even if prices were doubled or trebled.

BIBLIOGRAPHY

R. Rothon, ed. (1995) *Particulate-filled Polymer Composites*, Longman.
J.S. Falcone, Jr. (1992) Fillers, in *Encyclopedia of Chemical Technology*, Kirk-Othmer, 4th edn, Vol. 10, ed. Mary Howe-Grant, Wiley.

Keywords: calcite, limestone, chalk, marble, particle size, surface treatment, calcium stearate, titanate, zirconate, PVC.

See also: Flame retardants: the approaches available.

Carbon black

John Accorsi and Michael Yu

INTRODUCTION

There are many additives employed in the plastics industry. Among the most versatile is carbon black. It is used to provide colour, opacity, protection from ultraviolet light, electrical properties, thermal conductivity, and even reinforcement. Advanced production methods have enabled carbon black suppliers to develop a wide range of carbon black grades that, in turn, provide plastics processors with additive selections geared to specific end-use properties. Selection of the proper carbon black is critical to successful end-use performance. Unlike some plastics additives, carbon black is not simply added to the mix. It must be dispersed into a resin system, and the quality of the dispersion is essential to performance. This chapter discusses the fundamentals of carbon black, its selection in plastics applications, and dispersion equipment and techniques.

PROPERTIES

Carbon black is a particulate form of industrial carbon produced by the thermal cracking or thermal decomposition of a hydrocarbon raw material. Two crystalline forms of carbon – diamond and graphite – exist in nature. While it more closely resembles graphite than diamond, carbon black exhibits a microstructure with hexagonal layers that are farther apart than those of graphite and have no vertical orientation. Therefore, carbon black is said to have a semi-graphitic structure. Three to four hexagonal layers combine in crystallites or bundles that join to form primary particles of carbon.

Plastics Additives: An A–Z Reference
Edited by G. Pritchard
Published in 1998 by Chapman & Hall, London. ISBN 0 412 72720 X

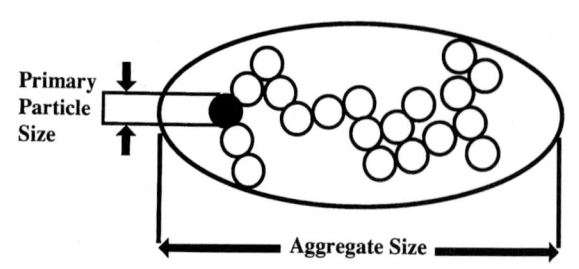

Figure 1 A typical carbon black aggregate.

Carbon black can be produced using several methods, including the lampblack, channel black, thermal black, and acetylene processes. However, the vast majority – over 90% – of today's carbon black is manufactured using the oil furnace process, a highly efficient method that permits rigid control of chemical and physical properties. The oil furnace process yields carbon black in fluffy or low-density powder form. Many grades are subsequently converted to pellets (beads) for ease of handling.

Carbon blacks are primarily classified on the basis of three of their properties – surface area/particle size, structure and surface chemistry.

SURFACE AREA/PARTICLE SIZE

Typical furnace blacks are composed of spherical particles ranging in diameter from 10 to 100 nm and in surface area from 25 to 1500 m^2/g. These particles fuse together in clusters that form the characteristic units of carbon black, called aggregates (Figure 1). The properties of both the particles and the aggregates they comprise are important factors in controlling carbon black performance.

Since the surface area of carbon black is expressed in square metres per unit weight, the smaller the primary particles the larger the surface area. Surface area affects the 'jetness' (intensity of black colour), the UV absorption, and the conductivity of the carbon black. Greater surface area increases both jetness and UV resistance, because more surface is available for the absorption of visible and UV light.

STRUCTURE

The second important property of carbon black is 'structure'. Structure is determined by the number of particles per aggregate, and by aggregate size and shape. A carbon black grade whose aggregates are composed of many primary particles, with considerable branching and chaining, is referred to as a 'high-structure' black. In contrast, if the aggregates consist of relatively few primary particles forming a more compact unit, the carbon black is considered to have low structure (Figure 2). Structure

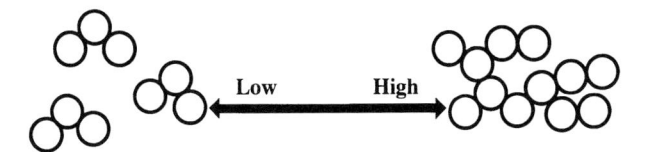

Figure 2 Carbon black structure.

affects the dispersion and viscosity characteristics of carbon black. High-structure blacks are more easily dispersed because the greater distances between aggregate centres weakens the attractive forces. The larger spaces that exist within aggregates of high-structure carbon blacks can incorporate more resin, thus increasing compound viscosity.

SURFACE CHEMISTRY

All carbon blacks have some level of chemisorbed oxygen complexes, i.e. carboxylic, phenolic, quinonic, or lactonic groups on their surfaces which constitute the volatile content. Oxygen complexes are acidic and they control the pH value of carbon black. As volatile content increases, pH decreases. Some carbon black grades are chemically oxidized to increase the volatile content for certain applications. For example, in resistive plastics, increasing the volatile content tends to insulate the aggregates and so more energy is needed for electron transfer. Therefore, after-treated grades of carbon black are suitable for applications requiring dielectric properties.

QUALITY AND OTHER PROPERTIES

While carbon blacks are generally classified on the basis of surface area/particle size, structure and surface chemistry, other properties may also be considered when selecting and using the proper carbon black. Carbon blacks can contain varying amounts of ash and grit as a result of the furnace production process. Special clean carbon black grades containing low levels of ash and grit are available for plastics applications. Compared with a black containing higher amounts of ash and grit, a clean black is easier to disperse, produces smoother extrusions, and gives better mechanical properties.

APPLICATIONS

Carbon black applications encompass numerous products ranging from rubber tyres to printing inks. In plastics applications, specialized carbon blacks are preferred over grades used for commodity rubber applications, due to their better cleanliness and higher quality. Before selecting a special carbon black for a given plastic system, the end use should be carefully considered. The attributes that the carbon black must supply, such as colour, UV protection, or conductivity, should be defined in order to identify both the appropriate grade and the proper loading level of the carbon black.

COLOUR

The most noticeable quality of carbon black is its intense jetness. Carbon black has always been the most widely used black pigment because of its colouring ability, cost-effectiveness, and performance. Carbon blacks are available in a wide range of jetness levels to meet colouring requirements in all types of plastics. Simple colouring requires only enough carbon black to achieve opacity – usually in the range 0.5–3.0% loading. Grade selection is largely dependent upon the degree of jetness required by the end product.

In addition to jetness, there are other important appearance considerations for plastics that can be controlled by the selection of an appropriate carbon black grade, such as undertone, tint strength and gloss. Undertone is a subtle, secondary colour underlying jetness. The undertone exhibited by carbon blacks can range from a warm brown to a cool blue. Tint strength is the relative ability of the carbon black to darken a coloured resin. Gloss level can range from a very shiny appearance to a matt finish.

Carbon black appears black because of its natural ability to absorb energy from all wavelengths of light. Small particle carbon blacks are better absorbers than large particle blacks because of the greater surface area available for increased light absorption. Therefore, small particle carbon blacks have jetter appearance. In general, to obtain the maximum light absorption needed for optimal jetness and tint strength, a carbon black with high surface area should be chosen. Low-structure carbon blacks exhibit increased jetness and higher gloss than higher structure blacks.

CONDUCTIVITY

Carbon black is far less resistant to the flow of electricity than the plastics in which it is dispersed. Therefore, carbon black can be used to lower the resistivity of plastics, imparting anti-static, semi-conductive, or conductive properties. The loading level of a carbon black grade is an important factor in the conductivity of the compound. In general, higher loadings increase conductivity. However, because carbon black grades can vary significantly in their inherent conductivity, some grades require higher loadings than others to achieve the same level of conductivity in the resulting compound. End-uses for conductive carbon blacks range from electronic packaging, business machines, and hospital equipment requiring anti-static properties, through semi-conductive shielding for power cable, to electromagnetic interference (EMI) shielding and video disks, where maximum conductivity is needed.

The structure of a carbon black affects the conductivity. High-structure carbon blacks have many primary particles and considerable branching and chaining which creates more potential paths for electron transfer through the resin. This translates into decreasing resistivity. Particle size also plays a role in determining conductivity. Smaller particle size and smaller aggregates mean that there are more aggregates per unit weight of carbon black distributed throughout the fixed volume of resin. This means smaller distances between aggregates and greater ease of electron transfer from one aggregate to another. Therefore, fine particle size and, consequently, smaller aggregates will lower the electrical resistivity when dispersed in plastics, with other properties remaining constant.

Increasing the surface oxidation of the carbon lowers the electrical conductivity in plastics, assuming other properties are constant. For applications where conductivity is important, grades of carbon black with lower volatile content should be selected.

UV PROTECTION

Exposure to ultraviolet light accelerates the physical and chemical degradation of many types of plastics, including polyolefins, polyvinylchloride, polystyrene, and to a lesser extent acrylonitrile-butadiene-styrene, polyamide and acrylic. Absorption of UV light by the plastics provides the energy to break key chemical bonds near the surface and form free radicals, which react with oxygen and attack other molecules, setting up a chain reaction. Plastics end-uses requiring UV stabilization include jacketing for communications or power cable, plastic pipe, agricultural film, geomembranes and automotive parts.

By absorbing and scattering UV light, and reacting with free radicals, carbon black aggregates significantly reduce the effect of UV radiation on plastics. Primary particle size is the key parameter that determines UV absorption efficiency. As particle size decreases, UV absorption increases until a size of about 20 nm is reached. At sizes smaller than 20 nm, absorption levels off. Volatile content also affects UV protection, as chemisorbed oxygen complexes react with free radicals, inhibiting any chain reaction caused by UV radiation. High-volatile-content carbon blacks offer an additional increment of UV protection.

Higher loadings of carbon black offer a greater degree of UV light stability, but only up to the level of opacity. Adding carbon black above this level will not significantly improve UV resistance. Carbon black loadings from 2.0 to 3.0% by weight are generally sufficient for most systems. The desired service life of the finished product determines the selection of a carbon black and its loading.

HANDLING

The pelleted (beaded) form of carbon black is generally preferred by most plastics compounders. Handled in a gentle manner, pelleted grades are free-flowing solids which can be conveyed using mechanical systems such as bucket elevators and screw feeders, or with low-velocity, dense-phase pneumatic systems. It should be noted that pellet attrition can result in the creation of dusty fines which can cause handling, processing, and housekeeping problems. Binders are added to some grades of carbon black to increase the hardness of the pellet.

Pelleted grades of carbon black also have higher bulk density than fluffy grades. Higher density increases the rate of incorporation into a polymer, an important factor in short residence time processes used in the plastics industry. Plastics generally have high viscosities, creating sufficient shear stresses in the mixing equipment to break the pellets and disperse the carbon black.

Fluffy grades are often used when a dry pre-mix is made using granular resins (PVC) and other additives.

Pelleted carbon blacks are available in plastic or paper bags, intermediate bulk containers (IBCs) or railcars. Fluffy grades are generally available only in paper bags.

DISPERSION

The dispersion of carbon black in a resin system is a critical factor in its performance. Optimal dispersion means that the carbon black is separated into discrete aggregates and the surface of each aggregate is completely covered with resin. A good dispersion ensures that the carbon black is uniformly distributed so that its maximum benefit can be realized.

The dispersion process involves several steps. Incorporation displaces occluded air and covers agglomerates (clusters of aggregates) with resin. In the next step, shearing force is applied to the mixture to break down agglomerates into discrete aggregates. Finally, each separated aggregate adsorbs enough resin to cover its surface completely. However, perfect dispersion is not always necessary or even desirable, e.g. conductive applications. A number of carbon black properties – such as surface area, structure and density – can influence the ease of dispersion. Greater surface area (resulting from finer particle size) requires more energy for wetting, and therefore makes dispersion more difficult. Lower structure blacks are also harder to disperse because closer packing of the aggregates creates stronger attractive forces. Similarly, denser carbon blacks, such as those in pelleted form, possess stronger inter-aggregate forces than fluffy blacks.

Achieving a good dispersion depends on selection of a carbon black grade that is as low as possible in surface area and as high as possible in structure. However, because the carbon black properties that produce the best dispersion frequently are not ideal for a given end use, it is necessary to determine a balance between dispersion qualities and end-use requirements. Other factors in a successful dispersion include the formulation and type of equipment to be used.

MASTERBATCH PROCESS

Carbon blacks are incorporated into polymeric media by one of two methods. The proper amount of carbon black may be dispersed directly into the polymer using a suitable mixing device. Far more common, however, is the masterbatch process in which carbon black is first dispersed into a carrier resin in concentrations of 30–50%. This concentrate is then let down (diluted) via the 'salt and pepper' procedure attained by introducing a mixture of concentrate and polymer pellets into the mixing unit. The ratio of compound to pellets is adjusted to provide for the final carbon black loading.

MIXING EQUIPMENT

Plastics processors have several options for mixing equipment to achieve adequate dispersion of carbon black in resin systems.

- Intensive mixers are generally used for the blending of dry ingredients prior to actual dispersion.
- Batch internal mixers such as Banbury mixers provide sufficient shearing force for good dispersion. This versatile equipment is used for compounding a wide range of resins. A typical charge consists of two layers of resin with a layer of carbon black sandwiched in between. When the rotors are activated and the internal pressure increased, the temperature will rise quickly due to high shearing force.
- Continuous internal mixers are similar to batch internal mixers, but offer the advantage of continuous operation and high throughput. The resin and carbon black can be pre-blended or the ingredients can be fed separately, provided that accurate measuring and/or metering is employed. Pelleted carbon blacks are required for this type of equipment.

Compounding extruders are a class of equipment that includes several distinct designs. Single screw extruders do not generate sufficient shearing force to disperse pelleted or high-surface-area blacks adequately, although they may be used with low-surface-area fluffy blacks. Mixing units that incorporate both rotational and axial screw movement generate sufficient shearing force to disperse pelleted carbon blacks. Twin screw extruders can be configured to achieve adequate dispersion; however, it is important to control residence time to minimize heat generation.

MEASUREMENT

The most effective way to measure the quality of carbon black dispersion is by microscopic techniques. A thin, translucent film prepared from the dispersion is viewed or photographed microscopically to determine the size and number of agglomerates remaining undispersed (Figure 3).

Other dispersion evaluation methods include the pressure rise test, tape quality test and blown film test. The pressure rise test involves extruding a given amount of plastics compound through a screen pack and monitoring pressure build-up as undispersed carbon black agglomerates and other contaminants plug the screen. For the tape quality test,

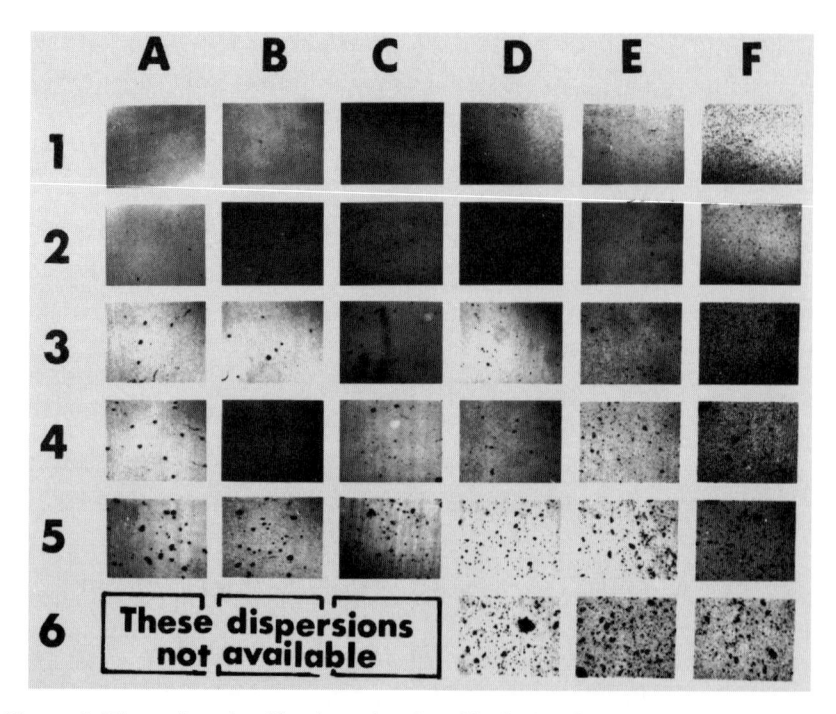

Figure 3 Dispersion classification of carbon blacks in plastics.

a thin (2–25 mil) tape made of the resin compound is extruded and compared with a standard of surface smoothness. A similar test consists of examining blown film prepared from the resin compound under strong light.

CONCLUSIONS

As a key additive in many types of resins, carbon black can impart a variety of desirable characteristics, including colour, protection from UV radiation, and conductivity. Within each of these broad categories, different grades of specialty carbon blacks can provide specific properties – such as a particular intensity of black colour, degree of UV resistance, or level of conductivity – or a unique combination of features.

There are a number of variables to consider in selecting the best carbon black grade for a given plastics application. Although each end-use may call for different carbon black characteristics, there is such a wide range of grades available that it is usually possible to identify one with the particular combination of properties needed. Careful consideration of the resin as well as accurate formulation and proper dispersion are also important in achieving desired end-use performance.

The information provided here is intended as a broad overview of carbon black as an additive in plastics. Additional information, including performance properties of specific grades of carbon black and appropriate health and safety recommendations, is available from carbon black suppliers.

BIBLIOGRAPHY

Donnet, J.-B., Bansal, R.C. and Wang, M.-J. (1993) *Carbon Black*, 2nd edn, Marcel Dekker, Inc., New York, Basle, Hong Kong.
Yu, M.C., Bissell, M.A. and Whitehouse, R.S. (1995) *SPE, ANTEC Tech Papers*, 41, 3246.
Accorsi, J.V. and Romero, E. (1995) Special carbon blacks for plastics, *Plastics Engineering*, April 1995.
Yu, M.C., Menaslii, J. and Kaul, D.J. (1994) *SPE, ANTEC Tech Papers*, 40, 2524.

Keywords: colour, opacity, UV absorber, electrical conductivity, structure, aggregate, surface area, dispersion, viscosity, masterbatch, mixing.

See also: Conducting fillers for plastics: (1) Flakes and fibres;
Conducting fillers for plastics: (2) Conducting polymer additives;
Fillers;
Fillers: their effect on the failure modes of plastics.

Compatibilizers for recycled polyethylene

P.S. Hope

INTRODUCTION

There are strong environmental pressures on industry to recycle waste polymers, particularly those used in packaging applications. Polyethylenes (PE), in the form of high density polyethylene (HDPE) moulding grades, and low density polyethylene (LDPE) or linear low density polyethylene (LLDPE) film grades, currently command more than half of the polymer recyclate market. They form the major polymeric part of post consumer waste streams, along with significant amounts of polypropylene (PP), polystyrene (PS), polyvinyl chloride (PVC) and polyethylene terephthalate (PET), and small amounts of other polymers such as polyamides, polycarbonates and barrier polymers. Despite undergoing sorting processes the PE recovered from post consumer waste inevitably contains significant amounts of these other polymers.

The mechanical properties of recycled PE blends can be significantly impaired, in particular their impact and ultimate tensile elongational behaviour. For example, the effect of adding minor levels of polymeric 'contaminants' on the Charpy impact strength of HDPE is shown in Figure 1. While adding LDPE hardly affects impact strength, PS, PVC and PET cause a rapid fall off even at low addition levels (<5%), requiring the introduction of compatibilizers to recover performance. Unless otherwise stated, impact provides the yardstick by which the performance of the compatibilizers discussed in this chapter is evaluated. However compatibilizers are not always required. For example, thin walled articles such as household detergent containers fail by ductile

Plastics Additives: An A–Z Reference
Edited by G. Pritchard
Published in 1998 by Chapman & Hall, London. ISBN 0 412 72720 X

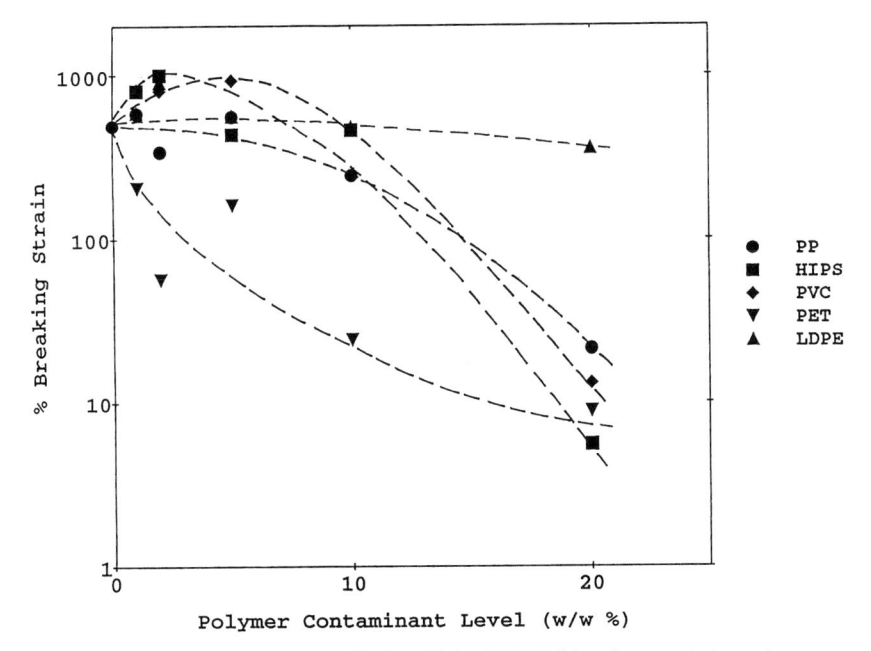

Figure 1 Charpy impact strengths (23°C) for HDPE blends containing minor proportions of PP, LDPE, HIPS, PVC and PET. After Hope, P.S., Bonner, J.G. and Miles, A.F. (1994) *Polymer Blends and Alloys*, Blackie Academic and Professional.

tearing mechanisms, and their impact performance tends to be relatively insensitive to the incorporation of other polymers, as a result of which significant levels (up to 10%) can at times be tolerated in recycling without property loss.

COMPATIBILIZATION MECHANISMS

Immiscible polymer blends form dispersions of two or more phases, in the same way as oil and water, resulting in an uneven and poorly bonded structure in the final product. It is well known that polymers with suitable molecular structures can be used as compatibilizers, which when added at the melt blending stage act as emulsifiers, produce finer and more even phase dispersion. They tend to locate preferentially at the phase boundaries and increase the adhesion between phases. A third effect is to stabilize the dispersed phase particles against growth during annealing, again by modifying the phase–boundary interface. In practice it is likely that all these effects will occur to some extent with addition of a particular compatibilizer, leading to greatly improved end-use performance. Examples illustrating the emulsifying and stabilizing effects of compatibilizers are shown in Figures 2 and 3.

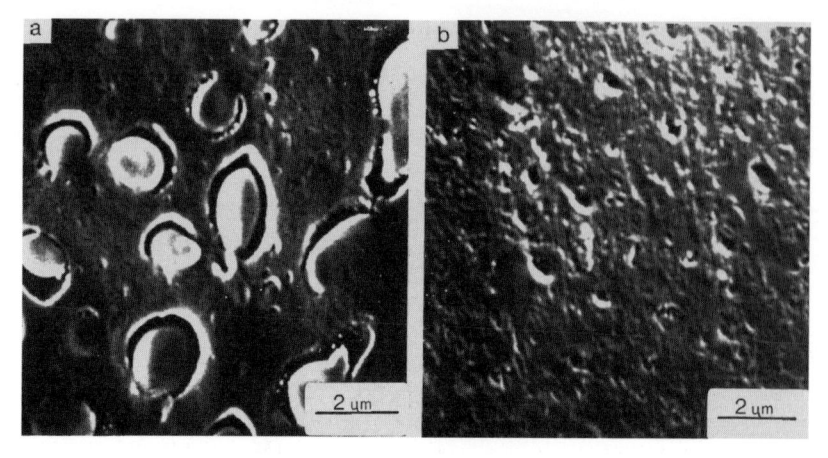

Figure 2 SEM images of LLDPE/PP 80/20 blends (a) unmodified (b) with 2 pph EP random copolymer. After Hope, P.S., Bonner, J.G. and Miles, A.F. (1994) *Polymer Blends and Alloys*, Blackie Academic and Professional.

Compatibilizers are usually block of graft copolymers, but can also be polymers which have been chemically modified to contain functional or reactive chemical species. Block and graft copolymers containing segments chemically identical to the blend components are natural choices as compatibilizers, given that miscibility between the copolymer

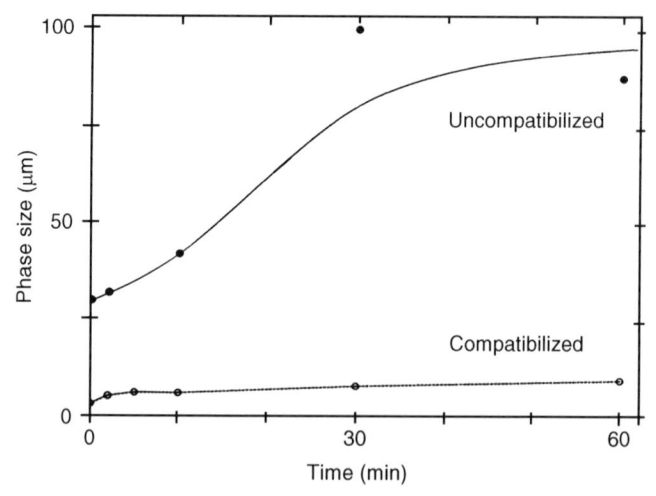

Figure 3 Effect of compatibilization on annealing behaviour of PE/polyamide blends (up to 1 hour at 285°C). After Capaccio, G., Gardner, A.J., Hope, P.S. and Wilkinson, K. (1990) *Die Makromolekulare Chemie*, Macromolecular Symposia, **38**, 267–273. Reproduced by permission of Huthig and Wepf Publisher, Zug, Switzerland.

segments and the corresponding blend component is assured, provided the copolymer meets certain structural and molecular weight requirements. For example, the ultimate tensile properties of PE/PS blends can be improved by adding PE-PS copolymers, for which it has been shown that:

1. block copolymers are more effective than graft copolymers;
2. diblock copolymers are more effective than triblock or star-shaped copolymers;
3. tapered diblock copolymers are more effective than pure diblock copolymers.

Copolymers with blocks of chemical composition identical to those of the two homopolymers are often not commercially available, or only available at high cost. Commercially used compatibilizers are therefore often partially random copolymers.

Functional polymers are also used as compatibilizers. Usually a polymer chemically identical to one of the blend components is modified to contain functional (or reactive) units, which can interact with the second blend component. This may be via chemical reaction, or via physical interaction such as the formation of ionic bonds. The functional modification may be achieved in a reactor or via an extrusion-modification process. Examples include the grafting of maleic anhydride or similar compounds to polyolefins, the resulting pendant carboxyl group having the ability to form a chemical linkage with polyamides via their terminal amino groups. Functionalized polymers (usually maleic anhydride (MAH) or acrylic acid grafted polyolefins) are commercially available. Another approach is the formation *in situ* of graft of block copolymer by chemical bonding reactions between reactive groups on component polymers; this may also be stimulated, for example, by addition of a free radical initiator during blending.

COMPATIBILIZERS FOR POLYETHYLENE BLENDS

Table 1 lists polyethylene blend systems for which commercially available materials have been reported as providing a compatibilizing action. Note that the list of suppliers is indicative, and similar polymers may be available from alternative sources. In most cases the compatibilizing effect reported included establishment of a stable, fine morphology and achievement of improved impact or ultimate extensional properties compared to the un-compatibilized blend. The systems are discussed briefly below, where mention is also made of non-commercial systems described in the technical literature. Readers who wish to investigate further the systems described here are directed to Bonner and Hope (1993), which contains a comprehensive bibliography.

Table 1 Commercially available compatibilizers for polyethylene blends

Blends of polyethylene with:	Compatibilizer	Trade name	Supplier
PP	EP random copolymer	Dutral	Himont
	EPDM	Nordel	Dupont
	VLDPE	Norsoflex	Orkem
PS	S-EB diblock copolymer	Kraton	Shell
	SEBS triblock copolymer	Kraton	Shell
PVC	EVA copolymer	EVA	Exxon
	Chlorinated PE	CPE	Dow
	Ionomer	Surlyn	Dupont
PET	SEBS triblock copolymer	Kraton	Shell
Polyamides	MAH-grafted PE or PP	Polybond	Uniroyal
	Ionomer	Surlyn	Dupont
Mixed polymers	Functional polymer	Bennet	High Tech Plastics

PE/PP blends can be compatibilized using ethylene-propylene (EP) random copolymers. A large range of effects are seen, including reduction of dispersed phase size and improvement of impact strength, although findings can appear to be contradictory and attention to detail is important. Some success has been reported using ethylene propylene diene rubbers (EPDM), in which the EPDM exists in the blends both at the PE/PP interface and as discrete particles, giving a toughening effect, and for specially synthesized EP copolymers containing PP grafts. Functional polymers such as maleic anhydride grafted EP copolymers and maleated SEBS triblock copolymers have also been used. Very low density polyethylenes (VLDPE) can effectively compatibilize LDPE/PP and LLDPE/PP blends at addition levels as low as 2%; this is of particular importance for recycling as VLDPE are relatively inexpensive.

Compatibilization of PE/PS blends can be achieved using ethylene-styrene graft and block copolymers, styrene-ethylene butene (S-EB) graft and block copolymers, styrene-ethylene butene-styrene (SEBS) triblock copolymers and hydrogenated butene-styrene graft and block copolymers. Again a large range of effects are reported, including reduced size of the dispersed phase, and increased elongation and yield, breaking and impact strength. The effect of a commercial compatibilizer (SEBS) on the impact performance of HDPE blends containing 20% by weight high impact polystyrene (HIPS) is shown in Figure 4.

Blends of PE with PVC have been compatibilized using graft copolymers of ethylene and vinyl chloride, chlorinated polyethylenes (CPE) and ethylene-vinyl acetate (EVA) copolymers. Compatibilized

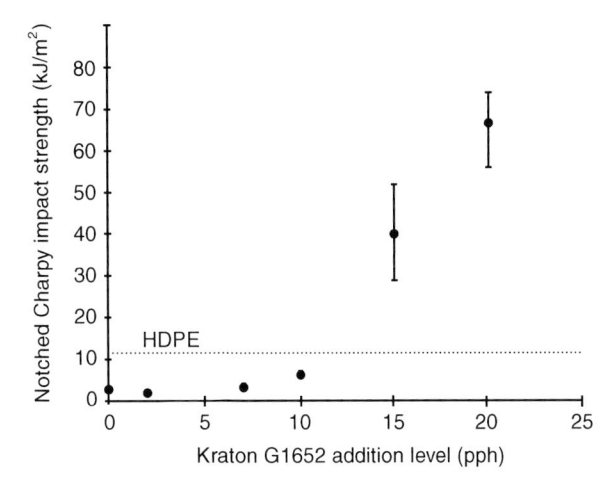

Figure 4 Effect of adding SEBS triblock copolymer compatibilizer on the Charpy impact strength of HDPE/HIPS 80/20 blends. After Hope, P.S., Bonner, J.G. and Miles, A.F. (1994) *Polymer Blends and Alloys*, Blackie Academic and Professional.

blends display finer morphologies, but changes in individual properties (elongation, toughness, modulus and tensile strength) vary according to the compatibilizer chosen. More recently, ionomers such as partially neutralized ethylene-methacrylic acid copolymers have been shown to be more effective than EVA copolymers, and direct addition of peroxides to promote graft copolymer formation during blending has also proved effective.

PE/PET blends have been studied for a long time, the impetus being recycling of PET carbonated drink bottles, and more recently HDPE milk bottles. Commercially available SEBS triblock copolymers can increase elongation and impact performance of HDPE/PET blends. Commercially available EVA copolymers have also been considered, but more recently ethylene-glycidyl methacrylate (E-GMA) copolymers have been investigated, and found to give improved impact strengths. E-GMA functions as a reactive compatibilizer, the ethylene backbone giving some miscibility with polyethylenes, while the GMA undergoes a reaction with the PET.

A commonly used method of inducing compatibility between PE and polyamides is chemical modification of the PE to contain pendant carboxyl groups, which form chemical linkages to the polyamide via the terminal amino groups. The concept has been employed to produce commercial compatibilizers such as MAH grafted PP or PE. Other approaches reported in the literature include the use of ionomers, acrylic acid/butyl acrylate/styrene terpolymers and nylon 6-polybutene multiblock copolymers.

For polymeric mixed waste a combination of compatibilizers can provide a solution, although this may prove too expensive. An alternative approach is the addition of free radical initiators (such as peroxides) during processing, to stimulate the generation *in situ* of compatibilizing copolymers. A 'universal compatibilizer' containing a range of functional groups, as well as elastomers to improve toughness, is marketed for this purpose. Another system reported recently generates compatibilizing copolymers directly from waste polymers by ozone activation in a heated reactor to produce peroxides, followed by monomer addition.

COMPATIBILIZER SELECTION

It is helpful to define compatibilizers as additives which improve the properties of polymer blends so that they possess a commercially desirable set of properties. This highlights the importance of defining the level of property (or properties) which need to be achieved for the

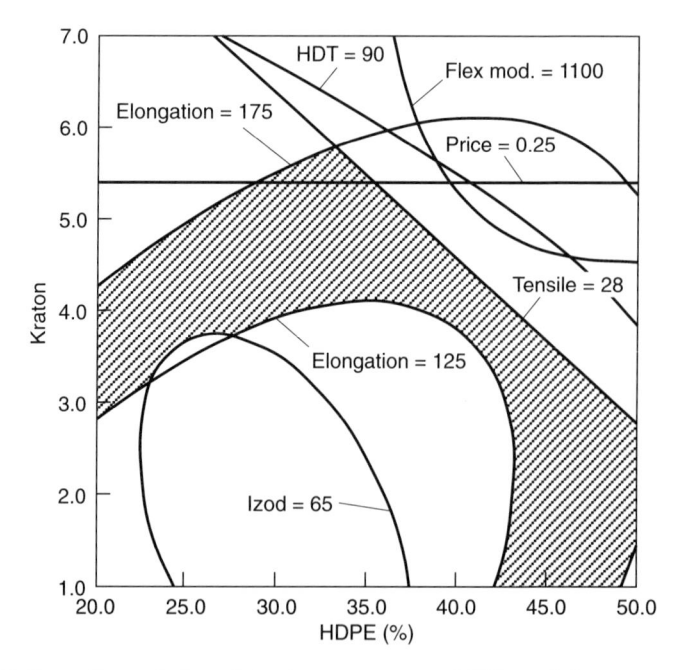

Figure 5 Operating window (at mixer screw speed 300 rpm) for compatibilized HDPE/PP blends to achieve desired property targets within maximum cost constraint. The vertical axis refers to the quantity (phr) of Kraton FG1901X compatibilizer (i.e. a maleated SEBS triblock copolymer). The diagram should be regarded as schematic rather than quantitative, hence the absence of units. See text for further commentary. After Adewole, A.A., Dackson, K. and Wolkowicz, M.D. (1994) *Proceedings of ANTEC '94*, pp. 3044–3049.

application under consideration before attempting to select compatibilizers. It is also important to keep cost firmly in mind, as margins are likely to be tight when recycling commodity polymers such as polyethylenes. Clearly there are many cost–performance options to be addressed and a great deal of expensive experimentation could be required to achieve an optimized system.

Some workers have used statistical experimental designs to produce 'response surfaces', quantifying the effect of different blend composition/compatibilizer combinations on a given end-use property. By superimposing cost information, it is possible to determine operating windows for a desired balance between properties and cost. The example in Figure 5 shows such an operating window for HDPE/PP blends compatibilized using Kraton FG1901X, a maleated SEBS triblock copolymer. The shaded area describes compositions which meet specified performance in terms of flexural modulus, Izod impact, elongation at break, tensile strength and heat distortion temperature (HDT), at or below a maximum acceptable price. Note that a range of solutions exist, and that the window can be substantially affected by the mixing regime employed, exemplified here by screw speed.

BIBLIOGRAPHY

La Mantia, F.P. (ed.) (1993) *Recycling of Plastic Materials*, ChemTech, Ontario.
Bonner, J.G. and Hope, P.S. (1993) Compatibilisation and reactive blending, in *Polymer Blends and Alloys* (eds M.J. Folkes and P.S. Hope), Blackie Academic and Professional, Glasgow, pp. 46–74.
Hope, P.S., Bonner, J.G. and Miles, A.F. (1994) *Plastics, Rubber and Composites Processing and Applications*, **22**(3), 147–158.

Keywords: blend, block copolymer, compatibilizer, functionalized polymer, graft copolymer, polyethylene, recycle.

See also: Polymer additives: the miscibility of blends;
Recycled plastics: additives and their effects on properties.

Conducting fillers for plastics: (1) Flakes and fibers

Richard G. Ollila and Donald M. Bigg

INTRODUCTION

Flakes and fibres are used as fillers in plastics to convert the inherent insulating polymeric material into electrically conductive composites. These composite materials have been studied and evaluated to create products to transmit electricity, to dissipate static charges built up by friction, or to provide electromagnetic interference (EMI) shielding. In this section, the focus is on recent developments in flake and fiber fillers that are used to provide EMI shielding for electronic devices that are found in homes, factories, offices, commercial establishments and transportation systems that ever increasingly rely on electronic broadcast systems for communications.

These electronic devices are usually housed in plastic structures to take advantage of many attributes of polymers, including light weight, reduced manufacturing costs, low maintenance costs, and corrosion resistance. However, as most polymers are ordinarily transparent to high frequency electromagnetic waves, the electronics inside must be shielded from EMI by using electrically conducting materials. To satisfy government regulations to limit electromagnetic radiation emitted from these devices, studies have been conducted to develop the knowledge of materials that would provide adequate EMI shielding for plastics and how to use them.

Much of the early work was reported in the works of Bhattacharya and Bigg [1], among others. They and others examined a wide range of metal and metal alloy particles including aluminum, iron, steel, copper, silver,

Plastics Additives: An A–Z Reference
Edited by G. Pritchard
Published in 1998 by Chapman & Hall, London. ISBN 0 412 72720 X

ELECTRICAL PROPERTIES

The excellent electrical conducting properties of metals are well known. Metals have resistivities in the range of 10^{-6} ohm cm compared with the resistivities of polymers which are on the order of 10^{15} ohm cm. Early studies showed there was a correlation between the electrical resistivity of the composite material and the amount of finely divided metal particles blended into it. More importantly, it was determined that a critical volume fraction of metal was required to make the plastic conductive. This critical volume or concentration is called the percolation threshold. This is the concentration at which each particle in the matrix makes contact with at least two other neighboring particles and creates a three-dimensional network in the matrix.

This phenomenon is shown in Figure 1 for aluminum flakes. Figure 1 shows the effect of shear processing in the percolation threshold. These studies also showed that for randomly dispersed silver spheres in a Bakelite composition, the percolation threshold (where the composite became electrically conductive) approached 40%. This amount of metal in the composite makes it hard to mold, and reduces its mechanical integrity for most applications.

However, additional studies showed that the percolation threshold was also dependent on the geometry of the filler particles, and that fibers or flakes with high aspect ratio (length to diameter or thickness) have much lower percolation thresholds (less than 6% to 8% by volume). In addition, the electrical conductivity is diminished by the thickness of the oxide coating on certain metals, such as copper and aluminum. Figure 2 shows the relationship between volume resistivity in a composite and the amount of EMI shielding attenuation achieved. To achieve at least 30 dB of attenuation the shielding material must have a resistivity of at least 1 ohm cm. For most consumer applications, 30 dB is considered acceptable. Military requirements frequently specify 50 dB or more. Based on the data in Figure 2, this necessitates an order of magnitude decrease in the volume resistivity of a plastic composite. This change greatly increases the difficulty in using plastic composites. This section summarizes recent efforts to meet those more stringent goals.

Over the years studies of various metals, shapes, and sizes have been reported. Table 1 shows the characteristics of a few metals.

Some interesting results have been reported. While copper is one of the best electrical conductors (lowest resistivity), it has the disadvantage that, along with its copper-based alloys like brass, it reacts unfavorably with

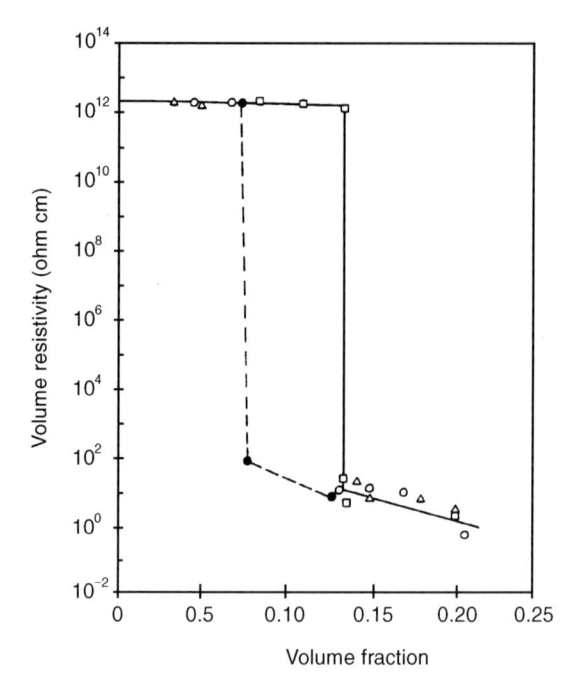

Figure 1 Volume resistivity versus volume fraction of aluminum flakes in various polymers. Open symbols represent injection-molded composites with 50:1 aspect ratio flakes. Filled symbols represent compression-molded composites 16.7:1 aspect ratio flakes. (□, ●) polypropylene, (△) polycarbonate, (○) ABS. From Li, L. and Chung, D.D.L. (1994) *Composites*, **25**(3).

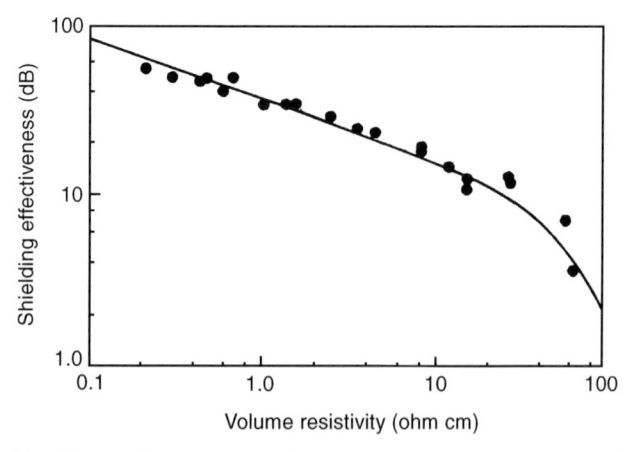

Figure 2 Shielding effectiveness of 3 mm-thick composites at 1 GHz versus volume resistivity. Solid line is theoretical curve. From Li, L. and Chung, D.D.L. (1994) *Composites*, **25**(3).

Table 1 Characteristics of metal fibers and flakes used in polymer composites

Metal	Form	Production method	L (mm)	d (μm)	L/d	Density (g/cc)	Resistivity (10^{-6} ohm cm)
Stainless steel	Fibers	Drawing	1.6	30–56	300–500	8.02	74.0
Aluminum	Flakes	Melt spinning	1.25	30*	42	2.71	2.8
Aluminum	Fibers	Chatter machining	3	60	50	2.71	2.8
Copper	Fibers	Chatter machining	3	60	50	8.90	1.7
Brass	Fibers	Chatter machining	3	60	50	8.40	7.0
Iron	Fibers	Chatter machining	3	60	50	7.90	9.8
Nickel	Fibers	Drawing	1.0	20	50	8.90	8.5

*Thickness

some polymers, most notably polycarbonates, which are favored for many electronic housing applications; therefore, they have not been used much as fillers. Very small aluminum particles have an oxide coating which causes them to behave as an insulator in a polymeric composite; however, large flakes and fibers are excellent fillers for EMI shielding. Iron and steel fibers lack corrosion resistance in hostile environments. Metal-coated fiber glass tends to agglomerate when mixed and, therefore, it is difficult to distribute uniformly in the matrix. Significant breakage often occurs during processing. Silver, also with excellent electrical conducting properties, is usually too costly for most EMI applications.

In current practice, the most commonly specified EMI shielding materials are nickel fibers, carbon fibers, stainless steel fibers and aluminum flakes. The choice is dependent on several factors including the level of shielding required, the polymer matrix, the complexity of the component or the housing, manufacturing considerations, mechanical properties of the structure, and cost. Recent studies have been directed toward finding the best fillers for unusual applications and for more compact electronic devices, as requirements are published to reduce the weight of both military and civilian vehicles, and as the demand for light-weight personal electronics devices grows, and as miniaturized medical instruments and therapeutic devices are being developed.

The Composite Materials Research Laboratory headed by Chung at the State University of New York at Buffalo, New York, has recently conducted some interesting evaluations of a wide variety of EMI shielding filler materials for plastics. In all cases, the polymeric material they studied was polyethersulfone (PES). This is a very attractive polymer for the molding of housings for electronic devices. Its properties are shown in Table 2.

In the course of their studies, Chung and her colleagues evaluated the effectiveness of aluminum flakes, coated and uncoated carbon fibers, nickel fibers, stainless steel fibers, combinations of flakes and fibers, and low melting-point metal alloys.

For low temperature applications, Chung and her colleagues showed that a 2% by volume solder-like material, 60% tin and 40% lead, could enhance the EMI shielding effectiveness of nickel coated carbon fibers when the Sn–Pb (melting temperature 188°C) was melted *in situ* during the fabrication of the composite [2]. For a PES composite having 20% by volume nickel-coated carbon fibers, the shielding effectiveness at 1 GHz increased from 19 dB to 45 dB. For an uncoated carbon fiber composition, and for compositions when the Sn–Pb material was added but not melted, the increase in shielding effectiveness was only a few dB, and in the case of the unmelted fibers being added to the uncoated carbon fibers, the shielding effectiveness declined by

5 dB. They attribute the increase in shielding effectiveness of the melted Sn–Pb alloy *in situ* to its wetting of the nickel coating and subsequent inter-fiber adhesion characteristics. This permits a more effective three-dimensional network to be formed and enhances the electrical conductivity of the matrix.

A similar study was conducted to determine the shielding effectiveness of this class of solder-like materials as flakes in a PES composite. The 60Sn–40Pb flakes were very small (approximately 2 μm thick and with a mesh of −200 ± 400). This composite was formed by melting the powder *in situ* by hot pressing the composite at 240°C and a pressure of 6.5 MPa. The resulting composite had a shielding effectiveness of 30.5 dB at 1 GHz. However, the mechanical and shielding effectiveness were very sensitive to the hot pressing temperature and pressure. This implies a requirement for a very tightly controlled manufacturing process.

The authors suggest, based on this information, that a class of low temperature alloys such as In–Pb, Sn–Zn, Pb–Sn–Ag, Sn–Pb, Sn–Pb–Sb, Sn–Ag, Bi–Pb–Sn–Sb could be used as shielding materials. Other thermoplastics such as polyimide, polyetherimide, polyamideimide, polyphenylene sulfite, and PEEK are possible matrices as well. Recently a product which is a thermoformable EMI-shielding material has been introduced. It is made of solder-like low melting temperature bismuth–tin alloy and a non-woven EVA polymer. At 1 GHz its shielding effectiveness is claimed to be greater than 40 dB [3].

In a comprehensive study of the electrical and mechanical properties of electrically conductive polyethersulfone (PES) composites, Chung and Li provided comparative data for aluminum flakes, carbon fibers, nickel fibers, and stainless steel fibers for a range of filler volume fractions [4]. The study contains extensive tables which provide data on the EMI shielding effectiveness between 1 GHz and 2 GHz, and the electrical resistivity of the composites. It also contains tensile strength properties (strength, modulus and elongation) data. In addition, the critical filler volume fraction (percolation threshold) was calculated and measured for the various fillers. Lastly, the authors determined the effect of heating

Table 2 Properties of PES

T_g (glass transition temperature)	220°–222°C
Density	1.37 g/cm^3
Particle size	100–150 μm
Tensile strength	45.93 ± 1.12 MPa
Tensile modulus	2.64 ± 0.19 GPa
Elongation at break	(3.1 ± 0.3)%
Electrical resistivity	>10^{10} ohm cm
Coefficient of thermal expansion	55 × 10^{-6} K^{-1}

(140°C for up to 144 hours) on the properties of the composites. For all fillers, an EMI shielding effectiveness of greater than 50 dB was achieved; for stainless steel and nickel fibers it occurred at a filler volume of 20%; for aluminum flakes, at 30%; and for carbon fibers, at 40%.

However, the aluminum flake-filled composite had better mechanical properties than the others, as evidenced by good bonding between the flakes and the PES matrix. Part of this was attributed to the relatively rough surface of the flakes produced by using rapid solidification manufacturing techniques. Also, the oxide on the flakes may contribute to the bonding between the flakes and the matrix. Previous studies showed the uniqueness in the relationship between the strength and filler concentration for various polymer types and individual filler materials. The combination of these factors (excellent EMI shielding, good mechanical properties, and stability of properties when exposed to high operating temperatures for long periods) led the authors to conclude that a PES composite containing 40% by volume of aluminum flakes would be an attractive material for packaging electronic components in automotive applications.

Another interesting study was Chung's evaluation of very small diameter high aspect ratio fibers as an EMI-shielding filler material in PES [5]. In this case, nickel and carbon filaments were evaluated and compared with nickel fibers (20 μm diameter and 100 μm long). The nickel filaments were made by electroplating nickel onto a carbon core. The typical filament was estimated to be 100 μm in length and 0.5 μm in diameter. The carbon filament was about 100 μm and had an aspect ratio of about 600. The results of the experiments showed that to achieve an EMI shielding effectiveness of 40 dB or greater, the volume of filler had to be 20% or greater. Again, in these studies the composites were prepared by dry-blending and compression molding. However, at filler loadings of less than 15%, the nickel filaments and the carbon filaments were more effective than the nickel fibers, with shielding effectiveness of about 38 dB compared with about 10 dB for the nickel fiber. These results indicate that it is possible to achieve relatively high levels of EMI shielding using small diameter filaments in plastics, thus opening the door to provide EMI-shielding for sub-miniature components in electronics devices.

As was mentioned earlier, copper with its excellent electrical conducting properties has not been used for EMI-shielding because of its tendency to oxidize and to degrade certain polymers. However, in recent years, a thin, low-melting metal coating has been developed for copper to prevent oxidation and to act like a solder at molding temperatures to increase the three-dimensional network that provides the electrical conductivity within the polymer matrix. The fibers are about 50 μm in diameter with an aspect ratio of about 100. According to the manufacturer, they provide more effective shielding than equivalent stainless steel fibers in the 15–750 MHz range [6].

DESIGN CONSIDERATIONS

When designing a plastic housing that requires EMI-shielding, it is the designer's responsibility to select the optimum filler material based on performance, compatibility with the environment, structural considerations, aesthetics and cost. In most cases, cost will dictate the choice, once viable alternatives are developed.

The cost of EMI shielding materials depends on several factors – raw materials, processing costs (finely divided particles or small, large aspect ratio fibers are difficult to make), packaging, and ease of blending with the plastic matrix. On a unit weight basis aluminum and copper are low-cost, followed by stainless steel, nickel, and carbon fibers. To arrive at a true cost one must consider not only weight, but the volume and weight percentage of material required to achieve the desired shielding effectiveness in the component. Table 3 contains some comparative cost data for some commonly used metal fillers in a PES composite.

SAFETY, HEALTH AND ENVIRONMENTAL CONSIDERATIONS

In addition to flakes and fibers providing adequate EMI-shielding for the electronic component, the materials must meet government regulations designed to protect the user of the device, the workers who manufacture it, and the environment. All materials sold in the United States are required to have a Material Safety Data Sheet (MSDS) that contains information about the material, its toxicity effects on humans, animals, and the environment, and any precautions that must be taken when working with or handling the material.

As most of the fillers used for EMI-shielding are finely divided powders or fibers, special care must be taken not to breathe these particles or to expose the skin to them because exposure can lead to short- or long-term health problems, either because the material is an irritant or because it is toxic. Aluminum flakes are relatively large and pose little or no dust problem. Many producers of stainless steel or nickel fibers package the fibers in mat or coated fiber form to reduce the potential exposure to the compounder or blender of the fiber into the matrix. Nickel in powder form may be carcinogenic, so special precautions must be taken when handling small fibers and powders that can be inhaled or can come in contact with the skin. Protective clothing (i.e. respirators and gloves) are mandatory when nickel is being handled. Stainless steel fibers and powders when inhaled will produce flu-like symptoms, and in the long-term may mask other lung diseases when the lungs are X-rayed.

Table 3 Comparative costs of EMI shielding composites

Material	Size	L/d	Density (gm/cc)	Filler (vol %)	Filler (wt %)	Filler (cost $/lb)	Cost of materials for composite ($/lb)
Aluminum flake	1.3 × 1.0 × 0.03 mm	40	2.7	30	46	$3.00–5.50	$3.00–5.80
Carbon fibers (uncoated, pitch-based)	400 μm	40	1.6	30	33	$10.00–20.00	$5.30–10.60
Nickel fibers	1000 μm	50	8.9	20	62	~$25	$16.60–17.80
Stainless steel wool fibers	1600 μm	30–50	8.0	20	59	$5.00–9.00	$4.20–7.80
Stainless steel fibers	4–6 mm	400–600	8.0	1.5	10	$23.00–28.00	$5.00–8.20

Notes: 1. Material required to achieve SE = 50 dB or greater @ 1–2 GHz.
2. Polymer is polyethersulfone (PES) with $\rho = 1.37$ gm/cm^3 and cost between $3.00 and $6.00/lb.
3. Does not include compounding and molding costs.
4. Data from industry survey and References 2, 4 and 5.

SOURCES OF MATERIALS

Suppliers of metallic flakes and fibers can be found in compilations of industrial material and services such as the Thomas Register. Included in the Thomas Register are Transmet Corporation, Columbus, Ohio, which produces aluminum flakes of uniform size and shape; the N.V. Bekaert S.A. and the Interational Steel Wool companies, which produce stainless steel fibers in mat and chopped fiber forms; the National-Standard Company, which produces nickel fibers; Toshiba Chemical, which produces coated copper fibers, and the 3M Company, which produces thermo-formable EMI-shielding products. In addition, AMOCO and Ashland Chemical Company among others produce carbon fibers, and there are several companies that specialize in supplying metal-coated carbon and metal fibers for customized applications.

REFERENCES

1. Bigg, D.M. (1986) Electrical Properties of Metal-Filled Polymer Composites, in *Metal-Filled Polymers*, 1st edn (ed. S.K. Bhattacharya), Marcel Dekker, Inc., New York, pp. 165–225.
2. Li, L., Yih, P. and Chung, D.D.L. (1992) Effect of the second filler which melted during composite fabrication on the electrical properties of short fiber polymer-matrix composites, *Journal of Electronic Materials*, **21**(11), pp. 1065–71.
3. Yenni, D.M., Jr. and Baker, M.G. (1996) One Step Thermo Formable EMI Shielding, *The International Journal of EMC^{TM} (ITEM) Directory and Design Guide*.
4. Li, L. and Chung, D.D.L. (1994) Electrical and mechanical properties of electrically conductive polyethersulfone composites, *Composites*, **25**(3), pp. 215–24.
5. Shui, X. and Chung, D.D.L. (1994) Submicron nickel filaments as a new filler material for electrically conducting polymer-matrix composites, *7th International SAMPE Electronics Conference*, Vol. 7, pp. 39–43.
6. Miller, B. (1991) New compounds raise the ante in shielding efficiency, *Plastics World*, **49**(12), pp. 57–59.

Keywords: aluminum, automotive applications, carbon, conducting fillers, copper, cost, electromagnetic interference (EMI), electronic devices, fibers, flakes, metal, nickel, polyethersulfone (PES), safety, health and environment, solder-like alloys.

See also: Carbon black;
Conducting fillers for plastics: (2) Conducting polymer additives.

Conducting fillers for plastics: (2) Conducting polymer additives

Asoka J. Bandara

ELECTRICALLY CONDUCTING PLASTICS

This entry concerns the use of conducting polymers and similar substances as additives enabling insulating polymers to form electrically conducting plastics or composites.

In the frequency range 30 MHz to 1 GHz, electromagnetic emissions are a major concern, as they cause interference in broadcast communications. The United States Federal Communications Commission (F.C.C.) created a large commercial market for new composite materials when it issued regulations which limited the field strength of devices emitting radiation in the pre-stated frequency range. The most common method of isolating electromagnetic radiation is to attenuate the signal at the device's structural housing. This is commonly achieved by fabricating the housing with conductive materials, which usually contains conductive metal fillers of either a particulate or fibrous nature. However, attractive features of the plastics host material such as low density, corrosion resistance and ease of formation of complex housing shapes are lost or reduced when metallic fillers are introduced. The introduction of metal fillers into polymers to produce conducting polymer composites has stimulated interest in a different strategy: the use of electrically conducting polymers (ECPs) as the conductive component in composite materials.

A similar and related requirement exists for anti-static materials. Anti-static agents are substances used to reduce the tendency of polymeric products to acquire or maintain a surface charge. The build-up of surface charge can often result in a spark discharge, which can be hazardous in

Plastics Additives: An A–Z Reference
Edited by G. Pritchard
Published in 1998 by Chapman & Hall, London. ISBN 0 412 72720 X

environments containing combustible dust, flammable vapours or high oxygen levels. More importantly, high levels of static discharge can also destroy delicate electronic components, causing the loss of information encoded in computer storage devices, or generating interference with electronic communication. In order to alleviate static charge build-up, a decay mechanism must be provided for the charge. The most common method used is that of charge drainage from the plastic surface to ground, and this usually requires the plastic material to be semiconductive.

Although there may be some overlap in individual cases, composites which minimize electrostatic discharge (ESD) require different properties from those used to provide protection from electromagnetic interference (EMI) or radio frequency interference (RFI). The major difference is the level of electrical conductivity required for the composite to serve its specific purpose. In the latter cases a conductive filler is added to give an overall volume conductivity in the range 10^{-3}–10^{-6} S cm^{-1} so that incident electromagnetic radiation can be dissipated as small but measurable leakage currents to ground. ESD materials usually operate with much lower volume conductivity in the range 10^{-6}–10^{-9} S cm^{-1}.

A major draw-back to the use of ECPs as conductive fillers is the instability associated with many of the key properties of these polymers. A number of electrically conducting polymers, such as polyacetylene, are unstable in air over long periods. Thermal stability can be problematical in some ECPs, and would therefore render these polymers unsuitable for use in hot environments.

More critically, some polymers rely on (oxidative) doping processes to bring their electrical conductivity into the range of semiconductors. In general these doping reactions are not well understood, and the stability of the high levels of conductivity achieved in this manner with respect to time is questionable. One the other hand, the choice of ladder polymers as conductive fillers in two-phase composites is supported by their attractive properties. As a result of their extended π-electron conjugation along the polymer backbone, they tend to exhibit good thermal stability and good resistance to chemical degradation and quite often possess intrinsic electrical conductivities in the semiconducting range. Such polymers were first synthesized in the mid-1960s with the goal of achieving thermal stability by employing a double-stranded molecular architecture. This approach not only led to polymers of increased thermal stability, but also resulted in an increase in polymer chain rigidity, which in turn markedly contributes to the structural order and mechanical properties of ladder polymers. However, the influence of rigid ladder structure on the solid state properties (particularly the optical and electronic properties), remained largely unexplored until as recently as the early 1980s. At this time, in parallel with work carried out on single strand polymers such as polyacetylene, polythiophene, polyaniline and polypyrrole,

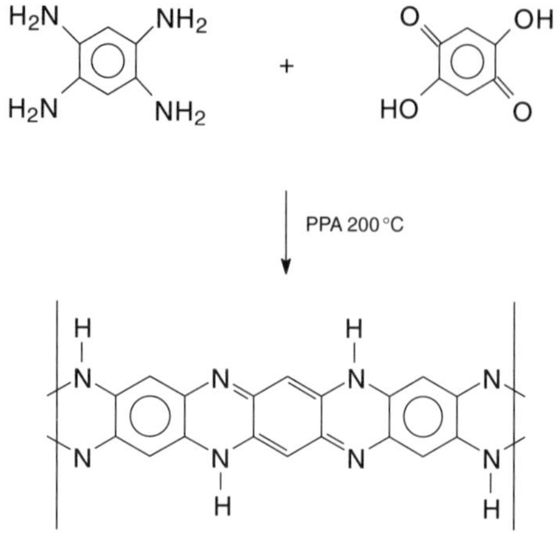

Figure 1 A general reaction scheme for the synthesis of a polyquinoxaline.

conjugated ladder polymers were treated through chemical and electrochemical doping to improve their electrical conductivity. During this period the search for electrically conducting ladder polymers became focused on a few select sub-classes of ladder polymers, namely the imidazoisoquinolines, the polyphenothiazines, the polyphenoxazines and the polyquinoxaline-type ladder polymers. These types are usually synthesized in polycondensation reactions under a variety of conditions. A general reaction scheme for the synthesis of a polyquinoxaline is presented in Figure 1.

The joint goals of highly conducting organic polymers together with a high degree of processibility are mutually exclusive, because the very factors that contribute to high conductivity (rigidity of structure and high levels of π-electron delocalization), also contribute to the insolubility and intractability of ladder polymers.

Conductive polymer composites can be defined as insulating polymer matrices which have been blended with filler particles such as carbon black, metal flakes or powders, or other conductive materials to render them conductive. Although the majority of applications of polymers in the electrical and electronic areas are based on their ability to act as electrical insulators, many cases have arisen more recently when electrical conductivity is required. These applications include the dissipation of electrical charge from rubber and plastic parts and the shielding of plastic boxes from the effects of electromagnetic waves. Consequently, materials scientists have sought to combine the versatility of polymers with the electrical properties of metals. The method currently used to increase the electrical conductivity of plastics is to fill them with conductive additives such as metallic powders, metallic fibres, carbon black and intrinsically conducting polymers such as polypyrrole.

In the case of intrinsically conducting polymers, the structural features which contribute significantly to the conductivity of the polymer (linearity, planarity of structure, regularity of structure, and high molecular weight) also contribute greatly to the processing problems associated with these polymers. Conducting polymers are usually insoluble and possess high softening temperatures. In ladder polymers, these problems are accentuated by their highly conjugated, fused ring structures. As a result, with few exceptions, ladder polymers are rarely processed as films cast from solution.

Composites consisting of a conducting polymer filler (in the powder form) and an insulating polymer matrix provide a convenient solution to these processing problems. These conductive composites have an advantage in that the mechanical and physical properties of the composite can be influenced by the choice of the host insulating matrix.

EFFECT OF INTERACTION BETWEEN FILLER PARTICLES AND HOST MATRIX

The addition of a conductor or semiconductor to an insulator affects the electrical properties of the composite according to the filler loading and the proximity of the conductive particles to other conductive particles. Three situations are possible: no contact between the particles, close proximity, and physical contact between the particles.

No contact between conductive particles

When the conductive particles are isolated, the conductivity of the composites changes only slightly or not at all. The composite remains an insulator, although its dielectric properties may change significantly.

Conductive particles in close proximity to each other

When the conductive particles are in close proximity to each other, electrons can cross the gap between particles, creating a current flow. The ability of an electron to cross a gap under a given voltage field increases exponentially with decreasing gap size. Gaps as large as 10 nm can be jumped.

Electron transport across an insulator gap can occur by one of two methods: hopping or tunnelling. Tunnelling is a special case of hopping where electrons can tunnel from the valence band of molecules or ions on one side of the gap to the conduction band of molecules or ions on the other side of the gap without any energy exchange. Hopping consists of the same type of electron flow between conductors across an insulating gap. However, in this case the electron must have its energy level increased to that of an appropriate level from which it can jump across the gap. Hopping, therefore, required an activation energy.

Physical contact between conductive particles

Under this condition the composite conducts through the particle network by the conduction mechanism of the conductive particles. Composites using metal particles as the conductor would exhibit band-type conduction. Band conduction and hopping conduction can be differentiated by the A.C. and D.C. behaviour of the composite. A composite which conducts by a hopping mechanism will exhibit a higher A.C. conductivity than D.C. conductivity. It will also show an increase in conductivity with an increase in frequency. Highly loaded composites which show linear current–voltage characteristics (i.e. show Ohmic behaviour) contain filler particles that are in actual physical contact with each other, whereas those composites with lower filler loadings exhibit non-Ohmic behaviour because they rely upon the hopping mechanism for electron transport. The total non-Ohmic behaviour of composites is claimed to be due to extended space charge distributions near the electrodes. Space charge distributions are claimed to be generated by local polarization of the matrix material. It is probable that metal particles and fibrous conductive fillers conduct by hopping, although actual physical contact between the particles is also possible. Their much larger size makes them less likely to be isolated from other conductive particles and more likely to penetrate a thin film.

NETWORK FORMATION BETWEEN FILLER PARTICLES

Composites consisting of an insulating matrix and a conductive filler will show a transition from insulator to conductor over a very narrow range of filler concentration. A typical curve of resistivity versus filler concentration (Figure 2) will show that the composite remains an insulator at low concentrations [1].

At a critical volume fraction, the resistivity of the composite falls sharply to a level at which the composite can readily conduct electricity. Increases in filler concentration above the critical loading do not appreciably reduce the resistivity. The sharp change from insulator to conductor is due to the formation of a network among the filler particles. Network formation has most frequently been treated as a percolation process. The percolation model refers to a means of continuous network formation through a lattice, taking into account the relative amounts of the two materials comprising the network. It is a statistical representation and is most frequently analysed by Monte Carlo techniques.

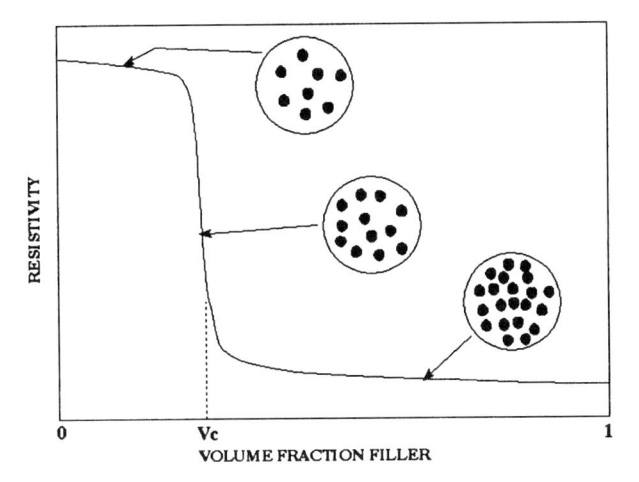

Figure 2 A schematic graph of resistivity versus filler concentration.

The probability of a continuous network being formed by filler particles in a matrix is related to the statistical average number of contact each particle makes with neighbouring particles and the average number of contacts per particle that are sterically possible.

CONDUCTIVE COMPOSITES AND THEIR APPLICATIONS

Conducting polymer composite materials are becoming common in many applications such as electrodes, conductive coatings and adhesives, positive temperature coefficient materials, and heaters.

Screening of electromagnetic waves using composites is based on cellular foamed plastics which incorporate one or more films of electrically conductive organic polymers. In some cases the composites employ two different conductive polymers possessing two different electrical conductivities. For example a melamine resin base material coated with a film of a polypyrrole 100 μm thick showed a conductivity of 150 S cm^{-1} and exhibited a shielding capacity [2] of 45 dB over the frequency range 1–1500 MHz.

A lightweight composite material with high electrical conductivity can be prepared by incorporation of polypyrrole in the pore of a porous divinylbenzene-crosslinked polystyrene host polymer, prepared by an emulsion polymerization method. The recorded conductivity of the composite was 0.5 S cm^{-1} and the shielding effectiveness (SE) of the composite reach 26 dB at 1.0–2.0 GHz. Polypyrrole–poly(ethylene-co-vinylacetate) composites which possess good electrical and mechanical properties have been made in two steps: first, a concentrated emulsion was generated by stirring a toluene solution of pyrrole, to which was added poly(ethylene-co-vinylacetate) in a small amount of an aqueous solution of sodium dodecylsulphate. Then the oxidative polymerization of the monomer was carried out by introducing an aqueous solution of FeCl$_3$. After polymerization, flexible and stretchable films were obtained by hot-pressing the composite powders. The conductivity of the composite films exhibited a percolation threshold in the range 5–20% w/w and reached a maximum value of 5–7 S cm^{-1}.

The electrical resistivities of polyaniline-poly(vinyl alcohol) composites are in the range of 60–1000 ohm. The electrochemical behaviour and mechanical properties of the composites meant that they were suitable materials for the dissipation of electrical charge.

Composites of high strength which were suitable for EMI shielding and charge dissipative applications can be made by (a) melting or dissolving a non-conducting polymer, (b) dispersing monomers which form conducting polymers in the polymer, then adding oxidizing agents to cause oxidative polymerization of the monomers, and moulding the reaction mixtures within 30 minutes. A solution of 100 parts of a polyamide in methanol was mixed with 20 parts pyrrole, treated with ammonium bisulphate and formed into a 100 μm film in 2 minutes. The film showed a volume resistivity of approximately 6200 ohm cm and a tensile strength of 730 kgf cm^{-2} (71.6 MPa) [3].

A composite using polyaniline as the conductive component and bisphenol A-based polycarbonate as the insulating matrix can be prepared electrochemically. The conductivities of the composites were close to that of pure polyaniline prepared by the same method. Also composites prepared in this manner have different properties from those prepared from a simple mechanical mixture of the two materials. A polypyrrole–polyvinylacetate composite film was prepared by casting a mixed solution of polyvinylacetate, FeCl$_3$, and polypyrrole on a substrate. Polymerization of pyrrole, which was suppressed in the mixed solution by controlling the oxidation potential of the solution, was accelerated upon rapid evaporation of solvent when cast on a substrate. The high electrical conductivity of the film, even when only 5% pyrrole was incorporated, was explained by the formation of a network structure of polypyrrole [4]. The composite film prepared by this methods showed an enhanced anion doping–undoping process because of the porous structure originating from the evaporation of solvent.

A latex can be used to form conductive blends of thermoplastic rubber with π-conjugated polymers such as polyaniline, polypyrrole, and poly(3-methoxythiophene) made from Fe(III), Fe(II), Cu(II)–H$_2$O$_2$ *in-situ* aqueous oxidation systems. The composite films were then obtained by hot pressing. Suitable polymerization conditions involved using HCl, aniline and the oxidizing agent in 6:2:1 mole ratios respectively. Composite films containing 50% polyaniline indicated an electrical conductivity of $1 \times 10^2\,\mathrm{S\,cm^{-1}}$, tensile strength of 13 MPa, and good storage stability in air [5].

Melt-processible conducting polymer blends based on intractable conducting polymers have been prepared. The conducting polymers were in the form of fibrils or fine conducting needles with high aspect ratio (length much greater than diameter). The microfibrils were incorporated into polymer mixtures which were thermoformed into conductive shaped articles. The blends were electrically conductive with excellent mechanical properties and the conductivity of the final product could be controlled by adjusting the fibril–matrix ratio [6].

Composites based on polyaniline by diffusion–oxidation of aniline-swollen polymeric matrices using $FeCl_3$ as a oxidant had promising room temperature conductivity, transmittance at 400–800 nm and stability in air. Morphology of the composite films was dependent upon the polymerization time, concentration of $FeCl_3$, and the matrix used [7].

The *in-situ* preparation has been described of poly(isothianaphthene)–PMMA, i.e. poly(methylmethacrylate) composite films by mixing 1,3-dihydroisothianaphthan-2-oxide with PMMA in dichloromethane, spreading the solution and removing the solvent, and then treating the resulting film with a cyclohexane solution of thionyl chloride. The polymer achieved electrical conductivities in the range 10^{-4}–$10^{-7}\,\mathrm{S\,cm^{-1}}$. The films also exhibited good environmental stability and good mechanical properties. A process for depositing a uniform, smooth and coherent film of polypyrrole onto and into commercially available polytetrafluoroethylene (PTFE) tape showed that PTFE substrate can be oriented. Growth of polypyrrole on this surface can lead to an oriented conductive polymer film. Adhesion was promoted by the use of the perfluorosulphonate ionomer Nafion® to create an interface between the PTFE and the polypyrrole.

The temperature dependence of the conductivity of low density polyethylene–polypyrrole composites in the range 77–450 K depends upon the insulating polymer and the conductive polymer used. A positive temperature coefficient (PTC) phenomenon was tentatively explained by the interruption of the percolation path [8]. The microwave absorption characteristics of composites of polypyrrole with paper, cotton cloth and polyester fabrics have been investigated. Reflectivity measurements in the range 2–18 GHz and plane wave modelling revealed impedance characteristics with a common transition region. The relationships between substrate material, polymer loading and electrical performance have been considered. An electrical model was successful in predicting the performance of both the Salisbury screen and Gambian multi-layer designs of radar-absorbing material [9].

REFERENCES

1. Ruschan, G.R. and Newnham, R.E.J. (1992) Compos. Mater., 26, 18, 2727.
2. Naarman, N., Krueckan, F.E., Moibius, K.H. and Otta, K. German Patent 3821478 (19/01/1989).
3. Masuda, Y., Ofuku, K., Suzuki, K. and Kawagoe, T. Japanese patent 04149267 (22/05/1992).
4. Whang, Y.E., Han, J.J. and Miyata, S. New Material, 281, Ed. Joshi S.K. Narosa, New Delhi.
5. Liu, H., Maruyama, T. and Yamamoto, T. (1993) Polym. J., Tokyo, 25, 363.
6. Heeger, A.J. and Smith, P. World patent 9220072 (12/11/1992).
7. Wan, M. and Yang, J. J. (1993) Appl. Polym. Sci., 49, 1639.
8. Meyer, T. (1974) Polym. Eng. Sci., 14, 706.
9. Wright, P.V., Wong, P.C., Chambers, B. and Anderson, A.P. (1994) Adv. Mater. Opt. Electron., 4, 253.

Keywords: electromagnetic radiation, EMT, RFI, ladder polymers, polyquinoxalines, polypyrrole, polyaniline, hopping, tunnelling, band conduction, composites, percolation.

See also: Carbon black;
           Conducting fillers for plastics: (1) Flakes and fibres

Coupling agents

Geoffrey Pritchard

INTRODUCTION

A coupling agent is a chemical which improves the adhesion between two phases in a composite material. The term 'composite' is used here to denote a material which has two or more distinct constituents, not chemically bound to each other. The two examples of composite materials that are especially relevant here are:

1. resins containing glass fibre reinforcement;
2. thermosetting resins and thermoplastics containing particulate fillers.

There is a separate entry in this book dealing with the surface treatment of particulate fillers in plastics, and so the main discussion here will be about the surface treatment of glass fibres.

The most important coupling agents to-day are the organosilanes [1, 2], applied to the surface of glass filaments in the form of an aqueous or non-aqueous solution, to promote adhesion to resins. The organosilanes completely dominate this market, although there have been earlier coupling agent technologies based on the chemistry of chromium complexes. Zirconates and titanates have much more varied applications, mostly related to fillers.

COUPLING AGENTS FOR GLASS FIBRES IN REINFORCED PLASTICS

It is necessary to distinguish carefully between coupling agents, which have a single purpose in fibre reinforced plastics, namely to promote adhesion and thereby facilitate the load transfer which is of the essence

Plastics Additives: An A–Z Reference
Edited by G. Pritchard
Published in 1998 by Chapman & Hall, London. ISBN 0 412 72720 X

of reinforcement, and sizes, such as those based on poly(vinyl acetate), that are sometimes applied to the fibres as well. 'Size' refers to a more complex formulation which has several functions to fulfil. One of the most important of its roles is to bind together a large number of delicate filaments into more robust strands, so that they can more easily withstand the textile-type processing operations which convert continuous filaments into woven fabrics or so that they can survive as short fibre bundles in thermoplastics processing machinery. A common constituent of sizes is poly(vinyl acetate). The same poly(vinyl acetate) is also used in glass fibre technology as a binder (adhesive) to bond short chopped filament bundles together to form chopped strand mat, one of the commonest forms of glass fibre reinforcement.

Without the adhesion provided by the coupling agent between the two phases in fibre reinforced plastics, there is little that the fibres can do to protect the matrix from mechanical stresses. A failure of the interface between fibre and matrix will sometimes be visible under a low power microscope but it is more easily reflected in the mechanical properties of the reinforced plastics materials. In the case of unidirectionally re-inforced laminates, affected properties will include transverse flexural strength, transverse tensile strength, and shear strength (see some mechanical effects in Figure 1). Sometimes interfacial bond weakness is not evident immediately on applying a load, but is induced gradually by prolonged immersion in hot water, which can chemically attack the

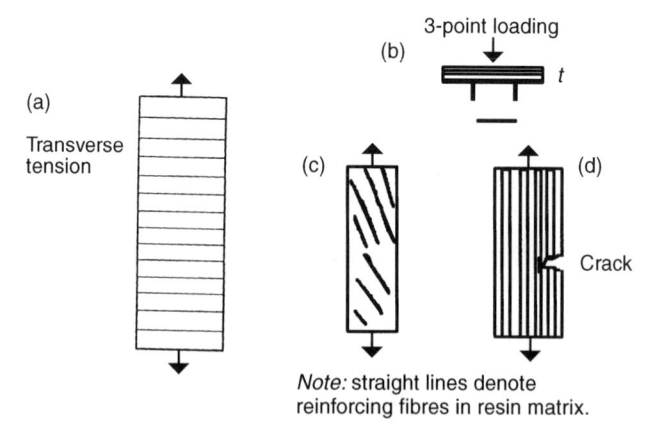

Figure 1 Mechanical properties of glass fibre reinforced plastics. Many mechanical properties are improved by appropriate coupling agents applied to the glass fibres, notably: (a) transverse tensile strength; (b) short beam shear strength; (c) 10 degrees off-axis tensile strength. Sketch (d) shows a crack propagation normal to the fibre reinforcement, facilitated by excessively good adhesion between fibres and matrix.

bond between the coupling agent and the fibres or resin, or can cause the resin to swell. High temperatures can also induce strain in the interfacial bond, because of the large difference between the thermal expansion coefficients of glass and those typically found in organic plastics.

It is possible for the adhesive bond between fibres and resin to be too strong. This means that the progress of a crack through the material is insufficiently hindered by the interface, i.e. the loss of propagation energy is slight, and this means a reduced impact strength.

ORGANOSILANES

A silane is a chemical compound containing a hydrogen–silicon bond. Such chemicals are also referred to as silicon hydrides, e.g. SiH_4, or $SiCl_3H$. Organosilanes have organic groups replacing the hydrogen atoms in silanes, but they still tend to be called simply silanes in reinforced plastics circles, although they do not always have the silicon–hydrogen bond. There is in theory a wide choice of organosilane chemicals available for use as coupling agents (Table 1). The main limitation is that at least one of the groups must be reactive towards the functional groups of the polymer, and at least one of the others must be reactive towards the fibre or mineral surface, so as to produce a chemical 'bridge'. An epoxy resin could readily react with an aminofunctional silane, or an unsaturated polyester resin could react with a silane containing a compatible, ethylenically unsaturated functional group such as methacrylate or vinyl. The organosilane can therefore be designed for a specific matrix resin, but it is more useful for it to be multi-purpose, having reactive groups for more than one resin type. Sometimes the length of the resin-functional group on the silane is extended, e.g. three or four extra methylene groups might be inserted between the silicon

Table 1 Some coupling agents used to promote adhesion of glass fibres to resins

Coupling agent structure	Name
$NH_2(CH_2)_3Si(OCH_3)_3$	Trimethoxy 1-aminopropylsilane
$CH_2=CHSi(OC_2H_5)_3$	Vinyl triethoxysilane
$CH_2=CHSi(OCH_2CH_2OCH_3)_3$	(γ-methacryloxypropyl)trimethoxysilane
$H_2NCH_2CH_2NH(CH_2)_3Si(OCH_3)_3$	3-(2-aminoethylamino)propyltrimethoxysilane
$CH_2=CHCH_2Si(OCH_3)_3$	Allyltrimethoxysilane
$CH_2=CHCOO(CH_2)_3SiCl_2CH_3$	3-acryloxypropylmethyldichlorosilane
$CH_2=CHSi(CH_3)_2Cl$	Vinyl dimethyl chlorosilane
$CH_2=CHSiCl_3$	Vinyltrichlorosilane
$CF_3(CH_2)_2SiCl_3$	(3,3,3-trifluoropropyl)trichlorosilane
$Cl-(CH_2)_3-Si(OCH_3)_3$	γ-(chloroproyl)trimethoxysilane
$H_2N(CH_2)_2NH(CH_2)_3Si(OCH_3)_3$	γ-(β-aminoethylaminopropyl)trimethoxysilane

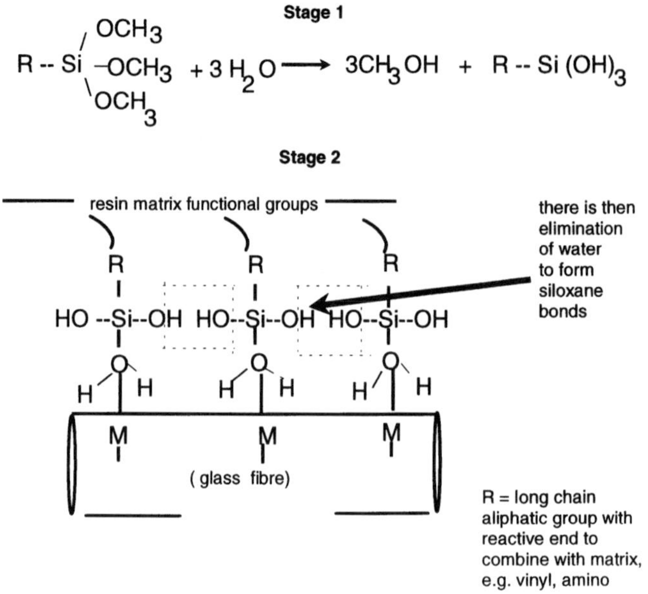

Stage 1

$$R-Si(OCH_3)_3 + 3H_2O \longrightarrow 3CH_3OH + R-Si(OH)_3$$

Stage 2

Figure 2 Idealized mode of action of silane coupling agents. M refers to Si, Al, Ca, Fe etc.

atom and the amine or methacrylate group, so that it is long and flexible enough to reach into the resin and react at the appropriate reactive sites.

The bonding of the silane to the glass surface is the other requirement and this is achieved through incorporating in the organosilane a hydrolysable group, such as chloro- or methoxy-, which participates in hydrolysis reactions with the inevitable adsorbed moisture on the fibre surface, and subsequently react with the silanol groups on the glass surface (Figure 2).

(In Figure 2, the letter M refers to silicon, or aluminium, magnesium, or another cation in the surface layer. The oxane bonds, M–O–Si, are hydrolysed by prolonged exposure to water, but they reform on drying.)

Under basic conditions the hydrolysis of alkoxy groups takes place during the silane application process, and generally occurs in a stepwise fashion. Except for aminosilanes, however, it is usual to employ acid conditions. The rate of hydrolysis of alkoxy groups decreases with an increase in their size, for instance FTIR evidence indicates that vinyl trimethoxysilane hydrolyses almost eight times as quickly as the triethoxy equivalent. The glass surface is likely to become covered with several monolayers of hydrolysed silane.

Organosilanes vary in their effectiveness, judged by such criteria as resistance of the glass reinforced resin to prolonged immersion in hot water, its resistance to debonding under stress, the magnitude of the

Organosilanes 193

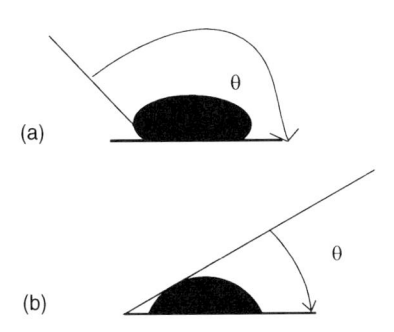

Figure 3 Contact angle between a liquid and a fibre surface: (a) poor wetting, large contact angle; (b) better wetting, smaller contact angle.

short beam shear strength etc. Some organosilanes, e.g. chloropropylsilanes, are effective with hot cured epoxy resins, but not with low temperature cured epoxy systems.

There has been some controversy about the extent to which the mechanism of action is intrinsically chemical, i.e. whether silanes simply act as chemical bridges between fibres and matrix, or whether they lower the contact angle (see Figure 3), thus improving the wetting capacity, altering the surface energy of the glass to make wetting thermodynamically more favourable. Certainly organosilanes are not generally very effective if they do not possess suitable chemical groups, but this does not rule out the possibility of additional mechanisms playing a part.

Water has a crucial role in the action of organosilanes. According to Plueddemann [1], in the silane series

$$(CH_3O)_3SiCH_2CH_2CH_2NHCH_2CH_2NHR \cdot HCl$$

if R is less than a three carbon atom group, the compound is soluble in water, either directly, or after hydrolysis. Between 3 and 10, the silanes are insoluble, but are rendered soluble by hydrolysis. Those with 12 to 18 carbon atoms are soluble, but rendered insoluble by hydrolysis in a concentrated homogeneous solution; higher homologues are always insoluble. Aqueous solutions of coupling agents are stable for several hours, but become unstable as condensation occurs.

Other coupling agents, based on chromium complexes (Figure 4) have been used in the past to promote adhesion between glass fibres and resins, but at present organosilanes are the preferred choice.

Some polymers are lacking in the reactive functional groups necessary to take advantage of the silane mechanisms outlined above. Polypropylene and other polyolefins are prime examples. They are therefore modified by a graft polymerization process so as to attach functional molecules to the polymers. Maleic anhydride is widely promoted as a suitable agent for use with polypropylene.

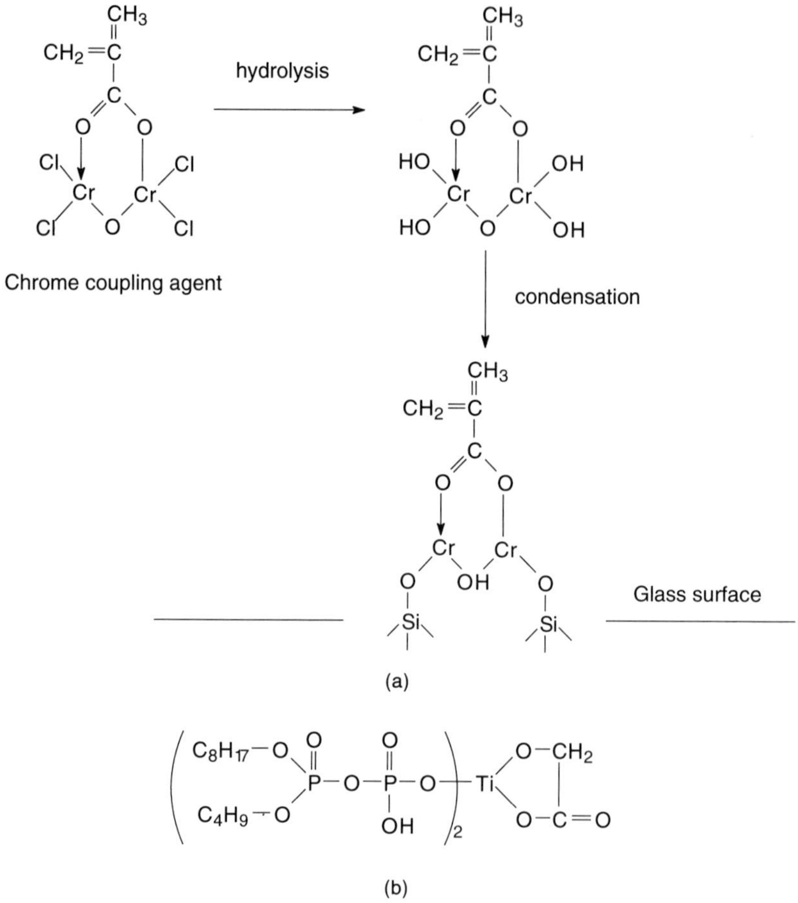

Figure 4 Structures of some non-silane coupling agents: (a) a chrome complex; (b) an organotitanate. From [1] and [3].

FIBRES OTHER THAN GLASS

There are many polar resins which would adhere to carbon or graphite fibres if the fibre surfaces were to contain similar polar groups such as carbonyl and carboxyl. High strength, low heat temperature PAN-based carbon fibres already have these surface chemical groups, and therefore there is no great need for coupling agents. High modulus, high heat treatment temperature carbon fibres have much greater orientation of their six membered carbon rings, which become aligned with their basal planes facing outward, with very little scope for surface polar groups. In the early days of carbon fibres, many surface treatments and coupling agents were evaluated, but the most favoured procedure was simply to subject the fibres to moderate oxidation, either chemically using acid/dichromate, nitric acid or hypochlorite, or electrolytically. The recent growth in use of high molecular weight, highly oriented polyethylene fibres would not have been possible without a similar surface treatment of the polyethylene by a plasma process.

FILLERS

The reasons for applying coupling agents or surface treatment agents to the surfaces of particulate fillers are usually quite different from those applicable to reinforcing fibres. A surface treatment may be intended to prevent the agglomeration of fine particulates during preliminary processing operations, as this is detrimental both to the viscosity of the polymer containing the filler, and also to the mechanical strength of the finished product. It may be used to improve the rheology of resins or to improve the electrical or chemical resistance of the filled resins. There is often little incentive to promote adhesion between a filler particle and a resin matrix, either because other considerations ensure that this is not a problem anyway, or because soft filler particles such as clay, alumina trihydrate and talc will surely fail along weak cleavage planes at much lower stresses than the resin fracture stress, so surface treatment will not enhance the strength of the filled resins.

The reader is referred to the entry in this book, entitled ' Surface treatments for particulate fillers in plastics', for further information.

TITANATE AND ZIRCONATE COUPLING AGENTS

Titanate (and zirconate) coupling agents as well as other chemical agents have been developed for use with particulate fillers rather than fibres. It should be sufficient to say here that titanates have several diverse applications: they are applied to the surfaces of minerals, pigments, carbon black, metals, metal oxides and fibres. They can be effective in preventing or reducing the unwanted agglomeration of particles, in promoting adhesion, changing the degree of hydrophobicity, reducing the viscosity of liquid resins containing particles (including the viscosity of clay-filled SMC moulding compounds), or otherwise altering the rheology of a system or its thixotropy; increasing the maximum achievable filler loadings, and reducing the ageing rate of filled plastics and plastics coated metals. They have been reported to give satisfactory performance as fibre and whisker surface treatments in advanced composites. Figure 4 gives some examples of the structures of non-silane coupling agents.

REFERENCES

1. Plueddemann, Edwin P. (1982) *Silane coupling agents*, Plenum Press, New York.
2. Arkles, B., Steinmetz, J.R., Zazyczny, J. and Zolotnitsky, M. (1991) Stable, water-borne silane coupling agents, Paper 2D, *46th Annual Conference, Composites Institute*, The Society of the Plastics Industry, Inc., Washington, D.C., USA.
3. Monte, S.J., Sugarman, G. and Seeman, D.J. (1977) Titanate coupling agents–update 1977, Paper 4E, *32nd Annual Conference, Reinforced Plastics/Composites Institute*, Society of the Plastics Industry, Inc., Washington, D.C., USA. Also subsequent papers in same conference series: Papers 2N and 2C (1978); 16E, (1979); 23F (1980); 18D (1981); 22B (1982); 3E (1983) by Monte and Sugarman.

Keywords: fibre, surface, organosilane, titanate, zirconate, chrome complex, surface energy, contact angle, interface, mechanical properties, adhesion, maleic anhydride.

See also: Fibres;
Surface treatment of filler particles;
Fibres: the effects of short glass fibres on the mechanical properties of thermoplastics;
Reinforcing fibres

Curing agents

Geoffrey Pritchard

The term 'cure' refers to the hardening, setting or crosslinking process in thermosetting resins. During crosslinking, a thermosetting resin undergoes several changes, usually (but not always) starting as a liquid, becoming progressively more viscous, undergoing a solidification process known as gelation, then becoming a soft, rubbery or cheese-like substance, and eventually becoming a hard, relatively brittle solid.

The crosslinking process consists of forming covalent bonds, either between small molecules, such as phenol and formaldehyde, or between existing linear macromolecules, as in the case of unsaturated polyesters or epoxy resins. The process of crosslinking existing linear macromolecules is analogous to the vulcanization of natural rubber.

Once the gelation stage is passed, some of the material is insoluble and infusible, making chemical analysis of the extent of reaction extremely difficult, so the determination of the extent of cure [1] frequently relies on the measurement of changes in physical, electrical and mechanical properties. The disadvantage of this approach is that complete reaction of the chemical groups involved cannot be assumed simply on the grounds that there is no further change in these properties. It may be necessary to increase the temperature in order to make the functional groups accessible. The process frequently stops short of completion in practice.

References to 'cure' of thermoplastics are inappropriate, as they concern non-reactive conditioning processes such as solvent removal by prolonged heating, annealing etc.

CURING AGENTS

Crosslinking is a chemical change that can in a few instances occur spontaneously, although it may be too slow or too rapid for commercial use. More

Plastics Additives: An A–Z Reference
Edited by G. Pritchard
Published in 1998 by Chapman & Hall, London. ISBN 0 412 72720 X

commonly, crosslinking has to be deliberately initiated, and the rate of reaction controlled, by the introduction of reactive substances called curing agents. Other chemicals known as accelerators (often of complex composition) may be needed as well for the cure reaction to occur at an acceptable rate, or at all. The curing agents must satisfy certain general requirements. They must be readily dispersed or dissolved in the resin; they must not react immediately, i.e. there should be a pot life to facilitate the processing operations, but they must eventually react completely or almost completely, and they must not adversely affect the properties of the final cured product, e.g. by promoting easier thermal or UV degradation.

Radiation, e.g. visible or ultraviolet light, or radiation from a cobalt-60 source, is an alternative to the use of curing agents for free radical cure reactions.

UNSATURATED POLYESTER RESINS

These resins are low molecular weight polyesters containing ethylenic unsaturation – almost invariably fumarate. They are supplied as solutions in styrene, or occasionally as solutions in another reactive vinyl monomer such as vinyl toluene.

The cure process proceeds by free radicals being first generated, and then allowed to facilitate the formation of styrene bridges between fumarate groups. Figure 1 shows schematically the process of cure as a function of time.

In the early stages, particularly when the process is being carried out at room temperature, there may be an induction period, associated with the clean-up of retarders, inhibitors etc. Among many inhibitors which delay or prevent crosslinking, mention should be made of hydroquinone, quinone, titanium dioxide in the anatase form (but not rutile), sulphur,

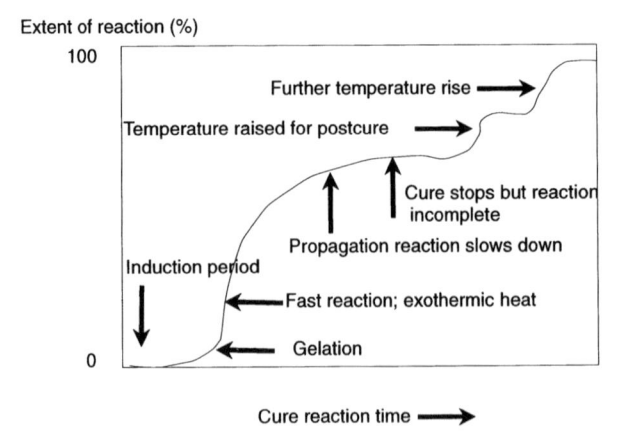

Figure 1 The crosslinking of an unsaturated polyester resin as a function of time.

oxygen, quaternary ammonium salts, aryl phosphites, and some copper compounds.

A common cause of inhibition is oxygen, whether stirred or dissolved into the resin, or merely contacting the resin surface. This allows the formation of much more unreactive peroxy radicals:

$$R^{\bullet} + O_2 \rightarrow RO_2^{\bullet}$$

As the reaction accelerates, it reaches the stage called gelation, when the crosslinks extend throughout the polymer, forming an infinite network, although only a small part of the overall reaction has occurred. After gelation, the propagation reactions continue, but termination reactions are sterically hindered by the growing three-dimensional network, and so the overall crosslinking process becomes much faster, with considerable exothermic heat evolved. This is known as the Trommsdorf effect. Eventually, the network becomes so dense that the propagation reaction is also sterically hindered, and the crosslinking process slows down. It cannot proceed further without a higher temperature, which enables more reactive sites to becomes accessible. Raising the temperature at this stage is called postcure. Often the reaction is incomplete. Ambient temperature cures for polyesters typically take about 24 hours, although the reaction will still not be complete, and residual styrene will persist until it either reacts or, more likely, diffuses out of the cured product.

High temperature cures, in contrast, can take anything from about one or two minutes to a few hours, depending on the resin, the curing agents and the temperature. The very long cure times in presses and autoclaves associated with some advanced composite resin systems are not customary with polyesters, which has been to their advantage, as such procedures constitute a disincentive to widespread use outside the defence/aerospace markets.

Characterizing a polyester curing agent requires a measurement of the time to reach gelation, under specified conditions, as well as of the heat evolved and of the time to reach the apparent end of the crosslinking process. It is also useful to know the extent to which the cure can really be called complete. More detailed examination of a cure system would consider its tolerance to contaminants and trace inhibitors, especially common substances such as moisture, trace metals and oxygen, pigments, thixotropic agents, release agents etc. The inhibition of cure at the surface by oxygen, which is observed with some curing agents, can lead to poor weathering and chemical resistance. Therefore surfaces sometimes have to be protected during cure.

Although many cure systems have been tried, only two ambient temperature cure systems for polyester resins have ever been widely used:

1. the benzoyl peroxide–tertiary amine system;
2. the hydroperoxide–cobalt(II) soap system.

Curing agents

The peroxides and hydroperoxides are not used in concentrated form, but are supplied already dispersed in inert diluents such as phthalates.

In the first system, the amine was once believed to form an amine oxide and to cleave the peroxide:

$$2(C_6H_5COO)_2 + (CH_3)_2N \cdot C_6H_5 \rightarrow$$
$$2C_6H_5COO^{\bullet} + (C_6H_5CO)_2O + (CH_3)2 \cdot NO^{\bullet}C_6H_5$$

but more recently benzoic acid has been identified as a by-product in the crosslinked resin, and the reaction may proceed by a single electron transfer from the nitrogen atom of the amine as follows:

[Benzoyl peroxide–amine reaction scheme]

Benzoyl peroxide–amine reaction

The crosslinking is very rapid at first with the benzoyl peroxide–amine system, unless the concentration of tertiary amine (N,N-dimethyl aniline or N,N-dimethyl para-toluidine) is very low. But the reaction fades early, probably because of the retarding effect of the benzoic acid. The mechanical properties of the finished product tend not to match those obtained with the cobalt system. Also there is a yellowing of the resin as cure proceeds. Consequently, the hydroxide–cobalt(II) system is more widely used. One proposed mechanism is:

$$ROOH + CO^{2+} \xrightarrow{fast} RO^{\bullet} + OH^- + CO^{3+}$$
$$ROOH + CO^{3+} \xrightarrow{slow} ROO^{\bullet} + H^+ + CO^{2+}$$
$$ROOH^{\bullet} \longrightarrow R^{\bullet} + O_2$$

where ROOH is any one of a number of hydroperoxides, such as cumene hydroperoxide, or methyl ethyl ketone hydroperoxide (MEKP). In practice, MEKP is a mixture of several hydroperoxide compounds, and their proportions can be altered, giving various levels of reactivity. The cobalt compounds used as accelerators are the naphthenate and the octoate, dispersed in white spirit or styrene. Many resins are sold with the cobalt accelerator already added, leaving the fabricator to mix the resin with the hydroperoxide. One reason is simply convenience, but it also minimizes the likelihood that excess cobalt will accidentally be added. Excess cobalt retards cure by using up some of the free radicals in the following reaction:

$$RO^{\cdot} + CO^{2+} \rightarrow RO^- + CO^{3+}$$

For cure at higher temperatures, several special peroxide systems have been advocated. The rate of decomposition of peroxides varies considerably with temperature and with the nature of the peroxide groups R in the structure R–O–O–R. For economic reasons the most rapid possible cure is required, consistent with the reaction approaching close to 100% completion. However, some peroxides are undesirable because although they provide their free radicals quickly, they tend to decompose prematurely, causing storage and handling problems. Tertiary butyl peroxy-benzoate, tertiary butyl hydroperoxide, cumene hydroperoxide and cyclohexanone hydroperoxide among others are, or have been, extensively used (Table 1).

Hydroperoxides and peroxides react violently to various stimuli, e.g. impact, heat, and contact with certain organic substances, including accelerators. Because of the use of inert diluents to reduce the hazards, care must be taken to distinguish between the weight of substance employed and the smaller quantity of active peroxide actually contained

Table 1 Peroxide activities

Peroxide	Half-life (hours) at	Temperature (°C)
Benzoyl peroxide	13	70
	2.1	85
	0.4	100
Lauroyl peroxide	13	60
	0.5	100
t-butyl perbenzoate	1.7	120
	0.55	130
t-butyl peroctoate	1.7	89
t-butyl hydroperoxide	29	160
Dicumyl peroxide	0.3	145
2,5-dimethyl hexane-2,5-di-hydroperoxide	19	145

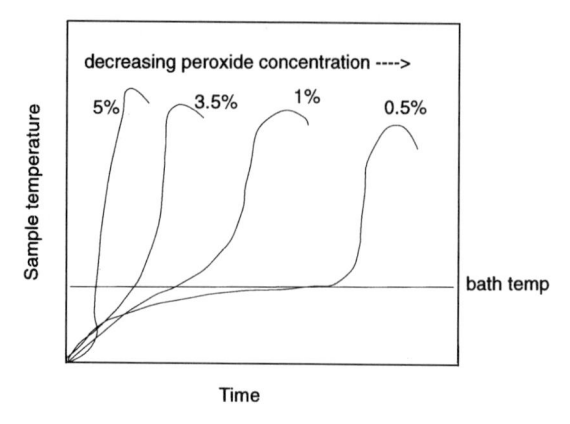

Figure 2 Typical data from an exotherm test at constant temperature. The temperature of the sample rises above the temperature of its surroundings, to an extent dependent on sample geometry.

in it. Manufacturers' advice must be obtained about the storage, handling and use of all these compounds.

The cure of thermosetting resins in thick sections is difficult because the heat generated by the reaction can cause cracking or even raise the temperature above the decomposition temperature of the resin. The amount of heat evolved, and the time taken to generate it, can be compared for various peroxide and accelerator combinations using standard tests such as the SPI exotherm test, which produces curves of the kind shown in Figure 2. Crosslinking is carried out in constant temperature baths, using sample tubes of standard size and geometry. The temperatures reached in Figure 2 would of course be much higher with a larger bulk.

Sometimes it is advisable to reduce the exothermic heat evolved during the cure of an unsaturated polyester resin. This can be done by adding a small quantity of a moderating agent such as alpha methyl styrene to the styrene. The unsaturation in the alpha methyl styrene is sterically hindered and rather unreactive. Fillers also have a benign influence on exotherm temperatures.

The peak exotherm temperature is affected by the nature and concentration of accelerator used; the higher the concentration, the higher the exotherm temperature, but reducing the accelerator concentration too drastically could stop the cure reaction altogether.

VINYL ESTER RESINS

Vinyl ester resins are cured by similar methods to polyesters [2]. They were originally developed to produce anti-corrosion laminates for markets such as the chemical process industry. They consist essentially of epoxy resins, or else epoxy novolaks, reacted with acrylic or methacrylic acid so as to introduce terminal unsaturation. The unsaturation is confined to the chain ends, where it is highly reactive with styrene. The MEKP/cobalt system is used for room temperature curing, but cobalt, hydroperoxides and tertiary amines are often used together in vinyl esters, because this has an accelerating effect. The concentrations typically used are in the range MEKP 1.5 to 2%, cobalt octoate 0.2 to 0.5% of 6% active solution, and 0 to 1.2% of 10% N,N'-dimethyl aniline solution. If a longer pot life is required, addition of 2,4-pentanedione is recommended to slow down the process without reducing the concentration of the curing agents, which might affect the ultimate degree of cure. Benzoyl peroxide, t-butyl perbenzoate or peroctoate can be used for high temperature curing during fabrication by filament winding, resin transfer moulding or pultrusion.

There are certain differences between the cure of vinyl ester resins and that of unsaturated polyesters. MEKP with a high dimer content is more reactive with vinyl esters than the standard varieties, but less reactive with polyesters.

EPOXY RESINS

Epoxy resins consist of substances containing two or more epoxy rings per molecule. The curing reactions are essentially ring-opening reactions of the epoxy group, i.e. reactions which convert epoxy groups to hydroxyl groups. There is a very wide choice of substances to carry out the ring opening reaction. Some are only suitable for ambient temperature use, others for elevated temperatures, while still other varieties are chosen because they have specific advantages such as improved safety in use, easier processing etc. Epoxy resins have more varied applications than unsaturated polyesters, being favoured for coatings and adhesives, as well as laminates and castings. Different types of application impose different requirements on curing agents. Curing agents for epoxy resins can be divided into the catalytic kind, of which only a small quantity is required to initiate a self-sustaining reaction, and the chain-extending kind, which is needed in stoichiometric quantities. The first type has something in common with polyester cure, because the small quantities of curing agent used do not have a large effect on the final properties, whereas in the second case, the final product may consist of chemical sequences derived from the original epoxy resin and from the curing agent, in roughly comparable quantities. Clearly, in the second case, the curing agent has a major effect on ultimate properties.

An alternative division is between ambient temperature cure systems and hot cure systems. Other important considerations include whether the curing agents are liquids, or solids which can be dissolved in acceptable solvents for ease of dispersion, or whether they are high melting solids which require to be added at high temperatures. This can be inconvenient.

Several epoxy curing agents are suspected of being potential health hazards, whether by skin contact, inhalation or some other process, and manufacturers' advice should always be sought when a new curing agent is used for the first time.

Catalytic processes

The catalytic process uses Lewis acids, such as boron trifluoride as the etherate, or a tertiary amine, or the more reactive imidazoles. Amine complexes of boron trifluoride are stable in epoxy resins until heated, when rapid homopolymerization of the resin takes place. The reaction mechanism for tertiary amines has been represented by Garnish [3] (see Figure 3).

If the alcohol in Figure 3 is replaced by a mercaptan, very rapid cure occurs.

Chain extending curing agents

The chain extending reactions require much larger quantities of curing agent than the catalytic ones and they rely on the multifunctional nature of the curing agent to produce a crosslinked network rather than simply producing a longer linear structure. A few advanced epoxy resins are themselves multifunctional, notably TGDDM (the tetra glycidyl derivatives of diaminodiphenyl methane). The simplest example of chain extension is the ambient temperature use of an aliphatic primary or secondary amine (Figure 4).

Aromatic amines can be used instead, but they require elevated temperatures. Dicyandiamide, although not aromatic, requires heating as well. It is a latent curing agent capable of providing a long shelf life at ambient temperature, and is used at about 125°C in conjunction with traces of an accelerator such as benzyldimethylamine or an imidazole.

Anhydrides are also high temperature curing agents. The reaction with epoxy resins is slow in the absence of an accelerator, and requires quite large quantities. Typical substances used include methyl nadic anhydride and hexahydrophthalic anhydride. (Methyl nadic anhydride is a convenient liquid at room temperature, whereas nadic anhydride is a solid.)

It can be assumed that the first step is the formation of a carboxylic group after anhydride ring opening by water or adventitious hydroxyl groups, which, once introduced, can be regenerated by various means.

Amides can induce the cure of epoxy resins. Polyamides are used in epoxy adhesives. Other epoxy curing agents are listed in Table 2.

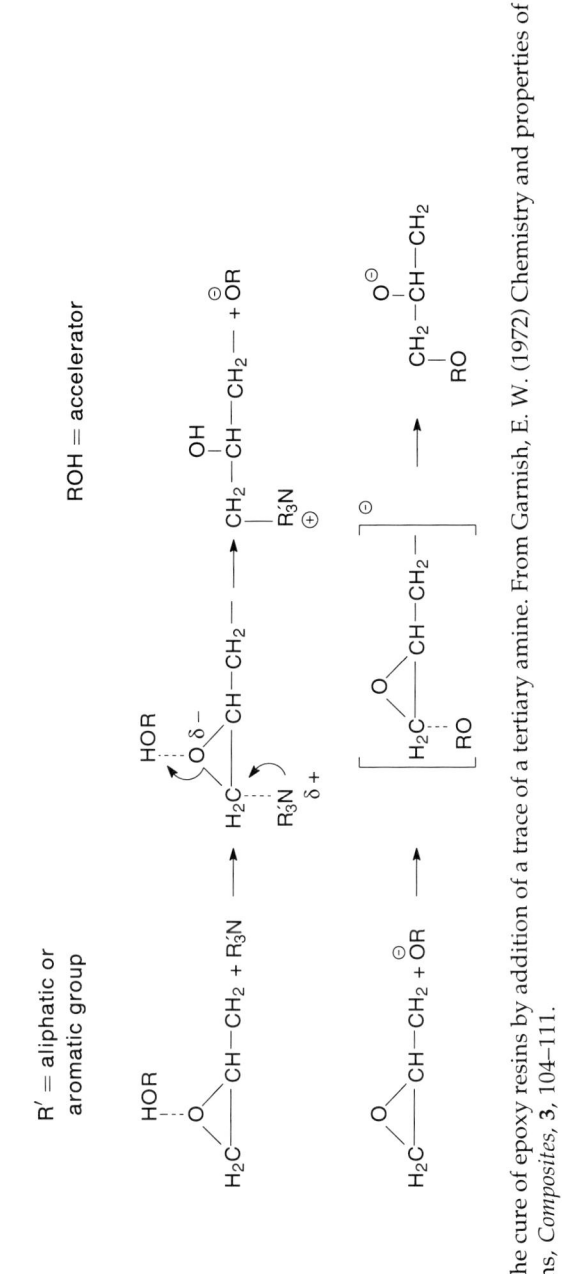

Figure 3 The cure of epoxy resins by addition of a trace of a tertiary amine. From Garnish, E. W. (1972) Chemistry and properties of epoxy resins, *Composites*, **3**, 104–111.

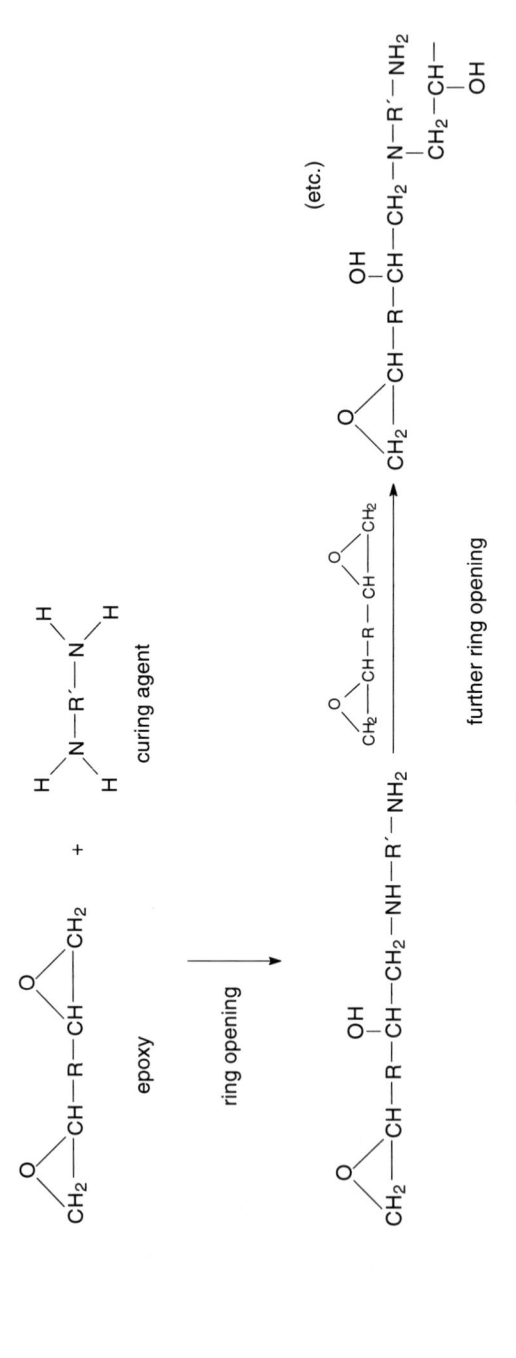

Figure 4 The cure of epoxy resins by primary amines.

Epoxy resins

Table 2 Some curing agents for bisphenol A (DGEBA type) epoxy resins.

Substance	State (ambient temp)	Typical cure cycle
Phthalic anhydride	Solid	48 h at 100°C then 24 h at 120°C
		6 h at 150°C 2 h at 200°C
Methyl nadic anhydride	Liquid	1 h at 120°C then 20 h at 220°C
Pyromellitic anhydride	Solid	18 h at 225°C
Hexahydrophthalic anhydride	Solid	1 h at 180°C then 1 h at 210°C
4,4'-diaminodiphenyl methane	Solid	1 h at 120°C then 3 h at 180°C
Boron trifluoride complex	Solution	90 min at 120°C then 6 h at 185°C
Triethylene tetramine	Liquid	Ambient then 2 h at 60°C
Benzyldimethylamine (N,N-dimethylbenzyl amine)	Liquid	2 h at 100°C 1 h at 200°C
Triethanolamine	Liquid	2 h at 100°C
Piperidine	Liquid	16 h at 120°C
4,4'-dimethyldiaminosulfone (DDS)	Solid (with accelerator)	8 h at 160°C
2,5-dimethyl-2,5-hexane diamine	Liquid	16 h at ambient then 1 h at 60°C
$CH_2=C(CH_3) \cdot CO \cdot N^- - N^+ - -(CH_3)_2CH_2CH(OH)CH_2$ (aminide)	Solid	3 h at 130°C 5 h at 160°C

Many commercial curing agents are actually much more complex than those given above. Some are eutectics of various amines, or they may be adducts of amines and epoxies. (The motivation is to overcome specific problems such as storage stability, or health and safety disadvantages.) Others are mixtures of two or three complex substances together with an accelerator. One example is a blend of 4,4'-methylene dianiline with polymethylene dianilines; it is slow acting and gives a stable pot life. Solid polyamides can be mixed with an epoxy resin to form a system with several months' stability at ambient temperature, but which can be used to activate high temperature cure (a few minutes at 170°C). Modified phenolic resins can also provide long term storage stability combined with hot cure capacity. Scola [4] has given a useful detailed review of epoxy resin curing agents, which although written many years ago, nevertheless still conveys something of the wide variety of substances investigated and used.

PHENOL–FORMALDEHYDE RESINS

Phenol is reacted with formaldehyde under either acidic (with excess of phenol) or alkaline (with excess of formaldehyde) conditions, to form a novolak or a resole respectively. Resoles are used extensively for making plywood and waferboard as well as for reinforced plastics; the cure is effected by heat alone. Novolaks require the addition of further formaldehyde for crosslinking to occur. This is achieved in practice by formulating moulding powders to include a solid curing agent, such as paraformaldehyde or hexamethylene tetramine. Both these substances decompose to give formaldehyde on heating.

When phenol–formaldehyde resins are used in glass laminates, organic acid catalysts such as p-toluene sulfonic acid promote the hardening. Alternatives have been developed in recent years, partly because acid catalysts can cause corrosion in processing equipment, but also to obtain better control over the storage stability and pot life. It has been claimed that alkaline catalysts produce products with less smoke on burning. Neutral catalysts have been developed as well.

LATENT ACID CATALYSTS

Uncatalysed resole phenolic resins have excellent storage stability, but acid catalysed ones have only a limited life at room temperature. Latent acid catalysts have been developed to overcome this problem. They are used in reaction injection moulding (RIM) and in resin transfer moulding (RTM) processes. Examples are phenyl hydrogen maleate, phenyl trifluoroacetate and butadiene sulfone. At elevated temperatures these catalysts generate the corresponding acids, which catalyse the resole reactions. A cycle time of one to two minutes at 150°C is achievable. The choice of a suitable latent catalyst can also reduce the peak exotherm temperature.

BISMALEIMIDES

Bismaleimides undergo a wide variety of reactions, leading either to chain-extended or crosslinked products. The process of cure is less easily defined with these resins than with polyesters and epoxies, because of the wide range of processes, some of which are still being elucidated. Bismaleimides have one important structural feature which makes it easy for them to undergo crosslinking: electron deficient maleimide

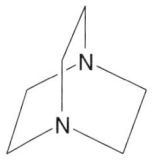

Figure 5 The structure of DABCO (diazobicyclo-2,2,2-octane).

double bonds. These bonds are very reactive and the resins can be crosslinked by heat alone at 165–200°C. The reaction can be further assisted by the addition of either free radical or ionic catalysts. Two cure accelerators suggested for commercial bismaleimide resins are imidazole, and DABCO (diazobicyclo-2,2,2-octane). The structure of DABCO is shown in Figure 5.

Anionic polymerization of bismaleimides can be achieved with butyl lithium at 40°C or with sodium t-butoxide at 20°C. Alkali tertiary butoxides have been used as anionic initiators at temperatures as low as $-72°C$ [5].

The maleimide bond is a reactive dienophile, and bismaleimides react with dienes to form ring systems by means of a Diels–Alder mechanism. Chain extension reactions can be carried out too, in which maleimide groups are reacted with allyl groups attached to aromatic rings, e.g. diallyl bisphenol A. Strong nucleophiles such as thiols and aromatic primary diamines undergo Michael addition reactions with bismaleimides, leading to crosslinking [6].

OTHER RESINS

Melamine–formaldehyde resins can be cured with substituted ammonium compounds. Silicone resins can be catalysed with cobalt octoate or naphthenate or with the corresponding salts of zinc, iron or lead. Polyurethanes can be cured with isocyanates.

REFERENCES

1. Chattha, M.S. and Dickie, R.A. (1990) *J. Appl. Polym. Sci.*, **40**, 411–416.
2. Anderson, T.F. and Messick, V.A. (1980), Chapter 2, in *Developments in Reinforced Plastics – Volume 1*, ed. G. Pritchard, Elsevier Applied Science Publishers, London.
3. Garnish, E.W. (1972) Chemistry and properties of epoxide resins, *Composites*, **3**, 104–111.
4. Scola, Daniel A. (1984) Chapter 5, in *Developments in Reinforced Plastics – Volume 4*, ed. G. Pritchard, Elsevier Applied Science Publishers, London.
5. Haguvara, Y. *et al.* (1990) *J. Polym. Sci.: Polym. Chem. Ed.*, **28**, 185–192.
6. Pritchard, G. and Swan, M. (1993) The crosslinking of eutectic mixtures of bismaleimides, *Eur. Polym. J.*, **29**(2/3), 357–363.

Keywords: free radical, initiator, accelerator, crosslinking, gelation, inhibition, peroxide, hydroperoxide, amine, anhydride, amide, Lewis acid, boron trifluoride, pot life, exotherm.

MANUFACTURERS AND SUPPLIERS OF CURING AGENTS

Akzo Nobel Chemicals BV
ARCO Chemical
Elf Atochem
Eastman Chemical
Anchor Chemical
Shell
Aldrich Chemical Company
Ciba Geigy
BP Chemical

Diluents and viscosity modifiers for epoxy resins

Salvatore J. Monte

WHY USE MODIFIERS?

Epoxy resins crosslink into thermosetting materials by reacting with various hardening agents such as amines, anhydrides, polyamides and catalytic curing agents. The desired properties in the ultimate finished products are obtained by selecting the appropriate combination of resin and hardener. Excellent chemical resistance, good electrical properties and toughness are common to nearly all epoxy resin systems.

Epoxy resin systems of the bisphenol A–epichlorohydrin type and epoxy novalac type generally lack flexibility. There are a number of proprietary flexible, low viscosity epoxy resins used to modify the above types to provide better impact resistance, elongation, or flexibility. These flexible epoxy resins are true epoxies and react completely with epoxy curing agents and becomes a permanent part of the cured system. However, they do not contribute towards lowering costs.

It is often necessary and desirable to alter an epoxy resin formulation for one or more reasons:

- to alter viscosity of the resin; increase the level of filler loading;
- to improve pot life and reduce exotherm;
- to improve certain physical properties such as impact and peel strength;
- to flexibilize; reduce surface tension; improve resin wetting action;
- to reduce the cost of the formulation.

Table 1 summarizes some diluents and viscosity modifiers for epoxy resins. They may be classified as:

Plastics Additives: An A–Z Reference
Edited by G. Pritchard
Published in 1998 by Chapman & Hall, London. ISBN 0 412 72720 X

Table 1 Viscosity and color of epoxy resins, diluents and viscosity modifiers

Description and features	Typical viscosity @ 25°C (cP)	Max. color (Gardner)
Resins		
Standard diglycidyl ether of BPA (bisphenol A)	11000–5000	1
Acrylate modifier epoxy – high reactivity with amines	85–115	2
Acrylate modified epoxy – higher reactivity with amines	800–1100	2
Acrylate modified epoxy – highest reactivity with amines	2700–3700	2
Diglycidyl ether of BPF	3000–4500	
Diluents and viscosity modifiers		
C_8–C_{10} Alkyl glycidyl ether – efficient viscosity reduction	3–5	1
C_{12}–C_{14} Alkyl glycidyl ether – efficient viscosity reduction	6–9	1
C_{12}–C_{13} Alkyl glycidyl ether - efficient viscosity reduction	6–9	1
Butyl glycidyl ether – efficient viscosity reduction	1–2	1
Cresyl glycidyl ether – low volatility	5–10	2
p-tert Butyl phenol glycidyl ether – low odor	20–30	1
2-Ethyl hexyl glycidyl ether – good viscosity reduction	2–4	1
1,4 Butanediol diglycidyl ether – superior difunctional viscosity reducer	13–18	1
Neopentyl glycol diglycidyl ether – low volatility	13–18	1
Cyclohexane dimethanol diglycidyl ether – good mechanical properties	55–75	1
Trimethylol ethane triglycidyl ether – high crosslink density	200–330	4
Trimethylol propane triglycidyl ether – high crosslink density	125–250	3
Dibromo neopentyl glycol diglycidyl ether – flame retarder	275–500	5
Aliphatic polyglycidyl ether – flexibilizer	200–320	1
Polyglycol diepoxide – flexibility and diluency	45–90	2
Polyglycidyl ether of castor oil – water resistant/flexibilizer	300–500	8
Glycidyl ester of neodecanoic acid – superior balance of properties	5–10	1
Cumyl phenyl acetate – reactive diluent – low toxicity	225–425	9
Alkyl naphthalenes – non-reactive diluent – low cost	5–50	
Alkyl phenanthrenes – non-reactive diluent – low cost	20–30	10
Tetra (2,2 diallyloxymethyl)butyl, di(ditridecyl)phosphito titanate – CA	40–60	2
Neopentyl (diallyl)oxy, tri(dioctyl)pyrophosphato titanate – coupling agent	3000–4500	6

Safety and handling notes

These products and the auxiliary materials normally combined with them are capable of producing adverse health effects ranging from minor skin irritation to serious systemic effects. Exposure to these materials should be minimized and avoided if feasible through the observance of proper precautions, use of appropriate engineering controls and proper personal protective clothing and equipment, and adherence to proper handling procedures. Each of these preventive measures depends upon responsible action by adequately informed persons referring to the Material Safety Data Sheet for each of the products to be used, stored or transported.

- reactive diluents;
- viscosity modifiers;
- plasticizers, extenders and non-reactive diluents;
- organometallic esters.

Reactive diluents

The most widely used reactive diluents are based on derivatives of glycidyl ether. To be effective, the diluent should react with the curing agent at almost the same rate as the resin, contribute substantial viscosity reduction at low concentrations, and be nonreactive with the resin under normal storage conditions.

Butyl glycidyl ether is most acceptable because maximum viscosity reduction is obtained with a minimum concentration (Figure 1). It contains reactive epoxide groups which react with the resin and is therefore

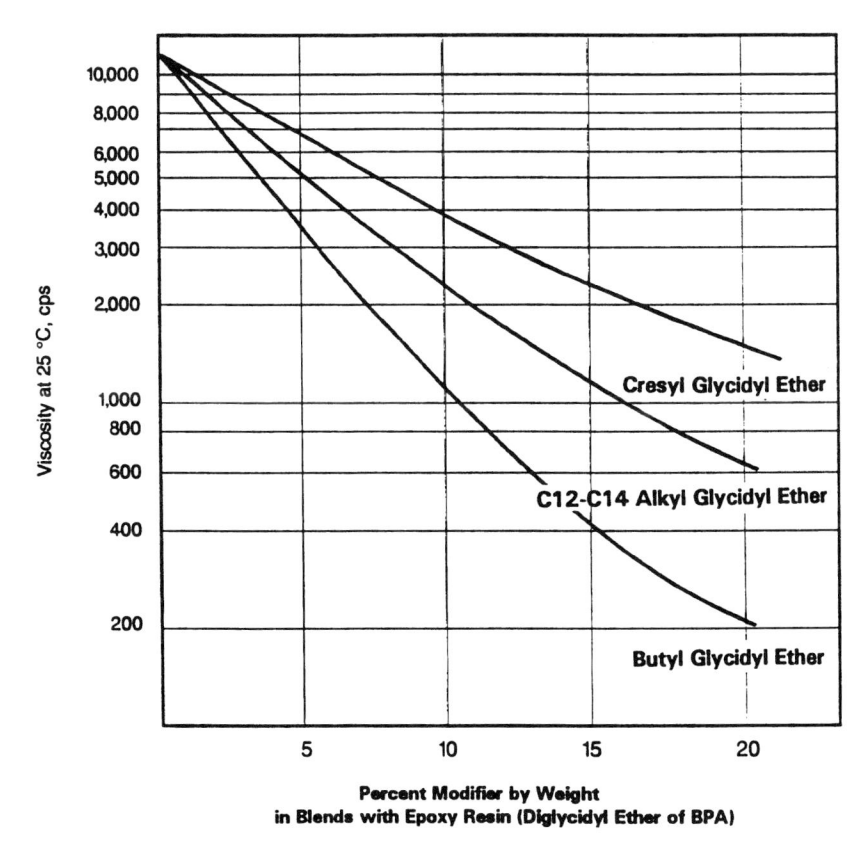

Figure 1 Viscosity reduction with various glycidyl ether derivative reactive diluents.

incorporated in the resin portion of the formulation. A number of proprietary epoxy resins are marketed with certain percentages of this reactive diluent to give a lower initial viscosity epoxy resin. The amount of curing agent used with such systems is calculated on the total epoxide equivalent of the blend. Reactive diluents generally decrease the properties of cured epoxy compounds.

Viscosity modifiers

These materials are used to improve thermal and mechanical shock, increase elongation, and obtain higher impact strength and flexibility. Usually, there is some sacrifice of physical strength, electrical properties and chemical or solvent resistance, and elevated temperature performance. Flexible epoxy resins or monofunctional epoxide compounds are examples of reactive epoxide-type modifiers. They can be used at ratios up to 1:1 to obtain a flexible and rubbery cured epoxy compound. They are shelf-stable when blended with the resin.

Modifiers which may be reactive as curing agents are often used. Among these triphenyl phosphite, liquid polysulfide polymers and various polyamides. High molecular weight aliphatic polyamines, which are also widely used, cure the resin slowly at room temperatures and must usually be heated to reduce their viscosity for easy blending with the epoxy.

The polysulfide polymers react slowly with the epoxies when used alone. One to three parts of an active catalytic amine, or amine salt are used to accelerate cure. Triphenyl phosphite reduces viscosity and somewhat reduces the ultimate cost of a compound. Although reactive with the epoxy, it is not effective as a curing agent by itself. A polyfunctional amine is necessary to effect a satisfactory cure.

Acrylate polymers and certain polyvinyl butyrals, silicone fluids, titanate esters and fluorocarbon compounds are used as flow control agents in powder coatings to modify the surface tension of the film in the melt stage, preventing crate formation and improving substrate wetting.

Plasticizers, extenders and non-reactive diluents

Materials which contain no reactive groups (such as epoxies) do not react with the resin and function primarily as diluents and plasticizers. The first requirement of such modifiers is that they be compatible to some extent with the resin before and after cure and not migrate excessively from the cured compound.

Material such as pine oil, dibutyl phthalate and glycol ethers have been used sparingly due to their adverse effects on the cured resin properties.

These materials are low viscosity, relatively low priced liquids which reduce the viscosity of the resin and improve castings. They permit higher filler content in epoxy resin compounds due to their wetting ability.

Monomeric plasticizers such as dioctyl phthalate and tricresyl phosphate, commonly used as plasticizers for vinyl resins and rubber, have poor compatibility with epoxy resins and are seldom used. They tend to separate out during cure and after storage and the small amount which is retained is of little value for improvement in flexibility.

The non-reactive diluents that have been used rather extensively are the phenolic pitches, which are high boiling extracts of cracked petroleum, and coal tar pitches which are hydrocarbon in nature. These materials are practically solid at room temperature and must be heated prior to use. They are black, and limited to applications where color and transparency are not factors. Their major use has been in road surfacing compounds since they can be used in high concentrations to reduce costs. The coal tars have some plasticizing effect upon the resin, such as improved flexibility, impact resistance, and low moisture absorption.

Some light, low boiling aromatic hydrocarbon solvents are used with epoxy resins to obtain fluidity. These aromatics are of assistance during the application of thin films. The solvent volatilizes during cure and has no effect on the ultimate finished properties of the compound.

The polarity or aromaticity of a plasticizer determines the degree of compatibility and, therefore, decides whether it will stay in the resin after cure and on storage. Unless there is some attraction between the resin molecules and the plasticizer, the latter will eventually work its way out and cause 'sweating' or exudation. On the other hand, if there is a strong affinity between the two, the plasticizer at use levels below 25 phr will stay in the resin matrix even under adverse conditions. A compatible plasticizer has high affinity for the epoxy resin and is a solvent for it. Although this is a physical phenomenon and no chemical reaction is involved, the chemical structure of the plasticizer determines how much attraction there is between the plasticizer and the resin.

Organometallic esters

Although used primarily as adhesion promoters, certain organometallic esters (titanates, zirconates, aluminates, zircoaluminates) and in some cases even silanes are used at catalytic levels in filled or reinforced epoxy resins to modify viscosity and achieve novel compositions.

ACKNOWLEDGEMENTS

Sam Sumner (Shell Chemical Co.) and Paul D. Sharpe (Kenrich Petrochemicals, Inc.) provided significant technical input.

Keywords: epoxy; reactive diluent; modifier, non-reactive diluent; coupling agent; flexibilizer; viscosity.

BIBLIOGRAPHY

Handbook of Epoxy Resins, Lee & Neville, McGraw Hill (1967).
Heloxy® Epoxy Functional Modifiers, Bulletin SC:1928-93, Shell Chemical Company (1993).
Formulating Powder Coatings with Epon® Resins, Bulletin SC:586-82, Shell Chemical Company (1982).
Monte, S. J. Ken-React® Reference Manual – Titanate, Zirconate and Aluminate Coupling Agents, Kenrich Petrochemicals, Inc., Summer 1993, 2nd Revised Edn.
Kenplast® G-Flexibilizer – Diluent for Epoxy Resins, Bulletin KPI 701, Kenrich Petrochemicals, Inc. (1970).

Dyes for the mass coloration of plastics

Lynn A. Bente

INTRODUCTION

In the few short years since 1856 when mauve was discovered by Perkin there has been a furious effort to develop commercially viable dyes for coloration of all media. Plastic resins ever since the invention of phenolics have been colored with both dyes and pigments. Dyes, as opposed to pigments, are soluble in or have an affinity for the media being colored. As dyes are solubilized in the polymer this molecular dispersion has the ability to develop much brighter and cleaner colors than pigments that derive their color from a crystal matrix. The trade-off for this brightness is reduced light and heat stability from pigments.

In recent years heavy metal pigments have been legislated out of use due to their supposed toxicity. Whether we agree or not with this concept, we no longer have them available for our use. This has necessitated an increase in the total number of colorants that can replace the high chroma colorants of lead and cadmium in specialty resins.

Dyes are classified and discussed by structure. To aid in this identification the AATCC and The Society of Dyers and Colorists have published the *Color Index*. This publication details dye classifications by structure, generic name and an identifying Constitution Number. Other information listed is solubility parameters, heat and light stability and chemical resistance. Recently, the larger dye manufacturers have chosen not to disclose a great deal of information concerning new dyes.

Plastics Additives: An A–Z Reference
Edited by G. Pritchard
Published in 1998 by Chapman & Hall, London. ISBN 0 412 72720 X

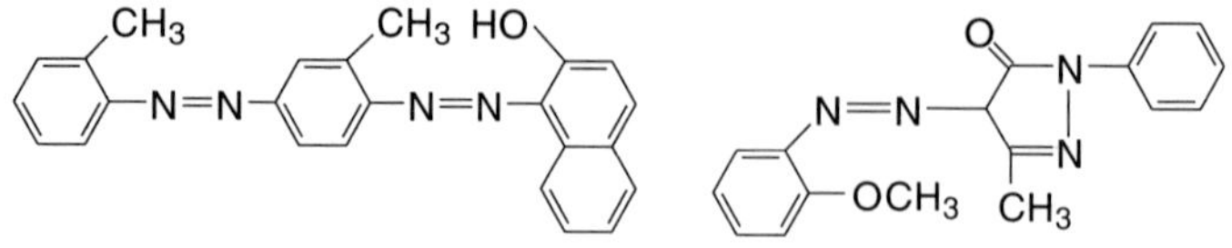

Figure 1 (a) Solvent Red 24; (b) Solvent Yellow 72.

AZOS

The general structure of a typical azo dye is shown in Figure 1. All azo dyes contain the bond $-N=N-$. One of the largest families of dyes, the azo dyes have found their way into many polymer applications. This is mainly due to their extreme brightness and low cost. Many of these dyes are staples of the color palette, especially for styrene, acrylics, some polyesters and minor ABS applications. Heat stability needs to be monitored but the light stability is acceptable for many low end applications.

The range of color of the azo family is normally yellow, orange and red with variations of each. Many of the violets and blue dyes have come out of production due to the greater heat and light stability of the violet and blue anthraquinone dyes.

In recent years, legislation in some countries has restricted the manufacture and use of azo dyes because of the possible toxicity of the intermediates and the potentially hazardous degradation products.

The most typically used azo dyes (Color Index Construction Number) are:

Solvent Yellow 14 (12055)	Solvent Red 1 (12150)	Solvent Black 3 (26150)
Solvent Yellow 16 (12700)	Solvent Red 23 (26100)	
Solvent Yellow 18 (12740)	Solvent Red 24 (26105)	
Solvent Yellow 72 (N.L.)	Solvent Red 26 (26120)	
Solvent Orange 7 (12140)	Disperse Red 1 (11110)	

As the melting points of these dyes are low (<180–200°C), processing temperatures need to be kept to a minimum and sublimation can occur in a number of them. Solubility of these dyes is very high and this allows them to disperse easily. Migration can occur at elevated loading, and use in plasticized resins is not suggested. Also they are not suggested for olefinic resins but azo and disazo dyes (i.e. Solvent Red 210) have been used in polypropylene holiday ribbon for many years.

Recently azo dyes have been attached to polymer backbones to make oligomeric type dyes. Problems with this type of colorant arise as you are adding as much of the polymer as you are of the dye.

More and more azo dyes are now produced in China and India. The purity of the dye needs to be monitored as to remaining insolubles such as intermediates or iron which can dull the color.

Metallized azo dyes are utilized for their brightness and clarity of color. Complexed with either chrome or cobalt these dyes have been used in styrene and acrylic. They have some enhanced heat and light stability over the standard azo dyes. Solvent Yellow 82 is a commonly used dye of this type.

Not many new azo dyes have become commercially available for plastics in the past few years. This may in part be due to the question of toxicity along with the inherent lack of stability of this type of molecule. There do exist a few higher stability azo dyes that can be well used in plastics, e.g. Solvent Red 195 and Disperse Yellow 241. Structures of these dyes are not disclosed.

ANTHRAQUINONE

Anthraquinone dyes have become the workhorses of the plastic colorant industry. From automobile taillight red to blue drinking cups these dyes are used heavily. The reason is quite simple. AQ dyes are relatively inexpensive, have very good solubility in aromatic resins, and good to excellent heat and light stability especially in the transparent modes. As compared with azo dyes they are less bright but their enhanced stability more than compensates for this. Another advantage with the AQ dyes is that a full range of colors, with similar solubilities and stabilities, can be developed by changing the pendants on the AQ molecule. A typical structure is shown in Figure 2.

AQ dyes are used for a broader spectrum of resins than that of the azo dyes. They are heavily used in styrene, ABS, SAN, polycarbonate, acrylics, cellulosics, polyesters etc. and are well known for excellent weatherability in the transparent mode such as red taillights, but in tint applications the light stability is greatly reduced. Heat, photo and chemical stability vary considerably as the pendants on the base AQ structure are substituted. Awareness of these differences is essential in color formulating. Most AQ dyes are not recommended for polyamide applications as these resins react with amine pendants and can remove the color. We are

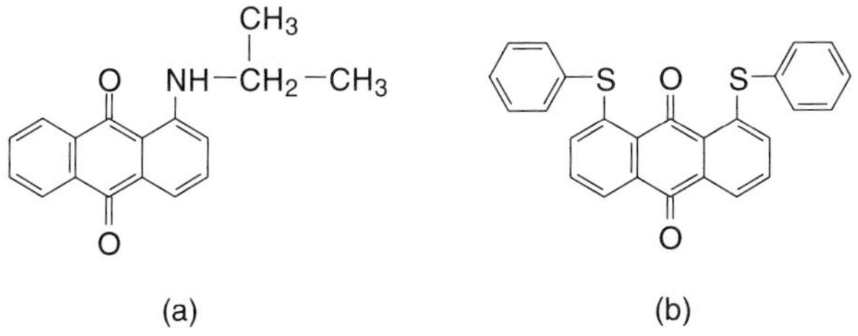

Figure 2 (a) Solvent Red 169; (b) Solvent Yellow 163.

also beginning to observe these restrictions in high butadiene ABS resins. It appears that the butadiene forms a highly reactive peroxide upon degradation and this then may react with particular chromophores.

A typical spectrum of anthraquinone dyes (Color Index Constitution Number) is

Solvent Yellow 163	(58840)	Solvent Violet 13	(60725)
Solvent Yellow 167	(N.L.)	Solvent Violet 14	(61705)
		Solvent Violet 36	(N.L.)
Solvent Red 111	(60505)	Disperse Violet 1	(61110)
Solvent Red 168	(N.L.)	Disperse Violet 26	(62025)
Solvent Red 169	(N.L.)		
Solvent Red 172	(N.L.)	Solvent Green 3	(61565)
Solvent Red 207	(N.L.)	Solvent Green 28	(N.L.)
Solvent Blue 35	(61554)	Solvent Blue 59	(61552)
Solvent Blue 36	(61551)	Solvent Blue 97	(N.L.)
Solvent Blue 58	(N.L.)	Solvent Blue 104	(N.L.)

(N.L. = not listed)

Many AQ dyes can be used in peroxide curing systems. Dyes such as Solvent Yellow 163 and 167, Red 168, 169, 172 and 207, and Solvent Violet 38 have all been found to be peroxide resistant. Solvent Red 111 has been used for taillight red for many years due to excellent solubility and the bright clean edge glow it produces in acrylic resins. Sublimation can be a problem though, especially in injection molded applications.

Solvent Yellow 163 and 167 are beginning to find applications in the coloration of polyamide 6,6. This non-reactive nature appears as a result of the replacement of the amine bridge with a sulfur atom. Solvent Red 168 is a bit bluer in shade than the Solvent Red 111 in acrylic, but has slightly increased heat stability and less sublimation problems. It also can be used in taillight applications.

QUINOPTHALONE

These are an abbreviated line of dyes with the kind of structures shown in Figure 3, commercially consisting of the following.

Dye	Constitution Number
Solvent Yellow 33	47000
Disperse Yellow 54 (Solvent Yellow 114)	47020
Disperse Yellow 64 (Solvent Yellow 105)	47023
Solvent Yellow 157	Not listed

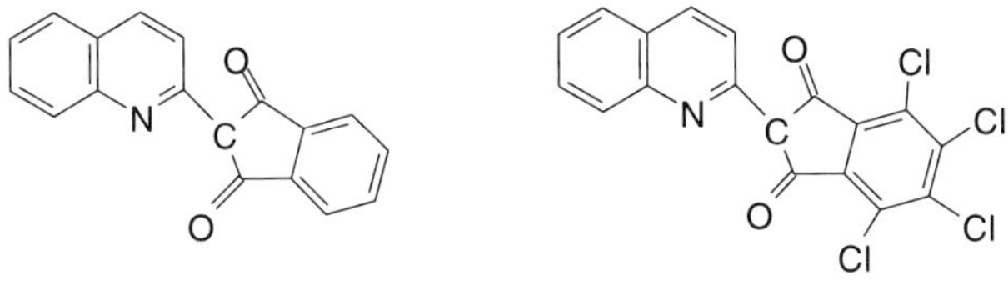

Figure 3 (a) Solvent Yellow 33; (b) Solvent Yellow 157.

The two main dyes are the Solvent Yellow 33 and Disperse Yellow 54. They are used in styrene, SAN, ABS and PET among others. The Solvent Yellow 33 is weak compared with other dyes but has good stability. Owing to the −OH group attached to the Disperse Yellow 54 this dye finds better stability and has become a workhorse dye for ABS and polycarbonate. Disperse Yellow 64 is a slightly redder shade and can be used in all of the same applications as the Disperse Yellow 54.

The Solvent Yellow 157, being the tetra chlorinated sister, has high heat stability with a melt point of 403°C. It is mostly used in Japan and is now being introduced into the U.S. The suggested resins would include polycarbonate, PBT, PET, PMMA and R-PVC.

PERINONE

This class is small in number compared with the others. Though commercially there are only a few available, they are some of the largest volume dyes used in the plastics industry today.

The most commonly used ones are as follows.

Dye	Constitution Number
Solvent Orange 60	Not listed
Solvent Red 135	Not listed
Solvent Red 179	Not listed
Solvent Red 180	Not listed

Perinones are used across the board in mass coloration of resin systems. Solvent Orange 60, Figure 4, is best known for application in amber taillight lenses of acrylic, styrene or polycarbonate. It has excellent heat stability (melting point 232°C) and good weatherability in the transparent applications. While not as bright as Disperse Orange 47, it is not as red a shade either and produces a nice mid shade orange. Solvent Red 135, the chlorinated sister to Orange 60, has similar stabilities and is used for the red taillight turn signals. It has a similar shade to Solvent Red 111, which

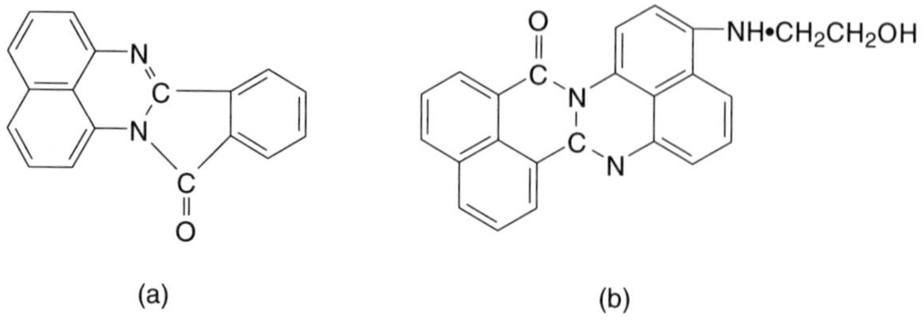

Figure 4 (a) Solvent Orange 60; (b) Solvent Red 180.

has historically been utilized for this application, but does not sublime when used in injection molding applications.

Solvent Red 179 and 180 are much less used but due to their good heat and light stability they find applications in a wide range of non-olefinic resins. Solvent Red 180 performs very well in flame retardant ABS.

Newly developed perinones are being offered for mass coloration of nylon. As these dyes are non-reactive to the polyamide carrier, they give good performance in these resin systems. Structures have not been disclosed. Perinones show excellent light and heat stability in most resins.

VAT DYES

Vat dye chemistry has been well known for many years. Vat dyes are high molecular weight structures that are well known for their light and heat stability. Owing to their increased size and weight many of these are also listed as pigments. This cross naming has been shown in the Color Index for many years (i.e. Vat Orange 7 is also known as Pigment Orange 43). They are not only bright colorants; many are also transparent. Bright vivid colors can be achieved with a selected palette of these colorants. As they are borderline dyes they can find applications in polyolefins and also aromatic resins such as ABS, SAN etc.

Vat dyes have been used for many years in textile applications. These colorants are not normally prepared in a manner that allows for incorporation into polymers. Special production methods must be accomplished to allow the reduction of diluents and also the control of particle size.

Vat dyes are not distinguished by any one molecular substructure. They stem from a wide variety of useful structures, as shown in Figure 5.

- Vat Blue 4 (Pigment Blue 60) This can be used as a red shade non-copper phthalocyanine blue replacement. It has excellent heat and light stability with a melting point in excess of 300°C. It can be used in ABS, SAN, PMMA and olefinic applications where warping can be a problem.

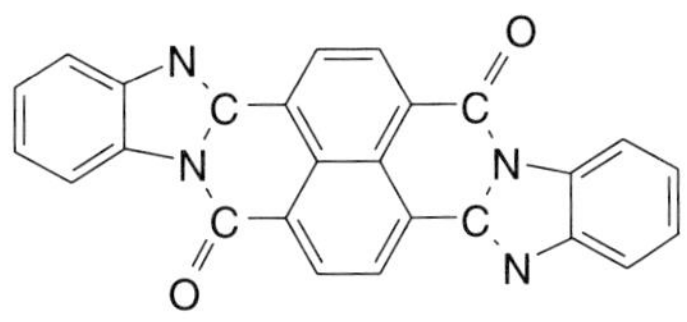

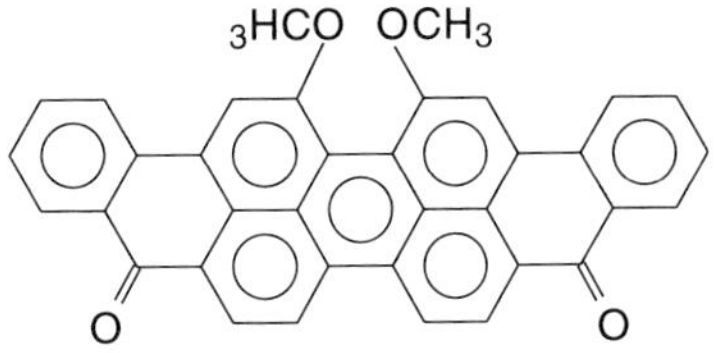

Methine and polymethine 223

Figure 5 (a) Vat Orange 7; (b) Vat Green 1.

- Vat Blue 6 (Pigment Blue 64) This chlorinated sister of Vat Blue 4 is greener in shade with good heat and light stability. It is used in olefinic applications where nucleation can be a problem. Melting point is >300°C.
- Vat Violet 1 (Pigment Violet 31) This is so bright that it is used as a fluorescent dye in polystyrene, SAN, and ABS. It is not suggested for olefinic formulations due to migration problems.
- Vat Orange 3 (Pigment Red 168) This is bright orange but not as bright as Vat Orange 7. It has good transparency and low migration tendencies.
- Vat Orange 7 (Pigment Orange 43) This has excellent brightness along with excellent heat and light stability. With a melting point of 460°C it finds use in a wide variety of resins including ABS.
- Vat Yellow 2 This is bright mid shade yellow with good transparency. It can be used in transparent olefinic applications.

SULFUR DYES

These are not normally thought of for use in resin coloration. Sulfur dyes have been used for years in textile applications due to good heat and light stability. They are normally thought of as dull colors, but selected grades of good purity have medium brightness with excellent stability. Unusual for the IR absorption bands, these dyes are now being studied for entry into the plastics markets. Sulfur Black 1 could probably be the first entry. It is believed that it is similar to the structure in Figure 6.

METHINE AND POLYMETHINE

This is another abbreviated family of colorants. Typical commercial examples of these dyes would be the Solvent Yellow 133, and Disperse Orange 47. They are known more for their brightness although they can exhibit good light stability. Disperse Orange 47 is an extremely bright red shade yellow used in many resins such as ABS, SAN, PET and polycarbonate. Light stability is borderline in ABS and this dye should be

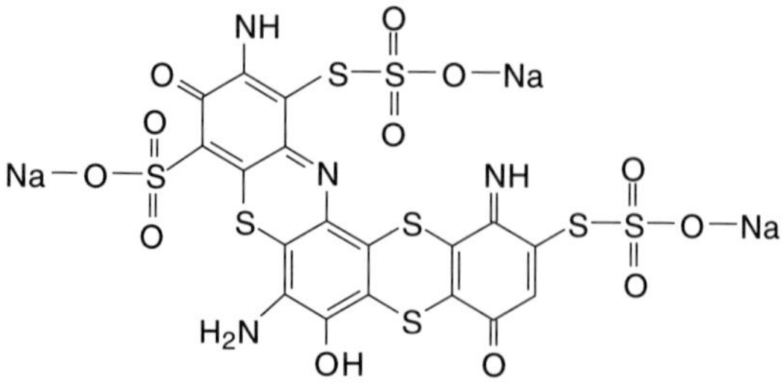

Figure 6 Sulfur Black 1.

avoided in outdoor applications. Solvent Yellow 133 is a green shade yellow, used in polyester resins.

AZINE

They are used in resin applications where jet black and transparent smoke colors are essential. Employed in phenolics, polyamides, olefinics, ABS etc, they exhibit good heat and light stability. Melt points are in excess of 180°C. Known as Solvent Black 5 and Solvent Black 7, these macromolecules have found use due to good stability and economy of price. Bluish black hues developed with these dyes show relatively good jetness in mass tones and transparency. They can also be used in olefinics and polyamides. The heat stability of both dyes is excellent.

Solvent Yellow 130 is known as an extremely bright yellow with good heat stability. It is used in resins ranging from styrene to ABS, PET, polyamides and has some minor uses in polycarbonate.

BENZODIFURANONES

This relatively new chemistry is being investigated due to excellent brightness along with good solubility in ABS and other aromatic resins. Colors sweep from orange to deep reds. Disperse Red 356 would be a prime illustration of this type of dye. Not much is known of the stability parameters in all resins. They are currently under investigation as toners for those difficult scarlet hues in ABS and polycarbonate. A typical structure is shown in Figure 7.

PHOTOCHROMIC DYES

Photochromic dyes, dyes that change from colorless or nearly colorless to bright hues on exposure to UV radiation, have entered the resin markets over the past three to four years. Conceptually invented for the ophthalmic lens market, they have found use in coatings and resin applications.

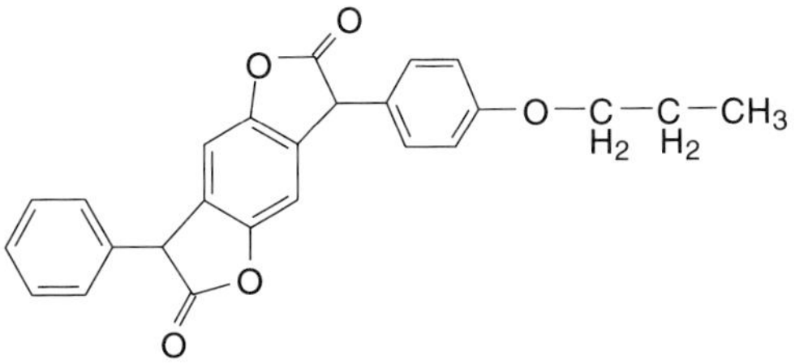

Figure 7 Disperse Red 356.

Heat stability has been improved and applications in olefinics, urethanes, styrene and some aromatic resins have been accomplished for inks, toys and novelty items. Light stability can be enhanced with specialty stabilizers such as hindered phenols. Color is developed due to a change in the extended conjugation and the charge separation within the molecule. With acids or very polar media the color change can become permanent.

The total amount of UV radiation, together with the effects of any unreacted intermediates, singlet oxygen, other radicals and acids can cause fatigue of the molecule. Commercial applications could include ophthalmic lenses, agricultural films, UV source detection, UV curing system testing, and security tracers.

BIBLIOGRAPHY

Christie, R.M. (1994) Pigments, dyes and fluorescent brightening agents for plastics: an overview, *Polym. Int.*, **34**(4), 351–361.

Rosen, R. (1994) Colours for a brighter future, *Polym. Paint Col. J.*, **184**(4353), 344–348.

Keywords: anthraquinone dyes, azine dyes, azo dyes, benzodifuranone dyes, color index, methine dyes, quinopthalone dyes, perinone dyes, photochromic dyes, sulfur dyes, vat dyes.

See also: Pigments for plastics.

Fibers: the effects of short glass fibers on the mechanical properties of thermoplastics

Harutun G. Karian, Hidetomo Imajo and Robert W. Smearing

INTRODUCTION

Glass fibers are added to plastics when it is necessary to improve their mechanical properties. The fibers may be short, or long. This entry is about short fibers exclusively. Short fibers may be used to improve both thermosetting resins and thermoplastics. This entry does not discuss thermosetting resins.

It should be realized that it is quite difficult to mix fibers with thermoplastics because of the exceptionally high viscosity of most thermoplastics, even at elevated temperatures. Furthermore, most thermoplastics have to be processed in machinery such as injection molding, extrusion or similar equipment. This presents considerable problems when glass fibers are involved, because of the effects of the machinery on the fibers. Fiber length is much reduced.

Another essential concept to grasp at the outset is that glass fibers have to be coated with special surface treatments. Sometimes this coating is loosely described as a size. The details will be discussed in the following sections.

For over 35 years, commercial glass fiber manufacture has exhibited continuous growth in volume to meet increasing demands in the market place for short glass fiber reinforced (SGFR) composites in many

Plastics Additives: An A–Z Reference
Edited by G. Pritchard
Published in 1998 by Chapman & Hall, London. ISBN 0 412 72720 X

different areas of end use, e.g., automotive, aviation, home appliance and a wide variety of consumer goods.

SGFR composites of thermoplastics are rapidly replacing metals in high performance demanding applications. The incentives to replace metals by plastics are multifold in purpose. Obviously, the cost per unit volume is greatly reduced by using cheaper and less dense thermoplastic resin. This enhancement of mechanical properties provides cost-effective options in the replacement of metals for many under-the-hood automotive applications [1] under extreme environmental conditions: ambient temperatures ranging from −40°C to 150°C, sunlight and moisture.

GENERAL FEATURES OF COST-EFFECTIVE COMPOSITES

The primary task of the glass fiber manufacturer is to design the fiber glass sizing chemistry to suit the polymer type used. The surface of glass fiber filaments is treated by a recipe containing (1) a coupling agent and (2) a film former. The coupling agent is generally an organosilane to promote bonding at the glass fiber–polymer interface. The film former acts like glue by holding the filaments together and producing chopped strands that have good integrity for metering to the feed port of the compounding extruder. The protective coating of film former prevents friability or fuzz generation in any handling equipment upstream from the compounding extruder.

The compounding extrusion and injection molding processes feature the same theme: efficient distributive mixing of feed ingredients (polymer resin, stabilizers, modifiers, mineral fillers and SGFR) into a homogeneous molten blend with minimum dispersive mixing. In the mixing section of the extruder, it is necessary to exert a degree of shear stress on the chopped glass fiber strand to overcome the adhesive forces that hold individual filaments together. This provides the necessary contact surface for efficient wetout by the polymeric melt containing chemical coupling additives.

Enhanced ultimate strength, rigidity and fracture toughness [1] are dependent on optimum interphase design coupled to minimum attrition of glass filament lengths during a sequence of process steps. Here we will describe the physicochemical nature of the interphase region between short glass fibers and the molecules of the polymer matrix. Interphase design is identified as the central gem affecting SGFR mechanical properties over the service lifetime of a molded part.

GLASS FIBER VARIABLES

Ming-Liang Shiao [2] listed three key glass fiber variables that influence ultimate mechanical properties: fiber diameter, average fiber length and fiber glass content.

E glass is used almost exclusively by all glass fiber manufacturers today. Chopped glass fiber strands consist of 800–3000 separate filaments. Upon a degree of attrition in the feed system and/or extrusion step, the strand bundles open up into separate filaments of glass fiber to provide the entire fiber surface for wetout by the molten polymer. Each filament has a diameter between 9 and 25 μm. The effect of glass fiber diameter on mechanical properties depends on the particular polymer type and sizing package.

Commercially, the diameter of glass fibers for polyamide resins is generally 9–10 μm. Shiao [2] observed an increased yield stress of SGFR nylon 6,6 for this diameter compared with 13 μm. Fiber glass types, having a 13–14 μm diameter, are generally sized so as to be compatible with chemically coupled polypropylene. However, both tensile strength and fracture toughness for SGFR polypropylene are improved by using the much smaller 9–10 μm diameter fiber. Conversely, there is information in the public domain that tensile and flexural strength are adversely affected by 17 μm size fibers.

After a sequence of compounding and injection molding processes, the fiber length is reduced from 3–5 mm to less than 1 mm for an average length depending on cumulative fiber damage. Shiao [2] and Ramani et al. [3] provide supporting evidence for a model of maximum volume packing of randomly oriented stiff rods to represent the effects of fiber–fiber interactions. This leads to the following inverse relationship between fiber glass content and length:

$$v_f = (constant) d_f / L_{ave} \qquad (1)$$

where d_f is fiber diameter and L_{ave} is the average fiber length for a given fiber volume fraction v_f. Shear forces act on the brittle fiber ends to cause a certain amount of fracture depending on v_f. Obviously, this behavior limits the effective length of glass fibers to a minimum amount dictated by this fundamental relationship. This is an unavoidable consequence of mixing processes, once the chopped fiber strand is dispersed into individual filaments. The objective is to prevent any further breakage of fiber filaments beyond this amount.

Shiao [2] investigated the brittle–ductile transition of nylon 6,6 composites as a function of fiber glass content. At low levels of fiber glass (0–5 wt%), the stress concentration at fiber ends is quite high. The inherent brittleness is due to the relatively small fiber glass diameter. With little overlapping of stress fields around glass fibers in a dilute medium, the notched Izod values are quite small. At higher loadings of glass fibers, the composites become more ductile and exhibit corresponding increased notched impact values. By decreasing the mean fiber end spacing below a critical threshold value (six times the fiber diameter), the stress fields around individual fibers overlap strongly to modify the deformation characteristics of the polymer matrix. Therefore, a matrix toughening mechanism results from plasticity around glass fibers, not unlike rubber particle incorporation.

In comparison with SGFR polyamide resins, Shiao described different trends of impact behavior with increase in fiber glass content for other thermoplastic materials. Semi-crystalline acetal exhibited a corresponding decrease in notched impact. Unmodified amorphous resin based materials exhibit an opposite trend. Relatively tough polycarbonate and the much more brittle styrenic resins (polystyrene and styrene-acrylonitrile) exhibit an increase in notched impact with increase in fiber glass content. However, rubber modified ABS composites have a lower impact behavior at increased fiber glass content.

SURFACE TREATMENT OF FIBERS

Glass fiber manufacturers introduce a sizing package [4,5] in aqueous medium which generally consists of the following key ingredients:

1. organo-silane as coupling agent to promote fiber-matrix adhesion: 0.1–0.5 wt%;
2. film former to keep fiber strand from being friable during the extrusion feed step: 1–5 wt%;
3. lubricant as process aid: 0.1–0.2 wt% fatty acid amide or polyethylene glycol;
4. Antistatic agent: 0.1–0.2 wt% ammonium chloride or quaternary ammonium salts;
5. Wetting agents and other specific property modifiers.

Garrett [4] indicated that about 90% of the coupling agents on glass fibers consisted of one of the three types of silanes:

1. γ-aminopropyl triethoxy silane: polypropylene and polyamide;
2. γ-glycidoxypropyl silane: acrylics, polysulfides, polypropylene;
3. methacryloxy silane: polyethylene, polypropylene, polyester and styrenics.

Garrett narrows the possible list of film formers to polyvinyl acetate (PVA), urethane, epoxies or polyester. Because of the need for dilute aqueous solution coating of glass fibers, he points out that there are severe constraints on emulsion particle sizes for proper selection of effective film formers.

COMPATIBILIZER BRIDGE: ADDITION OF CHEMICAL COUPLING PROMOTER

The glass fiber reinforcement of nylon 6,6 exemplifies a typical polar thermoplastic. Nylon end groups interact strongly with the aminopropyl silane to generate a three-dimensional network of H-bonds. In such cases, there is no need for any bridging groups to be introduced into the composite to facilitate adhesion between surface treated fiber glass and the polymer matrix.

Polyolefin resins are non-polar in nature. Hence, SGFR composites require a combination of suitable glass fiber sizing recipes and added chemical coupling promoter to enhance stress transfer necessary for strength and rigidity. This promoter is bifunctional in nature with grafted monomer on the polymer backbone. A number of monomers are typically used to functionalize polyolefin resins, e.g. acrylic acid, maleic anhydride, fumaric acid and himic anhydride.

Maleic anhydride grafted polypropylene, MA-g-PP, type compatibilizers constitute the majority of chemically coupled composites produced today. An entire list of available maleic anhydride grafted polypropylene additives supplied globally is given in Table 1. The particular choice of a compatibilizer and level is made generally on the basis of cost-effectiveness.

Figure 1 provides a schematic diagram of chemical coupling in SGFR polypropylene composites. This sequence of inter-connected structures is dependent on a strong compatibilizer bridge formed between the polymer matrix and surface treated glass fibers. In this conceptual model, added MA-g-PP molecules form the framework of this bridge as chemical coupling promoter.

Scanning electron micrographs (SEM) of cryogenically fractured surfaces can provide a vivid portrayal of interphase design differences. Figures 2 and 3 depict the degrees of chemical coupling from low to high levels, respectively. Note the much smoother fiber surfaces, characteristic of poor adhesion, in Figure 2.

CHARACTERIZATION OF FIBER DAMAGE DUE TO PROCESSING

In order to minimize short fiber damage and maximize thermoplastic composite properties, one must optimize the compounding and injection molding processes. Single screw and corotating-intermeshing twin-screw extruders are two types of compounding equipment used to mix a variety of ingredients with the thermoplastics.

In compounding SGFR thermoplastic composites, twin-screw extruders have a definite advantage over single-screw extruders. The higher initial cost of investment for a twin-screw extruder is offset by its capability of attaining higher throughput rates for the same bore diameter. The fiber glass can be metered downstream to a side feed port. This option provides a high degree of distributive mixing for an intimate coating of the individual glass filament surface by molten

Table 1 Commercial grades of maleic anhydride grafted polypropylene

Manufacturer	Product name	Form	Total grafted monomer content (wt%)	Melt flow rate (g/10 min) @ 230°C–2.16 kg [ASTM D-1238]	Molecular weight, $\bar{M}_w$
Uniroyal	PB 3001	Pellet	0.075	5	320 000
	PB 3002	Pellet	0.15	7	261 000
	Pb 3150	Pellet	0.40	50	139 000
	PB 3200	Pellet	0.80	90–120	110 000
Nippon Hydrazine	FTH 100	Powder	3.0	High	
Hercules	Hercuprime G	Powder	3.9	High	39 000
Hoechst Celanese	Hostaprime HC5	Powder	4.0	High	30 000
Du Pont	Fusabond P:				
	MD-9508	Pellets	0.1	30	160 000
	MZ-135D	Pellets	0.1	12	200 000
Eastman Kodak	P-1824-003	Pellets	0.2	13	200 000
Exxon	P0X1-1015	Pellets	0.3		
Elf Atochem NA	Orevac CA 100	Pellets	1.1	High	
Honan Petrochemical	PH-200	Powder	4.2	High	41 000
Sanyo Chemical	Youmex 1001	Pellets	4.0	High	15 000
	Youmex 1010	Pellets	9.6	High	4 000
Aristech	Unite MP320	Pellets	0.2	22	180 000
	Unite MP620	Pellets	0.4	34	150 000
	Unite MP800	Pellets	1.0	200	100 000

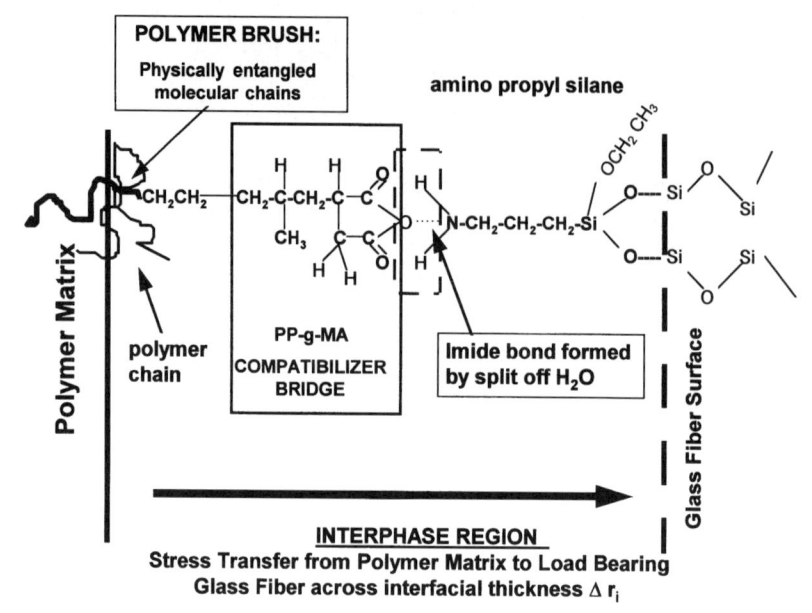

Figure 1 Chemical coupling reinforcement.

polymer. In spite of high shear rates, the residence times for such twin-screw extrusion processes are quite short in duration, e.g. 10–30 seconds. The degree of dispersive mixing is correlated with shear strain, which is the product of shear rate and residence time:

$$\text{degree of mixing} = \text{shear strain} = \text{shear rate} \times \text{residence time} \quad (2)$$

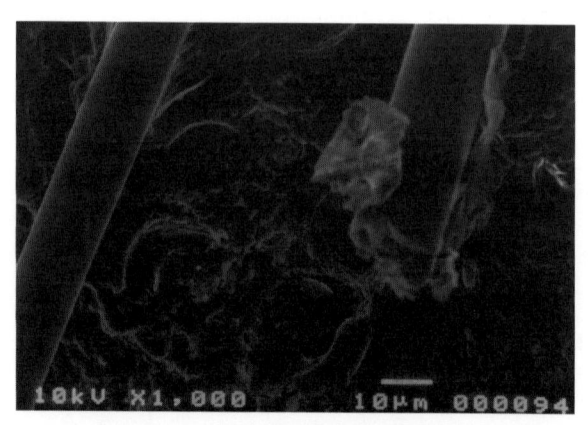

Figure 2 Low chemical coupling level.

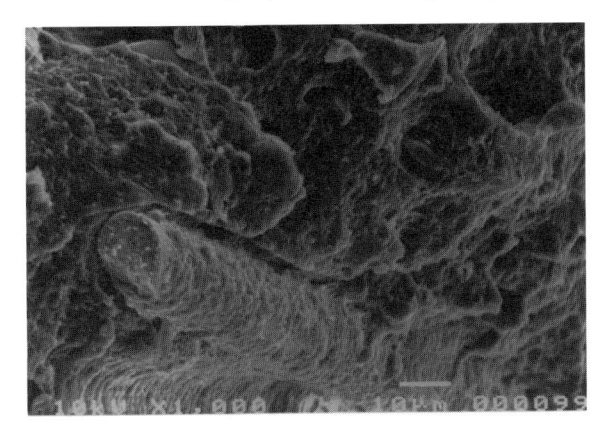

Figure 3 High chemical coupling level. SEM images were obtained courtesy of Mr Jerry R. Helms, Thermofil, Inc.

Consequently, high screw speed is offset by low residence time to provide an optimum degree of mixing without incurring severe fiber damage.

Ramani *et al.* [3] studied twin-screw compounding of SGFR syndiotactic polystyrene, a semi-crystalline thermoplastic resin. Three different mixing arrays featured combinations of gear tooth elements and reverse screw bushings to regulate the intensity of mixing. They were able to characterize fiber damage as a cumulative function of localized distributive mixing and wetout of glass fibers along incremental slices of axial molten flow, i.e. along the combined compounding and injection molding streamlines. A comparison of results between 30 and 40 wt% fiber glass content indicated greater ultimate damage for the higher level reinforcement.

TRENDS IN MECHANICAL PROPERTIES DUE TO FIBER GLASS CONTENT

Table 2 summarizes recent studies of SGFR nylon 6,6 composites as a function of fiber glass content. Internal data were obtained from materials compounded via a 63 mm single screw extruder. The grade of fiber glass used was a typical commercial type used for polyamides (9.7 μm diameter and 4.7 mm chopped strand length).

The trends in mechanical properties are consistent with Shiao's results [2].

1. Tensile strength versus fiber glass content is linear.
2. Impact behavior increases with increased fiber–fiber interactions due to higher v_f.
3. There is a decrease in fiber length from 4.7 mm to 0.2–0.4 mm.

Table 2 Summary of mechanical properties for SGFR nylon 6,6 composites

Fiber glass content (wt %)	v_f	Tensile strength (MPa)	Notched Izod impact (J/m)	Flexural strength (MPa)	Flexural modulus (GPa)	HDT @ 1.82 MPa (°C)	Average fiber length, L_{ave} (mm)	$L_{ave}v_f/d_f$
10	0.048	103.9	37.3	169.1	4.08	241	0.436	0.216
20	0.101	139.9	62.9	222.6	5.96	248	0.337	0.351
30	0.181	176.2	116.5	270.6	8.72	252	0.261	0.487
40	0.230	198.1	145.1	305.4	10.73	253	0.178	0.422
50	0.309	236.0	194.4	366.3	13.49	254	0.208	0.663

4. The average fiber length is inversely related to glass fiber volume fraction.

Recently, Yu et al. [6] provided a comprehensive review of existing computational models to analyze trends in mechanical properties, e.g., yield stress σ_c or tensile strength. The following analytical expression for the modified rule of mixture model is most commonly used to relate stress values to contemporary concepts of glass fiber structure and polymer property relationships:

$$\sigma_c = v_f \sigma_f [1 - L_c/2L_{ave}] C_o + v_m \sigma_m \tag{3}$$

The fundamental problem with this type of model is the presence of so many unknown empirical quantities that need to be evaluated for a given glass fiber loading: L_c (critical fiber length), L_{ave} (average fiber length), effective fiber strength σ_f, and C_o (fiber orientation factor). v_m is the matrix volume fraction.

The yield strength of the polymer matrix σ_m can be obtained from published data. The orientation factor C_o values range from 0 (completely random fibers) to 1 (completely oriented). Generally, the latter quantity is not actually measured but used as a 'fitting' parameter to equation (3).

The critical fiber length determination is coupled to the popularly known Kelly–Tyson expression (4) based on a shear-lag theory for metal filaments:

$$L_c = d_f \sigma_f / 2\tau \tag{4}$$

where τ is the interfacial shear strength of the base polymer. However, due to a gage length constraint [5] on strength value for SGFR, the effective σ_f value approximates to about half of the value (3.4 GPa) associated with continuous fibers. Generally, one needs to make a guess of either orientation parameter or effective fiber strength in order to calculate the other parameter. In retrospect, Garrett [4] makes a critique of the inability of theoretical models to predict mechanical behavior a priori. He concludes that theoretical tools for material development are lagging seriously behind the practical 'art' of developing better sizing chemistry.

The tensile strength versus fiber glass content relationship for polypropylene based composites is dependent on the efficacy of the compatibilizer bridge between surface treated glass fibers and the polymer matrix. Figure 4 depicts a series of plots for an array of materials which encompasses a spectrum of chemical coupling:

1. unmodified polypropylene resin (P);
2. low level of functionalized polypropylene (P1);

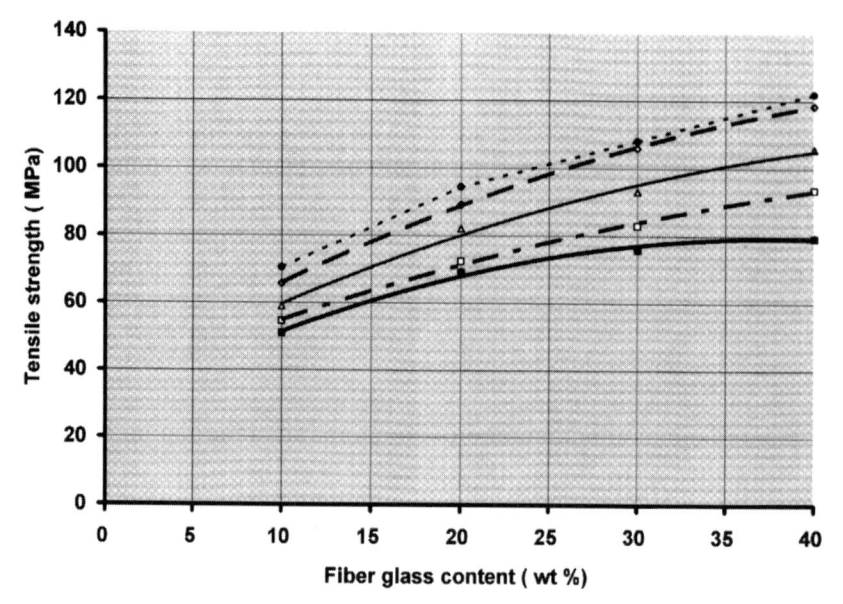

Figure 4 Tensile strength versus fiber glass level for chemical coupling spectrum: ■, P; □, P1; △, P6; ◇, P7; –●–, perfect coupling.

3. high level of functionalized polypropylene (P6);
4. mega coupled polypropylene (P7);
5. perfect coupling limit defined from calorimetric data [7].

The functionalized polypropylene can be any one of the grafted monomers described in Table 1. The degree of chemical coupling is dependent on promoter type and proprietary usage levels.

Flexural fatigue behavior described by S–N type plots can be used to characterize the endurance of SGFR composites subjected to oscillating stress loads S. The fracture of a rotating cooling fan in an automobile is a good example of this type of failure mode. A simple way to envision such a failure mode, is to take a flexural bar and bend it back and forth a number of times or cycles (N) until it breaks. Using the cantilever beam method at an oscillation frequency of 30 Hz, a typical bending stress (S) versus cycle-to-failure (N) plot exhibits an asymptotic convergence to a constant stress value plateau after a culmination of repeated oscillations. The stress limit at greater than 10^6 cycles to failure approximates to infinite service lifetime.

In Table 3, a comparison is made between two levels of chemical coupling for flexural strength, modulus and fatigue data at 20 and 40 wt% fiber glass levels. The mega coupled SGFR polypropylene (P7)

Table 3 Flexural behavior of chemically coupled PP

Fiber glass loading (wt %)	Chemical coupling level	Flexural strength (MPa)	Flexural modulus (GPa)	Flexural fatigue: stress limit @ infinite life
20	P6	102	3.42	36.1
20	P7	131	4.52	44.9
40	P6	139	7.44	46.7
40	P7	184	8.74	56.3

exhibits exceptional flexural behavior by measured values being greater than P6 chemically coupled composites.

If the applied load is held constant, the corresponding long term flexural behavior can also be characterized by flexural creep measurements. Trends in creep rupture times correlate with the results for fatigue stress limits. Tensile strength, creep and fatigue measurements gage the relative effectiveness of chemical coupling on stress transfer to loading bearing glass fibers. The tensile behavior due to P7 chemical coupling is much better than P6.

At elevated temperatures, the role of glass fiber reinforcement becomes a crucial factor for deciding what base material to use. Before the advent of more efficient chemically coupling combinations, there was a wide performance gap between SGFR composites based on polyamide and polypropylene resins. Obviously, that gap has shrunk significantly by improved interphase design for chemically coupled polypropylene materials.

Instead of upgrading SGFR polypropylene to suit engineering resin performance, another strategy is to reduce the cost of SGFR polyamide by incorporating a portion of less expensive polypropylene resins. Hence, there has been a major effort to develop compatibilized blends of these two resins at various composition ratios. The particular choice of fiber glass type to use has been shown to depend on the polyamide–polypropylene ratio. Perwuelz et al. [8] found that the choice of fiber glass type (polypropylene or nylon 6,6) had a crucial effect on the heat distortion temperature. They conclude 'the glass fibers were encased in the polymer for which their sizing was designed'.

Santrach [1] observed that HDT values were dependent on the amount and types of film former and lubricant. Fiber glass that has been sized for polyamide gives HDT values of about 240–252°C for nylon 6,6 levels above 50% in the polymer phase of compatibilized blends with polypropylene. However, in polypropylene type fiber glass, the HDT is only about 150–155°C. These observations are consistent with Santrach's comments about the addition of glass fibers to crystalline resins giving

HDT values near the melting point of the given resin. On the other hand, he cites evidence that amorphous resins exhibit an HDT that corresponds to the glass transition temperature instead.

ROLE OF INTERPHASE DESIGN TO ENHANCE MECHANICAL PROPERTIES

A differential scanning calorimeter (DSC) was used to quantify the degree of chemical coupling by a method [6] derived from the following fundamental concepts concerning polymer mobility in the interphase region surrounding the glass fibers.

Liptov's model

Heat capacity jump ΔC_p (composite) at the glass transition temperature T_g relates to interfacial thickness Δr_i of the interphase region surrounding the glass fiber. The abrupt rise in heat capacity value is greatly reduced by a portion of polymer molecules that are immobilized by inter-molecular entanglements. An interaction parameter λ is obtained from DSC measurements of $\Delta C_p (T_g)$ for the composite and unfilled based polymer materials. The calculated λ values range between 0 and 1 at the lower and upper limits of chemical coupling, respectively:

$$\lambda = 1 - (\Delta C_p^{composite} / \Delta C_p^{polymer}) \tag{5}$$

This determination is followed by a calculation of the interfacial thickness for a given chemical coupling level:

$$\Delta r_i = d_f \{[1 + (\lambda v_f)/(1 - v_f)]^{1/2} - 1\}/2 \tag{6}$$

using values for fiber diameter d_f and fiber glass volume fraction v_f.

Polymeric brush model

Depending on the molecular weight of the grafted polymer backbone, there is a certain amount of entanglement with molecules of the polymer matrix. Since this intermolecular arrangement is like hair tangled up in a hair-brush, the structure is referred to as being a polymer brush as depicted in Figure 1.

Tensile strength–interfacial thickness correlation

Thermomechanical properties were determined for a vast array of 10–40 wt% SGFR composites, which encompass a wide spectrum of chemical coupling. The observed data coalesced into a single master plot shown in

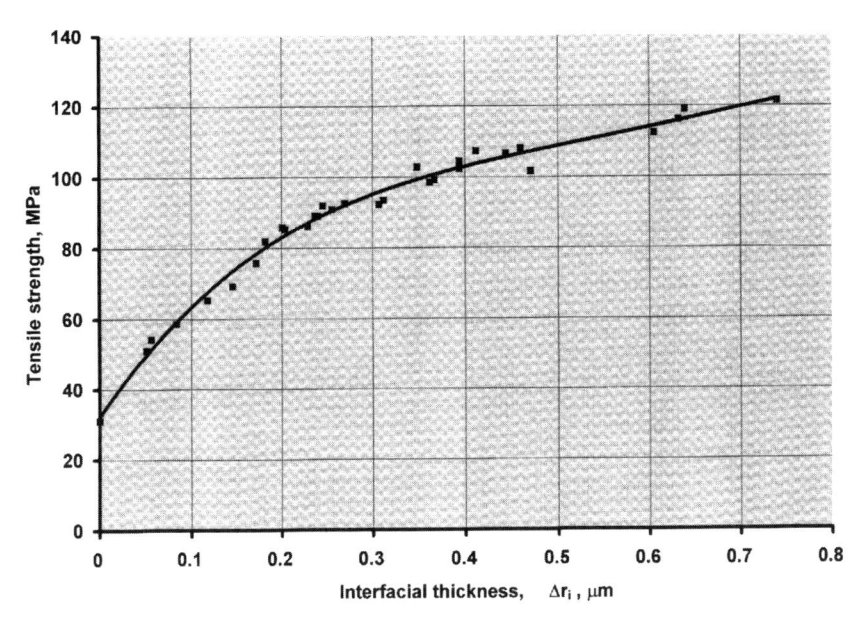

Role of interphase design to enhance mechanical properties   239

Figure 5 Correlation between tensile strength and interfacial thickness.

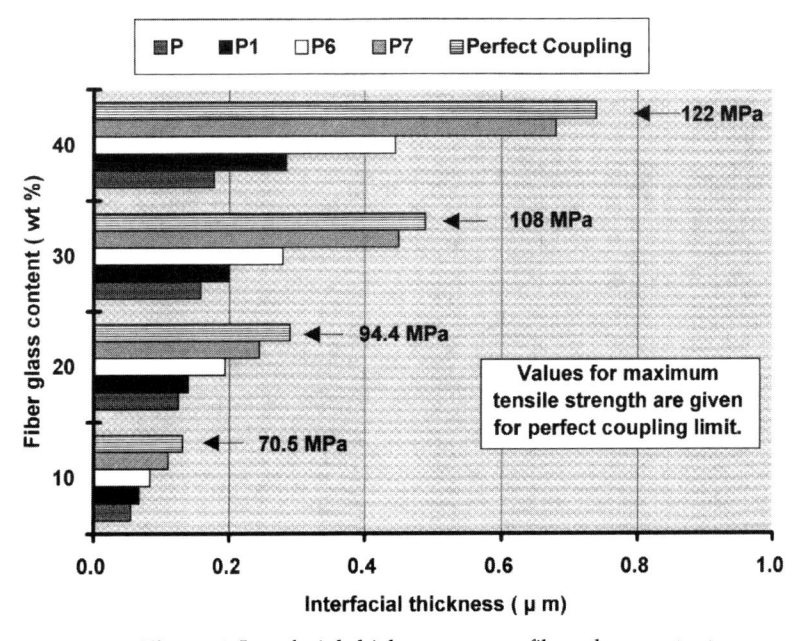

Figure 6 Interfacial thickness versus fiber glass content.

Figure 5. Hence, there is a strong inference that tensile strength is quantitatively dependent on the thickness of the interphase region.

Perfect coupling limit

By assuming that the entire polymer matrix forms a three-dimensional network via the compatibilizer bridge–entanglement combination, the corresponding maximum interfacial thickness at perfect coupling corresponds to setting $\lambda = 1$ in equation (6). By using Figure 5, we can make an estimate of the maximum attainable tensile strength value at a given fiber glass loading.

An estimate of tensile strength at the perfect coupling limit provides a useful bench mark in developing more cost-effective short fiber reinforced thermoplastics. Figure 6 provides a series of bar graphs of interfacial thickness at given fiber glass loadings for a series of composite materials representing the entire spectrum of chemical coupling.

REFERENCES

1. Santrach, D. (1982) Industrial Application and Properties of Short Glass Fiber-Reinforced Plastics, Polymer Composites, 3(4), 239–43.
2. Shiao, Ming-Liang (1993) The Role of Matrices and Glass Fibers on the Deformation and Fracture of Short Glass Fiber Reinforced Thermoplastics, Ph.D. Dissertation Thesis, University of Massachusetts, Mass., USA.
3. Ramani, K., Ban,. D. and Kraemer, N. (1995) Effect of Screw Design on Fiber Damage in Extrusion Compounding and Composite Properties, Polymer Composites, 16(3), 258–66.
4. Garrett, D.W. (1990) The Fiber–Matrix Interface: Fact, Fiction and Fantasy, SPE/APC Retec-10/16-18/90, Los Angeles, CA, USA.
5. Fraser, W.A., Ancker, F.H., Dibenedetto, A.T. and Elbirli, B. (1983) Evaluation of Surface Treatments for Fibers in Composite Materials, Polymer Composites, 4(4), 238–48.
6. Yu, Z., Brisson, J. and Ait-Kadi, A. (1994) Prediction of Mechanical Properties of Short Kevlar Fiber–Nylon-6,6 Composites, Polymer Composites, 15(1), 64–73.
7. Karian, H.G. (1996) Designing Interphases for Mega-Coupled Polypropylene Composites, Plastics Engineering, 52(1), 33–5.
8. Perwuelz, A., Caze, C. and Piret, W. (1993) Morphological and Mechanical Properties of Glass-Fiber Reinforced Blends of Polypropylene/Polyamide 6,6, J. Thermoplastic Composite Materials, 6(7), 176–189.

Keywords: short glass fiber, reinforced thermoplastics, chemically coupled, size, interface, polypropylene, nylon 6,6, flexural strength.

See also: Reinforcing fibres.

Fillers

Geoffrey Pritchard

Editor's note: This entry will not discuss all the particulate fillers used in plastics, because several are covered elsewhere in the book. It will outline some general principles and make brief references to specific fillers which are not allocated their own articles elsewhere.

INTRODUCTION

The popular perception is that fillers are used mainly to reduce costs, but many fillers have very beneficial effects on the properties of plastics. They can improve mechanical properties, especially the modulus; they can modify the electrical characteristics, change the density, improve the fire resistance, and reduce smoke evolution on burning. The use of fillers in thermosetting resins is often advantageous to reduce the excessive exothermic heating caused by the hardening reaction, which might otherwise crack the resin. On the other hand, the actual process of addition of particulate fillers to resins and plastics incurs an economic penalty, especially as fillers increase the viscosity of liquid resins during processing operations, making some fabrication processes more difficult, even when the adverse effects have been mitigated to some extent by surface treatment of the filler particles. The incorporation of fillers at high loadings can increase the likelihood of voids. The tensile and flexural strength and the ultimate elongation at break can be adversely affected, although this depends on the nature of the filler and the polymer, and on the filler particle size.

Plastics Additives: An A–Z Reference
Edited by G. Pritchard
Published in 1998 by Chapman & Hall, London. ISBN 0 412 72720 X

ENTRIES ON FILLERS ELSEWHERE IN THIS HANDBOOK

Several fillers are allocated their own separate entries in this book. The interested reader is referred to the following entries:

Calcium carbonate;
Carbon black;
Conducting fillers for plastics;
Hollow microspheres;
Mica;
Rice husk ash.

There is an account of 'Surface treatments for particulate fillers in plastics'.

The subject of rubber as a filler is dealt with in the entry entitled 'Surface-modified rubber particles for polyurethanes'. Hydrated alumina, otherwise known as alumina trihydrate (ATH), is considered under the heading 'Flame retardants: inorganic oxide and hydroxide systems'. Some mechanical property considerations are dealt with in the entry entitled 'Fillers: their effect on the failure modes of plastics'.

It is usual to distinguish between fillers and fibres, the latter being traditionally a separate subject, although this distinction becomes difficult to maintain at very short fibre lengths. Nevertheless it can be noted that there are separate entries on 'Reinforcing fibres' and (more relevant to fillers) on 'Fibres: the effect of short glass fibres on the mechanical properties of thermoplastics'.

GENERAL CHARACTERISTICS OF PARTICULATE FILLERS

Fillers can be crystalline or amorphous. Examples of crystalline fillers include calcium carbonate and anatase (titanium dioxide) whereas solid glass beads are amorphous. Many, but not all, fillers are extracted from the earth's crust by mining or quarrying operations; examples include calcium carbonate, talc, bentonite, wollastonite (calcium metasilicate) and titanium dioxide. Some fillers are extracted along with impurities that can seriously affect the colour, electrical properties and toxicity of plastics unless they are removed. Others, such as wood flour, have organic origins. The use of wood flour itself has been rather limited because of compatibility problems.

PARTICLE SHAPE AND SIZE

Filler particles can have a wide variety of shapes, and where there is some regularity, they can be conveniently categorized as spherical, plate-like, needle-shaped or fibrous. Others are more irregular than any of these words suggest. Spherical particles such as glass beads have the important property that they do not increase resistance to the flow of filled plastics in the mould as much as the other shapes do (an increase in viscosity is an undesirable consequence of adding fillers) and spherical particles do not constitute such severe stress concentrations as angular ones. ATH particles are plate-like and they, too, greatly increase the viscosity of liquid resins such as polyesters and epoxies. This constitutes a limitation to the quantity of filler that it is practicable to add, although it is still possible using special methods to achieve 200 parts by weight of ATH per 100 parts by weight of resin. Larger filler particles have slightly less effect on the viscosity, and certain combinations of large and small particles can provide optimum viscosity control, but the largest ones reduce the strength of the product, and large particle agglomerations formed from the grouping together of several small ones are particularly undesirable, especially as the boundaries between the constituent particles can act as initiation cracks which help to explain the drastic reduction in elongation at break often observed (Table 1). Surface treatments are applied to filler particles for various reasons, and often it is in order to reduce agglomeration, and aid dispersion, rather than to improve bonding to the resin as would be the case with reinforcing fibres. There is usually not much need to improve adhesion between filler particles and the matrix, partly because the very large difference between the thermal expansion coefficients of filler and resin ensures that the resin exerts a compressive stress on the particles, and partly because many filler particles already have numerous polar groups on their surfaces, which usually improve adhesion to the resin.

It is possible to identify some filler types simply by their highly uncharacteristic shapes, using optical microscopic examination of a very small number of filler particles. If we wish then to measure the size of a given particle, we have to overcome the problem that irregular particles have by definition no particular dimension which defines their size, as spherical

Table 1 The tensile properties of alumina trihydrate filled epoxy resins. From Phipps, M.A. (1995) The strength of alumina trihydrate filled epoxy resins, Ph.D. thesis, Kingston University, U.K.

Filler content (phr)	Tensile strength (MPa)	Tensile elongation at break (%)	Tensile modulus (GPa)
0	77	4.4	3.2
10	60	3.4	3.2
20	53	1.8	3.6
50	52	1.4	5.0
100	49	0.8	7.2

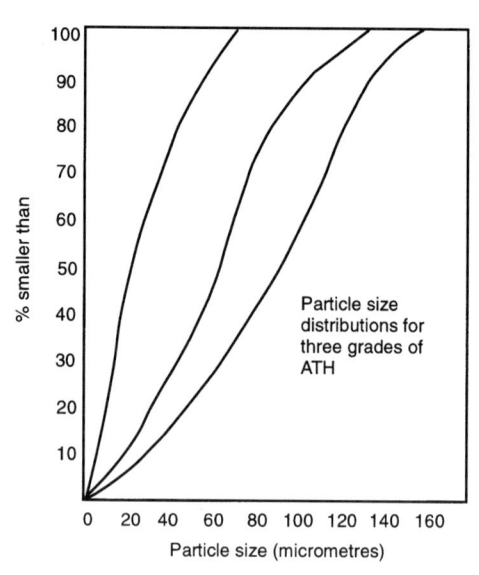

Figure 1 Particle size distributions of some ATH filler grades. Re-drawn and modified from: Wainwright, R.W. (1991) Mechanical properties of epoxy/alumina trihydrate filled compositions, Ph.D. thesis, Kingston Polytechnic, U.K.

particles do. Therefore the volume of the particle is first determined, and the diameter of a sphere of equal volume is taken as a practical descriptor of particle size. The diameter in this case is called the equivalent spherical diameter (esd). The likelihood that all the particles in a given sample have very similar sizes or esd values is extremely small. It is customary to quote not just an average particle size but a particle size distribution (Figure 1) which can be expressed graphically in various ways. One way is to plot the cumulative wt% filler against the esd.

The irregularity of a particle can be expressed quantitatively by calculating its surface area and dividing it by the surface area of a sphere with the same volume.

SURFACES

The surface area is usually deduced from the quantity of nitrogen adsorbed as a monolayer at the surface. It is an important property, affecting the rheology and ease of dispersion of the filler – two crucial filler properties, from the commercial standpoint.

Other surface properties must also be considered, such as its chemical nature. Metallic particles can have quite different effects on the electrical properties of plastics, depending on whether their surfaces are oxidized or not. Surfaces can be hydrophilic or hydrophobic and in some cases,

OTHER PROPERTIES OF FILLERS

such as silica, they are adjusted to become more or less hydrophobic, using appropriate surface treatments.

OTHER PROPERTIES OF FILLERS

The mechanical properties of individual filler particles – as opposed to those of filled plastics – can be quite difficult to determine, and often the data are not readily available. Hardness is an exception, being one of the more frequently quoted mechanical properties of fillers, and the materials are sometimes divided loosely into hard and soft fillers. Hardness values, together with density and refractive index, are listed for several fillers in Table 2. The density can be expressed either as true

Table 2 Some properties of fillers for plastics. (Many but not all of the data have been obtained from: Falcone, Jr., J.S. (1992) Fillers, in *Encyclopedia of Chemical Science and Technology*, Kirk-Othmer, 4th edn (ed. Mary Howe-Grant) Wiley, New York.)

Filler	True density (g/cm^3)	Refractive index	Hardness (Mohs scale)
Alumina trihydrate	2.42	1.57	3
Carbon black	1–2.3	–	2–4
Calcium carbonate	2.60–2.75	1.49	3
Kaolin	2.58–2.63	1.55–1.57	–
Quartz	2.65	1.55	7
Wood flour	0.65	–	–
Starch	1.5	–	–
Feldspar	2.6	1.53	6–6.5
Fumed silica	2.2	–	–
Diatomaceous earth	2.65	–	–
Crystalline silica	2.65	1.55	7
Fly ash	0.3–1.0	–	5
Solid E-glass beads	2.6	1.548 at at	5.5
Talc	2.7–2.8	1.57	1++
Wollastonite	2.9–3.1	1.63	4.7
Titanium dioxide (rutile)	3.9	2.7	6.2
Titanium dioxide (anatase)	4.2	2.55	5.7
Calcium sulfate, anhydrous	2.96	1.57	–
Aluminium	2.55	–	2.5
Barium sulfate	4.50	1.6	3.3
Magnesium hydroxide	2.39		
Mica	2.8–2.9	1.6	2.7

density, e.g. that obtained by the liquid displacement method, or as the apparent (bulk) density, which is a property of the filler in large quantities, taking into account the space between the particles. Apparent density is particularly relevant to bulk storage and transportation. True density cannot be measured by the classical liquid displacement method if the filler particles contain many pores. In this case a gas method such as the gas pycnometer technique is used.

The optical properties of fillers can be among the most important. Titanium dioxide is first and foremost a white pigment because, like magnesium oxide, its particles reflect visible light. Carbon black, in contrast, absorbs visible light and produces a black plastics material when used as a filler. It incidentally absorbs ultraviolet light as well and is beneficial in outdoor applications where colour is not important. Other fillers have intermediate characteristics between those of titanium dioxide and carbon black, and they produce filled plastics with different appearances. Glass transmits rather than absorbs or reflects light, and a glass bead filled product will be clear and colourless if the polymer itself is also clear, and the refractive index values of the two materials are appropriately matched.

Sometimes fillers are used together with fibres in reinforced plastics. To obtain the highest fibre volume fraction, the particle size of the filler should be less than the average fibre diameter.

THE MODULUS OF FILLED PLASTICS

A large number of equations have been proposed to predict a variety of mechanical properties of filled plastics. The principles underlying the derivations require detailed and lengthy study to be understood, and regrettably there is no room here to discuss the rational basis for the predictive equations mentioned; we shall simply indicate the kind of predictions that have emerged and moreover we shall confine the discussion to one property: the modulus. The original references can be found in any review of the effect of fillers on mechanical properties, such as the theses by Wainwright and Phipps, mentioned elsewhere in this article. Contributors include Nielsen, Paul, Narkis, Ishai, Bueche, Sato and Furukawa, Halpin, Chow etc.

The equations usually make several simplifying assumptions, so verification by comparison with real commercial fillers is not always easy, and different workers have used a variety of approaches, so comparison is difficult.

One early expression for the tensile modulus, E_c, of filled plastics was given for high modulus fillers by Nielsen. Nielsen proposed that:

$$\frac{E_c}{E_m} = \frac{1 + AB\phi}{1 - B\psi\phi}$$

where ϕ is the filler volume fraction, and ψ is a closely related term:

$$\psi = 1 + \left(\frac{1-\phi_\infty}{\phi^2}\right)$$

ϕ_∞ is the maximum possible volume fraction. The constant A is equal to $K_E - 1$ where K_E is the Einstein coefficient, which is effectively a measure of the increase in modulus for a given volume fraction (the subscripts f and m refer to filler and matrix respectively):

$$K_E = \frac{\left(\frac{E_f}{E_m} - 1\right)}{\phi}$$

The other constant, B, is given by $(C-1)/(C+A)$ where C is E_f/E_m.

Many other equations have also been offered. One of the simplest was a semi-empirical one by Narkis:

$$\frac{E_c}{E_m} = \frac{1}{K(1-\phi^{1/3})}$$

The symbols have the same meaning as in the previous equation except that K is a constant related to stress concentrations and is said to be between 1.4 and 1.7 for most fillers.

In an entirely different approach, Bueche proposed a model which assumed that fillers could act in a similar fashion to crosslinks:

$$E_c = E_m + \left(\frac{3sRTX}{u}\right)$$

where s is the specific surface area of the filler, u is the cross-sectional area of the polymer chain, R is the universal gas constant, T is the absolute temperature, and X is the ratio of filler volume fraction to matrix volume fraction.

Most of the equations assume spherical filler particles, but more recent work has taken their shape into account. Models have also been proposed for strength.

SPECIFIC FILLERS

The following discussion briefly mentions a range of fillers not allocated separate articles in this book. Some of them also belong to other additive categories such as pigments, thixotropes etc.

Silicates

Wollastonite (calcium silicate) consists of pure white, non-hydrous, needle-shaped crystals. The particle lengths are typically larger than the widths by a factor of between one and two, but the aspect ratio can be much higher, up to 15. The tensile moduli of individual crystals have been measured, and found to be almost as good as those of aramid fibres. Consequently wollastonite has a reinforcing property and it can compete with or partially replace other reinforcing fillers and fibres. As its cost is competitive this may happen more in the future. The mineral can be surface treated with an aminosilane to improve its adhesion to the polymer. If no such treatment is applied, the amount of reinforcement will be much smaller and there can even be a lower tensile strength for the filled polymer than for the unfilled equivalent.

Wollastonite is good for use where resistance to water or humid environments and to ultraviolet light is required. It is used in unsaturated polyesters, polyamides, polypropylene and vinyl plastisols. It has the disadvantage that it is vulnerable to acids.

Kaolin is clay, i.e. a hydrous aluminosilicate. There are two varieties: (a) a naturally occurring, hydrous form, and (b) the form obtained when the clay is heated at more than 600°C to remove the water. This is the calcined, anhydrous form. It is harder than the hydrous variety, i.e. 6–8 on the Mohs scale, instead of 2. The average equivalent spherical diameter (esd) of the natural form is 1.5–1.8 µm compared with a wider range (0.9–3) after calcining.

The particles are hexagonal platelets which tend to stack together producing larger particles. Surface treatment is needed to facilitate dispersion in a resin. The surface treatment can have side-effects, inhibiting reaction with epoxy and vinyl polymers. The particles are rendered hydrophobic by suitable treatment, for achieving low dielectric loss. Very fine kaolin particles can increase rather than decrease the strength of certain thermoplastics.

Talc is a naturally occurring hydrated magnesium silicate, $3MgO \cdot 4SiO_2 \cdot H_2O$. The variety used as a filler is lamellar. It is added to thermoplastics for the usual reinforcement reasons – to increase flexural strength and modulus, increase hardness, reduce creep, and increase the heat distortion temperature. There is some embrittlement, reflected in ultimate tensile strength and elongation and in reduced impact strength. There may also be an abrasive effect. Certain grades are found in nature in association with asbestos-type minerals.

Silica and glass

Silica is used in many different forms. The precipitated variety consists of small amorphous particles and is used to improve heat and wear resistance in phenolic brake linings. It is also added to rubbers, PVC and polyethylene. There is a market for it as an extender for titanium dioxide, and as an antiblocking agent. Pyrogenic silica is more expensive, being made by the hydrolysis of silicon tetrachloride in an oxygen/hydrogen flame. The product flocculates to form 'flocks' of up to 200 μm diameter, with a wide range of surface areas. There are silanol groups on the surfaces, which cause the particles to interact with each via their hydrogen bonds. Three-dimensional structures can build up, increasing the viscosity of a suspension in a liquid such as a thermosetting resin. The bonds are not permanent, and can easily be broken by moderate shear forces. This effect leads to the application of silica in the control of gelcoat thixotropy and the prevention of gelcoat 'sag'. Silanes can be used to render silica hydrophobic, and the increase in resin viscosity produced by adding it can be moderated by altering the nature of the surface treatment.

Pyrogenic silica can increase the dehydrochlorination temperature of PVC.

Microcrystalline quartz is obtained by pulverizing quartz sands and is a hard solid (7 Mohs). It increases the thermal shock resistance in brittle resins – some filled thermosetting resins are cracked by relatively few thermal cycles between, say, ambient temperature and 100°C – when added at high concentrations (typically 100–200 parts per hundred by weight). It can be surface treated with an aminosilane to enhance adhesion, when used in epoxy compositions to improve flexural modulus, electrical insulation or thermal properties, and in the case of unsaturated polyesters, it can be treated with a methacrylic silane.

Natural microcrystalline silica and fused silica are also used as fillers in thermosetting resins. Solid glass spheres are available in a range of diameters. They have the advantage of readily controlled geometry, unlike quarried minerals with irregular shapes and wide particle size distributions. They have no oil absorption characteristics, and lend themselves (as do their hollow microsphere analogues) to metallization. Whereas most fillers increase the modulus but lower the strength, they increase the strength as well, partly because their shape does not generate large stress concentrations. Their cost effectiveness in comparison with hollow microspheres depends on whether the cost is on a weight or a volume basis; they are cheaper on a weight basis.

Other oxides and hydroxides

Titanium dioxide is a white pigment – the most important white pigment used in PVC. The amount required can be between 1 and 10 parts titanium dioxide per 100 of PVC. The oxide has a very high refractive index (Table 2) and imparts good opacity and brightness. It is easy to disperse in PVC, and is not itself affected by the temperatures used in processing, although the various grades of titanium dioxide, together with the appropriate heat stabilizing systems, have different effects on the amount of yellowing detectable after processing. Further slight colour changes can sometimes be observed after natural weathering.

There are two crystalline forms: anatase and rutile. The latter has the higher refractive index and is more widely used, because the opacity is dependent on the difference between the refractive index of the filler and that of the polymer.

Titanium dioxide can be sufficiently abrasive to accelerate the wear on processing equipment and it is recommended that it is added late in the processing cycle to minimize damage. Without special surface treatment, titanium dioxide is photocatalytic towards the degradation of certain polymers, leading to weathering effects such as chalking and reduced gloss. Surface treatments with alumina, zirconia or similar substances provide a barrier between the titanium dioxide and the polymer, minimizing degradation. The surface treatment has implications for other properties such as ease of dispersion, and for opacity. Organic coatings are sometimes used in conjunction with inorganic ones in order to obtain the best balance of properties.

Magnesium oxide is not widely used but it has a low volume speciality use in polyester resins as a thickening agent in the production of sheet moulding compounds. More importantly, the hydrated form, magnesium hydroxide, has been evaluated in polypropylene and ethylene vinyl acetate copolymer as a combined filler and flame retardant, using surface coated particles. The flame retardant mechanism is similar to that of ATH, i.e. it evolves water by an endothermic reaction when heated, but with magnesium hydroxide there is a higher temperature of decomposition, 320°C instead of about 180°C. ATH would not survive the processing temperatures used with polypropylene, although ATH derivatives have now been produced, with higher decomposition temperatures.

Magnesium hydroxide is slightly hygroscopic.

Metals

Certain metallic fillers are essentially pigments, or in some way modifiers of surface appearance. They are used both in low-technology areas such as stopping compounds and in more specialized applications, sometimes for their effect on electrical conductivity and thermal properties. Aluminium particles are available in masterbatches and can impart characteristic surface appearances, such as lustre or glitter. The particles are dispersed in mineral oil or in phthalate plasticizers. Using various other metals, we can obtain gold, silver or coloured metallic effects. The larger the flakes of metal used, the brighter the appearance. Metal particles can be supplied with silver coatings on, for further enhancement of electrical conductivity.

There is a considerable mismatch between the thermal expansion coefficients of metals and those of organic polymers. The metal particles can sometimes shrink away from the surrounding polymer as the temperature falls from the moulding temperature. Improved bonding between metal and polymers would be beneficial. It is possible to obtain filled plastics with thermal conductivities up to 95% of those of metals, simply by using high aspect ratio aluminium flakes.

HEALTH AND SAFETY

There is an inherent danger to health in handling large quantities of fine, dust-like particles, and fabrication processes must be designed to minimize the likelihood of inhalation. Explosions are another risk associated with the handling of powders. There are also other problems specific to individual fillers. Potential users are advised to acquaint themselves with all the health and safety aspects of these materials.

BIBLIOGRAPHY

Enikolopian, I.N.S. (1991) *Filled Polymers: Science and Technology*, Advances in Polymer Science Series, Volume 96, Springer-Verlag.
Katz, H.S. and Milewski, John V. (1987) *Handbook of Fillers for Plastics*, Chapman & Hall, London.

Keywords: particle shape; size; surface area; silane; viscosity; silica; kaolin; wollastonite; titanium dioxide.

Fillers: their effect on the failure modes of plastics

S. Bazhenov

FAILURE MODES

Thermoplastic polymers filled with rigid inorganic particles display higher values of Young's modulus, better thermal stability, and lower wear under friction than do unfilled polymers. Unfortunately, fillers also lead to dramatic reductions in the fracture strain of the polymer [1] under tensile load and embrittlement. Some polymers, such as polytetrafluoroethylene (PTFE, trade mark Teflon®), yield uniformly. However, more usually we get non-uniform yielding with propagation of a neck through the entire length of a specimen as in polyethylene (PE) and polypropylene (PP). If a polymer yields with necking an increase in filler content leads to fracture during neck propagation, to quasibrittle fracture during formation of a neck and, finally, to true brittle fracture as illustrated by Figure 1. With brittle and quasibrittle failure modes fracture strain is roughly 100-fold lower than that of unfilled polymer.

In addition, filler may suppress necking and initiate yielding in crazes or, more exactly, in craze-like zones if the particles were treated by an anti-adhesive and their bonding with a polymer was weakened [2]. Thus, in filled polymers at least six modes of deformation behaviour are distinguished: (1) brittle; (2) quasibrittle fracture during neck formation; (3) fracture during neck propagation; (4) stable neck propagation; (5) micro-uniform yielding, and (6) yielding in crazes.

Crazes are similar to cracks, the opposite sides of which are connected with fibres of oriented polymer material. In filled polymers, crazes are caused by debonding of particles and by stretching of the polymer in

Plastics Additives: An A–Z Reference
Edited by G. Pritchard
Published in 1998 by Chapman & Hall, London. ISBN 0 412 72720 X

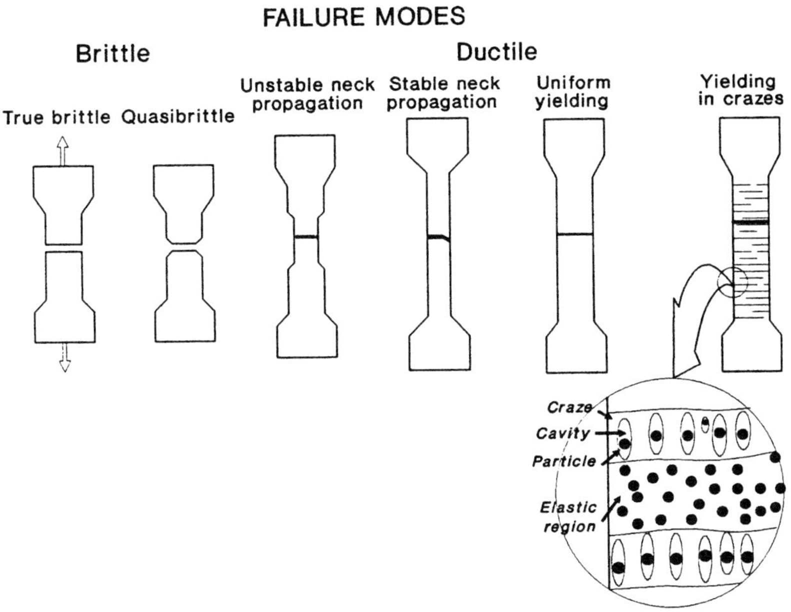

FAILURE MODES

Figure 1 Failure modes of filled polymers.

the space between the neighbouring particles. The thickness of the fibres is determined by the distance between the neighbouring particles.

The mechanical properties of elastic materials are characterized by ultimate fracture stress (strength), fracture strain, and Young's modulus (the ratio of stress to strain). The mechanical response of ductile materials is non-linear and their properties are additionally characterized by yield stress and draw stress (also called lower yield stress) if a polymer yields with necking. The cross-sectional area of a specimen reduces during yielding, and stress is calculated as a ratio of an applied force to the initial cross-section (engineering stress) or to the current cross-section of the specimen (true stress). Here engineering stresses are used.

The load versus elongation curve has specific features at each failure mode (Figure 2). A typical stress–strain curve of a necking polymer is characterized by the presence of a maximum. After reaching the yield point, the stress drops and remains constant while the neck propagates through the specimen. This is followed by a region of strain-hardening where the stress gradually increases. Quasibrittle composites reach a stress maximum, however, just past the yield point, as the stress decreases toward the draw stress, and the composite fractures. A curve with non-linear, monotonously increasing stress is typical for both uniform yielding and yielding in crazes. In brittle materials the stress–strain relationship is linear and the fracture strain is less than 2–3%.

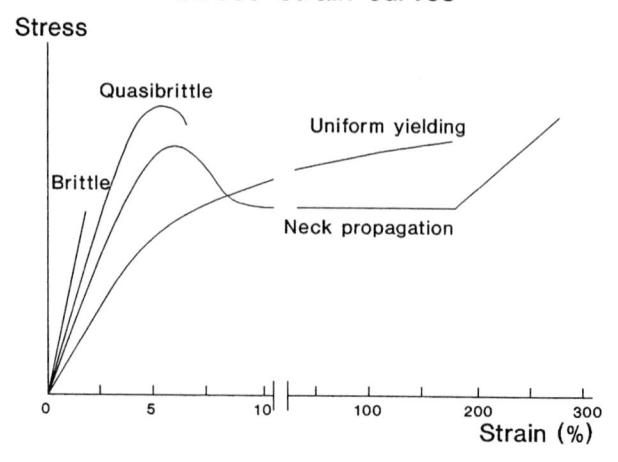

254   Fillers: their effect on the failure modes of plastics

Figure 2 Typical engineering stress versus strain curves for different failure modes.

STRENGTH AND FRACTURE STRAIN

Figure 3 illustrates the effect of particles on the relative strength of various polymers. The relative strength is comparatively insensitive to the polymer type and its adhesion to the particles, and depends on the volume fraction of particles. The decrease in strength is described by

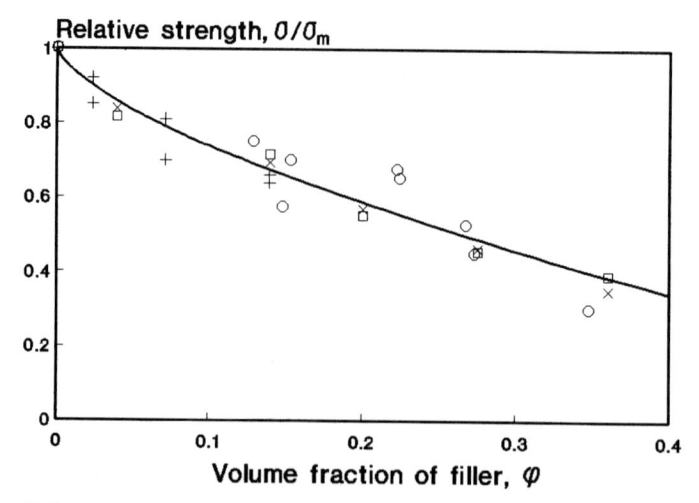

Figure 3 Relative strength of filled polymer, σ/σ_m, versus the volume fraction of filler ϕ. σ_m – strength of unfilled polymer. (O) – PTFE; (+) – PET/CaCO$_3$, diameter of particles 8 μm [5]; (□) – HDPE/Al(OH)$_3$, 1 μm [2]; (×) – HDPE/Al(OH)$_3$, 25 μm [2].

[Figure: Relative fracture strain, ε/ε_m vs Volume fraction of filler, φ — curves for PVC, PET, HDPE, PP, PTFE]

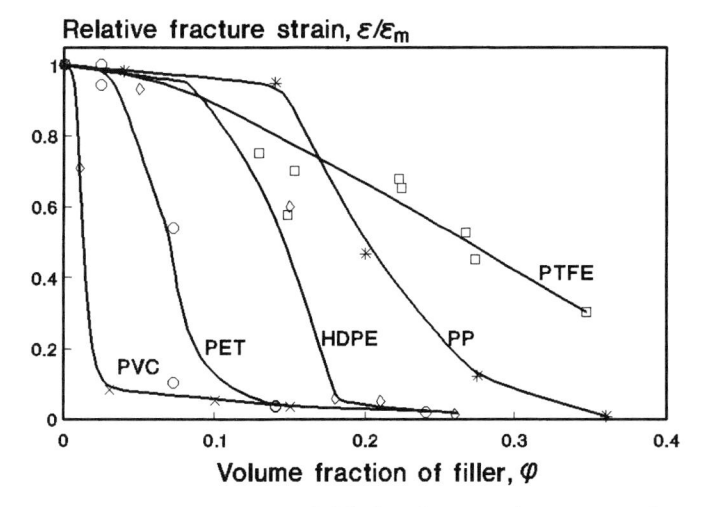

Figure 4 Relative fracture strain of filled polymer, $\varepsilon/\varepsilon_m$, versus the volume fraction of filler ϕ. ε_m – fracture strain of unfilled polymer. (∗) – PP/Al(OH)$_3$, 10 µm; (◇) – HDPE/Al(OH)$_3$, 10 µm [2]; (○) – PET/CaCO$_3$, 8 µm [5].

the 'two thirds' power equation derived by Smith [3] and widely used by Nielsen:

$$\sigma/\sigma_m = 1 - 1.21\phi^{2/3} \tag{1}$$

where σ_m is the strength of an unfilled polymer, and ϕ is the volume fraction of particles. Fillers lead to moderate reduction in strength, and at $\phi = 0.4$ the decrease in strength is approximately three-fold.

Figure 4 illustrates the effect of particles on the relative fracture strain of filled polymers. The fracture strain depends on the polymer matrix. The most typical dependence is step-like, and a moderate reduction in fracture strain at low filler contents is followed by a sharp drop of fracture strain in a comparatively narrow interval of filler contents. The magnitude of the strain decrease is approximately 100-fold, much higher than the decrease in strength. The drop in fracture strain is caused by the transition from a ductile to brittle fracture. Filled PTFE behaves differently and remains ductile up to $\phi = 0.35$.

YIELD STRESS AND ENGINEERING DRAW STRESS

Figure 5 illustrates the effect of particles on the relative yield stress σ_y of filled polymers. The yield stress is within the dashed fork region depending on adhesion between the polymer and the particles. The upper border of the fork is a constant, equal to the yield stress of the unfilled polymer σ_{ym} (the case of small particles of nanometre size, when the yield stress

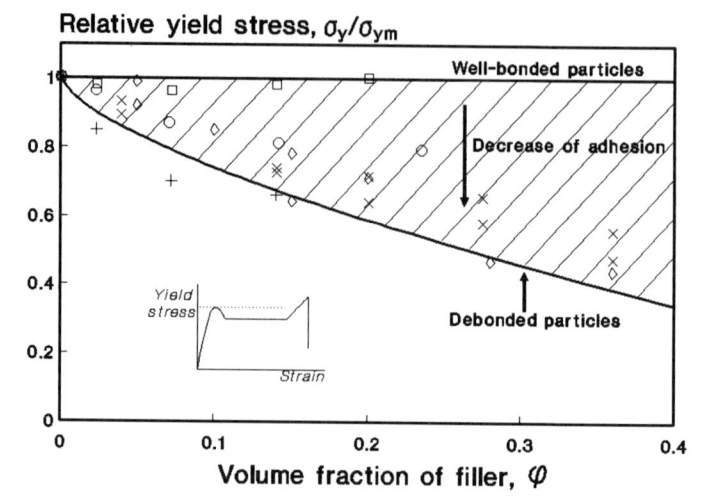

Figure 5 Relative yield stress of filled polymer, σ_y/σ_{ym}, versus the volume fraction of particles ϕ. (□) – PET/CaCO$_3$, 8 µm; (○) – PET filled by covered CaCO$_3$ particles [5]; (×) – PP/Al(OH)$_3$, 1 and 25 µm, (◊) – HDPE/Al(OH)$_3$, 10 µm [2].

may be higher than σ_{ym}, is not considered here). The yield stress of the filled polymer is equal to that of the unfilled polymer if adhesion is strong and at the yield point particles are well-bonded with the polymer matrix. Any reduction in adhesion leads to a decrease in the yield stress. The yield stress is described by the lower border of the fork if all particles debond before the yield point. If debonding is partial, the yield stress is inside the fork region, and is given by the equation:

$$\sigma_y/\sigma_{ym} = 1 - 1.21\alpha\phi^{2/3} \qquad (2)$$

where α is the fraction of particles debonded in the yield zone.

Figure 6 illustrates the dependence of the engineering draw stress on filler content ϕ. Like the yield stress, draw stress depends on adhesion between the polymer and the particles. The upper limit of the draw stress, equal to the draw stress of unfilled polymer, describes polymers filled with well-bonded particles. The lower border of the fork, corresponding to completely debonded particles in the yield zone, is a straight line with the slope approximately equal to minus one. If particles are debonded partially, the draw stress is given by:

$$\sigma_d/\sigma_{dm} = 1 - 1.25\alpha\phi \qquad (3)$$

where σ_{dm} is the draw stress of unfilled polymer, and α is the fraction of debonded particles. In contrast to the strength and the yield stress, the decrease in the draw stress is linear with an increase in ϕ [4].

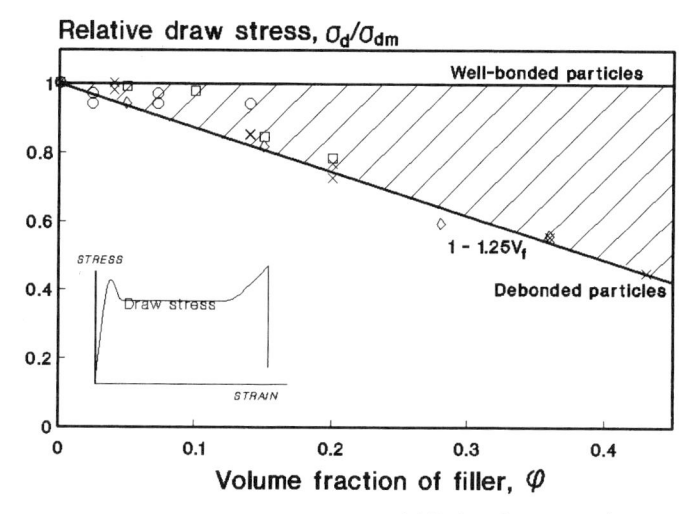

Figure 6 Relative engineering draw stress of filled polymer, σ_d/σ_{dm}, versus the volume fraction of particles ϕ. ($\square$) – PET/CaCO$_3$, 8 μm; (○) – PET filled by covered CaCO$_3$ particles [5]; ($\diamond$) – PP/Al(OH)$_3$, 10 μm; ($\times$) – HDPE/Al(OH)$_3$, 1 and 25 μm [2].

TRANSITIONS IN FAILURE MODES

Failure mode depends on matrix properties, particle content and adhesion between the polymer and the particles. Low amounts of filler do not change the failure mode of a polymer. An increase in filler content leads to a transition in failure mode. Sometimes, if particles were treated with an anti-adhesive, the transition is ductile-to-ductile from neck propagation to macro-uniform yielding in crazes [2]. However, more usual is the transition from ductile to brittle failure.

The embrittlement filler content is determined primarily by polymer ability to undergo orientation strain-hardening. The degree of strain-hardening is characterized by the ratio of the strength to the draw stress (or to the yield stress if unfilled polymer yields without necking). If a polymer does not strain-harden this ratio is equal to one. An increase in the degree of strain-hardening is described by an increase in this ratio.

Figure 7 shows the embrittlement filler content ϕ^* plotted against the ratio of the strength to the draw stress σ_m/σ_{dm}. ϕ^* increases with an increase in the σ_m/σ_{dm} ratio. The critical volume fraction for polyvinylchloride (PVC) is less than 0.05, and is not improved by toughening with rubber particles. If a polymer does not strain-harden, even a small amount of filler leads to loss of ductility. The critical volume fractions for polypropylene (PP) and for copolymers of polyethyleneterephthalate (PET), Kodar 6763, are higher, from 0.1 to 0.2. The critical filler content of

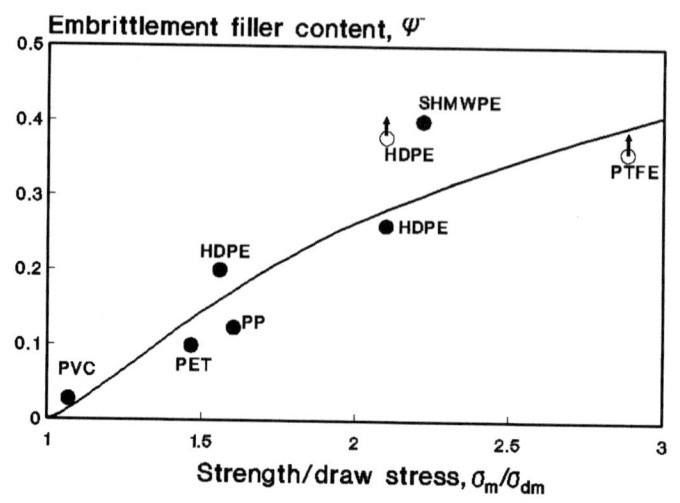

Figure 7 Embrittlement filler content of filled polymer ϕ^* versus the ratio of strength to draw (yield) stress of unfilled polymer, σ_m/σ_{dm} [2, 5].

HDPE is from 0.2 to 0.25 and depends on particle size and surface treatment [2]. In general, strain-hardening of polymers increases with molecular weight. The critical volume fraction for superhigh molecular weight polyethylene (SHMWPE), which is easily oriented, is the highest, 0.35 to 0.4. At the same time, PTFE and HDPE filled with particles treated by an anti-adhesive behave differently and remain ductile up to $\phi = 0.35$ (open points in Figure 7).

If particles in the yield zone are well-bonded and the unfilled polymer yields with necking, the fracture stress is higher than the draw stress at low filler contents. As a result, the applied force is enough to propagate the neck through the specimen, and the filler does not change the deformation mode of the polymer. However, when the fracture stress (equation (1)) decreases below the level of the draw stress, the neck is not able to propagate through the sample, and a ductile-to-brittle transition is observed. At the transition point the composite strength is equal to the draw stress $\sigma = \sigma_d$.

With well-bonded particles the draw stress σ_d corresponds to the upper border of the fork in Figure 6. Hence, σ_d is equal to that of the unfilled polymer, and the critical filler content ϕ^*, is found from equation (1) and the embrittlement criterion, $\sigma = \sigma_{dm}$ [4]:

$$\phi^* = \left(\frac{\sigma_m - \sigma_{dm}}{1.21\sigma_{dm}}\right)^{3/2} \quad (4)$$

where σ_m and σ_{dm} are the strength and the draw stress of unfilled polymer. For polymers with uniform yielding (e.g. PTFE) the critical filler content is described by equation (4), with σ_{dm} being the yield stress of the unfilled polymer. The critical filler content calculated from equation (4) is plotted as the solid line in Figure 7.

If adhesion between the polymer and the filler is weak and particles debond before the yield point, the degree of polymer strain-hardening affects the mode of the transition, which may be ductile-to-brittle or ductile-to-ductile. If the strength of a polymer is lower than its yield stress (low strain-hardening), an increase in filler content leads to embrittlement of the composite. Embrittlement filler contents for weakly bonded particles are higher than for well-bonded particles. If the strength of an unfilled polymer is higher than the yield stress (essential strain-hardening), particles will undergo a ductile-to-ductile transition from necking to yielding in crazes, and embrittlement is avoided (open points in Figure 7).

Thus, the main parameter that affects embrittlement is the degree of polymer strain-hardening and hence molecular weight. The second most important parameter is adhesion between the particles and the polymer. The decrease in particle diameter leads to debonding at higher strains. As a result, it may lead to embrittlement of a polymer in the same way as an increase in adhesion.

High filler contents necessarily lead to embrittlement of polymers with low ability to strain-harden, no matter whether adhesion is strong or weak. Examples are PVC and PETG. The opposite example is filled PTFE, which remains ductile up to high filler contents due to high strain-hardening ability and low (or even no) adhesion to fillers. PP and PE are between these two extremes. These polymers are produced with different molecular weights, strain-harden differently and, in addition, may be well or weakly bonded to the filler, depending on the surface treatment of the particles. As a result, PE and PP are brittle or ductile at high filler contents depending on their molecular weight, adhesion to the filler particles, particle size and shape.

REFERENCES

1. Tochin, V.A., Shchupak, E.N., Tumanov, V.V., Kulachinskaya, O.B. and Gai, M.I., *Mechanics of Comp. Mater.*, **20** (1984), 440.
2. Dubnikova, I.L., Topolkaraev, V.A., Paramzina, T.V. and Diachkovskii, F.S., *Visokomol. Soedin.*, **A32** (1990), 841.
3. Smith, T.L., *Trans. of the Soc. of Rheology*, **3** (1959), 113.
4. Bazhenov, S., *Polymer Eng. and Science*, **35** (1995), 813.
5. Bazhenov, S., Li, J.X., Hiltner, A. and Baer, E., *J. Appl. Polym. Sci.*, **52** (1994), 243.

Keywords: filled polymer, failure mode, fracture strain, embrittlement.

See also: Fillers.

Flame retardancy: the approaches available

G.A. Skinner

INTRODUCTION

All organic polymeric materials are based on carbon as the essential element, besides which most of them contain hydrogen and many have appreciable quantities of other elements such as nitrogen, oxygen, sulphur, fluorine and chlorine. The polymer backbones are based on carbon–carbon bonds, with a few having carbon–nitrogen and carbon–oxygen bonds at regular intervals because of the presence of heteroatoms in the main chains.

All of these organic polymers undergo thermal degradation if exposed to sufficient heat. The energy is absorbed until the carbon–carbon, carbon–nitrogen and carbon–oxygen bonds in the polymer backbones break when lower molecular weight volatile gases are generated. Although the mechanism is somewhat different in the presence of oxygen, giving rise to some new products, a gaseous mixture is still formed which is now inherently flammable. The precise nature of the degradation products is determined by the chemical composition of the polymer, the additives present and the degradation c]onditions. Although the stability of a given polymer will vary with its structure and composition nearly all polymers will ignite and burn at temperatures in the range 350°–450°C.

The role of the flame retardant is to make the polymer formulation less flammable by interfering with the chemistry and/or physics of the combustion process. They are only effective at the growth stage of a fire.

Plastics Additives: An A–Z Reference
Edited by G. Pritchard
Published in 1998 by Chapman & Hall, London. ISBN 0 412 72720 X

The emphasis is on reducing, rather than completely eliminating flammability, as virtually all flame retarded polymer formulations will ignite given extremely severe conditions of temperature and high oxygen concentrations. Nevertheless polymer formulations can be rendered relatively safe to normal conditions of usage by the right flame retardant or combination of flame retardants.

It should be stressed that not all polymer formulations contain flame retardants. Their usage is usually accompanied by an increase in cost due to their relative high price in some instances and/or problems associated with the processing of the formulations. As a result companies only use them when the polymer application demands it. In many applications legislation has been enacted which strictly controls the permissible flammability of the formulation.

In this brief introduction to flame retardants, it is intended to review the hazards of fires and the main approaches available to reduce these hazards. Mention will be made of some of the compounds exemplifying these approaches, together with their advantages and disadvantages, but a full discussion of individual flame retardant systems will be given in separate entries.

FIRE HAZARDS

Owing to the wide variations in the chemical composition of commercial polymers and the multitude of forms in which they are used, there is no universal flame retardant that is applicable in all formulations. What might be effective in a solid moulded item may be completely ineffective in a foam. Any flame retardant treatment of a textile fibre must be permanent to all the various washing procedures that the final garment might be subjected to whereas this is not such a stringent requirement in other polymer applications.

In assessing the performance of a flame retardant, there are specific fire hazards to be considered, associated with the burning of a polymer. These are:

1. polymer ignitability;
2. rate of flame spread;
3. rate of heat release;
4. formation of smoke and toxic gases;
5. corrosivity of acid gases.

Most of the flame retardants used up to and during the 1970s were assessed mainly on their ability to reduce the polymer's ignitability and rate of surface spread of flame. During the last couple of decades more attention has been focused on the rate of heat release and emission of smoke and toxic gases.

The hazards associated with smoke and toxic gases were recognized in the 1970s as a result of many fires occurring around the world in high rise buildings. Fatalities occurred on floors remote from the scene of the fires due to smoke decreasing visibility and hindering escape. Death resulted from inhalation of toxic gases such as carbon monoxide and suffocation from a lack of oxygen. The corrosivity of the acid gases released in a fire has been identified as a major problem where electrical and electronic components are concerned and where premature shutdown has occurred.

Ideally a flame retardant should reduce all the above-mentioned fire hazards. In practice this very rarely occurs. A polymer's ignitability may be reduced but the rate of heat release and/or smoke emission may be increased. Hence a flame retardant to be used in a specific formulation has to be carefully chosen to give the desired effect(s).

It is easy for the inexperienced scientist reading company literature to be misled. Most companies produce only one broad type of flame retardant, and their literature is designed to promote its beneficial effects. It is possible to get the impression that there are many ideal flame retardants on the market which are effective and can be used without giving rise to any problems. In reality this is not so, as there are advantages and disadvantages associated with all of them.

One conceivable approach to reducing the first hazard of some polymer formulations is to replace the polymer being used with a more thermally stable one. Even if the resulting formulation had all the desired properties, this is not usually a realistic approach because thermally stable polymers normally have a high aromatic or fluorine content. They tend to be difficult to process, and are relatively costly.

FLAMMABILITY TESTS

Large scale testing is expensive so there are numerous small scale tests used to assess the performance of flame retardant additives. They vary from simple ignitability tests on rods of plastics, such as the UL-94 and oxygen index tests, to the use of somewhat sophisticated instrumentation such as cone calorimetry, which gives information on all the fire hazards associated with the polymer specimen being tested. The relatively simple tests used in industry are geared to the particular hazard being assessed. Frequently there is no widely accepted test method and different countries and different industries have their own preferred methods.

FLAME RETARDANTS IN COMMERCIAL USE

Nearly all the present-day commercial flame retardants and smoke suppressants are based on the following elements, which are listed in alphabetical order:

aluminium	iron
antimony	magnesium
boron	molybdenum
bromide	phosphorus
chlorine	tin

They act by interfering with one or more stages of the combustion process. Some so-called flame retardants, such as antimony trioxide, are actually synergists and are only active in the presence of compounds of the other active elements.

ACTION OF FLAME RETARDANTS

Inhibition of vapour phase combustion

As previously mentioned, flame retardants act by interfering with one or more steps of the burning cycle. A schematic diagram of this cycle is given in Figure 1.

As shown in the burning cycle, a polymer burns via the vapour phase combustion of the volatile products produced during its thermal oxidative degradation. The vapour phase combustion is a free radical process which can be simplified and expressed as follows, where RCH_3 is representative of the hydrocarbon undergoing combustion:

$$RCH_3 + {}^{\bullet}OH \rightarrow RCH_2{}^{\bullet} + H_2O \tag{i}$$

$$RCH_2{}^{\bullet} + O_2 \rightarrow RCHO + {}^{\bullet}OH \tag{ii}$$

$$2HO^{\bullet} + CO \rightarrow H_2O + CO_2 \tag{iii}$$

$$H^{\bullet} + O_2 \rightarrow HO^{\bullet} + {}^{\bullet}O^{\bullet} \tag{iv}$$

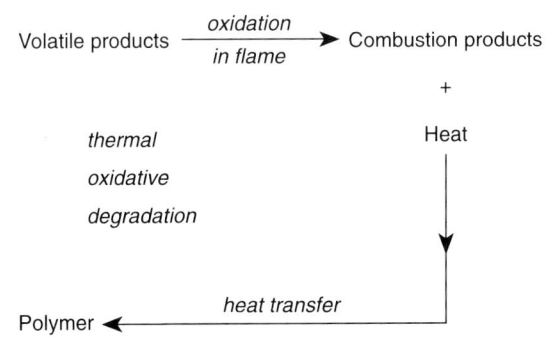

$$H_2 + {}^{\bullet}O^{\bullet} \rightarrow HO^{\bullet} + H^{\bullet} \tag{v}$$

Figure 1 The burning cycle.

Steps (iv) and (v) are thought to be the main chain branching steps in combustion. Several flame retardants act by inhibiting this vapour phase combustion. They produce decomposition products (at the same temperature as the polymer substrate decomposes) which interfere with the chain branching steps in combustion. Less reactive radicals are formed, which inhibit the combustion process.

Organochlorine and organobromine flame retardants emit hydrogen halide (HX) gas during their decomposition. This interferes with the flame reactions by promoting scavenging of the high energy H˙ and OH˙ chain branching radicals:

$$H^{\cdot} + HX \rightarrow H_2 + X^{\cdot} \qquad \text{(vi)}$$

$$HO^{\cdot} + HX \rightarrow H_2O + X^{\cdot} \qquad \text{(vii)}$$

These reactions compete with those occurring in the absence of halogen. The halogen atom produced is a low energy radical which is incapable of propagating the oxidation process. This reduces the rate of heat transfer back to the polymer, which decreases the burning rate and leads to extinction of the flame.

The relatively low flammability of chlorine-containing organic polymers such as poly(vinyl chloride), which contains 56.8% by weight of chlorine, is explained by this vapour phase inhibition. Incorporation of high concentrations of halogen into the polymer structure (the so-called 'reactive approach') is an obvious way of reducing the flammability of polymers but this is only possible in a few cases. The normal procedure is to add the halogen as an additive in the form of a highly halogenated organic compound.

Antimony, in the form of the trioxide, is used in many flame retardant formulations although on its own it is not a flame retardant. It acts as a synergist in the presence of halogen and reduces the amount of halogen needed to impart a given level of flame retardancy. It is believed to assist the transfer of the hydrogen halide to the vapour phase by the formation of antimony trihalide and to catalyse the recombination of hydrogen, oxygen and hydroxyl radicals to form water.

The halogen/antimony–halogen flame retardants have the advantage of imparting an appreciable degree of flame retardancy, when used at relatively low levels, regarding both the ignitability and flame spread hazards. For this reason they are very widely used. Bromine combinations are more effective than equivalent chlorine combinations and are frequently preferred despite being more expensive and more susceptible to photochemical degradation. All these halogen-containing formulations have the disadvantage of emitting more smoke than the untreated polymer formulations, a part of which is the corrosive hydrogen halide.

Dilution of the volatile products

An obvious way to slow down the reactions of the flame is by a simple dilution procedure where the diluent is inactive. One readily available diluent is water. What is needed is a compound that contains and releases large quantities of water at the same temperature as the polymer degrades into volatile products.

Alumina trihydrate, $Al_2O_3 \cdot 3H_2O$, is well suited to this application, as it starts to break down to give water (nearly 35% of its weight) and aluminium oxide at about 180–200°C. This reaction is responsible for the relatively recent high usage of the trihydrate as a flame retardant. In addition it has the advantages of being inexpensive, which reduces material cost, and does not result in the emission of any toxic or corrosive gases. Its main disadvantages are that high loadings of it are needed in the polymer formulation to impart a reasonable degree of flame retardancy, and that its degradation temperature is on the low side. This gives problems in the processing of some polymers when temperatures in this range are reached.

More recently magnesium hydroxide, $Mg(OH)_2$, has been investigated. It acts in an identical fashion to alumina trihydrate giving off nearly 31% of its weight as water at temperatures above 340°C. For polymers processed above 200°C, such as ABS, PVC and polypropene, the magnesium hydroxide can be incorporated without any premature decomposition occurring during melt processing. Unfortunately high loadings are still needed to obtain a reasonable degree of flame retardancy. Surface coatings have been used to improve processability and properties at these high loadings.

Promotion of char forming reactions

As is shown in the burning cycle, the burning of a polymer occurs by the combustion of the volatile products emitted during the oxidative thermal degradation of the polymer. If the degradation mechanism could be altered to produce more char and less volatiles the flame retardancy of the formulation should be decreased. Many flame retardants act in this way, promoting low energy, solid-state reactions which lead to the carbonization of the polymer and a carbonaceous char on the surface.

Many phosphorous additives act as flame retardants in this way in hydroxyl containing polymers such as cellulose. During the polymer degradation process, phosphorus acids are produced which lead to char via phosphorylation and dehydration reactions. Relatively low quantities of phosphorus compounds are needed to impart a reasonable degree of flame retardancy. Several boron additives behave in a similar way by the promotion of carbonaceous char through esterification and dehydration reactions.

Formation of a protective coating

Forming an intumescent coating is another way of providing a protective layer on the surface of the polymer. When heated at certain temperatures, these coatings swell and decompose to form an insulating barrier of carbonaceous char, which protects the polymer from continued combustion. The essential ingredients of these coatings are a carbonific such as dipentylerythritol, a dehydration catalyst such as ammonium pyrophosphate and a blowing agent such as melamine. They have the advantage of emitting innocuous products such as ammonia and water but they have never achieved much commercial success. Factors responsible for this include the high loadings required, difficulties of processing and a high price.

Some compounds act as flame retardants by forming a thermal shield on the polymer's surface without necessarily increasing the char content. The thermal shield insulates the polymer substrate, slowing down the thermal degradation process and subsequent formation of volatile gases. Compounds that normally behave in this manner are the high melting point inorganic compounds. This is recognized as being a secondary flame retardant role for alumina produced from the decomposition of alumina trihydrate and zinc borate, the latter forming a glass-like coating on the substrate.

Removal of heat of combustion

If the heat transferred back to the polymer from the combustion of the volatiles is sufficient to continue degradation, the burning cycle will be self-sustaining. If some of this heat is removed, the cycle will be broken and the combustion stopped because of a slowing of the rate of pyrolysis of the polymer.

Both alumina trihydrate and magnesium hydroxide reduce the flammability of formulations in this way as their breakdown reactions to release water are highly endothermic processes. Particulate fillers can be beneficial, as they act as heat sinks and conduct heat away, so that the polymer temperatures are lowered and no significant degradation occurs.

Smoke suppressants

As mentioned earlier, smoke is one of the hazards associated with the burning of polymeric materials. It has been identified as the cause of death in many instances. Any additive that will reduce the smoke generation of a given material is beneficial.

Much attention has been given to the thermal degradation of PVC, because it burns with the emission of a considerable amount of smoke. It has been found that compounds based on iron, molybdenum, tin and zinc act as smoke suppressants in this polymer by increasing the amount of char at the expense of the emission of volatiles.

SELECTION OF A FLAME RETARDANT

The primary concern when selecting a flame retardant for a given application is that it is effective to the extent required for the application. The end-use application will often determine the flame retardant selected. Assessment is made by exposing samples of the final formulation to a series of tests.

In addition the additive(s) should not adversely affect the physical or mechanical properties of the polymer formulation, must not colour the sample, and must not react with any other additive present. Ideally it must not give any problems during processing, must not lead to the emission of any toxic or harmful products, and must not interfere with attainment of desired product aesthetics and form.

In recent years there has been special interest in non-halogenated flame retardants because of their potential for providing reduced toxic emissions and less smoke generation. This trend has been promoted by governmental and regulatory bodies.

BIBLIOGRAPHY

Paul, K.T. (October 1993) Aspects of flame retardants. Paper presented at the conference *Enhancing Polymers using Additives and Modifiers*, RAPRA, Shawbury, England.

Skinner, G.A. (1981) Smoke – the hazard, the measurement and the remedy, *Journal of Chemical Technology and Biotechnology*, **31**, 445–452.

Cullis, C.F. and Hirschler, M.M. (1981) *The Combustion of Organic Polymers*, Clarendon Press, Oxford.

Conference Proceedings (January 1993), *Inorganic Fire Retardants – All Change?*, Industrial Division of the Royal Society of Chemistry, London.

Keywords: flame retardant, fire, smoke, mechanism, heat, flame spread, burning cycle, synergism, hazard, intumescent coating, char.

See also: Flame retardants: borates;
Flame retardants: halogen free systems (including phosphorus additives);
Flame retardants: inorganic oxide and hydroxide systems;
Flame retardants: intumescent systems;
Flame retardants: iron compounds, their effect on fire and smoke in halogenated polymers;
Flame retardants: poly(vinyl alcohol) and silicon compounds;
Flame retardants: synergisms involving halogens;
Flame retardants: tin compounds; Smoke suppressants.

Flame retardants: borates

Kelvin K. Shen and Roderick O'Connor

INTRODUCTION

Boron compounds such as borax and boric acid are well known fire retardants for cellulosic products [1]. However, the use of boron compounds such as zinc borate, ammonium pentaborate, boric oxide, and other metallo-borates in the plastics industry has become prominent only since the late 1970s. This entry will review the manufacturing, chemical and physical properties, end-use applications, as well as modes of action of major boron compounds as fire retardants in polymers. The subject is also mentioned in the section entitled 'Flame retardants: inorganic oxide and hydroxide systems'.

THE ORIGIN OF BORON

Boron is present in the earth's crust at a level of only 3 ppm, but there are areas where it is concentrated as borate salts (salt of oxidized form of elemental boron) in substantial volume for mining. The United States and Turkey supply 90% of global borate demand. The principal minerals in the deposits are tincal, kernite and colemanite (Table 1).

THE PRODUCTS

Table 2 lists properties of and applications for various types of boron-containing fire retardants.

Plastics Additives: An A–Z Reference
Edited by G. Pritchard
Published in 1998 by Chapman & Hall, London. ISBN 0 412 72720 X

Table 1 Principal borate minerals

Ore name	Formula	B_2O_3*	Source
Tincal	$Na_2O \cdot 2B_2O_3 \cdot 10H_2O$	36.5	United States, Turkey
Colemanite	$2CaO \cdot 3B_2O_3 \cdot 5H_2O$	50.9	Turkey
Kernite	$Na_2O \cdot 2B_2O_3 \cdot 4H_2O$	50.9	United States
Ulexite	$Na_2O \cdot 2CaO \cdot 5B_2O_3 \cdot 16H_2O$	43.0	Turkey, Chile
Hydroboracite	$CaO \cdot MgO \cdot 3B_2O_3 \cdot 6H_2O$	50.5	Argentina
Szaibelyite	$2MgO \cdot B_2O_3 \cdot H_2O$	41.4	Russia

*Theoretical maximum percent.

Borax pentahydrate ($Na_2O \cdot 2B_2O_3 \cdot 5H_2O$)

Commercially, borax pentahydrate is produced by conventional extraction and recrystallization from sodium borate mineral tincal ($Na_2O \cdot 2B_2O_3 \cdot 10H_2O$). Owing to its low starting dehydration temperature (approximately 60°C) and high water solubility (4.4% by wt. at room temperature), it has been used mainly as a fire retardant in cellulosic products. Its use as a fire retardant in urethane foams was also reported in several patents [2]. The anhydrous form of the material ($Na_2O \cdot 2B_2O_3$) is also commercially available, but its potential as a fire retardant in plastics has not been fully explored.

Boric acid ($B(OH)_3$)

Boric acid is produced by reacting borate mineral such as kernite ($Na_2O \cdot 2B_2O_3 \cdot 4H_2O$) or colemanite with sulphuric acid (equation 1):

$$Na_2O \cdot 2B_2O_3 \cdot 4H_2O + H_2SO_4 + H_2O \rightarrow 4B(OH)_3 + Na_2SO_4 \quad (1)$$

Owing to its low starting dehydration temperature (approximately 75°C) and significant water solubility (4.7%), boric acid is mostly used as a fire retardant in cellulosic products such as cellulosic insulation, cotton batting and medium density fibreboard. Depending on the fire standard to be met and co-additives to be used, the boric acid loading level is generally between 6 and 12% by weight. It has also been used as a fire retardant in flexible urethane foam.

Boric acid is also an important precursor of other boron-containing fire retardants.

Boric oxide (B_2O_3)

Boric oxide, also known as anhydrous boric acid, is produced commercially by calcining boric acid (equation 2):

$$2B(OH)_3 \xrightarrow{260-300°C} B_2O_3 + 3H_2O \quad (2)$$

Table 2 Boron-based fire retardant products

Chemical name	Starting dehydration temperature (°C)	Water solubility (wt % around 20°C)	Applications
Borax pentahydrate ($Na_2O \cdot 2B_2O_3 \cdot 5H_2O$)	65	4.4	Paper products and wood composite board
Boric acid (H_3BO_3)	70	4.7	Urethane foams, paper products, cotton batting and wood composite board
Boric oxide (B_2O_3)	–	2.6	Engineering plastics
Calcium borate – colemanite ($2CaO \cdot B_2O_3 \cdot 5H_2O$)	300	0.4	Rubber modified ashphalt
Ammonium pentaborate (($NH_4)_2O \cdot 5B_2O_3 \cdot 8H_2O$)	120	10.9	Urethane, epoxy and other coatings
Barium metaborate ($BaO \cdot B_2O_3 \cdot H_2O$)	200	0.4	Flexible PVC and coatings
Disodium octaborate tetrahydrate ($Na_2O \cdot 4B_2O_3 \cdot 4H_2O$)	40	9.7	Paper products
Melamine diborate ($C_3H_6N_6 \cdot 2B(OH)_3$)	130	0.7	Water-based coatings, and epoxy resins
Zinc borates ($2ZnO \cdot 3B_2O_3 \cdot 3.5H_2O$)	290	0.2	PVC, polyolefins, nylon silicone, urethane, unsaturated polyesters, and elastomers
($2ZnO \cdot 2B_2O_3 \cdot 3H_2O$)	190	0.2	PVC
($2ZnO \cdot 3B_2O_3$)	–	0.2	Engineering plastics

This material has been reported to be an effective fire retardant at very low loadings (0.5–2%) in engineering plastics such as high impact polystyrene, polyetherketone and polyetherimide [3]. Boric oxide, when used in conjunction with red phosphorus, is an effective fire retardant in fibreglass reinforced nylon 6,6. Boric oxide softens at about 325°C. It is, however, hygroscopic and can absorb water and revert to boric acid without losing its fire retardant efficacy.

Ammonium pentaborate ((NH_4)$_2$O·5B_2O_3·8H_2O)

It can be produced by reacting ammonium hydroxide and boric acid. It has been used as a fire retardant mostly in urethane foam and epoxy. This material releases its water of hydration beginning at about 120°C and releases ammonia beginning at about 220°C. Its usage in polymers, however, is limited by its high water solubility (10.9%) and low dehydration temperature.

Barium metaborate (BaO·B_2O_3·H_2O)

It can be produced from borax and barium sulphide, and finds use primarily as a corrosion inhibitor and fungistatic pigment in coatings. This material is also claimed to be a fire retardant in plastics, where it is mostly used as a partial replacement for antimony oxide in halogen-containing polymers. But its efficacy is not as good as that of zinc borates (see below). Its usage in plastics is also limited by its low starting dehydration temperature (200°C).

Calcium borate (2CaO·3B_2O_3·5H_2O)

This natural mineral – colemanite – is thermally stable up to about 300°C. Depending on its clay content, the B_2O_3 content of this material can vary from 34 to 44%. It has been used mostly in applications such as fire retardant rubber modified asphalt, where chemical purity and consistency are not so critical.

Disodium octaborate tetrahydrate (Na_2O·4B_2O_3·4H_2O)

This sodium borate is produced from a boric acid and borax solution. It is an amorphous material and thus can be dissolved into water rapidly (9.7% at room temperature). It is used mainly as a fire retardant for cellulosic materials.

Melamine diborate ($C_3H_6N_6$·2B(OH)$_3$)

This material is produced by reacting boric acid with melamine in water. It is a white powder which is virtually insoluble in water (0.7% by wt.).

It starts releasing water at 100°C and continues up through 200°C. Above 200°C, boric oxide is formed. At 300°C, the melamine starts to decompose. It is used mostly in water-based coatings and epoxy resins.

Zinc borates

This is the most important class of boron compound used as a fire retardant in the plastics industry. Zinc borates are commercially produced by reacting either boric acid or sodium borate with zinc oxide. Depending on the reaction conditions, a host of zinc borates with different mole ratios of $ZnO:B_2O_3:H_2O$ can be produced.

Zinc borate ($2ZnO \cdot 3B_2O_3 \cdot 3.5H_2O$)

This form of zinc borate was discovered in 1967 (equation 3). In contrast to other commercial zinc borates, it is stable up to 290°C and is suitable for use even in engineering plastics such as nylons [4] (Figure 1). It is available from several manufacturers (Table 3).

$$2ZnO + 6B(OH)_3 \rightarrow 2ZnO \cdot 3B_2O_3 \cdot 3.5H_2O + 5.5H_2O \qquad (3)$$

In halogen-containing polymers, the zinc borate can either partially or completely replace antimony oxide as a fire retardant synergist of halogen sources. The combination of antimony oxide and zinc borate can provide not only cost savings but also synergism in fire test performances (Figure 2). In addition, in contrast to antimony oxide, the zinc borate also

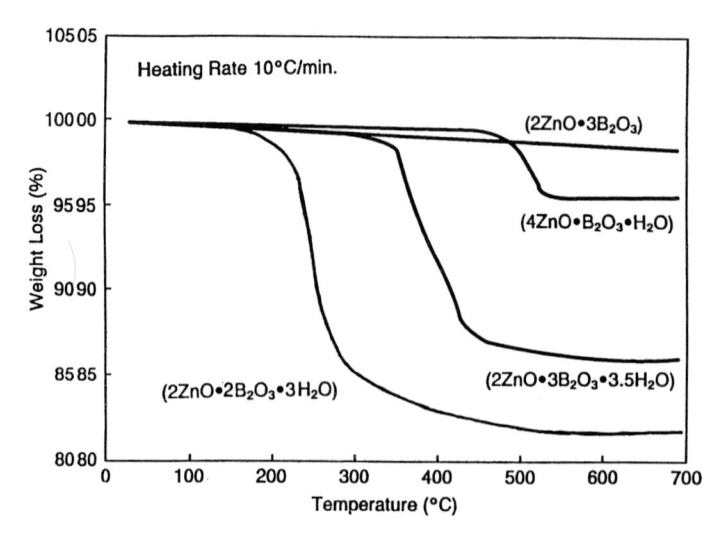

Figure 1 Thermogravimetric analysis of zinc borates.

Table 3 Major suppliers of boron-based fire retardants

Trade name	Producer	Composition
Ammonium pentaborate	U.S. Borax Inc.	$(NH_4)_2O \cdot 5B_2O_3 \cdot 8H_2O)$
Polybor® disodium octaborate tetrahydrate	U.S. Borax Inc.	$Na_2O \cdot 4B_2O_3 \cdot 4H_2O$
Boric acid	U.S. Borax Inc., NACC	$B(OH)_3$
Boric oxide	U.S. Borax	B_2O_3
Bulab® Flamebloc	Buckman Lab	$BaO \cdot B_2O_3 \cdot H_2O$
Colemanite	Etibank	Crude $2CaO \cdot 3B_2O_3 \cdot 5H_2O$
Neobor® borax pentahydrate	U.S. Borax Inc.	$Na_2O \cdot 2B_2O_3 \cdot 5H_2O$
Firebrake® ZB zinc borate	U.S. Borax Inc.	$2ZnO \cdot 3B_2O_3 \cdot 3.5H_2O$
Firebrake® 500 zinc borate	U.S. Borax Inc.	$2ZnO \cdot 3B_2O_3$
Melamine diborate	DSM	$(C_3H_6N_6 \cdot 2B(OH)_3)$
V-bor	NACC	$Na_2O \cdot 2B_2O_3 \cdot 5H_2O$
Zinc borate	Joseph Storey, Larderello	$2ZnO \cdot 3B_2O_3 \cdot 3.5H_2O$
ZB-467	Anzon	$4ZnO \cdot 6B_2O_3 \cdot 7H_2O$
ZB-223	Anzon	$2ZnO \cdot 2B_2O_3 \cdot 3H_2O$

functions as a smoke suppressant, afterglow suppressant and anti-tracking agent. Its loading level is typically in the range 3–25 phr (parts per hundred parts of resin).

Where halogens are not used, the recommended zinc borate usage level is in the range 5–200 phr, normally used in conjunction with alumina

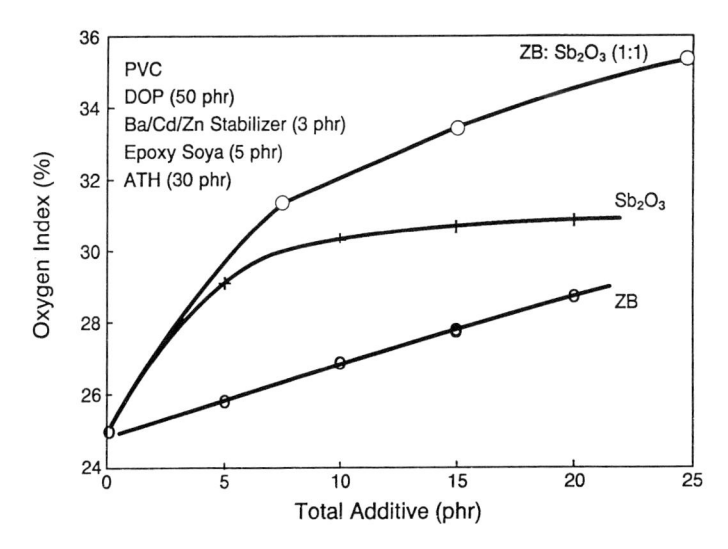

Figure 2 Oxygen index of flexible PVC formulations.

trihydrate, magnesium hydroxide or a silica source such as silicone polymers.

Major applications of zinc borate include flexible PVC (wire and cable, auto upholstery, conveyor belting, wall covering, floor tiles and tenting), rigid PVC (panels), polyolefins (wire and cables and electrical parts), nylon (electrical connectors and adhesives), SBR (flooring and conveyor belting), epoxy (coatings), thermoset polyester (panels), and silicones (insulator, coatings and wire insulations).

Zinc borate ($2ZnO \cdot 2B_2O_3 \cdot 3H_2O$)

This zinc borate has also been used in some PVC wire and cable formulations. Its application has been limited by its low dehydration temperature (starting at around 190°C).

Zinc borate ($2ZnO \cdot 2B_2O_3$)

This anhydrous zinc borate is produced by calcining zinc borate ($2ZnO \cdot 3B_2O_3 \cdot 3.5H_2O$). It is intended for use in engineering plastics such as polyether ketone and polysulphones, which are processed at temperatures above 290°C.

FIRE RETARDANT MECHANISM OF BORON COMPOUNDS

For cellulosic systems, endothermic release of water and melting to form a sodium borate or boric oxide coating from borate fire retardants (such as boric acid, borax pentahydrate and disodium octaborate tetrahydrate) are responsible for the fire retardancy. Chemically, borates can also react with the C6-hydroxyl group of cellulose to give a borate ester, blocking the release of flammable C-6 fragments. In addition, dehydration of alcohols is enhanced by boric acid, which accounts for the larger quantities of char produced when borate treated cellulose is burned.

Boric oxide or any precursors such as boric acid, which can form boric oxide under fire conditions, is well known to reduce afterglow combustions. It is generally believed that boric oxide, a low melting glass, can cover the char and inhibit oxygen permeation, and thus prevent oxidation of the char. Another school of thought believes that effectiveness of boric oxide in afterglow suppression is related to its high ionization energy/or electron affinity. Active sites for oxygen adsorption on the char surface may be deactivated by boric oxide via electron transfer [5].

Ammonium pentaborate, upon thermal decomposition, can evolve a large amount of water, ammonia, and boric oxide, which is a glass former. Ammonium pentaborate functions as both an inorganic blowing agent and a glass forming fire retardant (equation 4):

$$(NH_4)_2O \cdot 5B_2O_3 \cdot 8H_2O \xrightarrow{120-450°C} 2NH_3 + 9H_2O + 5B_2O_3 \quad (4)$$

Zinc borate in halogenated systems, such as flexible PVC, is known to increase markedly the amount of char formed during polymer combustion; whereas the addition of antimony, a vapour phase flame retardant, has little effect on char formation. The hydrogen chloride released from PVC thermal degradation can react with the zinc borate [6] to form zinc hydroxychloride and zinc chloride as well as boric oxide and boron trichloride (equation 5).

$$2ZnO \cdot 3B_2O_3 \cdot 3.5H_2O + HCl \rightarrow ZnCl_2 + Zn(OH)Cl + B_2O_3 + BCl_3 + H_2O \quad (5)$$

The zinc species remaining in the condensed phase can alter the pyrolysis chemistry by catalysing dehydro-halogenation and promoting cross-linking, resulting in increased char formation and a decrease in both smoke production and flaming combustion. The boric oxide released can promote the formation of a strong char and, more importantly, can stabilize the char; whereas the boron trichloride released, in a much smaller proportion relative to boric oxide, can function as a gas phase flame retardant. The water released at above 290°C can promote the formation of a foamy char which can act as a good thermal insulator.

Where halogen is not used, it was reported that zinc borate in conjunction with alumina trihydrate promotes the formation of a porous ceramic residue during polymer combustions. This residue is an important thermal insulator for the underlying substrate or unburned polymer. Both differential scanning calorimetric and differential thermogravimetric analyses, as well as char yield analysis suggest that partial replacement of alumina trihydrate with zinc borate in a halogen-free polymer such as ethylene-vinyl acetate can favourably alter the oxidative-pyrolysis chemistry of the halogen-free base polymer.

PRODUCERS AND TRADE NAMES

Table 3 lists major producers and trade names of boron-based fire retardants.

REFERENCES

1. Lyons, J.W. (1970) *The Chemistry and Uses of Fire Retardants*, Wiley-Interscience, NY.
2. Ishikawa, T. (1975) *Japan Kokai*, 76, 124, 195.
3. Lohmeijer, J.H. *et al.* (1990) European Patent Application, 0 364 729A1.
4. Shen, K.K. and Griffin, T.S. (1990) American Chemical Society Symposium Series 425, *Fire and Polymers*, p. 157.

5. Rakszawski, J.F. and Parker, W.E. (1964) *Carbon*, 2, 53.
6. Shen, K.K. and Sprague, R.W. (1982) *Journal of Vinyl Technology*, 3, 120.

Keywords: borates, flame retardant, boric acid, borax, char, boric oxide, smoke suppression, afterglow suppression, dehydration temperature.

See also: Series of entries with titles 'Flame retardants:...';
Flame retardancy: the approaches available;
Smoke suppressants.

Flame retardants: halogen-free systems (including phosphorus additives)

John Davis

INTRODUCTION

There are three essential conditions to be met if a polymer, once ignited, is to continue burning. There must be a supply of heat to the bulk polymer, a generation of fuel (typically volatile decomposition products) and there must be a flame. Halogen-based systems act by a well-documented flame poisoning mechanism in the vapour phase. The alternative halogen-free systems, which encompass a wide variety of additives, tend to act by mechanisms which disrupt heat flow and the supply of fuel to the flame. Here the mechanisms are not always understood in great detail but two broad types of flame retardant action can be defined.

First there are additives which act to remove heat by endothermic decomposition and/or the generation of copious quantities of inert gases to dilute the combustible polymer degradation products. Materials such as alumina trihydrate (ATH) and magnesium hydroxide, which in tonnage terms are by far the most widely used halogen-free flame retardants, work in this way. These additives are more fully described in the section 'Flame retardants: inorganic oxide and hydroxide systems'.

The second type of flame retardant action involves the formation of char and this is most often accomplished by phosphorus-containing additives. Char formation is a process which occurs mostly in the condensed phase and has several benefits. A good char layer is difficult to ignite and acts as a physical barrier. It hinders the escape of polymer degradation

Plastics Additives: An A–Z Reference
Edited by G. Pritchard
Published in 1998 by Chapman & Hall, London. ISBN 0 412 72720 X

products to the flame zone and restricts oxygen access. It can also insulate the underlying polymer against thermal degradation, thereby significantly reducing the rate of fuel generation. Modern techniques of characterizing fire properties such as the use of the cone calorimeter have stressed the importance of factors such as minimizing the rate of heat release and mass loss. Char-forming systems, because they interfere with polymer degradation itself rather than act on the combustion processes of the degraded polymer, tend to perform well against these criteria.

Char formation by phosphorus flame retardants is easiest to achieve in oxygenated polymers such as polyesters, polyamides and polyurethanes. In hydrocarbon-like polymers such as polyolefins and styrenics it is more common to find combinations of phosphorus compounds with other additives. Flame retardant systems based on elemental phosphorus, phosphates and organophosphorus compounds are considered separately below. Very often nitrogen compounds have been used as co-additives in these charring systems, usually because synergistic behaviour can be observed. However in some cases melamine and its derivatives can claim to exhibit some primary flame retardant effect, in the absence of phosphorus, which does not involve charring. Other halogen-free additives, including zinc borates, molybdenum oxides, tin and iron compounds have been promoted. However, they tend to be used as adjuncts to halogen flame retardants or in halogen-containing polymers and hence fall outside the scope of this section.

One other approach to flame retarding plastics is to incorporate a more fire-resistant polymer, i.e. to use a polymer blend or alloy. The obvious commercial example is the use of poly(phenylene oxide) in polystyrene. It is notoriously difficult to achieve flame-retardant PS with halogen-free additives but here the char-forming ability of the PPO makes the blend more fire-resistant so that the addition of a simple triaryl phosphate will suffice. There are also references to the use of small quantities of powdered PTFE as an anti-drip agent in a variety of formulations, though this somewhat goes against the 'halogen-free' description.

In assessing the relative performance of different flame retardant additives in plastics a number of small scale tests have been developed. These are carried out under well-defined conditions and attempt to measure reduced rates of burning or flame propagation and/or increased resistance to ignition. It should be noted that the performance of a flame retardant plastic in small scale tests does not necessarily reflect the hazards present under actual fire conditions.

The most widely quoted small-scale test is the Underwriters Laboratory UL-94 Vertical Burning Test, in which materials are classified as V-0, V-1, V-2 or are non-rated (NR). In the test the bottom of a vertically-mounted test strip is exposed to a flame twice, each time for 10 seconds duration, after which any burning behaviour is timed. To achieve the top V-0 rating, burn times after each flame application must be less than 10 seconds and the sample must not generate flaming drips. For V-1, burn times up to 30 seconds are permitted and for V-2, 30 second burn times plus some flaming drips are allowed. The classification is highly dependent on the thickness of the test strip. In practice V-0 or V-2 ratings at either 3.2 or 1.6 mm are the most frequently seen specifications.

Measurements of Limiting Oxygen Index, a candle-type burning test in atmospheres of varying oxygen content, is another test which has proved useful in formulation development. Description of other test methods have been summarized by Troitzsch [1].

RED PHOSPHORUS

The simplest phosphorus-based flame retardant is the element itself in the form of amorphous red phosphorus. This additive is effective in a wide range of polymeric systems including polyolefins, polyamides and thermosetting resins. As the name implies, it is manufactured as a dark red powder. However its use in powder form demands careful processing because, although not spontaneously flammable like white phosphorus, it is capable of being ignited by sparks or overheating. Flame retardant manufacturers have addressed these problems by offering red phosphorus in a masterbatch form in a variety of carrier polymers. These are safe to handle and can be let down on conventional compounding equipment. As the final loading of red phosphorus is typically less than 10% the masterbatch route does not present any economic difficulties either.

Red phosphorus is believed to act in both the solid and vapour phase. In the solid phase it oxidizes eventually to yield polyphosphoric acid species which, as a non-volatile film, acts as a barrier to polymer degradation products reaching the flame. In oxygen-containing polymers, or in the presence of co-additives, some useful char formation can be expected. Some flame poisoning by phosphorus-containing radicals is also thought to occur in the vapour phase. Separate studies have shown that PO$^•$ radicals can act as scavengers for H$^•$, one of the key radical species involved in flame propagation reactions.

One application where red phosphorus is widely used is for electrical switches and mountings moulded from glass-filled nylon. These are internal parts where the brick-red colour imparted by the phosphorus is unimportant. Other flame retardants are also effective in nylon but different end uses favour different additives depending on their performance characteristics. This is shown in Table 1 where red phosphorus is compared with a typical halogen system and two other halogen-free additives, magnesium hydroxide and melamine cyanurate. Among the

Table 1 Comparison of flame retardant suitability in nylon 6 and 6,6

	Halogen + Sb_2O_3	Red phosphorus	Melamine cyanurate	Magnesium hydroxide
Typical loading (wt %)	19 + 5	7	10	55
UL-94 rating	V-0 (0.8 mm)	V-0 (0.8 mm)	V-0 (0.8 mm)	V-0 (1.6 mm)
Smoke hazard	Toxic, corrosive	Low	Low	Low
Effective with glass fibres	Yes	Yes	No	Yes
Comparative Tracking Index	250 V	400–600 V	>600 V	>600 V
Preferred end-use	Glass-filled non-electrical	Glass-filled electrical	Non glass-filled	Nylon 6 only

factors considered are the loading required to achieve flame retardancy (in terms of UL-94 rating), whether the additive generates harmful smoke, whether it is effective in both unfilled and glass-filled systems and whether the compound is suitable for electrical applications (as defined by a CTI (Comparative Tracking Index) >400 V).

The decomposition temperature of magnesium hydroxide (~260°C) means that its use is restricted to nylon 6. It has to be used at high loadings and this inevitably impairs mechanical properties but, on the other hand, mould shrinkage is low, so parts with high dimensional stability can be obtained. Much more widespread is the use of melamine cyanurate. This additive, which is a white powder stable to almost 400°C, is extremely effective at low loadings, with little smoke hazard and providing good retention of mechanical and electrical properties. In part it achieves its flame retardant action by promoting dripping and this is also its Achilles' heel, for it is remarkably ineffective in glass-filled nylon. It is believed that the glass reinforcement hinders the dripping and that once alight the glass fibres have a wicking effect, acting to conduct molten polymer into the flame zone. Of the other candidates, halogen systems give a low CTI value and generate corrosive smoke on burning so are inappropriate for electrical applications. This makes red phosphorus the preferred additive in these areas.

A number of ways have been found to enhance the effectiveness of red phosphorus. For instance, the addition of phenolic resin can improve UL-94 ratings in both polyamides and polyesters. The result is better charring, involving dehydration of the phenolic by phosphorus acids and the formation of a more crosslinked network which retards dripping. The addition of EVA copolymers can have a similar effect. In epoxy resin systems red phosphorus is typically used with ATH. Small amounts of red phosphorus can replace large amounts of ATH giving a lower overall loading. This is a case of two flame retardants combining to give a better cost/performance ratio.

Intumescent flame retardants

Intumescent systems are a particular class of phosphorus-based flame retardant. They are characterized by the production of an expanded foamed char on exposure to heat which then protects the underlying polymer. Ideally the char volume should be large, with a thick continuous surface crust and an inner structure which should be similar to a closed-celled foam. The char structure is important in providing a good insulative effect and hence reducing thermal decomposition of the polymer to volatile flammable products.

Few single additives display intumescence so instead these materials are carefully formulated blends. They are particularly suited to flame retarding polypropylene although LDPE, HDPE and EVA plastics can also be protected. In their favour intumescent systems have the advantage of being effective at lower loadings than ATH or magnesium hydroxide and of producing less corrosive decomposition products than the halogen flame retardants. In addition smoke production is generally low. The chief drawback is their limited thermal stability which restricts the range of processing temperatures. Improvements here would considerably further the commercial acceptance of these materials. Typical applications are in injection-moulded parts, from white goods to fan housings and stadium seating.

Intumescent systems are usually based on ammonium or amine phosphates, of which ammonium polyphosphate (APP) has received the most attention, followed by melamine phosphate (MP) and ethylenediamine phosphate (EDAP). Other related materials which have been reported or used in mechanistic studies are melamine amyl phosphate and di-ammonium pyrophosphate.

All the phosphates can be viewed as sources of phosphoric acid or polyphosphoric acid and most of the development work in this area has concentrated on identifying the best co-additives to employ. In simple systems such as APP blended with polyhydric compounds like pentaerythritol it has been established that, above 200°C, the components react to form phosphoric ester bonds. Further elimination of water and ammonia leads to a carbon–phosphorus char. Without the pentaerythritol no char is formed. The char structure can be further improved by the addition of melamine which in this case can be viewed as a blowing agent, increasing the char volume. The formation of the char can therefore be seen as a sequence of overlapping reactions which must occur in a particular order for optimum performance [2].

In this context, melamine phosphate can be viewed as a combination of acid source and blowing agent. Formulation with a polyhydric compound such as dipentaerythritol (which is more stable than pentaerythritol) produces an effective system. Ethylenediamine phosphate,

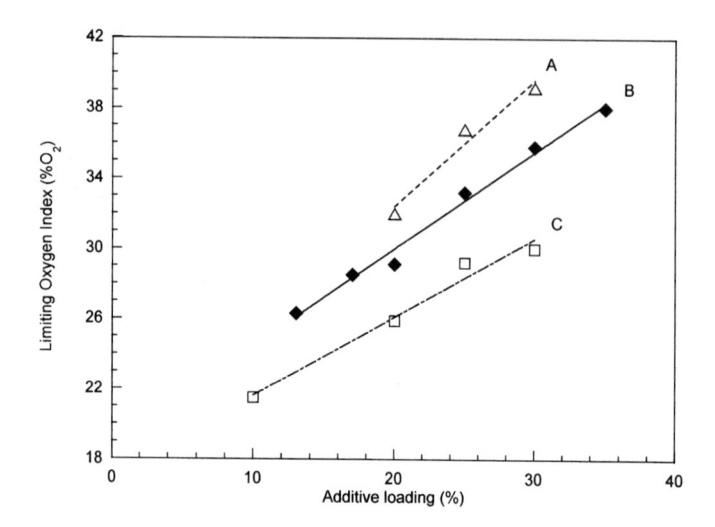

Figure 1 LOI of intumescent systems based on various phosphates: A – ammonium polyphosphate + poly(triazine-piperazine); B – ethylenediamine phosphate + melamine; C – melamine phosphate + dipentaerythritol.

the only phosphate to show some intumescent behaviour without the need for co-additives, is instead formulated with nitrogen compounds such as melamine. In this case synergistic behaviour can be clearly demonstrated.

The formulation of more recent APP-based systems also seems to be based on phosphorus–nitrogen synergism. Co-additives are polymeric in nature such as poly(triazine-piperazine) materials, essentially substituted melamine rings linked by piperazine groups, and poly(ethylene-urea-formaldehyde) condensates. The sequence of reactions leading to char formation in these systems is poorly understood. It is likely that both water and ammonia are evolved at certain stages, some phosphorus and probably some nitrogen remain incorporated in the char structure, and some phosphorus ends up as polyphosphoric acid.

The flame retardant effect in polypropylene of three intumescent systems, each based on a different phosphate, is shown in Figure 1. This demonstrates that effective systems can be developed from a variety of starting materials. The Limiting Oxygen Index increases linearly over the typical loading range of 15–40%. This greatly simplifies flame retardant evaluation because a single measurement at, say, 30% loading is sufficient to characterize the effectiveness of different formulations.

There is a reasonably good correlation between LOI and UL-94 test ratings. Thus it will usually be found that a flame-retarded polypropylene of LOI 28–30 will also meet UL-94 V-0 at 3.2 mm whilst a LOI of 30–32 is a good indicator of a V-0 pass at 1.6 mm. With lower flame retardant loadings the polypropylene will typically fail the UL-94 test because, unlike halogen flame retardants, there is no significant loading range over which a UL-94 V-2 rating can be expected. If a V-2 rated polypropylene is specified then this is more effectively accomplished with a low loading (<10%) of a halogen system. However, it should be remembered that this rating is only achieved because the halogen system promotes dripping of the ignited polymer and hence ensures rapid removal of the flame (and heat) from the test piece.

Once a certain level of flame retardancy has been achieved other factors serve to separate one system from another. For instance, APP suffers from poor water resistance because it can be hydrolysed back to soluble ammonium orthophosphate. Encapsulated grades have been developed to overcome this drawback.

The effectiveness of intumescent flame retardants is frequently reduced when fillers are added. Interactions can be either chemical or physical. Materials which are basic in character such as aluminium and magnesium hydroxides and calcium carbonate tend to interfere chemically with the phosphoric acid precursor in the intumescent system, presumably forming inorganic phosphates. Such antagonistic behaviour can be easily recognized by an almost complete lack of char formation.

For fillers such as talc, mica, titanium dioxide the interactions are more varied. Indeed, if the level of filler is relatively low, say 1–5% on top of a flame retardant loading of 25–35%, it is possible to observe a small enhancement of flame retardancy. This has been attributed to improved char mechanical strength or better thermal shielding caused by the filler reflecting more radiant heat. As the filler loading is increased from 5 to 20% the flame retardancy is progressively reduced – a lack of char swelling becomes apparent and the cellular structure of the inner char is lost. It is thought that the filler confers too much rigidity to the char as it begins to form, which retards and limits the blowing action and makes it prone to cracking when swelling does occur. The result is an inadequate physical barrier to the passage of volatiles and a poor thermal insulator.

Other plastics additives, especially colorants, need to be carefully screened. Certain organic pigments such as phthalocyanine blues or greens are known to interfere with intumescent systems.

ORGANOPHOSPHORUS COMPOUNDS

A vast number of organophosphorus compounds have been proposed as flame retardants. Alkyl and aryl phosphates are well-known plasticizers for PVC and also have a flame retardant effect. Organophosphorus compounds, sometimes also containing chlorine or bromine, are also

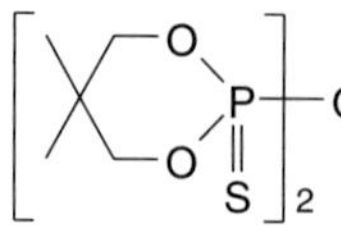

Figure 2 Organophosphorus flame retardants for plastics.

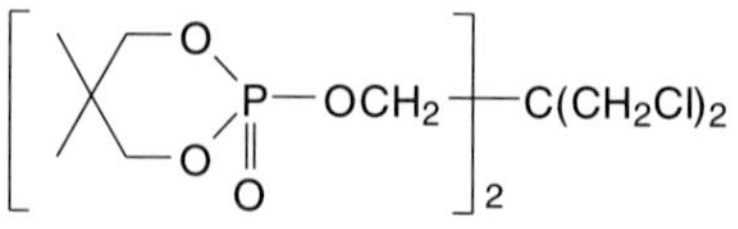

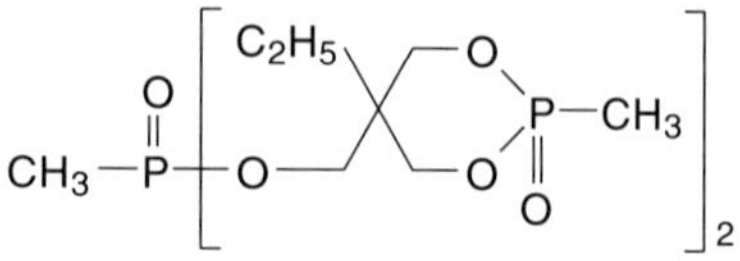

used in polyurethane foams, but their use in other plastics is relatively limited. The chief problem is one of thermal stability. At the processing temperature of thermoplastics many of these compounds will be liquid and pose problems with volatility and decomposition as well as handling difficulties for compounders used to dealing with solid additives.

One apparent exception is triphenyl phosphate which, although having a low melting point of 50°C, is stable enough to be processed into engineering thermoplastics such as polycarbonate/ABS blends. A vapour phase action is assumed as it boils undecomposed at 370°C.

A number of other compounds which have found use in niche applications show structural similarities in that they contain one or more cyclic 1,3,2-phosphorinane structures. This cyclic structure affords higher melting points (up to around 200°C) and lower volatility than comparable open chain compounds of similar phosphorus content and molecular weight [3]. The idea is that, on exposure to heat, the additive will not boil off but remain in the polymer to exert some char-forming action when it decomposes in the 300–400°C range.

Such compounds tend to be used when they have a very specific advantage over other flame retardant types – for example meeting requirements such as transparency or in applications such as films and fibres where particulate additives would be undesirable. A number of examples are shown in Figure 2. Compound I can be used in blown polyethylene film, showing better transparency and with less effect on mechanical properties than a halogen–antimony system. Compound II, which also contains chlorine, is effective at 10–15% loading in PMMA. Though a white powder melting at 190°C, it is completely soluble in the polymer and so excellent transparency is retained. A liquid additive such as compound III is effective at low levels in thermoplastic polyester – as little as 3.5% is required in PET to boost the UL-94 rating (at 1.6 mm) from V-2 to V-0.

MELAMINE

Besides the use of melamine as a component of intumescent systems it has also been advocated as a flame retardant in its own right. It has the advantages of being inexpensive, readily dispersible in most thermoplastics and is commercially available in grades of varying particle size. The flame retardant effect is largely due to a combination of heat sink effects and vapour phase dilution effects so in these respects it has some similarities with the hydrated filler flame retardants.

A number of processes contribute to the flame retardancy [4]. First, melamine sublimes at around 350°C. The heat of sublimation is around $-29\,\text{kcal mole}^{-1}$ and therefore has a considerable cooling effect, thus retarding ignition. The melamine vapour also dilutes the gaseous combustibles from polymer degradation. It is a poor fuel and amongst its combustion products nitrogen also acts as an inert diluent. In addition, the endothermic vapour phase dissociation of melamine to cyanamide may occur, leading to more cooling. Lastly, decomposition of cyanamide and melamine will release ammonia, which has long been recognized as having a flame retardant effect. As with most flame retardant systems the relative contribution of these different processes has not been established.

The use of melamine has been reported in polyolefins, polystyrene and polyamides. Loadings range from less than 10% in nylon 6 and 6,6 to 50% in polystyrene. In polyamides melamine has a tendency to give plateout or blooming and so its 1:1 salt with cyanuric acid, melamine cyanurate, is preferred in practice. Melamine cyanurate is more thermally stable but otherwise its flame retardant action is essentially the same. The use of melamine compounded with finely divided poly(phenylene oxide) and kaolin has given good results in EPDM rubber for cable insulation. This is a good example of an application imposing constraints on the choice of additive and where two different modes of flame retardant action have been utilized. Melamine acts primarily in the vapour phase and the PPO acts to bind the kaolin filler into a more robust coherent char to protect the polymer surface.

The practical development of halogen-free systems will continue to exploit the potential of additive blends, particularly where synergistic effects can be seen. Such blends offer the best opportunity of tailoring a combination of flame retardant mechanisms to achieve an efficient, cost-effective performance in each plastics application.

REFERENCES

1. Troitzsch, J. (1990) *International Plastics Flammability Handbook*, 2nd edn, Hanser, Munich.
2. Camino, G. and Costa, L. (1986) Mechanism of intumescence in fire retardant polymers. *Reviews in Inorganic Chemistry*, 8(1, 2), 69–100.

3. Wolf, R. (1986) Phosphorus-containing fire retardants for transparent plastics and film. *Kunstoffe*, **76**(10), 943–947.
4. Weil, E.D. and Choudary, V. (1995) Flame-retarding plastics and elastomers with melamine. *Journal of Fire Sciences*, **13**, 104–125.

Keywords: endothermic decomposition, alumina trihydrate, magnesium hydroxide, phosphorus, organophosphorus, intumescent flame retardants, melamine.

See also: Series of entries with titles 'Flame retardants:...';
Flame retardancy: the approaches available;
Smoke suppressants.

Flame retardants: inorganic oxide and hydroxide systems

S.C. Brown

INTRODUCTION

This category of flame retardants contains the two major additive flame retardants in use today – aluminium trihydroxide, Al(OH)$_3$ and antimony trioxide, Sb$_2$O$_3$. The broad discussion is best separated into two parts. First we shall consider metal hydroxides, which in general decompose endothermically to liberate water. They are normally smoke suppressants and work predominantly in the condensed phase of combustion. Secondly, we shall deal with metal oxides, which are only smoke suppressants in specific circumstances and find greatest use as 'synergists' in conjunction with other flame retardant additives, notably those containing the halogens chlorine and bromine. As such they are often vapour phase flame retardants. Two noteworthy additives, or rather families of additives, the zinc borates and zinc stannates, fall into both categories. These are, for convenience, discussed under 'Metal oxides' below, although borates will be discussed more fully in a separate entry.

METAL HYDROXIDES

Aluminium trihydroxide

Aluminium trihydroxide is often incorrectly called alumina trihydrate, but fortunately the abbreviation ATH serves for both. It is also known as Gibbsite or Nordstrandite or Hydrargillite. Some 40–50 million tpa are manufactured from bauxite ores by extraction using sodium

Plastics Additives: An A–Z Reference
Edited by G. Pritchard
Published in 1998 by Chapman & Hall, London. ISBN 0 412 72720 X

hydroxide and subsequent precipitation – the Bayer process. The vast majority of this Bayer ATH is then converted by calcination to aluminium oxide (alumina) and then electrolytically to aluminium metal. The earliest recorded data on the evaluation of ATH as a flame retardant were in the 19th century when Sir William Henry Perkin found it to be the best of the 25 additives that he studied to reduce the flammability of cotton flannelette. Little further work was carried out until the 1950s, when patents appeared which extolled the virtues of ATH as an additive for polyesters, neoprene butyl rubber and epoxies to improve the arc resistance of electrical components. In the 1960s Connolly and Thornton showed the improvements that could be obtained in Limiting Oxygen Index and reduced burning rates for polyesters containing ATH, and this led to substantial growth in its use in polyesters for building panels, electrical components, machine housings, automotive parts etc. [1]. Through the last quarter of a century ATH has made inroads into virtually all thermoset and rubber markets. ATH is little used in thermoplastics with the exception of PVC and 'thermoplastic' cable sheathing. Estimates of annual world-wide consumption of ATH as a flame retardant range from 100 000 tpa to >300 000 tpa depending on the survey consulted. Major markets include polyester and acrylic thermosets, ethylene copolymer cable sheathing compounds, PVC and latex-based flooring.

The key benefits of ATH are that it is cost effective and of low health hazard – it might be termed environmentally friendly by comparison with some other flame retardants. In addition to its primary function of flame retardancy it has useful secondary characteristics. For example, large quantities are used in synthetic marble whereby a clear thermoset polyester or acrylic is filled with a high whiteness ATH, of similar refractive index to the resin, to make high quality work surfaces or vanity ware with a visually appealing onyx effect. The smoke suppression benefits are also increasingly being valued in the marketplace. After-flow suppression and tracking-suppression are other key benefits. The wide range of demands and applications for ATH has resulted in the commercialization of several hundred different grades from perhaps twenty significant suppliers. Prominent among these suppliers are ALCOA (USA), ALCAN (Canada and UK), Martinswerk (Germany), Showa Denka (Japan) and Huber-Solem (USA).

Viscosity control

The first differentiating factor between commercially available ATH grades is median particle size, or d_{50}. In the 1970s most grades of ATH were derived by grinding from a normal coarse 'Bayer' ATH feed of anything from fifty to one hundred microns (µm). This feed was reduced to a series of products from around 30 µm down to 4 µm. Hammer mills, roller mills and ball mills were the favoured comminution devices. These simple ground products still dominate much of the business in the 1990s. ATH is effectively a filler in the sense that it is often less expensive than the polymer it displaces (at least on a weight basis), and by comparison with reactive flame retardants it needs to be used at high loadings. Thus most users will wish to maximize the ATH loading in their product and the constraint will often be process viscosity (which increases with increasing loading). Simple ground products do not have an ideal shape, surface chemistry or size distribution to give good viscosity. From the late 1970s a variety of 'low viscosity' products were launched which offered more suitable particle size distributions to enable particles to pack together more effectively at high loadings, hence giving lower viscosity, or allowing higher loadings for the same viscosity. The initial grades were generally bimodal mixtures at the appropriate proportions of two simple milled products. Later grades were more sophisticated (e.g. tri- or quadri-modal) and used more sophisticated feeds or other means than simple blending to achieve optimum size distributions. These products have enjoyed considerable success in thermoset polyester applications.

Another route to lower viscosity is by chemical modification of the ATH surface, and silane coatings are acknowledged to be technically the best modifiers. These coatings render the ATH surface more compatible with the host matrix. A second benefit of coatings, particularly silane coatings, is improved mechanical properties. It is debatable, and largely academic, whether the improvements in mechanical properties are due to a 'coupling' reaction between the polymer and the coating, combined with a bonding of the coating to the ATH surface, or whether the effect is simply due to improved dispersion of the ATH. Recent developments in viscosity performance have combined one or more of the above effects with grades of improved particle shape and surface roughness. A smooth blocky precipitated ATH has inherently less surface for any given d_{50} than a milled product and hence lower viscosity.

For products of 2 μm and below, various precipitation processes have always been preferred by the key end market – cable sheathing. In cable sheathing based on various rubbers and/or polyethylene co-polymers, a small particle size ATH is required in order to give maximum tensile strength. The major developments over recent years have been improvements in the chemical purity, particularly soluble ionic surface impurities such as sodium, and improvements in the bulk handleability of such fine powders (flowability etc.).

Other specialist grades of ATH exist with properties such as improved whiteness, improved thermal stability, 'synergistic' mixtures with other flame retardants/smoke suppressants. Coloured grades exist and tailored particle morphologies have been developed for some markets.

There are a variety of mechanisms by which ATH can contribute to flame retardancy and smoke suppression [2] but the dominating effect is endothermic decomposition to aluminium oxide, which absorbs around 1 kJ/g of heat energy:

$$2Al(OH)_3 \rightarrow Al_2O_3 + 3H_2O$$

This occurs in the condensed phase and interferes with heat feedback from the burning gases in the flame to the decomposing polymer beneath. It also promotes the formation of a layer of char which further protects and insulates unburned material. The smoke suppression effect may be viewed as a consequence of char promotion (that is carbon-rich particulates that would have otherwise become smoke, are locked up in the condensed phase as char). It is also likely that the very high surface area transition aluminium oxides formed during decomposition of ATH will adsorb many volatile species and fragments that could otherwise become smoke.

Magnesium hydroxide

Magnesium hydroxide, $Mg(OH)_2$, is manufactured in large quantities for applications including magnesia for refractory bricks and linings and as a water treatment chemical and pharmaceutical. It is most commonly extracted from brines and seawater by mixing with slaked lime or dolime:

$$CaO \cdot MgO + 2H_2O \rightarrow Ca(OH)_2 + Mg(OH)_2$$

$$Ca(OH)_2 + Mg(OH)_2 + MgCl_2 \rightarrow 2Mg(OH)_2 + CaCl_2$$

Other sources of magnesium hydroxide are via magnesium ores including magnesite ($MgCO_3$), dolomite ($CaCO_3 \cdot MgCO_3$) and serpentinite ($Mg_3[Si_2O_5](OH)_4$). It also occurs native in some places as the mineral brucite ($Mg(OH)_2$). Some other ores such as huntite ($Mg_3Ca(CO_3)_4$) and hydromagnesite ($3MgCO_3 \cdot Mg(OH)_2 \cdot 3H_2O$) can be used as flame retardants in their own right, or converted to magnesium hydroxide [3].

An early reference to magnesium hydroxide's use as a flame retardant was Belgian patent 645,879 in 1964. As a flame retardant and smoke suppressant, magnesium hydroxide has broadly similar performance to an equivalent loading of ATH and can in theory be used in the same applications, although in some systems one or other additive has a performance edge. In poly(ethylene vinyl acetate) (EVA), for example, the reactivity of $Mg(OH)_2$ to evolved acetic acid seems to contribute towards better performance in the critical oxygen index test. However, its main benefit as a flame retardant compared with ATH is that it decomposes at higher temperatures (350°C versus 200°C for ATH). Its main disadvantage is that the material normally produced from the brine or seawater route is crystallographically unsuitable for most polymer applications. This is manifested primarily as very small crystallite aggregates which give a high surface area for any given aggregate median particle size. The viscosity of such a material in a liquid polymer matrix during compounding will often be problematic. To improve this crystallography has, to date, proven relatively expensive and this has limited its penetration into traditional ATH markets. There are now a large number of routes, some patented, to narrow or eliminate this performance difference, some of which may be cost effective. The first route that gained significant commercial impact was due to Kyowa (Japan) whose patented route involved treating a basic magnesium chloride or nitrate in an autoclave. Another route is the Aman process used by Dead Sea Periclase (Israel) and hydropyrolyses concentrated brine to form a magnesium oxide which is then hydrolysed to the hydroxide. The Magnifin process used by Martinswerk (Germany) converts serpentinite into magnesium chloride, which is purified and roasted to the oxide and then hydrolysed. This route has the benefit of regenerating HCl, which is then recycled to treat further serpentinite:

$$Mg_3[Si_2O_5](OH)_4 + 6HCl \rightarrow 3MgCl_2 + SiO_2 + 5H_2O$$

$$MgCl_2 + H_2O \rightarrow MgO + 2HCl$$

$$MgO + H_2O \rightarrow Mg(OH)_2$$

Another problem with magnesium hydroxide is absorption of atmospheric moisture and carbon dioxide, which presumably causes surface formation of carbonate and further water pick-up. This has been a factor in persuading many manufacturers to offer only grades with a hydrophobic coating such as stearic or oleic acid. These coatings can further add to production costs but can also give additional benefits, most notably in dramatically improved tensile elongations. The main applications are for cable sheathing and some thermoplastic uses including polypropylene and polyamides. Total current world-wide consumption of magnesium hydroxide as a flame retardant may be around 10 000 tpa (outside of Japan). The Japanese cables market has often favoured magnesium hydroxide over ATH. Significant suppliers include Kyowa (Japan), Martinswerk (Germany) and Dead Sea Periclase (Israel).

Other hydroxides

Zinc–tin and zinc–boron materials are discussed along with their oxides in the metal oxides section below. Very few other metal or mixed metal hydroxides seem to be used commercially as flame retardants or smoke suppressants. Some small quantities of calcium–aluminium hydroxides and magnesium–aluminium hydroxides have been used intermittently as flame retardants for special applications.

Calcium hydroxide (Ca(OH)$_2$) may find some small use. One might expect that it could outperform Al(OH)$_3$ or Mg(OH)$_2$ because conversion of calcium hydroxide to oxide absorbs more heat energy than the corresponding reaction for aluminium or magnesium hydroxide. However, the reaction occurs at a higher temperature (450°C) and in any case in a fire it is believed that the calcium hydroxide is largely converted into the carbonate, an exothermic reaction.

METAL OXIDES

Antimony trioxide

The predominant metal oxide flame retardant is antimony trioxide, Sb$_2$O$_3$, which finds substantial usage in PVC (see Figure 1), in other halo-polymers, and (in conjunction with halogenated additives) in other systems. It is effective at low loadings, which means that its use can

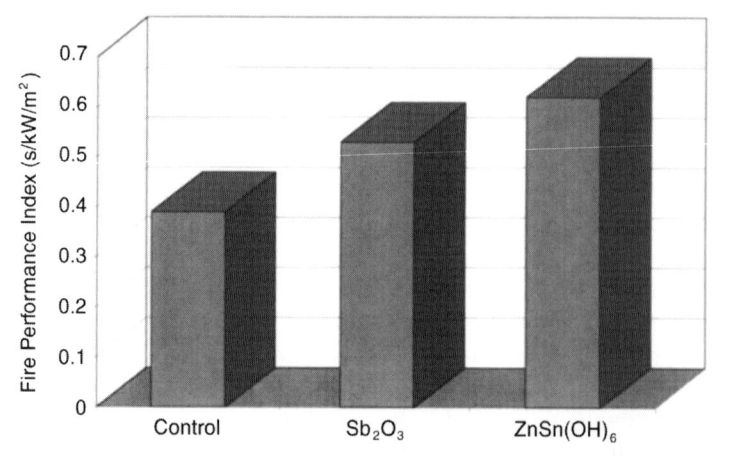

Figure 1 Inorganic oxides and hydroxides are frequently used in combination with each other and with other flame retardants. PVC (flexible) is perhaps the polymer most able to tolerate additions of various additives. This bar graph shows control data from a high performance flexible PVC containing ATH plus a flame retardant plasticizer and shows the benefit in Fire Performance Index (FPI) gained by adding a further 5 pphr of antimony trioxide or 5 pphr zinc hydroxy stannate. FPI is a measure of fire performance derived from cone calorimetry and is frequently quoted as being the best single cone parameter to assess the fire hazard of a material. High values indicate good performance.

often have a minimal effect on the mechanical properties of the host matrix. Antimony trioxide ore and antimony sulphide (stibnite) are found in China, Bolivia, Peru, Thailand, Russia, Turkey, South Africa, Mexico and to a lesser extent in the USA. Flame retardant grades of antimony trioxide can be prepared either from stibnite or antimony metal. Antimony trioxide obtained directly from the ore often contains unacceptable entrained impurities and may need to be sublimed to increase the purity [4].

Antimony trioxide has traditionally been a very cost effective flame retardant, although fluctuations in the supply from China in recent years has led to price volatility. There has also been growing concern in recent years over the health and safety implication of antimony compounds. World-wide use of antimony oxides as flame retardants may be around 50 000 tpa.

Some of the useful secondary features of antimony trioxide are its good tinting strength together with the fact that it is a partial stabilizer for PVC, and presumably also for some of the halogenated co-additives such as chlorinated paraffins. Antimony trioxide is not normally a smoke suppressant, and indeed its use can sometimes lead to increased smoke evolution. Colloidal versions of the pentoxide, Sb_2O_5, have been marketed as additives that do act as smoke suppressants and flame retardants. Producers of antimony trioxide and pentoxide flame retardants include Anzon, Campine, Laurel, Asarco and Amspec.

The mechanism by which antimony trioxide acts as a flame retardant is complex, and usually explained simply by stating that antimony trihalide is formed during combustion and that antimony trihalide is the effective flame retardant. Certainly a number of studies have suggested that the optimum ratio of antimony to halogen is around 1:3 but this is not always the case. A large number of different studies and different theories have been published on the mechanism(s) of antimony–halogen flame retardancy and many are very complex. It may be appropriate here simply to argue that volatile halogens are effective flame retardants because they readily participate within the free radical chain reactions which dominate a fire, but essentially slow down the rate of such reactions compared with the situation in their absence. Antimony reacts with the halogen in a variety of ways to slow and 'quench' these reactions still further. Obviously, to sustain and propagate fire these exothermic reactions must occur at a sufficient rate to sustain burning. It is the heat from this burning which causes further degradation into volatile flammable fragments of the burning polymer, and there is a threshold below which the flame dies because of insufficient rate of heat generation. From this argument it may be suggested that other metals could fulfil a similar role. Zinc, tin and possibly boron have been prime candidates.

Zinc stannates

(The reader is referred also to the entry: 'Flame retardants: tin compounds' for a more detailed discussion on tin compounds). Zinc stannates of note consist of zinc stannate (ZS), $ZnSnO_3$, and its precursor zinc hydroxy stannate (ZHS), $ZnSn(OH)_6$. Both are marketed as total or partial replacements for antimony trioxide. The early literature on $ZnSn(OH)_6$ utilized a production via the reaction of sodium stannate with zinc chloride. The oxide (stannate) can be prepared by thermal decomposition of this hydroxide (hydroxy stannate). Like antimony trioxide, they work best in halogenated polymers or in the presence of a halogenated additive, although there is growing evidence of effectiveness in non-halogenated systems such as EVA, EPDM and some nitrogen-containing polymers. Their major benefit is their low health hazard, and in addition to flame retardancy they are usually smoke suppressants and carbon monoxide suppressants. It is clear from the large body of research data now available that ZHS is primarily a flame retardant and smoke suppressant because of its tin and zinc content rather than its endothermic decomposition. Thus, unlike magnesium or aluminium hydroxide, ZHS is effective at low loadings. ZS is also effective at low loadings. These additives work in the vapour (gas) phase *and* the condensed (solid) phase. Fire tests in halogenated polyester showed that between one-third and two-thirds of both the zinc and the tin is retained in the char. This combined action is particularly advantageous for users wishing to pass the Underwriters Laboratory UL 94 vertical burning test. The fall of burning droplets is often a cause of failure for materials such as polyamides containing antimony trioxide, a gas phase flame retardant which does not significantly promote char. Char promotion from the solid phase action of zinc stannate largely eliminates this problem.

Zinc borates

(The reader is referred to the entry: 'Flame retardants: borates' for a more detailed discussion on this topic.)

Zinc borates form a family of compounds usually depicted as a combination of oxides and water, $(x)ZnO \cdot (y)B_2O_3 \cdot (z)H_2O$. There are at least twenty-five distinct members, but two examples of favoured flame retardant with good thermal stability are $4ZnO \cdot 6B_2O_3 \cdot 7H_2O$ and $2ZnO \cdot 2B_2O_3 \cdot 3H_2O$. In reality the structure of the members is usually a complex arrangement of zinc, boron, oxygen and hydroxyl ions, with or without some water of crystallization. Borate ores, including tincal ($Na_2O \cdot 2B_2O_3 \cdot 10H_2O$), kernite ($Na_2O \cdot 2B_2O_3 \cdot 4H_2O$), colemanite ($2CaO \cdot 3B_2O_3 \cdot 5H_2O$), ulexite ($Na_2O \cdot 2CaO \cdot 5B_2O_3 \cdot 16H_2O$), hydroboracite ($CaO \cdot MgO \cdot 3B_2O_3 \cdot 6H_2O$) and szaibelyite ($2MgO \cdot B_2O_3 \cdot H_2O$) are concentrated at only a few sites. The USA and Turkey supply over 90% of the world demand for borates, but there are also significant deposits in Chile (ulexite), Argentina (hydroboracite) and Russia (szaibelyite). Zinc borate is usually prepared by reacting zinc oxide with either boric acid or sodium borate. Depending on reaction conditions a variety of zinc borates can be obtained. The key suppliers are US Borax and Etibank (Turkey).

Major use as a flame retardant is in PVC and halogenated polyester, usually in conjunction with antimony trioxide. Some would regard it as a partial replacement for antimony trioxide but it does have other benefits, most notably as an excellent after-glow suppressant. It is often used in combination with other flame retardants such as ATH. It also finds some use in flame retarding cellulosics, where its greater resistance to water compared with borax or boric acid is needed. Total world use of zinc borate as a flame retardant may be around 10 000 tpa.

Zinc borate works predominantly in the condensed phase and is a good char promoter. Analysis of residues show that around 80% of the boron and perhaps 63% of the zinc can remain bound up in the char [5].

Molybdenum trioxide

Molybdenum trioxide, MoO_3, has been used as a partial replacement for antimony trioxide, almost exclusively in PVC. It has a condensed phase action. Studies in PVC have shown that after burning, over 90% of the molybdenum resides in the char. In the case of PVC it is also an effective smoke suppressant, probably by acting to catalyse dehydrochlorination to all-*trans* polyenes, which cannot then cyclize to the aromatic structures that are prevalent in smoke. A major drawback is poor colour (blue-grey), and for this reason the whiter ammonium octa-molybdate has gained greater commercial success.

Tin oxide

Tin oxide, SnO_2, has also been studied as a flame retardant and smoke suppressant in conjunction with chlorine or bromine compounds. Performance was found to be generally less effective than that of the zinc stannates, which also contain less tin, and are hence more cost effective.

Iron oxides

Iron oxides have been shown to be good smoke suppressants in various systems. They also seem to be effective flame retardants when used in conjunction with a halogen source. (They are discussed more fully in the separate entries: 'Smoke suppressants' and 'Flame retardants: iron compounds, their effect on fire and smoke in halogenated polymers'.)

In polyamide formulations, ferric oxide (Fe_2O_3), black iron oxide (Fe_3O_4) and yellow iron oxide ($Fe_2O_3 \cdot H_2O$) have been shown to be effective. As with molybdenum trioxide, the colour of iron oxides has been a barrier to commercial use. Iron oxides are not suitable smoke suppressants for PVC because of the (heat) destabilizing effect of iron.

Other oxides

Oxides other than those detailed above are mentioned from time to time in the literature of flame retardants and smoke suppressants. But they seem to have made no commercial impact. There would seem to be some evidence that, for example, copper oxides, titanium dioxide and alumina can act as smoke suppressants in some systems.

REFERENCES

1. Evans, K.A. and Godfrey, E. (1987) New developments in alumina trihydrate, *Flame Retardants 87*, London, 25–26/11/1987, Elsevier.
2. Mead, N.G. and Brown, S.C. (1993) *ATH – An Effective Flame Retardant that also Works Synergistically with Other Products*, London, 13/1/1993, Royal Society of Chemistry – Industrial Inorganic Chemicals Group.
3. Horn, W.E. (1996) Mineral hydroxides – their manufacture and use as flame retardants, *Flame Retardants 101*, Baltimore, 24–27/3/1996, Fire Retardant Chemicals Association.
4. Touval, I. (1996) Antimony flame retarder synergists, *Flame Retardants 101*, Baltimore, 24–27/3/1996, Fire Retardant Chemicals Association.
5. Ferm, D.J. and Shen, K.K. (1996) Boron compounds as fire retardants, *Flame Retardants 101*, Baltimore, 24–27/3/1996, Fire Retardant Chemicals Association.

Keywords: alumina trihydrate (aluminium trihydroxide), magnesium hydroxide, calcium hydroxide, antimony trioxides, oxides of zinc, tin, molybdenum, iron, smoke suppression.

See also: Series of entries with titles 'Flame retardants:...';
Flame retardancy: the approaches available;
Smoke suppressants.

Flame retardants: intumescent systems

Giovanni Camino

INTRODUCTION

Intumescent fire retardant additives undergo a thermal degradation process on heating, which produces a thermally stable, foamed, multicellular residue called 'intumescent char'. When these substances are added to a polymeric material which is later involved in a fire, they produce an intumescent char which accumulates on the surface, while the polymer is consumed, providing insulation to the underlying materials and partially protecting it from the action of the flame.

The intumescent char acts essentially as a physical barrier to heat and mass transfer between the flame and the burning material. Thus, the process of pyrolysis of the polymer that produces combustible volatile products to feed the flame is reduced by a decrease in temperature, caused in turn by a lower heat supply from the flame. The diffusion of the volatile products towards the flame is hindered with further reduction of the flame feed. Furthermore, whatever may be the role of oxygen in the combustion process, its diffusion towards the polymer burning surface is also hindered. This series of events can lead to an interruption in the self-sustained combustion process because the flame is starved.

Thus, the condensed phase mechanism of fire retardant intumescent systems aims at reducing the rate of pyrolysis of the polymer below the threshold for self-sustained combustion. This limits the production of volatile moieties and hence reduces undesirable secondary effects of volatiles combustion such as visual obscuration, corrosion and toxicity, which are typical of the widely used halogen containing fire retardants.

Plastics Additives: An A–Z Reference
Edited by G. Pritchard
Published in 1998 by Chapman & Hall, London. ISBN 0 412 72720 X

Moreover, intumescent char adhesion to the surface of the burning polymer prevents molten inflamed polymeric particles from dripping, thus avoiding a source of fire propagation which is typical of some materials.

INTUMESCENT COATINGS

The intumescent approach has been used for about 50 years in coatings for the protection of metal and wood structures [1,2]. The introduction of intumescent systems in the bulk of polymeric materials is relatively recent [3–5]. The early developments in intumescent additives for polymers were based on experience acquired in coating applications. Indeed, the empirical approach had led to a recognition of the need for compounds capable of supplying the charred residue (a 'carbonific') and of blowing it to a foamed cellular structure ('spumific') as components of formulations showing intumescent behaviour in coatings.

In an attempt to rationalize the considerable empirical knowledge already gathered about intumescent coatings, Vandersall suggested that compounds belonging to three chemical classes were necessary to obtain intumescent behaviour [1]:

1. an inorganic acid, either free or formed *in situ* by heating an appropriate precursor;
2. a carbon rich polyhydric compound;
3. a nitrogen containing compound, typically an amine or amide.

Examples of these compounds are given in Table 1.

Very little has been published on chemical–physical mechanisms of intumescence in coatings. There has been speculation that the char is

Table 1 Examples of components of intumescent formulations [1]

Inorganic acid source
1. Acids: Phosphoric, Sulphuric, Boric.
2. Ammonium salts: Phosphates, Polyphosphates, Sulphates, Halides.
3. Amine, amide phosphates: Urea, Guanylurea, Melamine, Product of reaction of ammonia and phosphoric anhydride.
4. Organophosphorus compounds. Tricresyl phosphate, Alkyl phosphates, Haloalkyl phosphates.
5. Amine sulphates: p-nitroaniline bisulphate.

Polyhydric compounds
Starch, Dextrin, Sorbitol, Pentraerythritol (monomer, dimer, trimer), Resorcinol, Triethylene glycol, Phenol-formaldehyde resins, Methylol melamine, Linseed oil.

Amines/amides
Urea, Methylolurea, Butylurea, Dicyandiamide, Melamine, Aminoacetic acid (glycine), Urea-formaldehyde resins, Polyamides.

formed by dehydration of the polyhydric compound (carbonific) due to the action of heat in the presence of acid (char promoter). The volatiles formed by thermal degradation of the amine/amide perform a blowing action (spumific) to which a relevant contribution may also come from thermal dehydration and degradation of the carbonific. It was also proposed that the amine/amide catalyses the charring dehydration reaction of the polyhydric compound.

It has to be noted that the random selection of a compound from each of the three classes does not ensure intumescent behaviour in their mixture. The three compounds must show a suitably matching thermal behaviour. For example, it is obvious that the inorganic acid must be available for the dehydration action at a temperature lower than that at which the thermal degradation or evaporation of the char source occurs extensively. Furthermore, the blowing gases must evolve at a temperature above that at which charring of the mixture begins, but before the solidification of the liquid charring melt occurs. The size distribution of the cells in the intumescent char is of paramount importance to the fire retardant performance because it affects the insulation properties of the char. Commercial intumescent coatings give chars characterized by mostly closed cells, ranging from 20 to 50 µm in diameter, with walls 6–8 µm thick [1].

A typical sequence of events taking place when an intumescent coating is exposed to heat was indicated by Vandersall [1] to be as follows.

1. The precursor liberates the inorganic acid between 150–215°C.
2. The acid esterifies the polyhydric compound with amide catalysis.
3. The mixture melts, prior to (or during) esterification.
4. The ester decomposes, forming a carbon–inorganic (e.g. C–P) residue.
5. Water and other gases released from the above reactions blow the carbonizing mass.
6. Gelation and solidification take place, leaving a multicellular charred solid foam.

As far as temperatures are concerned, Vandersall reported that in an ammonium polyphosphate-based paint system, the film coating softened at 190°C, forming a clear melt at 240°C. Bubbling started at 310°C, and carbonization and foaming began at 330°C. The mass then gelled and solidified.

Commercial intumescent coatings were generally developed using only a few of the compounds listed in Table 1. The inorganic acid must be high boiling and not strongly oxidizing; it is very often phosphoric acid in the form of the ammonium salt, the amine or amide salt, an organic ester, amide, or imide. A linear high molecular weight ammonium polyphosphate (APP) is more widely used [1]. Pentaerythritol and its oligomers are traditional polyhydric compounds and urea, melamine, dicyandiamide and their derivatives are commonly used amines/amides.

Several additional compounds may be added to the basic components to improve fire protection in intumescent coating formulations. For example, vitrifying agents can increase the thermal resistance of the intumescent char, whereas nucleating additives can help to control cell size in the char [1, 2].

THE INTUMESCENT APPROACH APPLIED TO PLASTICS

Principles

Plastics can be protected from fire by intumescent coatings like any other substrate. However, it is well known that, generally, the present state of the art advises against such an approach, owing to technical difficulties and to economic considerations. Therefore, an alternative approach was chosen: to promote intumescent behaviour in the bulk of plastics. A contributing factor was the desire to make use of the low impact on the environment and on health, typical of intumescent fire retardants.

The guidelines illustrated above for the development of intumescent coatings are generally useful also in the case of plastics. However, it has to be taken into account that, in contrast to coatings, which are generally processed and applied at relatively low temperature, any intumescent system built into a plastics composition must stand much higher processing and fabrication temperatures without premature decomposition, charring and foaming. Furthermore, the effect of the intumescent system on the required service properties of the plastics (and on the role of other additives such as stabilizers) may be very different from that when intumescent coatings are used.

The strategy for inducing intumescent behaviour in plastics depends first of all on the ability of the polymer to leave a substantial amount of residue by thermal degradation ('char') which is stable at the temperature of burning. In this case the polymer itself can become the char source.

Charring polymers

A typical example of a polymer which can act as a char source is cellulose, which has a polyhydric structure and is the polymer present in the largest amount naturally on earth. On heating, pure cellulose decomposes through two competitive routes. One involves dehydration of the carbohydrate units (dehydrocellulose) and successive condensation of the resulting unsaturated structures, leading to char. The char yield depends on the type of cellulose and on the conditions of pyrolysis; it ranges between 6 and 23%. The alternative reaction is depolymerization to laevoglucosan, which is the main volatile combustible product:

$$\text{cellulose} \longrightarrow \begin{cases} H_2O + \text{Dehydrocellulose} \rightarrow \text{Char} + H_2O + CO + CO_2 + \text{etc.} \\ \text{Tar, primarily; laevoglucosan} \\ \text{Flammable gases} \end{cases} \quad (1)$$

Following the empirical rules deduced from coatings experience, if an inorganic acid or precursor and an amine or amide of Table 1 are added to cellulose, a mixture should result, showing intumescent behaviour on heating. Indeed, the addition of phosphoric acid increases the char yield from cellulose, while laevoglucosan decreases it. The acid-induced charring process can be stoichiometrically described as follows [1]:

$$nC_6(5H_2O) \xrightarrow[H^+]{\Lambda} 6nC + 5nH_2O \quad (2)$$

As far as the mechanism of reaction 2 is concerned, the dehydration was shown to occur through phosphorylation of the alcoholic groups of cellulose, followed by pyrolysis of the resulting ester, which gives back the phosphoric acid and creates a double bond:

$$>CH-CH_2-OH + H_3PO_4 \rightarrow >CH-CH_2OP(O)(OH)_2$$
$$\rightarrow >C=CH_2 + H_3PO_4 \quad (3)$$

A mechanistic study of the thermal degradation of the model compound cellulose phosphate suggests that conjugated double bonds might be formed in cellulose through successive phosphorylation–pyrolysis (reaction 3). The charred residue would then be formed by crosslinking of the unsaturation, as suggested for the dehydrocellulose reaction. The phosphoric acid tends to polymerize on heating, with elimination of water giving polyphosphoric acid, which is a stronger phosphorylating agent than phosphoric acid. Although phosphoric acid is not consumed in reactions 2 and 3, an amount of acid larger than the catalytic quantity is required to induce the charring process, possibly because of poor mixing with cellulose. The fire retardancy effectiveness of phosphoric acid on cellulose was increased by combined addition of amines and amides [1, 2]. This was attributed to the formation of phosphoramides, which should be more efficient phosphorylating agents than either phosphoric or polyphosphoric acids.

Foaming of the char formed by cellulose heated in the presence of phosphoric acid and its physical structure have not been discussed so far.

Great progress has been made in the past 20 years in the mechanistic study of the 'charring' of polymers because of the availability of powerful new techniques for the chemical and physical characterization of insoluble, intractable, carbonized 'chars' (e.g. FTIR, solid state NMR, FTRaman, ESR, TEM, EPR etc.) and in general for the identification of liquid and gaseous degradation products, formed simultaneously with the char. This provides us today with a rational, flexible and effective approach for the study of the intumescent behaviour in charring polymers, which is no longer restricted to the polyhydric type carbon source polymers.

For example poly(2,6-dimethyl-1,4-phenylene oxide) (PPO) decomposes, leaving about 40% residue, on heating to 700°C in an inert atmosphere. On blending with high impact polystyrene (HIPS) which is the common way to overcome PPO processing problems, an inherently intumescent material is obtained. Indeed, an intumescent char is formed on burning PPO–HIPS blends, although HIPS, which volatilizes completely on thermal decomposition to 450°C, does not substantially modify the char yield of PPO. The foamed char was shown to have a dual structure, with large cells of about 1 mm radius and walls 5 µm thick, in which a 'secondary' structure of small closed cells is present. The large voids are formed by the decomposition of PS in the PPO–PS matrix of the blend, whereas the small secondary cells (<5 µm) are due to the decomposition of the rubbery domains.

There are cases in which the gaseous products evolved in the thermal degradation process of the polymer fail to produce a foamed char by a self-blowing action, as in cellulose. In these cases a detailed knowledge of the charring process may allow the timing of gas evolution to make it occur in the time–temperature range between gelation and vitrification of the crosslinking, charring mass. For example, it has been shown that as little as 0.2–1% of aromatic sulphonates confer the intumescent behaviour to poly(bisphenol-A carbonate), which when heated alone thermally degrades giving about 20% of char, stable to 700°C. The sulfonates seem to act by accelerating the degradation/charring process, thus attaining the rate of gas evolution (e.g. CO_2) and viscosity of the charring mass necessary for intumescence to occur, without increasing the char yield [6].

With few exceptions, no method of making charring polymers intumesce is yet available, because of the dependence of the degradation mechanism on the structure of the polymer. A detailed study of this process will supply the necessary knowledge.

Non-charring polymers

The environmentally friendly intumescence approach to fire retardancy is of paramount interest in the case of polyethylene (PE), polypropylene (PP) and polystyrene (PS) which together account for about 60% of all synthetic polymeric materials used today. These polymers decompose quantitatively on heating, to give volatile products, leaving a negligible residue.

Although attempts have been made to induce charring in polymers by means of reactive additives [3], this method has not yet reached a sufficiently advanced development stage to be of general practical use. A recent example of such an approach is supplied by nylon 6, in which the presence of the reactive amide group in the chain makes it more suitable than the above polyhydrocarbons for a chemical modification of the degradation process. Addition of APP to nylon 6, which when heated alone gives a negligible residue at 400°C, promotes the formation of an intumescent char on the surface of the burning polymer. This was shown to be due to the thermal degradation of a char precursor formed by grafting of the polyamide chains onto the polyphosphate through an ester bond:

$$\left[\begin{array}{c}\text{O}\\ \parallel\\ \text{P-O}\\ |\\ \text{ONH}_4\end{array}\right]_n \xrightarrow{-\text{NH}_3} \sim\text{P-O-P-O-P-O-P-O}\sim$$
with OH, ONH$_4$, ONH$_4$, OH substituents

$$\sim(\text{CH}_2)_5-\overset{\text{O}}{\underset{\parallel}{\text{C}}}-\text{NH}-(\text{CH}_2)_5-\overset{\text{O}}{\underset{\parallel}{\text{C}}}-\text{NH}\sim$$

$$\longrightarrow \sim\text{P-O-P-O-P-O-P}\sim \quad \text{O-(CH}_2)_5-\overset{\text{O}}{\underset{\parallel}{\text{C}}}-\text{NH}\sim$$
with OH, ONH$_4$, ONH$_4$ substituents

$$+ \text{H}_2\text{N}-\overset{\text{O}}{\underset{\parallel}{\text{C}}}-(\text{CH}_2)_5\sim \quad (4)$$

The approach most widely adopted so far to impart intumescent behaviour to otherwise non-charring polymers involves the use of an intumescent additive system capable of creating the intumescent char on the surface of the burning polymer in which it is incorporated. However, the chemical and physical requirements for the intumescence process cannot be taken for granted when the intumescent system is in the bulk of the polymer, instead of being heated alone or concentrated in a thin coating film. For example, mixtures of additives might react less effectively when they are diluted in the polymer matrix. Moreover, the large amount of volatile products evolved by the thermal degradation of the polymer matrix may interfere negatively with the blowing process of the intumescent char.

A patent review [4] showed that early intumescent formulations incorporated in polymers contained a precursor of phosphoric or polyphosphoric acid, a pentaerythritol type char source, and melamine, as typical formulations of intumescent coatings. Further developments tried to reduce the complexity of the additive system, for example by using a binary combination of the acid precursor with nitrogen-containing compounds, which also act as a char source. While the acid source is generally APP, typical examples of the second component are: products of condensation of formaldehyde with substituted ureas; products of reactions between aromatic diisocyanates and pentaerythritol or melamine; polymers containing the piperazine ring in the main chain, also combined with substituted s-triazine, hydroxyalkyl isocyanurate etc.

The ideal simple solution would be to reunite all the three functions (acid source, char-forming and blowing) in a single compound such as Compound I (dimelamine salt of 3,9-bishydroxy-2,4,8,10-tetraoxa-3,9-diphosphaspiro[5,5]undecane-3,9-dioxide) or II (melamine salt of bis(2,6,7-trioxaI-phosphabicyclo[2,2,2]octane-4-methanol) phosphate):

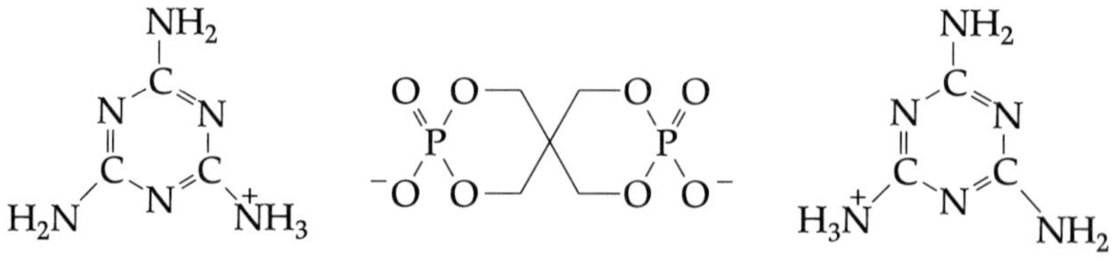

I

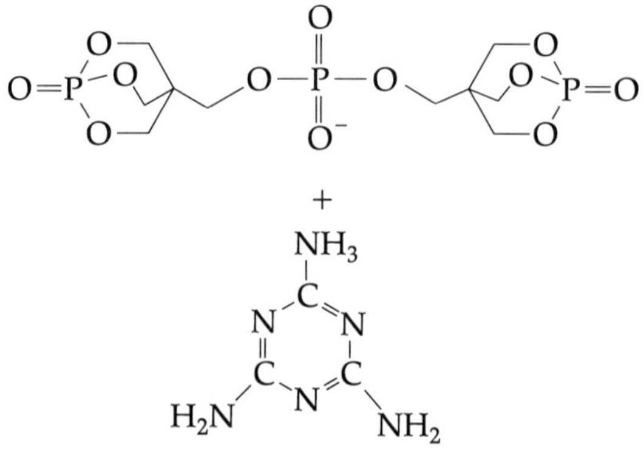

II

The use of a single additive overcomes the difficulty encountered by having two or three components of traditional intumescent systems, which have to diffuse within the polymer matrix until they collide and react. This approach might however reduce the fire retardant effectiveness, which generally depends on the ratio of the various chemical structures performing the different functions. For example, Compound II, in which the molar ratio of phosphorus to pentaerythritol to melamine is 1:0.7:0.3, was more effective in PP than Compound I, in which the ratio is 1:0.5:1. Indeed, the effectiveness of Compound I increased when it was used in conjunction with tripentaerythritol. The synthesis of simplified intumescence promoters should therefore be guided by a preliminary screening programme, based on the use of compounds each carrying one of the chemical structures to be combined in a new molecule. The dependence of the composition–effectiveness relationship on the type of polymer was often overlooked in the past, whereas it was shown that, at a given overall additive loading, the composition necessary to reach the maximum effectiveness for each polymer can vary greatly [4].

Inorganic intumescent systems

It has been pointed out recently that intumescent systems based on organic char sources may present some disadvantages. In particular, the charring reactions occurring in the intumescent process may be exothermic, thus spoiling the thermal insulating action typical of these fire retardant systems. Furthermore, the char obtained often lacks structural integrity, and may have low thermal resistance.

Low melting glasses or glass ceramics (melting, softening point <600°C) and inorganic glass forming systems have been evaluated as potential intumescent fire retardant protective layers to be used as additives for polymer (Table 2). A blowing agent can be added to the glass, or the glass itself can provide gaseous blowing products by its thermal degradation. Moreover, it has been suggested that any carbonaceous char from the polymer might be oxidized to CO by components of the glass. Finally, the gases from thermal degradation of the polymer can perform the blowing action.

PERSPECTIVES

A major problem of most intumescent additives developed so far is that they have to be used in relatively large amounts (20–30%), with adverse effects in terms of the properties of the material, its processability, cost etc. However, because of the interesting fire retardant behaviour of these systems, offering the possibility of a reduction in overall fire hazards, and because of their favourable environmental and health impact, a large effort is being made to develop new highly effective systems. Success in this research will largely depend on a parallel deepening of our understanding of the mechanisms of intumescence, which is not yet satisfactory.

Table 2 Inorganic intumescent glasses and glass-forming systems [7]

Type	Components (mol %)	
Sulphate	K_2SO_4	25
	Na_2SO_4	25
	$ZnSO_4$	50
Phosphate–sulphate	P_2O_5	36.6
	$ZnSO_4$	19.5
	Na_2O	18.3
	Na_2SO_4	9.7
	K_2SO_4	9.7
	ZnO	6.2
Borate–carbonate	B_2O_3	86.2
	Li_2CO_3	11.2
	$CaCO_3$	2.6

REFERENCES

1. Vandersall, H.L. *J. Fire Flammability*, **2**, 97 (1971).
2. Kay, M., Price, A.F. and Lavery, I. *J. Fire Retardant Chem.*, **6**, 69 (1979).
3. Camino G. and Costa, L. Mechanism of intumescence in fire retardant polymers. *Reviews in Inorganic Chemistry*, **8**, 69–100 (1996).
4. Camino, G., Costa, L. and Martinasso, G. Intumescent fire retardant systems. *Polymer Degradation and Stability*, **23**, 359–376 (1989).
5. Camino, G. Fire retardant polymeric materials in *Atmospheric Oxidation and Antioxidants* (ed. G. Scott) Elsevier, Amsterdam (1993) Chapter 10, 461–494.
6. Ballistreri, A., Montaudo, G., Scaporrino, E., Puglisi, C., Vitalini, D. and Cucinella, S. *J Polym. Sci., Polym Chemi. Ed.*, **26**, 2113 (1988).
7. Myers, R.E. and Licursi, E. *J. Fire Sci.*, **3**, 415 (1985).

Keywords: fire, combustion, coating, char, pyrolysis, polyhydric compound, amide, amine, inorganic acid, blowing action, foam, dehydration, phosphorylation, dehydrocellulose, intumescence, melamine, piperazine.

See also: Series of entries with titles 'Flame retardants:...';
Flame retardancy: the approaches available;
Smoke suppressants.

Flame retardants: iron compounds, their effect on fire and smoke in halogenated polymers

Peter Carty

INTRODUCTION

The purpose of this section is to give an up to date overview of the effects which some iron compounds have on flammability and smoke production when polymers burn in air. A glance at chemical abstracts indexes under iron, iron compounds or iron chemistry shows thousands of scientific papers appearing every year; however, very few of these deal with the use of iron compounds as flame retarding/smoke suppressing additives for polymers. The recent chemistry of iron has been reviewed by Silver [1] but again there are few comments on the potential uses of iron compounds in polymers.

The economic importance of iron is unsurpassed by any other element in the periodic table. Iron is the sixth most abundant element in the world and the most abundant metal. Iron and iron compounds are widely used in all aspects of economic activity. Iron compounds also occur widely in living systems, iron compounds being found at the active centres of many biological molecules where the Fe(II)/Fe(III) redox system and the ability of iron to form stable complexes with oxygen are used in many vital life processes. An adult human contains about 4 grammes of iron, most of which is present in haemoglobin. Pure iron(III) oxide is used to make high quality ferrites and other ceramic materials and it is also used as a light fast UV blocking pigment in paints and varnishes and also in colour printing. Iron oxides are essentially non-toxic and have been recognized as being

Plastics Additives: An A–Z Reference
Edited by G. Pritchard
Published in 1998 by Chapman & Hall, London. ISBN 0 412 72720 X

safe for selected FDA applications in the USA and iron oxides are not indicated in any toxic hazard assessments in the UK.

The first reports of iron compounds being used as smoke suppressants in plasticized and rigid PVC formulations began to appear in the 1970s and early 1980s. Ferrocene was the first iron compound to be investigated as a smoke suppressant for rigid PVC; however, its high volatility, colour, and the char oxidizing effect of iron(III) oxide (the final product of oxidation of ferrocene in this system) appeared to halt further commercial development of iron compounds for this use.

In a series of recent papers the author has shown that selected iron compounds (including a range of organo iron compounds) are particularly effective in suppressing smoke formation during the burning of polymers, especially in halogenated systems such as uPVC, pPVC, CPVC and blends of PVC with ABS [2,3]. In addition it has also been found that, contrary to work already published, some hydrated forms of iron(III) oxide have a good light and heat stabilizing effect in a range of PVC formulations including PVC plasticized with phthalate and phosphate plasticizers [4].

The detailed mechanisms of how these iron compounds reduce smoke from burning PVC and in blends of PVC with ABS have not yet been fully resolved but it is thought that the Lewis-acid activity of some iron(III) compounds formed *in situ* may play a part in the smoke suppressing effect.

Iron compounds appear to have considerable potential as smoke suppressants, although they have not yet reached full commercial development.

POLYMER COMBUSTION AND SMOKE PRODUCTION

The combustion of polymers is a very complex, rapidly changing chemical system which is not yet fully understood. Organic polymers undergo degradation processes with the formation of volatile organic compounds when they are heated above certain critical temperatures, which depend on their structures. If the gaseous mixture resulting from the mixing of these volatile organics with air is heated to a temperature greater than the ignition temperature of the mixture, combustion begins. This process (greatly simpified) is shown schematically in Figure 1.

The burning characteristics of polymers cannot be thoroughly evaluated by determining a few simple fire parameters in the laboratory. Fire and combustion is very complex and at least four major features interact during the overall burning process:

1. ignition;
2. rate of spread of flame;

Char formation, flammability and smoke formation

```
    POLYMER  ──thermal decomposition──▶  VOCs
        ▲                                  │
        │ heat                 air/ignition│
        │                                  ▼
              COMBUSTION
                  │
                  ▼
                SMOKE
```

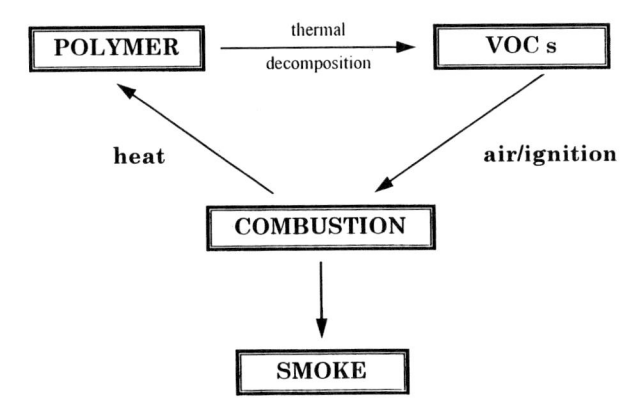

Figure 1 Polymer combustion cycle.

3. rate of heat development and release to the surroundings;
4. the formation of smoke and gases.

These processes all contribute to the hazards associated with burning polymers, but it is now recognized that the three most important factors are:

1. flammability;
2. the rate and amount of heat released;
3. the amount of carbon monoxide and smoke produced.

The main purpose of adding flame retardants to polymer formulations is to slow down or stop the burning process. All flame retardants function during the early stages of a fire, acting by interrupting the self-sustaining combustion cycle shown in Figure 1. It must be recognized that no organic polymer can be made fire-proof, even a polymer such as poly(tetrafluoroethylene) (PTFE) which has an LOI (Limiting Oxygen Index) value of 95 will burn in a real fire. Very detailed reviews of flame retardants and their use in polymers can be found in other entries in this book.

The most important commercial flame retardant systems are summarized in Figure 2, which shows qualitatively, the effects which flame retardants have on burning and on the production of smoke.

CHAR FORMATION, FLAMMABILITY AND SMOKE FORMATION

We can identify three general ideas linking char formation and combustion processes. These three empirical statements can be summarized as:

1. to reduce smoke by a factor of two requires an increase in char yield by a factor of three;

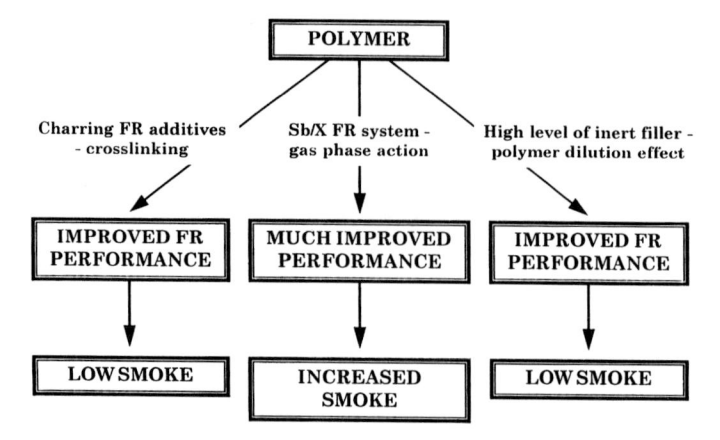

Figure 2 Flame retardants and their effect on smoke production.

2. to have a smoke value of D_{MAX} (see [5], p. 410 for details of this notation) of less than 100 requires a minimum char yield of 30%;
3. to achieve a UL V-0 flammability rating (see [5], p. 347 for full details of this test procedure), requires a minimum char yield of 30%.

While char formation is recognized as an important mode for achieving flame retardancy, little progress has been made to reduce smoke formation by incorporating char forming additives into polymer formulations.

The formation of char during the thermal decomposition of polymers can be a result of crosslinking reactions taking place in the solid phase. One of the beneficial effects of char formation is that polymer carbon is retained in the solid phase and, as a consequence, is not available for the formation of oxides of carbon, volatile organic compounds and of course smoke. Hence, substances which interact with the polymer to increase char formation during burning are in fact smoke suppressants, and only a few elements in the periodic table are known to react in this way. Zinc, molybdenum, iron and possibly tin compounds are known smoke suppressants and have been available commercially for use in PVC formulations.

It has been found that incorporating modest amounts of hydrated iron(III) oxide (FeOOH) into ABS/PVC blends increases char formation significantly when compared with equivalent formulations containing no iron. The formation of large amounts of char in the presence of basic iron(III) oxide results in a reduction in the amount of smoke produced when the polymer blend burns in air. In addition to the formation of char, it has also been found that this iron(III) compound improves the flame resistance of the blends and also stabilizes them against decomposition by both heat and light (Table 1).

Table 1 ABS/PVC/FeOOH blends, char yields and related flammability data (First published in *Polymer*, **35**(2), 344 (1994))

Formulation (phr)	Char (%) 350°C	500°C	650°C	800°C	Smoke D_{MAX} (g^{-1})	LOI
100 uPVC*	41.42	21.97	13.70	10.82	62	49.9
100 ABS†	96.51	2.6	<0.2	0	111	18.3
70 ABS/30 PVC	78.60	10.21	6.42	3.87	104	21.8
70 ABS/30 PVC/1 FeOOH‡	79.58	17.22	9.74	7.14	85	33.6
70 ABS/30 PVC/2.5 FeOOH	79.17	22.43	13.94	10.08	80	33.6
70 ABS/30 PVC/5 FeOOH	79.67	27.46	19.05	14.36	63	33.4
70 ABS/30 PVC/7.5 FeOOH	80.50	33.30	24.89	18.25	61	33.4
70 ABS/30 PVC/10 FeOOH	81.14	34.07	26.97	19.71	54	33.8
70 PVC/30 ABS	57.68	17.19	11.14	9.63	92	31.2
70 PVC/30 ABS/1 FeOOH	61.48	24.28	18.94	14.69	56	37.3
70 PVC/30 ABS/2.5 FeOOH	62.86	27.38	21.38	16.02	50	40.8
70 PVC/30 ABS/5 FeOOH	63.44	31.40	23.60	17.21	45	44.2
70 PVC/30 ABS/7.5 FeOOH	65.73	33.73	27.13	19.53	41	42.8
70 PVC/30 ABS/10 FeOOH	66.36	37.60	30.79	22.38	39	39.8

* PVC: 100 phr Corvic S67/111 (PVC resin); 5 phr tribasic lead sulphate (stabilizer); 1 phr calcium stearate (lubricant). † ABS: 100 phr Cycolac GSM. ‡ FeOOH: Bayferrox 3905, hydrated iron(III) oxide (Bayer UK Ltd).

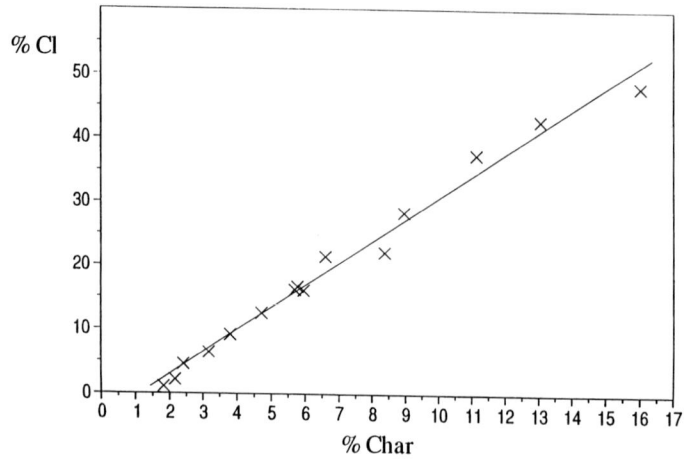

Figure 3 Percentage char versus percentage chlorine in ABS/PVC blends.

Recent work on char formation in ABS/PVC blends carried out in our laboratory suggests that char formation is dependent upon the PVC content and in the iron containing polymer blends, char formation depends not only on the PVC content, but also on the Fe/Cl stoichiometry (Figure 3).

There is a very high correlation ($r = 0.99$) between char formation and percentage of chlorine in the blend. This suggests that it is the PVC content of the blend which influences the char-forming crosslinking reactions. Across the range of ABS/PVC iron-free blends studied, even

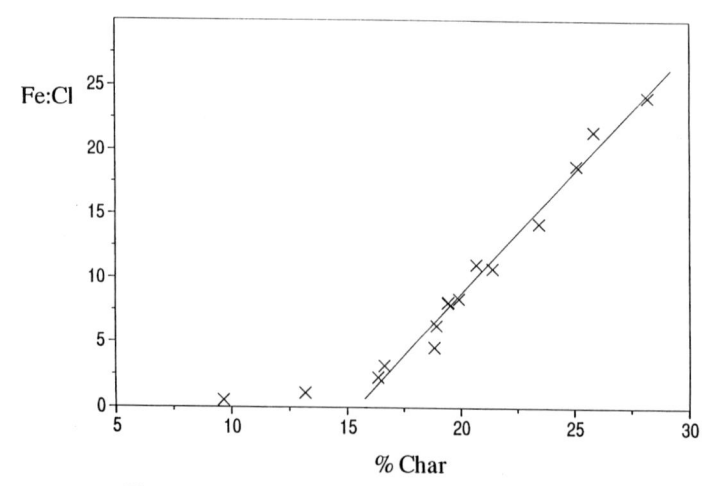

Figure 4 Fe:Cl ratio versus percentage char.

Char formation, flammability and smoke formation

down to formulations containing only 2 phr PVC, Figure 3 shows that char formation is very dependent on chlorine and therefore PVC content.

The incorporation of FeOOH into this polymer system has very dramatic effects on char formation and smoke suppression. Even with PVC levels of only 2 phr in ABS/PVC in the presence of 5 phr FeOOH, the char yield (when measured against an iron-free system) is increased by more than five times. There is another very high correlation between Fe/Cl ratio and char formation in these blends (Figure 4).

The author is of the opinion that although $FeCl_3$ or $FeCl_2$ (a very weak Lewis acid) formed *in situ* could be active in this system, a compound with an Fe/Cl ratio of about 1:1 is more likely to be the char forming/smoke suppressing intermediate. Work is currently in progress to clarify this.

Formulations containing carefully controlled amounts of FeOOH and PVC in blends of ABS/PVC have been investigated in some detail recently and the results obtained are shown graphically in Figures 5 and 6. Very small amounts of the smoke suppressing iron compound have large effects on char formation in 30 PVC/70 ABS, 70 PVC/30 ABS formulations all containing 5 phr of FeOOH and also in a range of ABS/PBC blends containing from 0 to 10 phr of FeOOH.

It is now thought that iron smoke suppressants reduce smoke in these ABS/PVC blends probably by a combination of two important processes:

1. by increasing char formation;
2. by changing the aromatic decomposition products formed when the polymer is thermally decomposed.

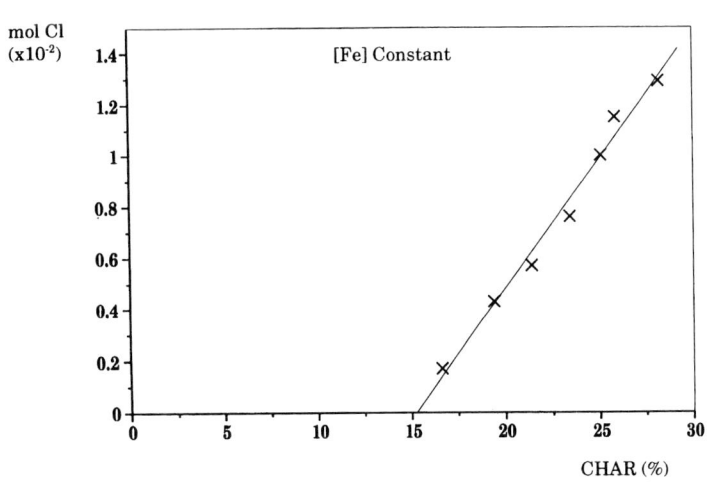

Figure 5 Mol Fe versus percentage char for PVC/ABS blends.

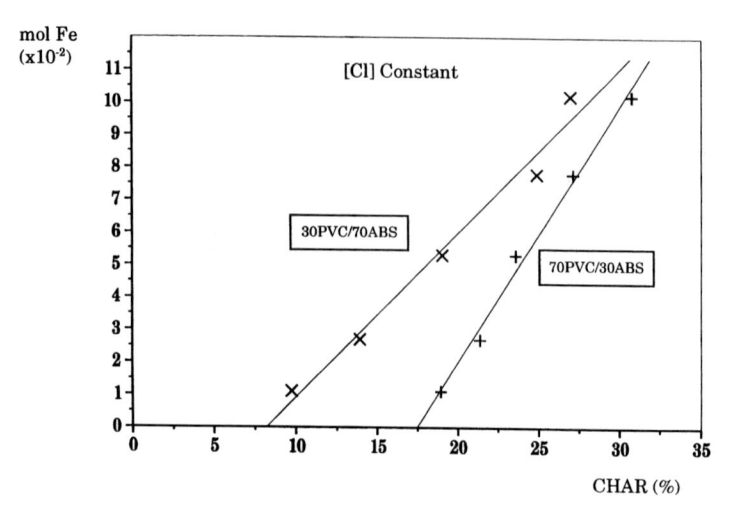

Figure 6 Mol Cl versus percentage char for PVC/ABS blends.

Blends of ABS with CPVC containing FeOOH are currently being investigated and early results suggest that in these related systems the action of the smoke suppressant is the same.

ACKNOWLEDGEMENTS

The author would like to thank Stewart White of Anzon Ltd who did the LOI and smoke density measurements described in this work, and Jan Eastwood, University of Northumbria at Newcastle, who typed the manuscript.

REFERENCES

1. Silver, J. (ed.) *Chemistry of Iron*. Blackie (Chapman & Hall), (1993).
2. Carty, P. and White, S. *Polymer*, **36**(5), 1109 (1995).
3. Carty, P. and White, S. Flammability studies – plasticised and non-plasticised PVC/ABS blends. *Polymer Networks and Blends*, **5**(4), 205–209 (1995).
4. Carty, P. and White, S. *Engineering Plastics*, **8**, 287 (1995).
5. Troitzsch, J. (ed.) *International Plastics Flammability Handbook, Principles, Regulations, Testing and Approval*. 2nd edn, Hanser Publications, Munich (1979).

Keywords: iron, smoke, flame retardant, char, PVC, ABS.

See also: Series of entries with titles 'Flame retardants:...';
Flame retardancy: the approaches available;
Smoke suppressants.

Flame retardants: poly(vinyl alcohol) and silicon compounds

Guennadi E. Zaikov and Sergei M. Lomakin

INTRODUCTION

In the modern polymer industry, the various existing types of polymer flame retardants based on halogens (Cl, Br), heavy and transition metals (Zn, V, Pb, Sb) or phosphorus–organic compounds reduce the risk from polymer combustion and pyrolysis, but may present ecological issues. The overall use of halogenated flame retardants is still showing an upward trend, but the above concerns have started a search for more environmentally friendly polymer additives. As a result it is quite possible that the future available flame retardants will be more limited than in the past.

One ecologically-safe flame retardant system, containing a char former, is polyvinyl alcohol (PVA) combined with a silicon-based inorganic system which can act in two ways as follows.

1. By the formation of a barrier (char) which hinders the supply of oxygen and reduces the thermal conductivity of the material; this is the role of the polyvinyl alcohol.
2. By trapping the active radicals in the vapour phase (and eventually in the condensed phase) – this is the role of the silicon inorganic system.

MECHANISMS

High temperature char-former

There is a strong correlation between char yield and fire resistance. Char is formed at the expense of combustible gases and the presence of a char

Plastics Additives: An A–Z Reference
Edited by G. Pritchard
Published in 1998 by Chapman & Hall, London. ISBN 0 412 72720 X

inhibits further flame spread by acting as a thermal barrier around the unburned material. The tendency of a polymer to char can be increased with chemical additives and by altering its molecular structure. Polyvinyl alcohol systems usually produce highly conjugated aromatic structures which char during thermal degradation and/or which transform into crosslinking agents at high temperatures. Decomposition of PVA occurs in two stages. The first stage, which begins at 200°C, involves mainly dehydration accompanied by formation of volatile products. The residues are predominantly polymers with conjugated unsaturated structures. In the second stage, polyene residues are further degraded at 450°C to yield carbon and hydrocarbons. The mechanism involved in thermal decomposition of PVA has been deduced by Tsuchya and Sumi [1]. At 245°C water is split off from the polymer chain, and a residue with conjugated polyene structure results:

$$(-CH-CH)_n-CH-CH_2- \;\;\rightarrow\;\; (-CH=CH)_n-CH-CH_2- + H_2O$$
$$\quad\;\;|\quad\quad\quad\;\;|\quad\quad\quad\quad\quad\quad\quad\quad\quad\quad\;\;|$$
$$\quad\;\;OH\quad\quad\;OH\quad\quad\quad\quad\quad\quad\quad\quad\quad\;OH$$

Scission of several carbon–carbon bonds leads to the formation of carbonyl ends. For example, aldehyde ends arise from the reaction:

$$-CH-CH_2\!-\!(CH=CH)_n-CH-CH_2- \;\rightarrow$$
$$\;\;|\quad\quad\quad\quad\quad\quad\quad\quad\quad\quad\;\;|$$
$$\;\;OH\quad\quad\quad\quad\quad\quad\quad\quad\quad\;OH$$

$$-CH-CH_2\!-\!(CH=CH)_n-CH + H_3C-CH-$$
$$\;\;|\quad\quad\quad\quad\quad\quad\quad\quad\quad\quad\;\;\|\quad\quad\quad\quad\;|$$
$$\;\;OH\quad\quad\quad\quad\quad\quad\quad\quad\quad\;O\quad\quad\quad\;OH$$

In the second-stage pyrolysis of PVA, the volatile products consist mainly of hydrocarbons, i.e. *n*-alkanes, *n*-alkenes and aromatic hydrocarbons [1].

Thermal degradation of PVA in the presence of oxygen can be adequately described by a two-stage decomposition scheme, with one modification. Oxidation of the unsaturated polymeric residue from dehydration reaction introduces ketone groups into the polymer chain. These groups then promote the dehydration of neighbouring vinyl alcohol units producing a conjugated unsaturated ketone structure [2]. The first-stage degradation products of PVA pyrolysed in air are fairly similar to those obtained in vacuum pyrolysis. In the range 260–280°C, the second-order-reaction expression satisfactorily accounts for the degradation of 80% hydrolysed PVA up to a total weight loss of 40%. The activation energy of decomposition appears to be consistent with the value of 53.6 kcal/mol (224 kJ/mol) which is obtained from the thermal degradation of PVA [2].

The changes in the IR spectra of PVA subjected to heat treatment have been reported [2]. After heating at 180°C in air, bands appeared at 1630 cm^{-1} (C=C stretching in isolated double bonds), 1650 cm^{-1} (C=C stretching in conjugated dienes and trienes), and 1590 cm^{-1} (C=C stretching in polyenes). The intensity of the carbonyl stretching frequency at 1750–1720 cm^{-1} increased, although the rate of increase of intensity was less than that of the polyene band at low temperatures. Above 180°C, although dehydration was the predominant reaction at first, the rate of oxidation increased after an initial induction period.

The identification of a low concentration of benzene among the volatile products of PVA has been taken to indicate the onset of a crosslinking reaction proceeding by a Diels–Alder addition mechanism [2]. Clearly benzenoid structures are ultimately formed in the solid residue, and the IR spectrum of the residue also indicated the development of aromatic structures:

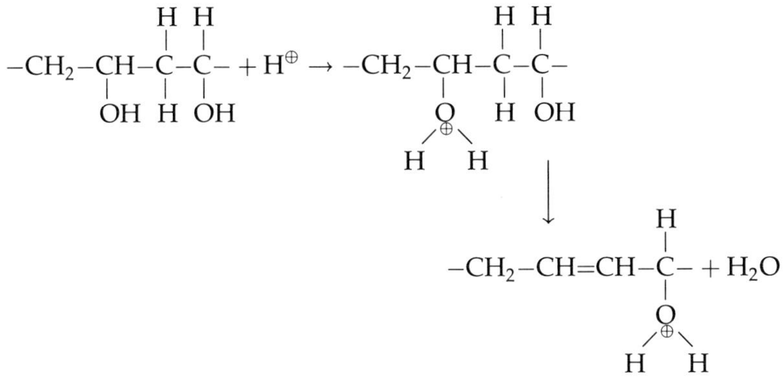

$$-CH_2-CH-\underset{OH}{\underset{|}{C}}\underset{H}{\underset{|}{\overset{H}{|}}}-\underset{OH}{\underset{|}{C}}- + H^{\oplus} \rightarrow -CH_2-CH-\underset{O^{\oplus}}{\underset{|}{C}}\underset{H}{\underset{|}{\overset{H}{|}}}-\underset{OH}{\underset{|}{C}}-$$

$$\downarrow$$

$$-CH_2-CH=CH-\underset{O^{\oplus}}{\underset{|}{C}}-\overset{H}{|} + H_2O$$

Acid-catalysed dehydration promotes the formation of conjugated sequences of double bonds (see (a) below) and Diels–Alder addition of conjugated and isolated double bonds in different chains may result in intermolecular crosslinking, producing structures which form graphite or otherwise undergo carbonization (see (b) below):

(a)

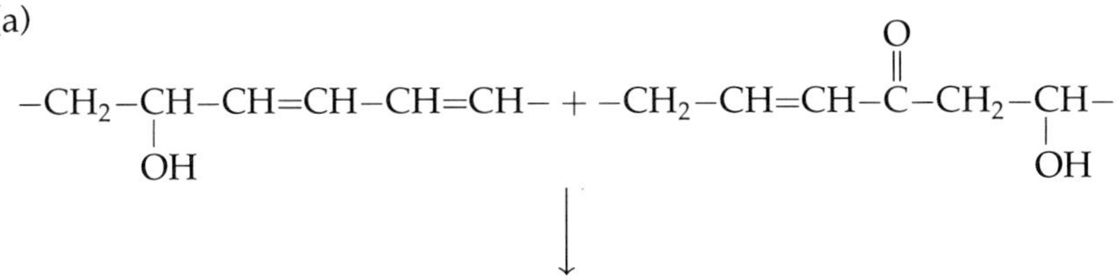

$$-CH_2-\underset{OH}{\underset{|}{CH}}-CH=CH-CH=CH- + -CH_2-CH=CH-\overset{O}{\overset{\|}{C}}-CH_2-\underset{OH}{\underset{|}{CH}}-$$

$$\downarrow$$

(b)

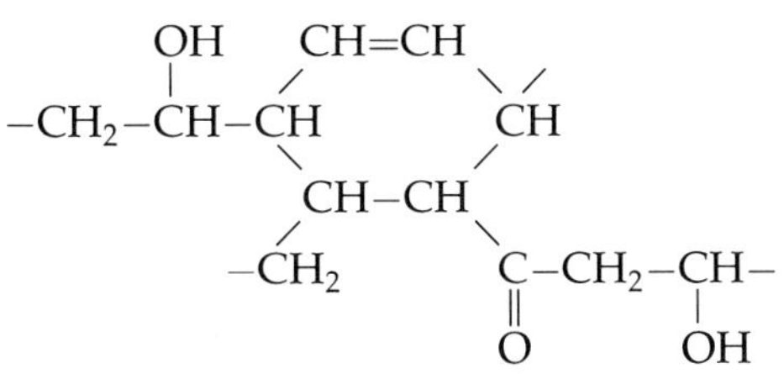

In contrast to PVA, when nylon 6,6 is subjected to temperatures above 300°C in an inert atmosphere it completely decomposes. The wide range of degradation products, which includes several simple hydrocarbons, cyclopentanone, water, CO, CO_2 and NH_3 suggests that the degradation mechanism must be highly complex. Further research has led to a generally accepted degradation mechanism for aliphatic polyamides:

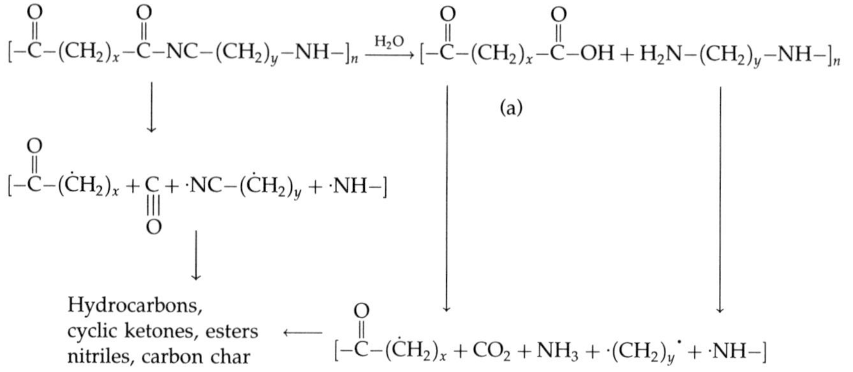

$$[-\overset{O}{\overset{\|}{C}}-(CH_2)_x-\overset{O}{\overset{\|}{C}}-NC-(CH_2)_y-NH-]_n \xrightarrow{H_2O} [-\overset{O}{\overset{\|}{C}}-(CH_2)_x-\overset{O}{\overset{\|}{C}}-OH + H_2N-(CH_2)_y-NH-]_n$$

(a)

$$[-\overset{O}{\overset{\|}{C}}-(\dot{C}H_2)_x + \underset{\underset{O}{|||}}{C} + \cdot NC-(\dot{C}H_2)_y + \cdot NH-]$$

↓

Hydrocarbons, cyclic ketones, esters nitriles, carbon char ⟵ $[-\overset{O}{\overset{\|}{C}}-(\dot{C}H_2)_x + CO_2 + NH_3 + \cdot(CH_2)_y{}^{\cdot} + \cdot NH-]$

1. Hydrolysis of the amide bond usually occurs below the decomposition temperature.
2. Homolytic cleavage of C–C, C–N, C–H bonds generally begins at the decomposition temperature and occurs simultaneously with hydrolysis.
3. Cyclization and homolytic cleavage of products from both of the above reactions occur.
4. Secondary reactions produce CO, NH_3, nitriles, hydrocarbons and carbon chars.

The idea of introducing poly(vinyl alcohol) into nylon 6,6 is based on the possibility of high-temperature acid-catalysed dehydration. This reaction can be provided by the acid products from the hydrolysis of nylon 6,6 which would promote the formation of intermolecular cross-linking and char. Such a system we have called 'synergistic carbonization' because the char yield and flame suppression parameters of the blend of poly(vinyl alcohol) and nylon 6,6 are significantly better than from pure poly(vinyl alcohol) or nylon 6,6 polymers.

It is well-known that nylons have poor compatibility with other polymers because of their strong hydrogen bonding characteristics. Nylon 6 is partially compatible with polyvinyl alcohol. 'Compatibility' here does not mean thermodynamic miscibility but rather ease of mixing blends to achieve small size domains.

The next step to improve the flame resistant properties of poly(vinyl alcohol)/nylon 6,6 system was the substitution of pure poly(vinyl alcohol) by poly(vinyl alcohol) oxidized by potassium permanganate (PVA-ox). This approach was based on the fire behaviour of the (PVA-ox) itself. Cone calorimetry showed a dramatic decrease in the rate of heat release and a significant increase in ignition time for the oxidized PVA in comparison with the original PVA.

The literature on the oxidation of macromolecules by alkaline permanganate presents little information about these redox systems. Hassan [3] investigated the oxidation of PVA by potassium permanganate in alkaline solution. It was reported that the oxidation occurs through the formation of two intermediate complexes I and/or II as follows:

$$(-CH_2-\underset{\underset{OH}{|}}{\overset{\overset{H}{|}}{C}}-)_n + OH^\ominus \underset{-H_2O}{\overset{K_1}{\rightleftharpoons}} (-CH_2-\underset{\underset{O^\ominus}{|}}{\overset{\overset{H}{|}}{C}}-)_n + MnO_4^-$$

(a) / (b) \

$$(-CH_2-\underset{\underset{O-Mn^{6+}O_4^{2-}}{|}}{\overset{\overset{H}{|}}{C}}-)_n \quad I \qquad (-CH_2-\underset{\underset{O^\ominus}{|}}{\overset{\overset{H----OMn^{6+}O_3^{2-}}{|}}{C}}-)_n \quad II$$

↓ OH$^\ominus$

$$(-CH_2-\underset{\underset{O-Mn^{6+}O_4^{2-}}{|}}{\overset{\overset{\ominus}{|}}{C}}-)_n$$

↓

$$(-CH_2-\underset{\underset{O}{\|}}{C}-)_n + Mn^{5+}O_4^{3-} \qquad\qquad (-CH_2-\underset{\underset{O}{\|}}{C}-)_n + HMn^{5+}O_4^{2-}$$

III   III

The reactions (a) and (b) lead to the formation of poly(vinyl ketone) (III) as a final product of oxidation of the substrate. Poly(vinyl ketone) can be isolated and identified by microanalysis and spectral data.

Silicon–inorganic systems

According to recent patent publications silicones may be considered as universal additives to improve the flammability properties of polymers and to decrease the harmful impact on environmental safety.

The addition of relatively small amounts of silicon inorganic additives (SI) to various polymers significantly reduces their flammability. A

Si–inorganic system (SI) inhibits gaseous phase combustion, and also affects char formation in the solid phase. This system can be incorporated into polypropylene or nylon 6,6.

The mechanism of flame suppression is postulated on the reaction of gaseous-phase inhibition by $SiCl_4$ and HCl, which can be produced only at temperatures above 300–500°C, exactly the temperatures realized on the surface of burning polymers:

350–500°C:
1. $2SnCl_2 + nSi = 2Sn + (n-1)Si + SiCl_4$
$SiCl_4 + 2H_2O = 4HCl + SiO_2$

410°C:
2. $2PbCl_2 + nSi = 2Pb + (n-1)Si + SiCl_4$

280–350°C:
3. $4CuCl + nSi = 4Cu + (n-1)Si + SiCl_4$

300°C:
4. $2CaCl_2 + nSi = 2Ca + (n-1)Si + SiCl_4$

400°C:
5. $4FeCl_3 + nSi = 4Sn + (n-1)Si + 3SiCl_4$

The position of silicon directly below carbon in the periodic table suggests that the chemistry of these elements will be similar. Like carbon, silicon has a valency of four. Silicon(II) is not stable, however, and the tendency to form divalent compounds increases with atomic weight so that compounds of the form SnL_2 and PbL_2 (where L is used to denote an arbitrary ligand) are relatively common. The Group IVA elements, other than carbon, do not form strong bonds with like atoms. The Si–Si bond, for example, is notably weaker (222 kJ/mole) than the Si–C bond (328 kJ/mole) (Table 1).

The silicon analogues of the halogenated carbons, in particular, would be expected to be effective flame inhibitors. This hypothesis was confirmed early on, at least with respect to silicon tetrachloride ($SiCl_4$). There are, however, sigificant differences in the behaviour of these compounds in the atmosphere. Unlike the halons, all the halosilanes readily hydrolyse in moist air. An important consequence is that these compounds will undergo rapid decomposition in the troposphere and would therefore be expected to have correspondingly lower potentials for ozone depletion and global warming than halons. Unfortunately, this beneficial property is offset by the fact that hydrogen halides are produced in the hydrolysis of halosilanes.

This effect is so pronounced that the presence of a single silicon–halogen bond in a molecule is sufficient to make its vapours corrosive

Table 1 Cone calorimeter data of nylon 6,6/PVA

Material, heat flux (kW/m^2)	Initial wt (g)	Char yield (% wt)	Ignition time (s)	Peak R.H.R. (kW/m^2)	Total heat release (MJ/m^2)
PVA, 20	27.6	8.8	39	255.5	159.6
PVA, 35	28.3	3.9	52	540.3	111.3
PVA, 50	29.2	2.4	41	777.9	115.7
PVA-ox KMnO$_4$, 20	27.9	30.8	1127	127.6	36.9
PVA-ox KMnO$_4$, 35	30.5	12.7	774	194.0	103.4
PVA-ox KMnO$_4$, 50	29.6	9.1	18	305.3	119.8
PVA (100°C) KMnO$_4$, 20	31.1	16.3	303	211.9	124.5
PVA (200°C) KMnO$_4$, 20	35.9	25.7	357	189.0	91.1
nylon 6,6, 50	29.1	1.4	97	1124.6	216.5
nylon 6,6 + PVA(8:2), 50	26.4	8.7	94	476.7	138.4
nylon 6,6 + PVA-ox(8:2) KMnO$_4$, 50	29.1	8.9	89	399.5	197.5

and dangerous to breathe. But for SI-flame retarded nylon 6,6 this harmful influence is not so important, because HCl forms in the zone of combustion only at temperatures above 500–600°C. It also takes part in flame inhibition and undergoes reactions with tin (apparently forming stannic chloride).

Several years ago, a list of results was published in Lyon's book on fire retardants [4] showing that the flame suppression efficiency of SiCl$_4$ is between that of CF$_3$Br and that of carbon tetrachloride. In an independent study flame velocity measurements were made for a series of additives, including some halosilanes and related compounds [5]. The figure of merit was the volume percent of inhibitor required to reduce the burning velocity of a premixed (stoichiometric) n-hexane flame by 30%. On this basis, it was determined that the flame inhibition activity of SiCl$_4$ was comparable to that of Br$_2$ but considerably more effective than CCl$_4$. The hierarchy for inhibition: SnCl$_4$ > GeCl$_4$ > SiCl$_4$ > CCl$_4$ was also found to apply to increases in the ignition temperatures of hydrocarbon/(O$_2$ + N$_2$) mixtures.

322 Flame retardants: poly(vinyl alcohol) and silicon compounds

The mechanism by which the halosilanes effect flame inhibition is similar, if not identical, to that applicable to the well known halogenated carbons.

ILLUSTRATIVE EXAMPLES OF EFFECTS

Table 1 shows typical cone calorimeter data for nylon 6,6 mixed with poly vinyl alcohol, both unoxidized and oxidized. The carbon residue (wt%) and peak heat release rate (PHR) in kW/m^2 indicate a substantial improvement in fire resistance by the oxidized PVA. The heat fluxes were 20, 35 and 50 kW/m^2; even at the highest flux, there was a 9% char residue.

One reason for this phenomenon may be the ability of PVA when oxidized by KMnO$_4$ (polyvinyl ketone structures) to act as a neutral (structure I) and/or monobasic (structure II) bidentate ligand:

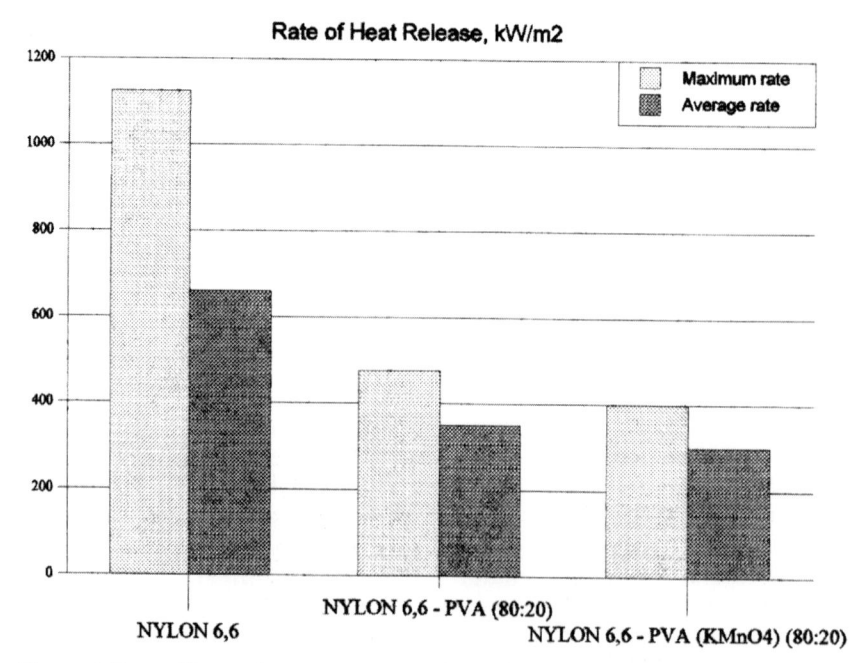

$$\underset{I}{+(CH_2-\underset{\underset{O}{\|}}{C}-CH_2-\underset{\underset{O}{\|}}{C})_n} \rightleftharpoons \underset{II}{+(CH_2-\underset{\underset{O}{\|}}{C}-CH=\underset{\underset{OH}{|}}{C})_n}$$

Figure 1 Rate of heat release versus time for: nylon 6,6; nylon 6,6/PVA (80%:20% wt.).

Spectroscopic data provide strong evidence for coordination of the ligand (some metal ions Cu^{2+}, Ni^{2+}, Co^{2+}, Cd^{2+}, Hg^{2+}) through the monobasic bidentate mode (structure II). From the above, we can propose this structure for the polymeric complexes:

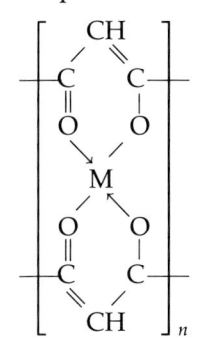

$$\left[\begin{array}{c} CH \\ \diagup \diagdown \\ C \quad\quad C \\ \| \quad\quad | \\ O \quad\quad O \\ \diagdown \diagup \\ M \\ \diagup \diagdown \\ O \quad\quad O \\ \| \quad\quad | \\ C \quad\quad C \\ \diagdown \diagup \\ CH \end{array} \right]_n$$

where M is a metal.

Figures 1–3 show the rate of heat release for various polymer/flame retardant compositions.

Table 2 shows further cone calorimetry data for polypropylene at a heat flux of $35\,kW/m^2$. The mix contained 3 wt% of silicon and 2 wt% of stannous chloride. All the results in the table show an improvement in fire performance with silicon.

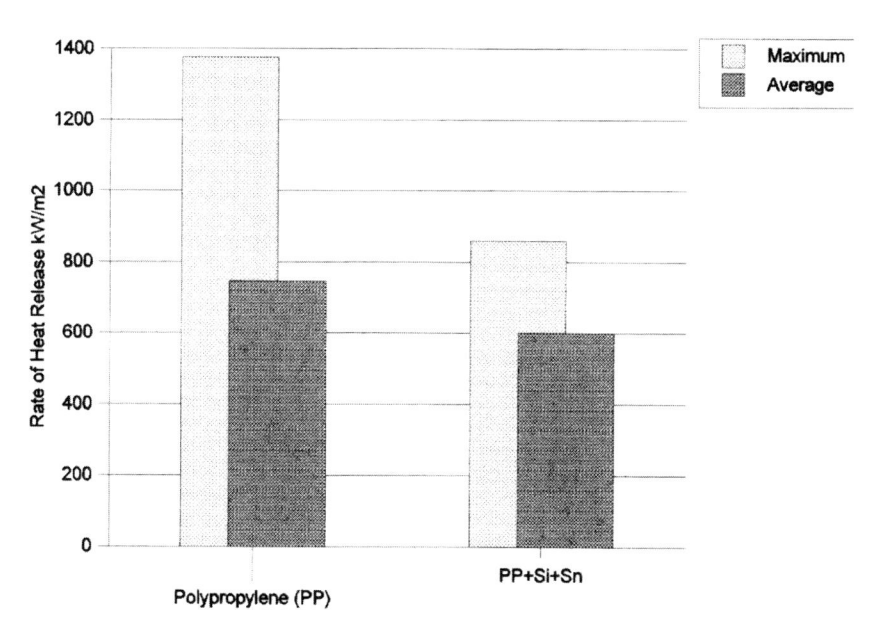

Figure 2 Rate of heat release for PP compositions at heat flux of $35\,kW/m^2$.

324 Flame retardants: poly(vinyl alcohol) and silicon compounds

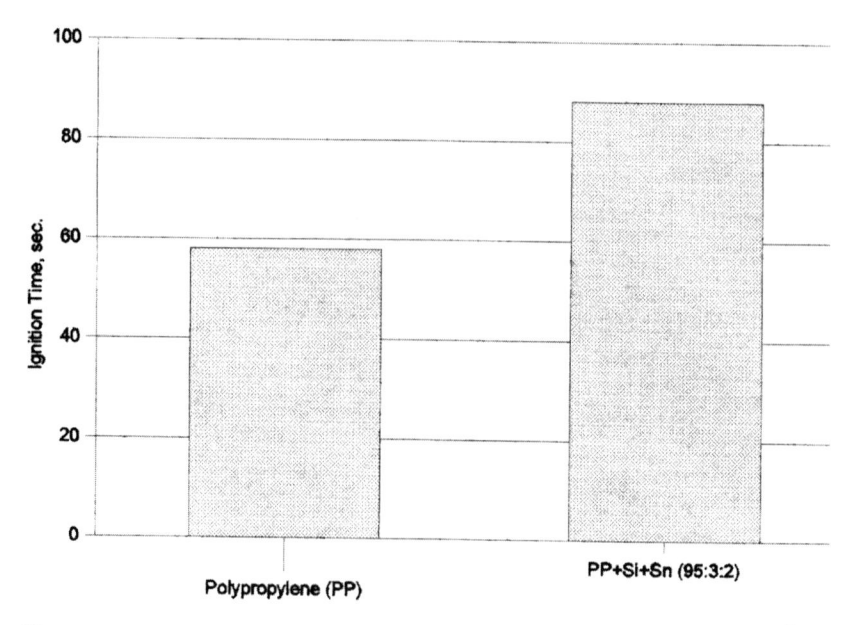

Figure 3 Ignition time delay for PP compositions at heat flux of 35 kW/m².

Finally, Figures 4 and 5 show the limiting oxygen index values and char yields (respectively) for several nylon 6,6 compositions.

CONCLUSIONS

It can be concluded that polyvinyl alcohol incorporated in nylon 6,6 reduces the rate of heat release and increases the char yield. Cone calorimetry data for permanganate-oxidized polyvinyl alcohol indicate an improvement in peak rate of heat release compared with polyvinyl alcohol alone, but the smouldering process is exothermic. Silicon/stannic chloride systems act as ecologically friendly flame retardants for both nylon 6,6 and polypropylene.

Table 2 Cone data of Si–PP system at a heat flux of 35 kW/m²

Cone data	Polypropylene	$PP + Si + SnCl_2$ (95:3:2)
Char yield (% wt)	0.0	10.1
Ignition time (s)	62	91
Peak RHR (kW/m²)	1378.0	860.1
Total heat release (MJ/m²)	332.0	193.7

Conclusions

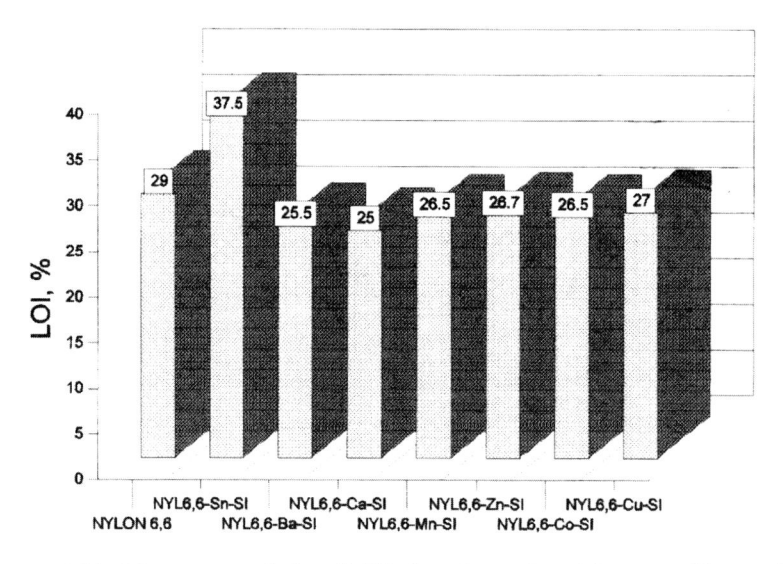

Figure 4 Limiting oxygen index (LOI) data for nylon 6,6 compositions with: $SnCl_2$–Si (2:3% wt), $BaCl_2$–Si (2:3% wt), $CaCl_2$–Si (2:3% wt), $MnCl_2$–Si (2:3% wt), $ZnCl_2$–Si (2:3% wt), $CoCl_2$–Si (2:3% wt), $CuCl_2$–Si (2:3% wt).

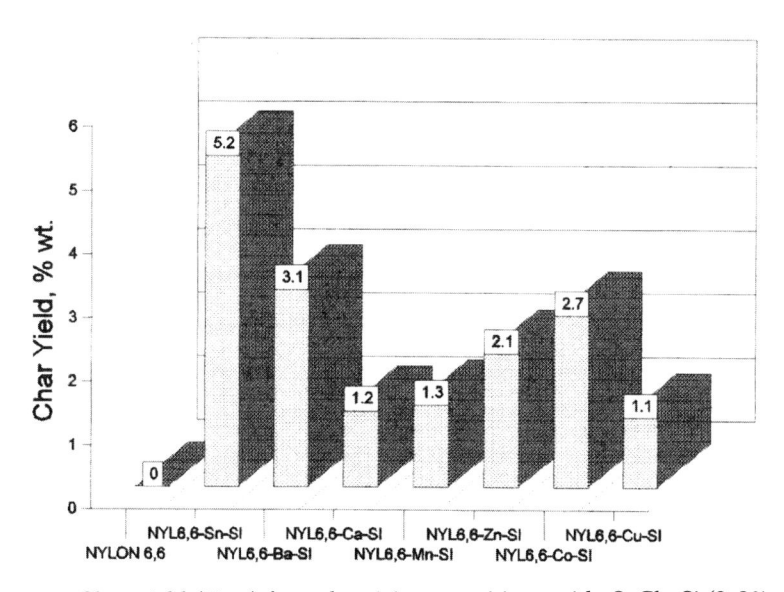

Figure 5 Char yield (% wt) for nylon 6,6 compositions with: $SnCl_2$–Si (3:2% wt), $BaCl_2$–Si (3:2% wt), $CaCl_2$–Si (3:2% wt), $MnCl_2$–Si (3:2% wt), $ZnCl_2$–Si (3:2% wt), $CoCl_2$–Si (3:2% wt), $CuCl_2$–Si (3:2% wt) (TGA analysis, 750°C, air).

REFERENCES

1. Y. Tsuchiya and K. Sumi, *J. Polym. Sci.*, **A-1**, 7, 3151, (1969).
2. C.A. Finch (ed.) *Polyvinyl Alcohol, Properties and Applications*, John Wiley, London–New York–Sydney–Toronto, 622, (1973).
3. R.M. Hassan, *Polymer International*, **30**, 5–9, (1993).
4. J.W. Lyons, *The Chemistry and Uses of Fire Retardants*, Wiley-Interscience, New York, 15–16, (1970).
5. M.E. Morrison and K. Scheller, The effect of burning velocity inhibitors on the ignition of hydrocarbon–oxygen–nitrogen mixtures, *Combustion and Flame*, **18**, 3–12, (1972).

Keywords: char former, cone calorimeter, ecology, flame retardants, halosilanes, heat of combustion, heat release rate, nylon 6,6, oxygen index, polymer combustion, polypropylene, polyvinyl alcohol, silicon, synergistic carbonization, total heat release.

See also: Series of entries with titles 'Flame retardants:...';
Flame retardancy: the approaches available;
Smoke suppressants.

Flame retardants: synergisms involving halogens

Ronald L. Markezich

INTRODUCTION

The following sections concern the use of a chlorinated flame retardant (CFR) in plastics. CFR consists of the Diels–Alder adduct of hexachlorocyclopentadiene and 1,5-cyclo-octadiene. Some experimental results are given, without the proposal of any mechanisms, which exemplify the flame retardancy improvements obtainable by various synergistic combinations.

ALTERNATE SYNERGISTS FOR USE WITH HALOGEN FLAME RETARDANTS

The synergist 'antimony oxide' has been used in combination with halogenated flame retardants for years to impart flame retardancy to plastics [1]. Today, many highly efficient antimony oxide/halogen systems are used to give flame retardancy properties to a wide variety of polymers. Complete or partial substitutes for antimony oxide in certain polymers have been reported [2]; they are ferric oxide, zinc oxide, zinc borate and zinc stannate. Most of these synergists are effective with polyamides and epoxies when using a chlorinated flame retardant.

The chlorinated flame retardant, CFR* (see introductory paragraph above for chemical composition), can also be used in nylons, epoxies and polybutylene terephthalate (PBT) to achieve flame retardancy using

---

* Dechlorane Plus®, Occidental Chemical Corporation.

Plastics Additives: An A–Z Reference
Edited by G. Pritchard
Published in 1998 by Chapman & Hall, London. ISBN 0 412 72720 X

Table 1 Sources of synergists

Material	Trade name	Supplier
Antimony oxide	Thermoguard™ S	Atochem
Zinc oxide	Kadox 911	Zinc Corp. of America
Zinc borate	Firebrake® 290	US Borax
Zinc stannate	Flametard S	Alcan Chemical
Zinc phosphate	Kemgard® 981	Sherwin Williams
Zinc sulfide	Sachtolith® L	Sachtleben Chemie
Iron oxide (Red) Fe_2O_3		J T Baker
Iron oxide (Black) Fe_2O_4	Akrochem® E-8846	Akrochem Corp.
Iron oxide (Yellow) $Fe_2O_3H_2O$	YLO-2288D	Harcros Pigments

synergists other than antimony oxide. Other synergists include zinc borate, zinc oxide, zinc stannate and the iron oxides. Synergists for nylon are listed in Table 1, together with some sources. The synergists can be used alone or in combination.

Mixed synergists can be used to lower the level of the total flame retardant package needed to give flame retardant materials. To achieve a UL-94 V-0 material at 0.4 mm, one can use 20% CFR with 10% antimony oxide or 16% CFR with 2% antimony oxide and 4% zinc borate.

Table 2 gives the results of using mixed synergists with CFR in FR-nylon 6,6. All these formulations are UL-94 V-0 down to 1.6 mm. Formulations 1 to 4 use 16% CFR with 4% zinc borate and 2% of a second synergist – antimony oxide, zinc sulfide, zinc phosphate or zinc stannate.

Table 2 FR-nylon 6,6 using different synergists in combination with zinc borate

Formulation (wt %)	1	2	3	4	5
Nylon 6,6	78	78	78	78	78
CFR	16	16	16	16	12
Antimony trioxide	2	–	–	–	–
Zinc borate	4	4	4	4	1.5
Zinc sulfide	–	2	–	–	–
Zinc phosphate	–	–	2	–	–
Zinc stannate	–	–	–	2	–
Iron oxide	–	–	–	–	1.5
Results					
UL-94 (3.2 mm)	V-0	V-0	V-0	V-0	V-0
(1.6 mm)	V-0	V-0	V-0	V-0	V-0
Tensile strength (MPa)	67	63	65	62	70
CTI, K_c (volts)	450	600	600	600	350

Alternate synergists for use with halogen flame retardants 329

Table 3 25% glass-filled FR-nylon 6,6 using various synergists

Formulation (wt %)	1	2	3	4	5
Nylon 6,6	49	48	55	54	54
Fiberglass	25	25	25	25	25
CFR	18	18	16	16	16
Antimony trioxide	8	–	2	–	–
Zinc phosphate	–	9	–	3	3
Zinc borate	–	–	2	2	–
Zinc oxide	–	–	–	–	2
Results					
UL-94 (3.2 mm)	V-0	V-0	V-0	V-0	V-0
(1.6 mm)	V-0	V-0	V-0	V-1	V-1
Tensile strength (MPa)	128	119	128	121	114
CTI, K_c (volts)	225	325	300	350	350

The most effective synergist package is the mixture of zinc borate with iron oxide, shown in Formulation 5 in Table 2. Only 12% CFR with 1.5% zinc borate and 1.5% iron oxide gives a V-0 material.

The major distinction in using the different synergists in these FR-nylon formulations is the CTI (comparative tracking index) values obtained, Formulation 1, with 2% of antimony trioxide and 16% CFR with 4% zinc borate, gives a CTI of 450 volts while the use of 2% zinc phosphate, 2% zinc stannate or 2% zinc sulfide with 4% zinc borate all give a CTI of 600 volts.

Table 3 shows the results of tests with a 25% glass filled nylon 6,6 formulation using CFR and different synergists. Formulations 1 and 2 both use 18% CFR with a single synergist; Formulation 1 uses 8% of antimony trioxide while Formulation 2 uses 9% zinc phosphate. The CTI of the zinc phosphate formulations is 325 volts while the antimony trioxide formulation only gives 225 volts. Formulations 3, 4 and 5 use mixed synergists. Formulation 3 only uses 2% antimony trioxide with 2% zinc borate and 16% CFR to give a UL-94 V-0 material down to 1.6 mm.

Synergists other than antimony trioxide can also be used with CFR to flame retard epoxy resins [3] as shown in Table 4. Formulations 1, 2 and 3 use 16 phr of CFR with 5 phr of the iron oxide. Yellow iron oxide is more effective than the red or black.

Table 5 gives some white formulations using zinc stannate in combination with zinc borate. Using 18 phr of CFR with 5 phr of zinc borate and 2 phr of zinc stannate gives a UL-94 V-0 material as 3.2 mm. Using 8 phr of zinc borate with 1 phr of zinc stannate only gives a UL-94 V-1 material. But the addition of 1 phr of yellow iron oxide and lowering of the concentration of zinc borate to 5% gives a UL-94 V-0 material at 3.2 mm (Formulation 3).

Flame retardants: synergisms involving halogens

Table 4 FR-epoxy using different iron oxides and high levels of CFR

Formulation (phr)	1	2	3	4	5
CFR	16	16	16	15.5	19.5
Red iron oxide	5	–	–	7	5
Black iron oxide	–	5	–	–	–
Yellow iron oxide	–	–	5	–	–
Results					
UL-94 (3.2 mm)	V-1	V-1	V-0	V-1	V-0
Total burn time (sec.)	66	53	34	55	20
Oxygen Index (O.I.)	–	–	27.5	–	28

Table 5 FR-epoxy using zinc stannate synergist

Formulation (phr)	1	2	3
CFR	18	18	18
Zinc borate	5	8	5
Zinc stannate	2	1	1
Yellow iron oxide	–	–	1
Results			
UL-94 (3.2 mm)	V-0	V-1	V-0
Total burn time (sec.)	38	42	37
Oxygen Index (O.I.)	26	26	28.5

FLAME RETARDANTS FOR PLASTICS COMBINED HALOGEN AND NON-HALOGEN SYSTEMS

Polyolefin wire and cable (W&C) formulations can be flame retarded using a mixture of a chlorinated flame retardant and inorganic salts such as magnesium hydroxide. The mixture of these two flame retardants shows a synergistic effect in the oxygen index test. These formulations give less smoke when tested in the NBS smoke chamber and also show less acid gas formation than without one of the additives. Using alumina trihydrate instead of magnesium hydroxide does not show a synergistic effect. The talc used was Mistron® ZSC from Cyprus Industrial Minerals and the magnesium hydroxide was Zerogen® 35 from the Solem Division of the J M Huber Corporation.

The ethylene vinyl acetate copolymer (EVA) used in these evaluations contained 9% vinyl acetate and had a melt index of 3.2. This and the other polyolefins used were obtained from commercial sources.

Flame retardants for plastics combined systems

Table 6 FR-EVA W&C Formulations

Weight %	1	2	3	4	5
EVA	47.9	47.9	47.9	47.9	47.9
CFR	25	20	15	10	5
Sb_2O_3	5	5	5	5	5
$Mg(OH)_2$	0	5	10	15	20
Talc	20	20	20	20	20
Agerite resin D	1.4	1.4	1.4	1.4	1.4
Luperox 500R	0.7	0.7	0.7	0.7	0.7
Results					
O.I (%)	28.1	28.75	28.41	27.42	27.25
Tensile elongation (%)	350	320	300	190	140

Table 6 gives some typical talc filled FR-W&C formulations using CFR as the flame retardant in combination with magnesium hydroxide. The total flame retardant level is constant at 25% with 5% Sb_2O_3. Figure 1 shows a graph of the % $Mg(OH)_2$ versus oxygen index. There is a synergistic effect between the chlorinated flame retardant and magnesium hydroxide which gives the highest O.I. with 20% CFR and 5% $Mg(OH)_2$.

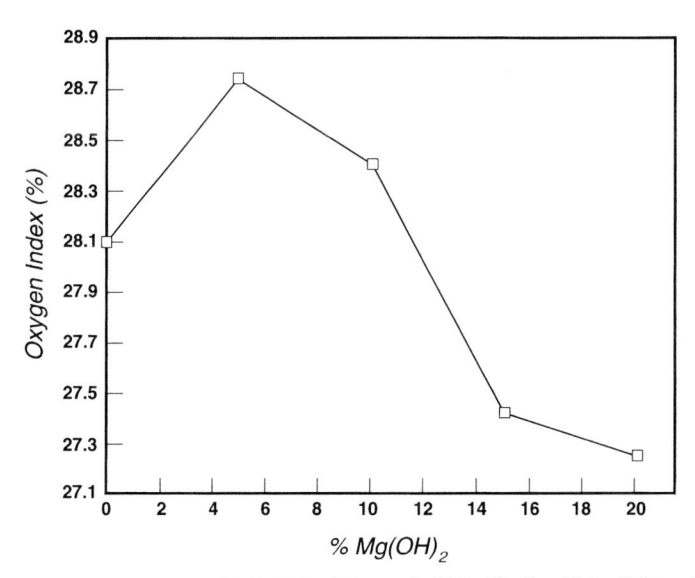

Figure 1 Oxygen index of FR-EVA/20% talc/5% Sb_2O_3 25% CFR minus % $Mg(OH)_2$.

Table 7 FR-EVA W&C formulation (no talc)

Substance	Weight, %	
	1	2
EVA	67.9	62.9
CFR	30 to 5	30 to 5
Sb_2O_3	0	5
$Mg(OH)_2$	0 to 25	0 to 25
Agerite resin D	1.4	1.4
Luperox 500R	0.7	0.7

Table 7 shows some talc-free formulations. Their oxygen indices are shown in Figure 2. There is a maximum oxygen index when the mixture contains 5 to 10% $Mg(OH)_2$ with 5% Sb_2O_3.

The use of alumina trihydrate, instead of magnesium hydroxide, in these talc filled EVA W&C formulations does not show the same synergistic effect as $Mg(OH)_2$ (Figure 3).

FR-EPDM formulations are shown in Table 8 and their oxygen indices are plotted in Figure 4. There is a large synergistic effect in the talc filled sample but the samples that do not contain talc do not show the same degree of synergy.

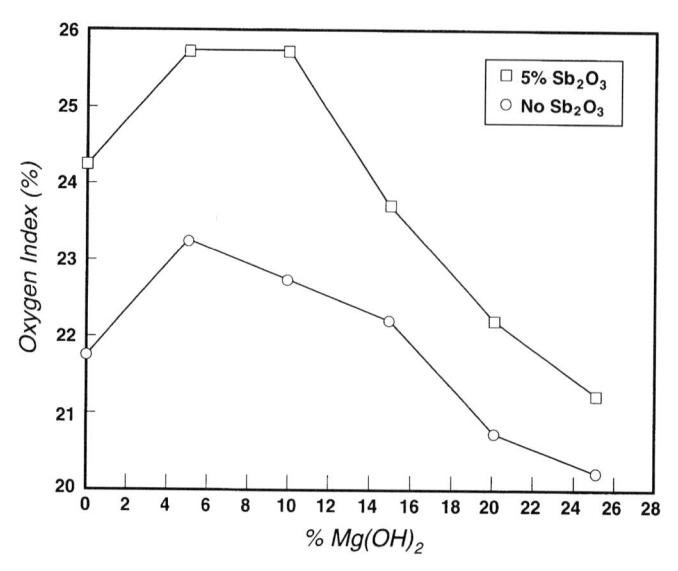

Figure 2 Oxygen index of FR-EVA (no talc) 30% CFR minus % $Mg(OH)_2$.

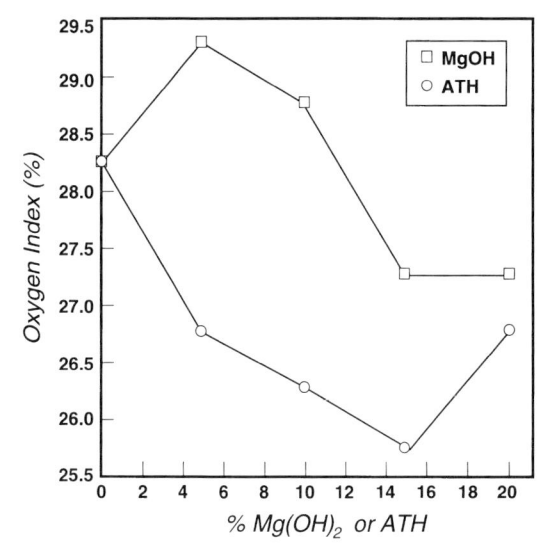

Figure 3 Oxygen index of FR-EVA/20% talc/5% Sb$_2$O$_3$ 25% CFR minus % ATH or Mg(OH)$_2$.

BROMINE/CHLORINE SYNERGISMS FOR FLAME RETARDANCY

The synergistic action between chlorinated and brominated flame retardants to impact flame retardant properties to plastics is not currently well known. In 1976, Gordon, Duffy and Dachs [4] reported the use of mixtures of the chlorinated flame retardant, CFR (see above), and DBDPO (dibromodiphenylene-oxide) to flame retard ABS. A more recent patent to Ilardo and Scharf [5] covers the use of mixtures of chlorinated and

Table 8 FR-EPDM W&C formulations

Substance	Weight, %	
	1	2
EPDM	40.5	53.5
LDPE	8	13
CFR	18 to 0	20 to 0
Mg(OH)$_2$	0 to 18	0 to 20
Talc	20	0
Sb$_2$O$_3$	5	5
Agerite resin D	1	1
Luperox 500 R	1	1
Zinc oxide	2	2
Paraffin wax	2	2
Vinyl silane A-172	0.5	0.5

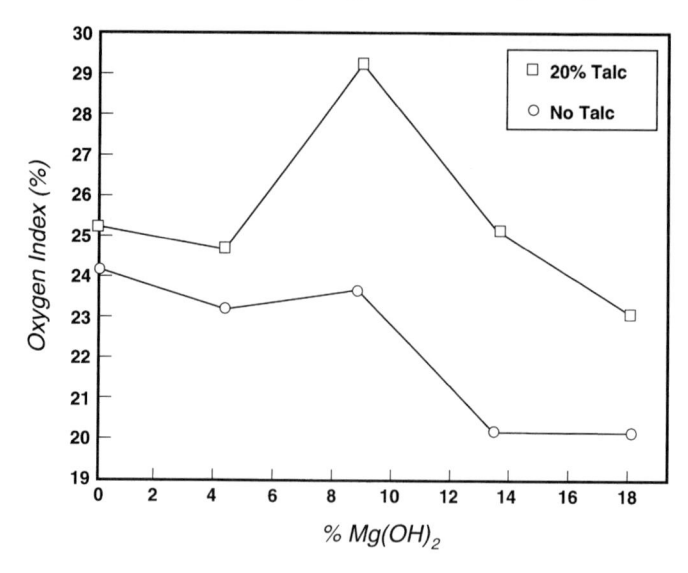

Figure 4 Oxygen index of FR-EPDM 18% CFR minus % Mg(OH)$_2$ with 5% SB$_2$O$_3$.

brominated flame retardants in polyolefins. Figure 5 represents data from this patent. The use of polydibromophenylene-oxide with the chlorinated flame retardant, CFR, gives a maximum oxygen index with a 1:1 mixture of the chloro and brominated flame retardants.

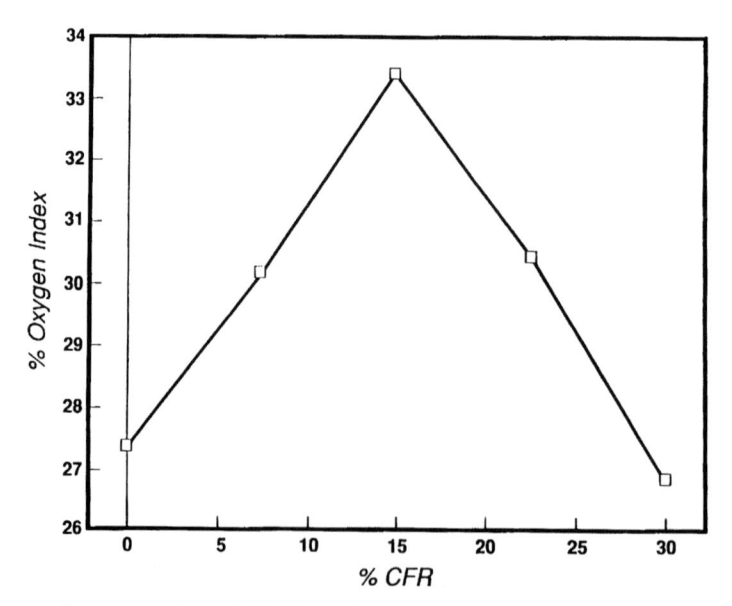

Figure 5 Oxygen index of FR-LDPE/30% total halogen CFR minus % polydibromophenylene-oxide.

Bromine/chlorine synergisms for flame retardancy 335

Table 9 FR-ABS formulations using different ratios of bromo- and chloro- FRs

Substance	1	2	3	4	5
ABS	78.1	76.8	75.55	74.28	73
CFR-2	16.9	12.7	8.45	4.22	–
Sb_2O_3	5	5	5	5	5
BFR-1	–	5.5	11	16.5	22
Oxygen Index	25.75	28.25	31.25	28.25	27.25
Results					
UL-94 3.2 mm	V-0	V-0	V-0	V-0	V-0
1.6 mm	NC	NC	V-0	NC	NC
Notched Izod (J/m)	64	82	97	107	97
% Cl	11	8.2	5.5	2.7	–
% Br	–	2.8	5.5	8.4	11
Total halogen %	11	11	11	11	11

High impact polystyrene (HIPS) can also be flame retarded using a mixture of chloro- and bromo- flame retardants, as discussed in a recent Japanese patent [6]. There has also been another recent patent to O'Brien [7] on the use of bromine/chlorine in flame retarded polyethylene wire insulations.

Table 9 gives several FR-ABS formulations using a mixture of chlorinated and brominated flame retardants. The oxygen indexes from this table are plotted in Figure 6. The highest oxygen index is obtained

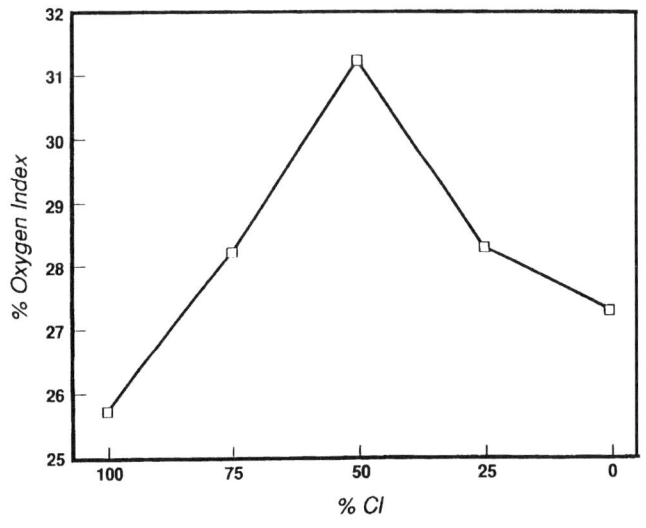

Figure 6 Oxygen index of FR-ABS/11% total halogen CFR-2 minus BFR-1.

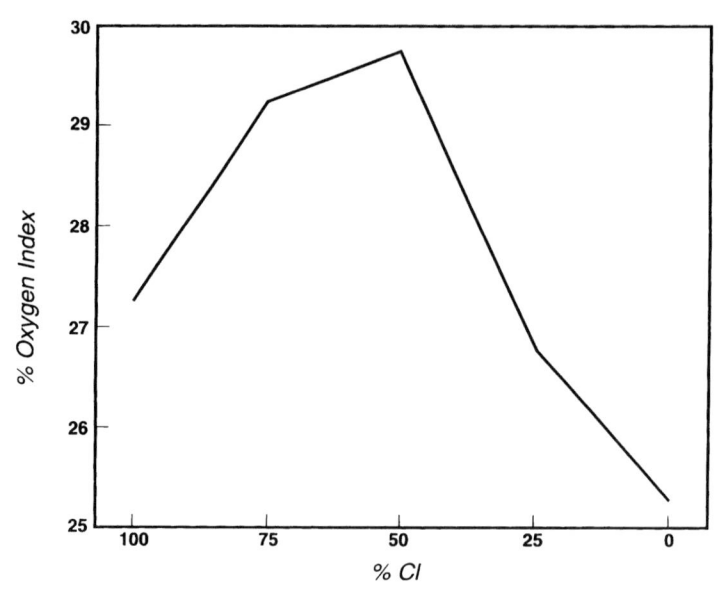

Figure 7 Oxygen index of FR-ABS 11% total halogen/CFR-2 minus BFR-2 with 6.1% Sb$_2$O$_3$.

when there is a 1:1 mixture of Cl and Br (Formulation 3). This is also the only formulation that is UL-94 V-0 at both 3.2 mm and 1.6 mm.

Figure 7 is a plot of oxygen index data comparing CFR-2 with BFR-2 in ABS with 6.1% antimony oxide added as a synergist. When 16.9% CFR-2

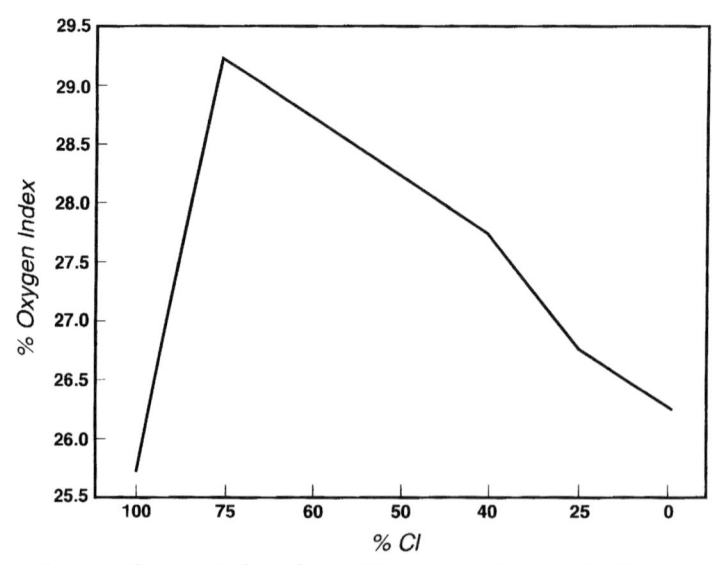

Figure 8 Oxygen index of FR-ABS 11% total halogen/CFR-2 minus DBDPO.

Table 10 FR-HIPS formulations using CFR and BFR-1

Formulation (wt %)	1	2	3	4
HIPS	78	77.17	76.34	75.5
CFR-2	18	15	12	9
Sb_2O_3	4	4	4	4
BFR-1	–	3.83	7.66	11.5
Results				
Oxygen Index	24.75	25.75	26.25	25.75
UL-94 (3.2 mm)	V-1	V-1	V-0	V-0
Notched Izod (J/m)	59	62	66	67
% Cl	11.7	9.75	7.8	5.85
% Br	–	1.95	3.9	5.85
Total halogen %	11.7	11.7	11.7	11.7

(11% Cl) is used, the oxygen index is 27.25, and the material is UL-94 V-0 at 3.2 mm. When 15.7 BFR-2 is used, the oxygen index is 25.25, and the material is NC (HB) in the UL-94 test. The highest oxygen index (29.75) is obtained when there is a 1:1 mixture of Cl and Br.

The use of DBDPO and CFR-2 to flame retard FR-ABS using 5% antimony oxide is shown in Figure 8. The maximum oxygen index of 29.25 is reached using 12.68% CFR-2 and 3.3% DBDPO (75% Cl and 25% Br).

The use of Br/Cl synergism to flame retard HIPS is shown in Table 10. A mixture of CFR and the brominated epoxy, BFR-1, is used. The highest

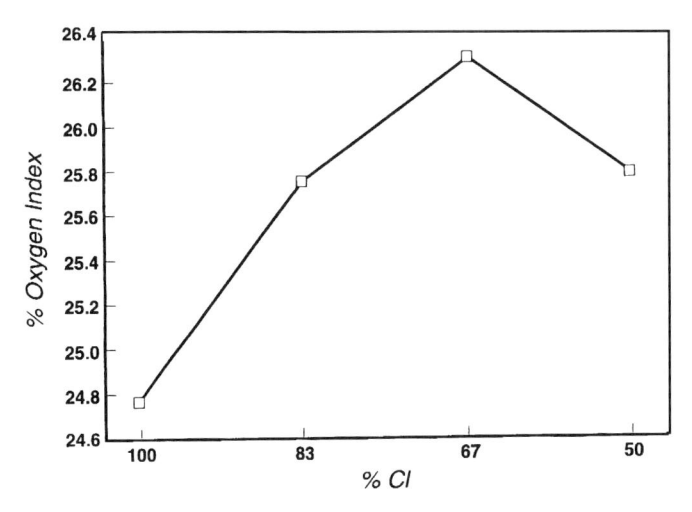

Figure 9 Oxygen index of FR-HIPS 11.7% total halogen/CFR minus BR epoxy.

Table 11 Materials used in bromine/chlorine synergism work

CFR	Diels–Alder adduct of hexachlorocyclopentadiene and 1,5 cyclooctadiene	Dechlorane Plus®	Occidental Chemical Corporation
CFR-2	Same chlorinated flame retardant but with a mean particle size of less than 2 µm	Dechlorane Plus®	Occidental Chemical Corporation
BFR-1	Brominated epoxy resins with 51% BR	YDB-406	Tohto Kasei
BFR-2	Bis(tribromophenoxy)ethane	FF-680	Great Lakes

oxygen index is obtained using 12% CFR and 7.66% brominated epoxy. This material is also UL-94 V-0 at 3.2 mm. A plot of the oxygen indexes from this table is given in Figure 9. Table 11 lists the materials used in the bromine/chlorine synergism work.

REFERENCES

1. K. Othmer *Encyclopedia of Chemical Technology*, Volume 10, 3rd Edition (1980).
2. R.F. Mundhenke and R.L. Markezich 'Flame Retarded Plastics Using a Chlorinated Flame Retardant and Mixed Synergists', *The Fire Retardant Chemical Association (FRCA) Fall Meeting*, October, 1993.
3. R.L. Markezich 'Flame Retardant Additives for Epoxy Resin Systems', *Epoxy Resin Formulators Division Winter Meeting*, February 1987.
4. I. Gordon, J.J. Duffy and N.W. Dachs, U.S. Patent 4,000,114 (1976).
5. C.S. Ilardo and D.J. Scharf, U.S. Patent 4,388,429 (1983).
6. M. Hirata, R. Fujihira, A. Suzuki and M. Machide, JP 05112693 A2 (1993).
7. D. O'Brien, U.S. Patent 5,358,881 (1994).

Keywords: Diels–Alder adduct, magnesium hydroxide, zinc, oxygen index, smoke, talc, polyolefin, polyamide, antimony trioxide.

See also: Series of entries with titles 'Flame retardants:...';
Flame retardancy: the approaches available;
Smoke suppressants.

Flame retardants: tin compounds

P.A. Cusack

INTRODUCTION

Tin compounds have been known as flame retardants since the mid-nineteenth century, when processes based on the *in situ* precipitation of hydrous tin(IV) oxide were developed to impart flame-resist properties to cotton and other cellulosic materials:

$$SnCl_4 + 4NH_3 + 2H_2O \rightarrow SnO_2 + 4NH_4Cl \tag{1}$$

$$Na_2Sn(OH)_6 + (NH_4)_2SO_4 \rightarrow SnO_2 + Na_2SO_4 + 2NH_3 + 4H_2O \tag{2}$$

More recently, tin salts have found use in flame-retardant treatments for woollen sheepskins and rugs. The active tin species are generally fluorostannate-based (e.g. SnF_6^{2-}), and these are electrostatically attracted to the protonated amino groups in the proteinaceous wool structure.

However, as far as plastics are concerned, commercial interest in the use of tin-based flame retardants has only developed over the past 10 years or so. Although it is estimated that over 600 000 tonnes of chemical additives are used worldwide annually as flame retardants for synthetic polymers, recent concerns about the toxic nature of certain additives have led to an intensified search for safer flame retardants. Hence, the generally low toxicity of inorganic tin compounds has been a major factor in their growing acceptance throughout the 1990s as flame retardants and smoke suppressants for plastics, elastomers and other polymeric materials.

TIN COMPOUNDS

Despite the fact that a wide range of inorganic and organo-tin compounds have been found to exhibit fire-retardant properties (Table 1), only a few

Plastics Additives: An A–Z Reference
Edited by G. Pritchard
Published in 1998 by Chapman & Hall, London. ISBN 0 412 72720 X

Table 1 Fire-retardant tin compounds

Compound	Recommended polymer(s)
ZnSn(OH)$_6$	Rigid and flexible PVC, polyester resins, elastomers
ZnSnO$_3$	Nylons, PET, epoxy resins, rigid & flexible PVC
SnO$_2$ (anhydrous or hydrous)	PVC, ABS co-polymer, polyolefins, nylons
Na$_2$Sn(OH)$_6$	Glass-reinforced polyester
CaSn(OH)$_6$	Flexible PVC
SnCl$_4$	Poly(methyl methacrylate)
SnCl$_2$	Nylon, PET, polyacrylonitrile
SnS$_2$	Rigid PVC
(NH$_4$)$_2$SnBr$_6$	Nylon
Sn(O·CO·C$_7$H$_{15}$)$_2$	Polystyrene
Sn(O·CO·CO·O)	Polystyrene
SnCl$_4$·2Ph$_3$PO	PET
SnCl$_2$·2Ph$_3$PO	PET
Bu$_2$Sn(O·CO·CH:CH·CO·O)	Polypropylene
Bu$_2$SnO	Polyacrylonitrile

have reached commercialization. Tin(IV) oxide, both in its anhydrous and hydrous forms, has been studied extensively as a potential substitute for antimony trioxide in halogen-containing polymer formulations. Although it has been found to exhibit good flame-retardant and smoke-suppressant activity, its relatively high price compared with Sb$_2$O$_3$ (particularly during the mid-1980s), has severely limited its usage. Similarly, an organotin derivative, dibutyltin maleate, has been found to impart both flame-retardant synergism and UV stabilization, when used in conjunction with an organobromine flame retardant. However, its commercial use has been limited to polypropylene fibres.

By far the most important tin-based fire retardants are zinc hydroxystannate (ZHS) and its anhydrous analogue, zinc stannate (ZS). Originally developed at ITRI during the mid-1980s, these two additives are now being marketed worldwide as non-toxic flame retardants/smoke suppressants for use in a wide range of polymeric materials.

Zinc hydroxystannate is manufactured commercially by the aqueous reaction of sodium hydroxystannate with zinc chloride:

$$Na_2Sn(OH)_6 + ZnCl_2 \rightarrow ZnSn(OH)_6 + 2NaCl \quad (3)$$

The product, which is precipitated as a white solid, is washed free of the sodium chloride by-product, and dried in air at a temperature of *ca.* 105°C.

Table 2 Properties of ZHS and ZS

Property	ZHS	ZS
Chemical formula	ZnSn(OH)$_6$	ZnSnO$_3$
CAS No.	12027-96-2	12036-37-2
Appearance	White powder	White powder
Analysis: Sn	41%	51%
Zn	23%	28%
Cl	<0.1%	<0.1%
free H$_2$O	<1%	<1%
Specific gravity	3.3	3.9
Decomposition temp. (°C)	>180	>570
Toxicity	Very low*	Very low*

* Acute oral toxicity, LD$_{50}$ (rats) >5000 mg/kg.

Zinc stannate is manufactured by controlled thermal dehydration of ZHS, usually at a temperature in the range of *ca.* 300–400°C:

$$ZnSn(OH)_6 \rightarrow ZnSnO_3 + 3H_2O \quad (4)$$

Although there is generally little difference in the effectiveness of ZHS and ZS, the latter is the preferred additive for polymers which are processed at temperatures above 180°C. Some important properties of ZHS and ZS are given in Table 2.

The major application areas for the zinc stannates (and, indeed, for other tin-based fire retardants) are in halogen-containing polymer formulations, where the tin additives are used as alternative synergists to Sb$_2$O$_3$. However, certain tin compounds, including ZHS, ZS and SnO$_2$, have been shown to exhibit beneficial flame-retardant and/or smoke-suppressant properties in halogen-free compositions, and both types of system are discussed here.

LABORATORY FIRE TESTS

Before discussing the fire-retardant properties of tin compounds, it is necessary to outline briefly some of the main laboratory test methods which are used to assess the combustion behaviour of materials.

1. Limiting Oxygen Index: The LOI test is a very widely used method for determining the relative flammability of polymeric materials. A numerical index, the LOI, is defined as the minimum concentration of oxygen, in a nitrogen/oxygen mixture, which is required to just support combustion of the test sample under the conditions of the test. Higher values of LOI indicate greater flame retardancy. The LOI test conforms to ISO 4589-2, ASTM D2863 and BS 2782 (Part 1, Method 141).

2. NBS Smoke Box: This test has long been the most widely used laboratory-scale method for measuring the smoke generated by burning materials. Test specimens may be burned in either the flaming mode or the non-flaming (or smouldering) mode, and the specific optical density of the resulting smoke is determined photometrically. Data from the NBS Smoke Box are usually presented graphically as cumulative specific optical density (Ds) versus time curves, and maximum corrected specific density values are quoted either directly as Dmc, or as Dmc/g, in which the values are normalized for sample mass loss during the test. The NBS Smoke Box conforms to ASTM E662, BS 6401, NFPA 258 and, in a modified version, to ISO 5659-2.
3. Cone Calorimeter: This instrument represents an important advance in laboratory-scale fire testing of polymeric materials and composites. Using the Oxygen Consumption principle, the method provides a reliable means of measuring the rate of heat release – the single most important property predicting fire hazard, because it governs both the rate of fire growth and its maximum intensity. Utilizing a truncated conical heater element to irradiate test samples at heat fluxes from 10 to 100 kW/m^2, the Cone Calorimeter can simulate a range of fire intensities, and has been shown to provide data which correlate well with those from full-scale fire tests. The Cone Calorimeter is the subject of ISO 5660-1, ASTM E1354, ASTM E1474 and BS 476 (Part 15).

Two key Cone Calorimeter parameters considered here are:
- peak rate heat release; the single property which most critically defines a fire is the heat release rate, in particular its peak value, which is indicative of the maximum intensity of the fire; and
- smoke parameter; defined as the product of the measured average specific extinction area and peak rate of heat release, this parameter gives an indication of the realistic amount of smoke that a material would produce under full-scale fire conditions.

HALOGEN-CONTAINING POLYMER FORMULATIONS

Although antimony trioxide has long been established as the most widely used flame-retardant synergist for use in halogenated polymers, recent concerns about its toxic nature, combined with severe fluctuations in price, have created a clear demand for alternative synergists. The effectiveness of ZHS, ZS and SnO$_2$ as flame retardants and smoke suppressants has been demonstrated in numerous halogen-containing polymers, including rigid and flexible PVC, unsaturated polyester resins, epoxy resins, chlorinated elastomers, alkyd resin-based paints, and various thermoplastics where halogen is incorporated as an additive-type flame retardant.

The relative fire-retardant efficiency of the tin additives compared with Sb$_2$O$_3$ is known to be dependent on a number of factors, which include:

Halogen-containing polymer formulations

Table 3 Fire test data for polyester resins containing different halogen additives*

Halogen additive[†]	Organic type	Synergist	LOI	Cone: Peak RHR (kW/m^2)[‡]
Cereclor 70	Aliphatic	None	21.6	376
		Sb$_2$O$_3$	29.0	234
		ZNS	26.0	195
		SnO$_2$ (hyd.)	24.2	224
HBCD	Alicyclic	None	23.8	359
		Sb$_2$O$_3$	29.1	238
		ZNS	30.1	254
		SnO$_2$ (hyd.)	26.5	–
DBDPO	Aromatic	None	24.5	369
		Sb$_2$O$_3$	33.4	191
		ZHS	26.2	247
		SnO$_2$(anhyd.)	29.3	282

* Resins contain 20 phr halogen additive + 4 phr synergist.
[†] Cereclor 70 = proprietary chlorinated paraffin (I.C.I.); HBCD = hexabromocyclododecane; DBDPO = decabromodiphenyl oxide.
[‡] Cone Calorimeter operated at 50 kW/m^2 incident heat flux.

- the chemical nature of the halogen source;
- the ratio of halogen to synergist;
- the nature of the host polymer;
- the presence of other additives in the formulation;
- the test method used to evaluate fire-retardant performance.

Comprehensive studies of polyester resin formulations have indicated that, in contrast to Sb$_2$O$_3$, tin synergists perform better when used in conjunction with aliphatic or alicyclic halogen compounds, than with aromatic halogen compounds (Table 3). Hence, whereas Sb$_2$O$_3$ gives the highest elevation in LOI and the greatest reduction in heat release rate when used in conjunction with the aromatic bromine flame retardant, DBDPO, ZHS gives superior performance in the LOI test with the alicyclic bromine additive, HBCD, and in heat release reduction when used with aliphatic chlorinated paraffin. Interestingly, although ZHS is generally superior to SnO$_2$ with regard to flame-retardant activity, SnO$_2$ gives a higher LOI value in the DBDPO system.

Investigations during the mid-1980s at City University, London, demonstrated that the optimum atomic ratio of halogen to tin for flame retardancy in the ABS/decabromobiphenyl/hydrous SnO$_2$ system, is much higher (i.e. *ca.* 9:1) than expected on a stoichiometric basis (4:1). Similarly, ITRI research on ZHS and ZS has shown that optimum flame retardancy often occurs at relatively high halogen:metal ratios, although

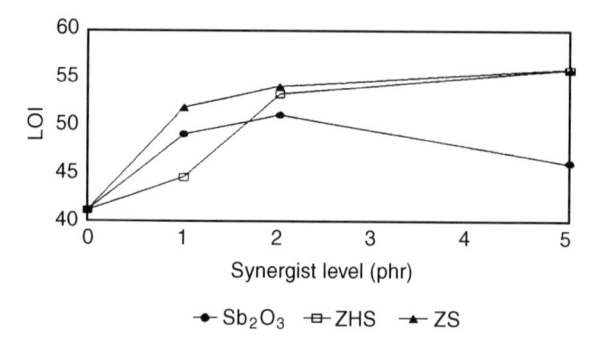

Figure 1 Effect of synergist level on the flammability of brominated polyester resin.

the precise ratio appears to be dependent on both the nature of the halogen source and the polymer itself.

Tin synergists are particularly effective in halogenated formulations in which the halogen is chemically bonded to the polymer chain. Hence, ZHS and ZS are superior to Sb_2O_3 in LOI tests on brominated polyester resin, in which the bromine is introduced as the reactive intermediate, dibromoneopentyl glycol (DBNPG), and on rigid PVC (Figures 1 and 2 respectively).

However, the most important performance benefit of the tin additives over antimony trioxide relates to their excellent smoke-suppressant properties. NBS Smoke Box tests have indicated that, whereas Sb_2O_3 increases smoke production for brominated polyester resin and flexible PVC formulations, the same incorporation level of ZHS (or ZS) gives significant reductions in smoke density (Figures 3 and 4).

Although tin compounds are themselves effective synergists when used in conjunction with halogen compounds, improved synergism has been observed in certain cases when further inorganic additives (or fillers) are incorporated into the polymer formulation. Examples of these ternary synergistic systems are given in Table 4.

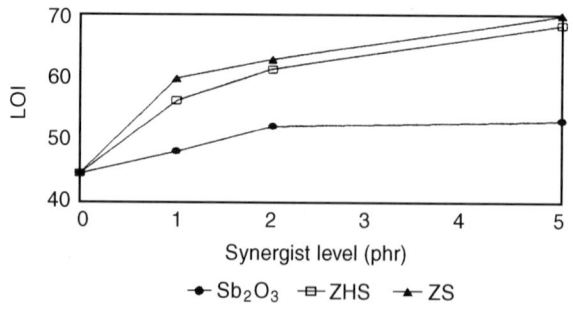

Figure 2 Effect of synergist level on the flammability of rigid PVC.

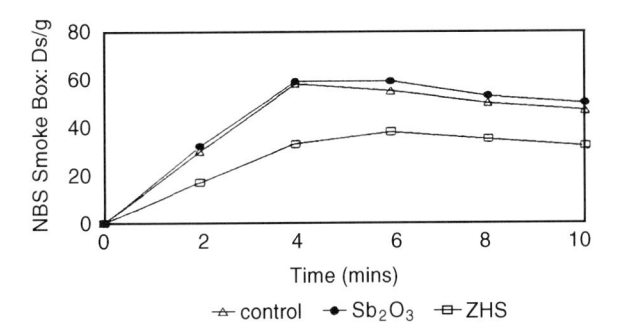

Figure 3 Effect of synergists on smoke density for brominated polyester resin.

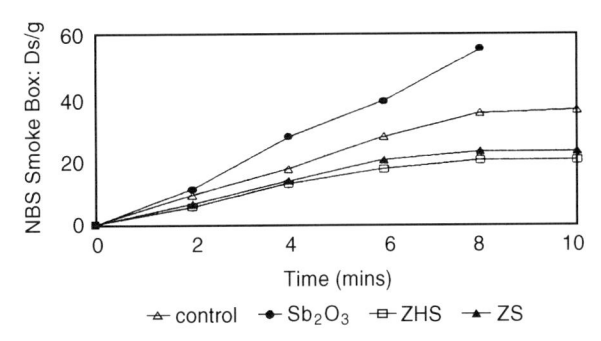

Figure 4 Effect of synergists on smoke density for a flexible PVC cable formulation.

Table 4 Ternary synergistic fire-retardant systems involving tin compounds

Tin compound	Inorganic additive/filler	Halogen source	Polymer
SnO_2	ATH*	–	PVC
SnO_2	MoO_3	–	PVC
SnO_2 (hydrous)	Fe_2O_3	DBB*	ABS
ZHS	ATH*	DBNPG*	Polyester resin
ZHS	Al_2O_3	DBNPG*	Polyester resin
ZHS/ZS	TiO_2	DBNPG*	Polyester resin
ZHS/ZS	ATH*	–	PVC
ZHS/ZS	$Mg(OH)_2$	–	PVC
ZHS/ZS	ZB*	–	PVC
ZHS/ZS	ZB*	Dechlorane Plus*	Epoxy resin
ZS	ZB*	Dechlorane Plus*	Nylons

* ATH = alumina trihydrate, $Al_2O_3 \cdot 3H_2O$; DBB = decabromobiphenyl: DBNPG = dibromo-neopentyl glycol; Dechlorane Plus = proprietary alicyclic chlorine additive (Occidental Chemicals); ZB = zinc borate, $2ZnO \cdot 3B_2O_3 \cdot 3.5H_2O$.

Table 5 Fire test data for chlorinated polyester resin samples*

Parameter	Control	2.5 phr Sb_2O_3	5 phr Sb_2O_3	2.5 phr ZHS	5 phr ZHS
Limiting Oxygen Index	26.8	39.2	42.9	35.6	38.0
Cone: Peak rate of heat release (kW/m^2)†	188	124	117	95	99
Cone: Smoke parameter (MW/kg)†	135	82	78	43	35
NBS Smoke Box (flaming mode): Dmc/g	50.0	40.2	–	24.3	–

* Resin contains 28% Cl as chlorendic anhydride.
† Cone Calorimeter operated at 60 kW/m^2 incident heat flux.

Differences in the relative performances of the synergists according to test method are clearly evident in a series of chlorinated polyester resins, in which the chlorine is chemically reacted into the polymer as chlorendic anhydride (Table 5). Whereas Sb_2O_3 is more effective than ZHS with regard to increasing the LOI of the resin, the tin additive gives a markedly superior performance, even at a lower incorporation level, in Cone Calorimeter evaluations of heat release rates and smoke generation. The outstanding smoke-suppressant properties of ZHS in this formulation are further confirmed in the NBS Smoke Box method.

With regard to practical usage, one of the major application areas for ZHS and ZS has been in flexible PVC wire and cable insulation, where the tin additives have been selected because of their combined flame-retardant/smoke-suppressant properties. The grades of ZHS/ZS used must have very low electrolyte levels and must not contain any free zinc oxide, since the presence of the latter compound can result in a phenomenon known as 'zinc burning', in which the PVC undergoes dehydrochlorination during processing.

HALOGEN-FREE POLYMER FORMULATIONS

In recent years, there has been much concern about the toxicity and corrosive nature of the smoke and gases generated during the combustion of halogen-containing polymers. As a result, the demand for plastics which comply with the specification 'low smoke zero halogen' flame retardancy has grown rapidly, particularly in application areas such as underground transport, mining and electrical installations.

Halogen-free compositions are often rendered flame-retardant by the incorporation of inorganic fillers, particularly alumina trihydrate (ATH) or magnesium hydroxide. Although these fillers are essentially non-toxic and relatively inexpensive, the high loadings necessary for effective

Table 6 Fire test data for halogen-free polyester resin samples

ATH (phr)	ZHS (phr)	LOI	NBS Smoke Box (flaming mode) (Dmc/g)
0	0	18.9	72.2
0	2	19.4	64.6
0	5	19.6	57.0
25	0	20.4	40.5
25	2	20.5	31.4
25	5	20.8	23.6
50	0	22.3	15.5
50	2	22.3	15.0
50	5	22.5	11.7

flame retardancy can lead to processing difficulties and a marked deterioration in the mechanical properties of the host polymer.

Although there have been relatively few studies of tin additives in non-halogenated plastics, it has been found that tin compounds, including ZHS, ZS and SnO_2, when utilized at low addition levels, can significantly improve the flame-retardant and, particularly, the smoke-suppressant properties of halogen-free polymer formulations.

For example, 2–5 phr levels of ZHS are found to reduce markedly the smoke density for unfilled and ATH-filled halogen-free polyester resins. Interestingly, however, the addition of the tin compound has no significant effect on the flammability of the resins, as adjudged by the LOI test (Table 6). Good smoke suppression using ZHS and other tin additives has also been observed in glass-reinforced polyester (GRP) and both flame- and smoke-retardant properties have been reported in non-halogenated elastomeric compositions.

FIRE-RETARDANT MECHANISM

Although much work has been carried out on the mode of action of flame retardants generally, the mechanisms associated with tin additives are only partially understood. It is clear that tin-based fire retardants can exert their action in both the condensed and vapour phases, and that the precise action in any particular system depends on a number of factors, including incorporation level, the amount and chemical nature of other additives present and, indeed, the nature of the polymer itself.

Thermoanalytical experiments have clearly shown that tin-based fire retardants markedly alter both the initial pyrolysis and the oxidative 'burn off' stages which occur during polymer breakdown. Hence,

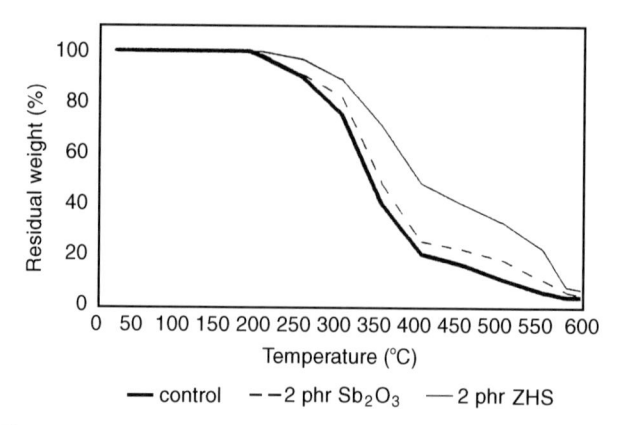

Figure 5 Thermogravimetric curves for brominated polyester resin samples in air.

whereas the incorporation of Sb$_2$O$_3$ into a brominated polyester resin only results in a slight change in its thermal decomposition profile, the addition of ZHS leads to dramatic changes in the thermogravimetric trace of the resin (Figure 5). The changes have been interpreted as being indicative of an extensive condensed phase action for the tin additive, in which the thermal breakdown of the polymer is altered to give increased formation of a thermally-stable carbonaceous char at the expense of volatile, flammable products. The consequent reduction in the amount of fuel supplied to the flame largely accounts for the beneficial smoke-suppressant properties associated with ZHS and other tin-based fire retardants.

Further insight into the mode of action of inorganic synergists in halogen-containing polymer formulations has been provided by quantification and analysis of the char residues remaining after combustion of the polymer in air. Data for such experiments on brominated polyester resins are presented in Table 7.

Table 7 Char yields and elemental volatilization data for brominated polyester resins

Resin	Synergist (phr)	Char yield	Elemental volatilization			
			Br	Sn	Zn	Sb
28% Br	None	24%	96%	–	–	–
28% Br	5Sb$_2$O$_3$	33%	94%	–	–	93%
28% Br	5SnO$_2$ (hyd.)	53%	85%	18%	–	–
28% Br	5ZS	53%	74%	43%	34%	–
10% Br	None	20%	96%	–	–	–
10% Br	2ZHS	39%	78%	64%	24%	–

Table 8 Boiling points of metal halides

Chlorides		Bromides	
Compound	b.p. (°C)	Compound	b.p. (°C)
SnCl$_4$	114	SnBr$_4$	202
SnCl$_2$	652	SnBr$_2$	620
ZnCl$_2$	732	ZnBr$_2$	650
SbCl$_3$	283	SbBr$_3$	280

It is found that the yields of involatile carbonaceous char are approximately doubled when additions of ZHS, ZS or hydrous SnO$_2$ are made to the polymer, this observation being consistent with condensed phase behaviour. Elemental analysis of the residues has been used to determine the degree of volatilization of the metals and of bromine during the combustion process. Although in the case of hydrous SnO$_2$ only a small fraction of the tin is volatilized, significant proportions of the tin and zinc are volatilized from the ZHS- and ZS-containing samples, which may be indicative of vapour phase flame-retardant action. The extent of bromine loss is significantly reduced for samples containing tin additives, particularly those containing ZHS and ZS, which is consistent with the formation of the relatively involatile metal halides, ZnBr$_2$ and SnBr$_2$ (Table 8). In line with this observation, ^{119}Sn Mössbauer spectroscopic studies of rigid PVC and Neoprene samples containing SnO$_2$ as a fire-retardant additive have shown that the SnO$_2$ is partially reduced to tin(II) species (SnCl$_2$ and SnO) and to metallic tin during thermal degradation and combustion. In such cases, the relatively involatile SnCl$_2$ is detected in the char, whereas the highly volatile SnCl$_4$ is not. Antimony, which undergoes almost complete volatilization from the brominated resin during combustion, shows little char-enhancing behaviour and operates primarily in the vapour phase by forming highly volatile antimony halides and oxyhalides.

The mode of action of tin additives in halogen-free polymer compositions has not yet been studied in any detail, but the near quantitative retention of tin in char residues is indicative of condensed phase activity. Thermal analysis of halogen-free polyester resin formulations containing ATH and ZHS has indicated that the tin compound exhibits significant char-enhancing properties (Table 9). Hence, a 5 phr addition level of ZHS is found to increase markedly the weight loss associated with char oxidation at the expense of the initial pyrolysis loss, when compared with the resin containing ATH alone. Further evidence of the char-promoting activity of ZHS is provided by the observed residual yield at 600°C, which is significantly greater than would be expected on the

Table 9 Thermoanalytical data for halogen-free polyester resin samples in air

ATH (phr)	ZHS (phr)	Pyrolysis stage* Weight loss	Pyrolysis stage* DTG_{max} (°C)	Char ordination stage† Weight loss	Char ordination stage† DTG_{max} (°C)	Residue at 600°C Found	Residue at 600°C Calc.
None	None	88.6%	363	11.3%	544	0.1%	0%
25	None	77.6%	342	9.9%	534	12.5%	13.1%
25	5	67.8%	336	14.1%	497	18.1%	15.7%

*Temperature range = ca. 260–450°C.
†Temperature range = ca. 450–570°C.

basis of the involatile inorganic materials (i.e. $Al_2O_3 + ZnSnO_3$) present in the char.

It has been suggested that certain metal oxides may act as dehydrogenation catalysts in halogen-free polymer systems and proprietary grades of ATH and magnesium hydroxide containing small amounts of char-promoting metal oxides are apparently being marketed. In addition to the above hypothesis, the highly endothermic dehydration of ZHS at temperatures above 180°C may partially account for its fire-retardant activity when used in halogen-free formulations.

RECENT DEVELOPMENTS

As part of a programme to develop improved activity tin-based fire-retardant systems, ITRI has patented a series of processes for producing novel additives, including 'Ultrafine ZHS/ZS powders' and 'coated fillers'.

Studies have been undertaken to develop Ultrafine (UF) ZHS powders for application in polyester resins. A laboratory produced UF grade of ZHS, with a particle size of ca. 0.1–0.4 µm, exhibits a number of performance benefits over commercial grades of ZHS, the latter having typical particle sizes of ca. 2–3 µm. In particular, the UF material does not settle out in polyester resins, it can be used in formulations where translucency is required, and its fire-retardant efficiency is markedly higher than that of standard grade ZHS (Figure 6).

ITRI's 'coated fillers' are developmental powders each of which comprise a coating of ZHS, ZS or SnO_2 on an inorganic filler, the latter including ATH, $Mg(OH)_2$, $CaCO_3$ and TiO_2. These additives have been shown to exhibit significantly enhanced flame-retardant and smoke-suppressant properties, compared with simple mixtures of the individual components, when evaluated in flexible PVC (Table 10) and other halogen-containing polymer formulations. The observed improvements are believed to arise from an improved dispersion of the active tin species

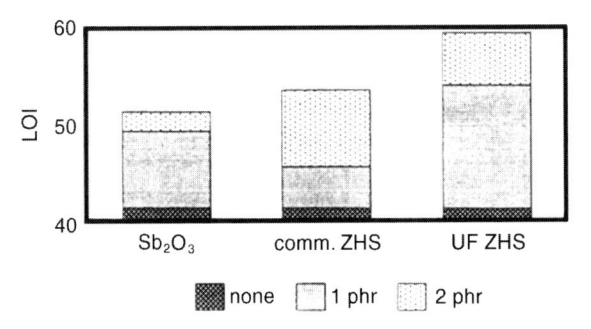

Figure 6 Effect of synergists on the flammability of brominated polyester resin, highlighting the exceptional performance of Ultrafine ZHS.

throughout the polymeric substrate. The observation that 20 phr of ZHS-coated Mg(OH)$_2$ outperforms 50 phr of uncoated Mg(OH)$_2$ is of particular significance since the marked reduction in filler loading necessary for a given degree of fire retardancy would be expected to result in markedly improved polymer processability of physical properties.

CONCLUSIONS

Certain tin compounds, in particular, zinc hydroxystannate, zinc stannate and tin(IV) oxide, are effective flame retardants in a wide range of polymeric materials and offer several advantages over many of the available alternatives:

- non-toxicity;
- combined flame retardancy and smoke suppression;
- marked reductions in heat release rates;
- synergist effects when used in conjunction with other additives and fillers.

Table 10 Fire test data for flexible PVC samples

Sample	LOI	Cone: Peak rate of heat release (kW/m^2)*	Cone: Smoke parameter (MW/kg)*
Control	25.4	239	210
20 phr Mg(OH)$_2$	26.8	220	120
18 phr Mg(OH)$_2$ + 2 phr ZHS	28.1	196	62
20 phr ZHS-coated Mg(OH)$_2$†	28.8	182	70
50 phr Mg(OH)$_2$	28.4	207	79
45 phr Mg(OH)$_2$ + 5 phr ZHS	32.4	188	56
50 phr ZHS-coated Mg(OH)$_2$†	33.9	162	40

* Cone Calorimeter operated at 50 kW/m^2 incident heat flux.
† Composition of coated filler = 10% ZHS on 90% Mg(OH)$_2$ by weight.

The tin additives exert their fire-retardant action in both the condensed and vapour phases, by promoting the formation of a thermally stable carbonaceous char and (in halogen-containing polymer formulations) by generating volatile metal halide species which assist in free radical scavenging reactions in the flame.

Although current usage of tin-based flame retardants is relatively small (perhaps in the hundreds of tonnes per annum), their numerous benefits combined with the continuing development of improved products should lead to a rapid growth in the market for these compounds in the years ahead.

REFERENCES

1. Blunden, S.J., Cusack, P.A. and Hill, R. (1985) Fire retardants in *The Industrial Uses of Tin Chemicals*, The Royal Society of Chemistry, London, pp. 172–209.
2. Touval, I. (1972) The use of stannic oxide hydrate as a flame retardant synergist. *Journal of Fire and Flammability*, **3**, 130–43.
3. Donaldson, J.D., Donbavand, J. and Hirschler, M.M. (1983) Flame retardance and smoke suppression by tin(IV) oxide phases and decabromobiphenyl. *European Polymer Journal*, **19**, 33–41.
4. Cusack, P.A. and Karpel, S. (1991) Zinc stannates: novel tin-based fire retardants. *Tin and Its Uses*, **165**, 1–6.
5. Chaplin, D. (1992) New and improved flame retardants of low hazard. *Proceedings of Flame Retardants '92 Conference*, Elsevier Applied Science, London and New York, pp. 198–210.

Keywords: tin, tin oxide, zinc hydroxystannate, zinc stannate, organotin compounds, antimony trioxide, alumina trihydrate, magnesium hydroxide, titanium dioxide, molybdenum trioxide, iron oxide, zinc borate, alumina, halogenated flame retardants, metal halides, thermal analysis, Mössbauer spectroscopy, fire-retardant mechanism, ultrafine powders, coated fillers.

See also: Series of entries with titles 'Flame retardants: ...';
Flame retardancy: the approaches available;
Smoke suppressants.

Hindered amine light stabilizers: introduction

Ján Malík and Gilbert Ligner

INTRODUCTION

The expansion of polyolefins into new areas of industrial and every-day use has achieved tremendous progress during the last three decades. This progress was in most cases allowed by the employment of various speciality chemicals, such as different catalysts, polymer additives etc. Parallel to the growth of polyolefins applications, a great advance was observed also in the field of additives. Perhaps the most exciting progress was reached in the area of light stabilization by the discovery and introduction of Hindered Amine Light Stabilizers (HALS).

SUCCESS STORY OF HALS

The success story of HALS began in 1959, when the preparation of a stable 2,2,6,6-tetramethylpiperidine-N-oxyl radical (Figure 1) was described. Soon it was found that this stable nitroxyl radical is a very efficient stabilizer which is able to inhibit chain oxidation of organic materials. A principal disadvantage of the stable nitroxyl radical was its deep red-brown colour – a feature that completely precluded nitroxyls from practical use in polymer stabilization.

In 1967 it was discovered that colourless parent piperidines (4-substituted derivatives of 2,2,6,6-tetramethylpiperidine) were also very effective light stabilizers in polymers.

Plastics Additives: An A–Z Reference
Edited by G. Pritchard
Published in 1998 by Chapman & Hall, London. ISBN 0 412 72720 X

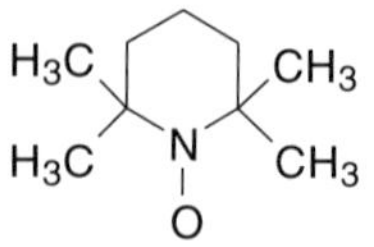

Figure 1 Stable 2,2,6,6-tetramethylpiperidine-*N*-oxide radical.

Finally, in 1974 the first commercial HALS structure (HALS-1) was launched to the market under the trade names Sanol LS 770 and Tinuvin 770.

Nowadays, there are approximately 30 HALS structures offered in the market, and the estimated world-wide HALS consumption for plastics in 1995 exceeded 7.5 thousand metric tonnes. Out of the HALS stabilizers produced, the most important structures in the polymer industry are HALS-1, HALS-3 and HALS-2. (Figure 2).

MECHANISMS OF HALS

Correspondingly to the industrial importance of HALS, the elucidation of their stabilization mechanism has attracted much attention during the 1970s and 1980s.

At the very beginning it was found that HALS do not absorb UV light, and so they cannot be classified as UV-absorbers or quenchers of excited states. The first mechanism proposed to explain HALS stabilization activity is known as the Denisov cycle. This mechanism attributes the key role in stabilization to HALS transformation products (nitroxyl radicals and hydroxylamino ethers), and it fits the generally accepted pathway of hydrocarbon degradation through alkyl- and alkylperoxy-radicals (Figure 3):

Such an antidegradation mechanism is certainly very attractive due to its 'catalytic' nature, but experimental works over the years showed that the situation is much more complex. Some of the other alternative HALS stabilization mechanisms include:

- complexation of transition metals;
- complexation of hydroperoxides and hydroperoxide decomposition;
- quenching of charge transfer complexes (CTC) of polymer–oxygen.

Based on the accumulated experimental evidence we can summarize by saying that HALS protect polymers by a combined mechanism where the principal activities are free radical scavenging, hydroperoxide deactivation, and formation of charge transfer complexes with oxygen. The stabilization action of HALS can be adversely affected by some other substances, like organic thioether costabilizers, halogen flame retardants, or an acidic environment. Several *N*-substituted HALS derivatives have

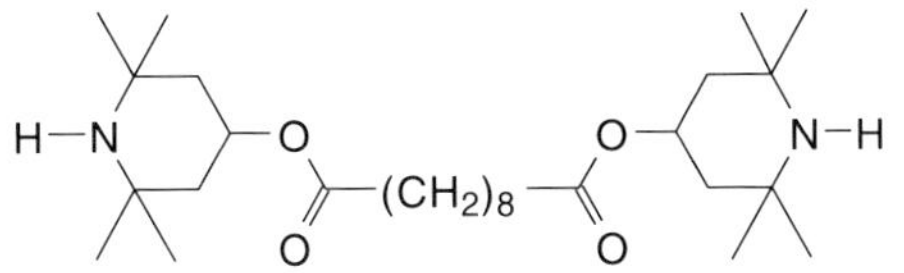

HAL-1

Tinuvin 770, Sanol LS 770, Mark LA-77, Ubaseb 770, Eversorb 90

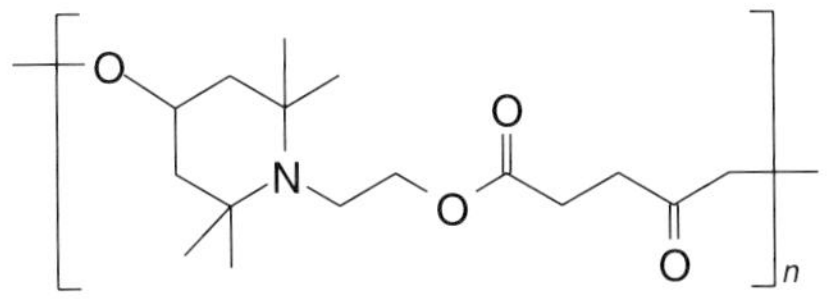

HALS-2

Tinuvin 622

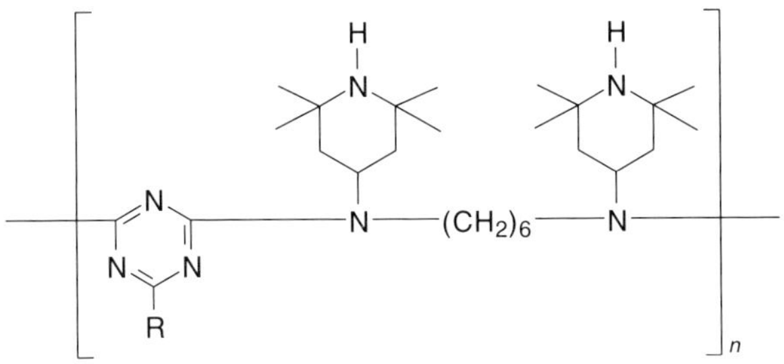

HALS-3

Chimassorb 944 (R: -NH-*t*-octyl)

Cyasorb UV 3346 (R: -N-morpholin)

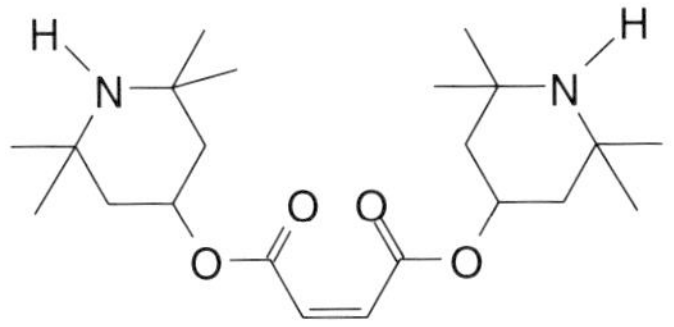

HALS-4

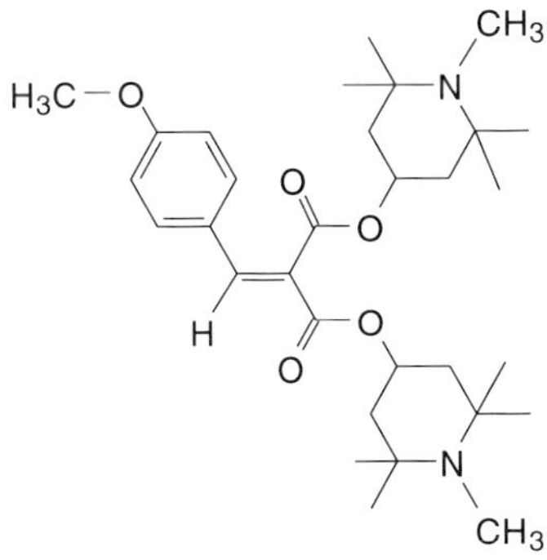

HALS-5

VP Sanduvor PR-31

Figure 2 Structures of HALS.

been developed to reduce these sensitivity problems originating from the basicity of the tetramethylpiperidine structure. *N*-methyl substituted HALS show practically the same basicity as the parent N-H HALS and their stabilization effect is comparable. *N*-acyl derivatives of HALS

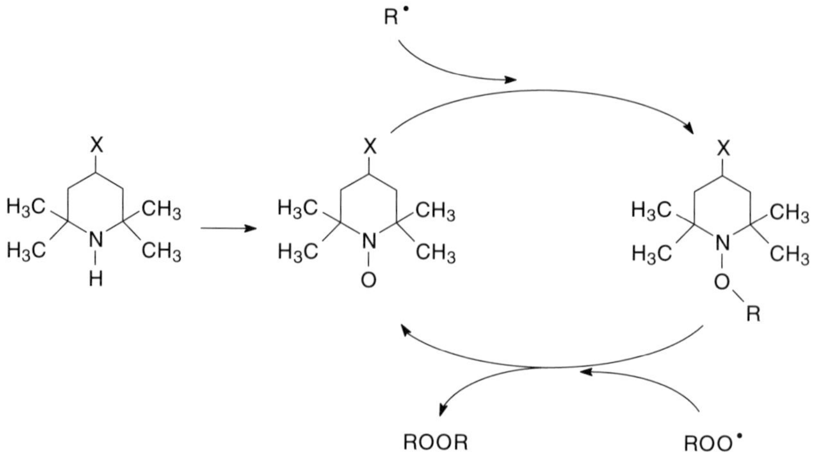

Figure 3 Mechanism of hydrogen degradation.

(N-COCH$_3$) have much lower basicity but, unfortunately, their light stabilization effectiveness in polyolefins is significantly reduced; N-acylated HALS are mainly used in the coating industry. N-OR derivatives of HALS (alkyloxyamines) also have low basicity but still good performance as light stabilizers.

Besides providing effective protection of polymers against photo-degradation, some high molecular weight and oligomeric HALS (e.g. HALS-3) have been shown to act as effective long term heat stabilizers. For this reason, in some relevant literature, it is possible to find the abbreviation HATS (Hindered Amine Thermal Stabilizers) or HAS (Hindered Amine Stabilizers). A detailed overview of the various mechanisms proposed for HALS action can be found in [1].

PHYSICAL ASPECTS OF POLYMER STABILIZATION BY HALS

Photo-oxidation of a stabilized polymer is accompanied by the loss of most of the effective light stabilizer. This depletion of the stabilizer is due to its chemical consumption in the stabilization reactions, and also due to physical losses of the stabilizer from the polymer by e.g. extraction, volatilization, blooming etc. The active participation of the stabilizer in the antidegradation chemical reactions is the actual reason why the additives are used in polymers but, obviously, the physical loss of the stabilizer is an undesirable process because it leads to the futile depletion of a relatively expensive chemical substance. The main factors that influence the physical losses of HALS are the rate of diffusion and the solubility of the stabilizer in the polymer matrix. These parameters are associated with the molecular characteristics of the additive (molecular weight, polarity etc.) as well as with the molecular and supermolecular characteristics of the polymer (polarity, crystallinity, glass transition temperature etc.). The physical aspects of polymer stabilization had been neglected for a long time, but the industrial development – also in the area of HALS – confirmed their importance.

The first commercial HALS-1 offered an excellent light stabilization effect in PP and HDPE. However, its performance in LDPE or in PP fibres did not fulfil the expectations. While in the case of low density polyethylenes a poor compatibility of HALS-1 with the polymer was blamed, in the case of PP fibres, the physical losses of HALS-1 were addressed as a source of the problem. Several polymeric and high molecular weight HALS structures have been proposed to reduce the problem of physical losses. Among the numerous structures tested, the best stabilization efficiency results have been achieved with the oligomers HALS-2 and especially HALS-3. These two oligomeric stabilizers were already commercialized in the early 1980s. Another recent development – the oligomeric HALS with a polymethylsiloxane backbone – is discussed in the next article, 'Hindered amine light stabilizers: recent developments'.

POLYMER STABILIZATION BY HALS

The sensible selection of a HALS stabilizer is governed by many variables, some of which are mentioned below:

- the molecular properties of HALS (e.g. the choice between low molecular weight HALS, oligomers, and polymeric stabilizers);
- the type of polymer (PP, LDPE, HDPE, engineering resins etc) to be stabilized;
- polymer processing conditions and type of final article (thick moulded articles, thin-walled films, fibres etc.);
- type and concentration of the other additives, pigments and fillers used;
- required lifetime under certain exposure conditions etc.

The above particulars must be considered in the selection of a suitable light stabilizer. Practical experience accumulated over the years enables us to give some general guidelines for the use of HALS for light stabilization of polymers. The concentration of HALS usually varies from 0.1 to 0.3% but for demanding applications the concentration can reach or even exceed 1% by weight.

Polypropylene

Thick injection moulded or compression moulded articles are advantageously stabilized with a low molecular weight HALS or with a combination of low molecular weight HALS with oligomeric HALS (e.g. HALS-1 with HALS-3). For fibre stabilization, oligomeric or high molecular weight HALS additives are recommended.

Polyethylene

Although polyethylene is less sensitive to UV-degradation than polypropylene, it requires good light stabilization for outdoor applications. LDPE and LLDPE resins are usually stabilized with oligomeric HALS (HALS-3, HALS-2) alone; in some cases combinations with benzophenone UV-absorbers are recommended (e.g. greenhouse films). HDPE resins can be well stabilized by low molecular weight HALS as well as by oligomeric HALS; but in general, oligomeric stabilizers are often preferred.

Styrenic polymers

HALS alone is not sufficient for light stabilization of styrenic polymers. The recommended light stabilization systems involve a UV-absorber (usually a benzotriazole type) with low molecular weight HALS, commonly in the ratio 1:1.

PVC

The light stability of PVC depends especially on its heat stabilization system. HALS are not very effective in rigid PVC; in plasticized PVC articles HALS can be used in combination with UV-absorbers.

HALS stabilizers are further used in engineering resins and coatings, but here the stabilization systems are usually composed of more components, and simple guidelines can be misleadings. More detailed information on such systems can be found in the specialized literature [2,3].

POSSIBLE DEVELOPMENTS AND NEW TRENDS

Since the commercial introduction of HALS, no fundamentally new light stabilizer structures have appeared that have surpassed their efficiency in polyolefin stabilization.

The protection against photodegradation conferred by today's HALS workhorses (HALS-1, HALS-2, and HALS-3) or their combinations in polymers (especially to polyolefins) in very high. Obviously, any new development in the HALS field should offer particular advantages that can justify its moving to the industrial scale.

Numerous attempts with bifunctional stabilizers (combining the HALS moiety with another functional group in one molecule) have not brought a serious improvement in stabilization performance.

More development has been seen in the area of the physical persistence of HALS. Owing to steadily increasing hygienic and eco-toxicological concerns, there is a continuous trend to improve physical persistence of HALS in the polymer matrix. Attempts with polymeric HALS stabilizers did not meet the expectations, because their limited compatibility and significantly reduced homogeneity of distribution of the stabilizing moieties lowered the stabilization performance. One of the more promising approaches is Polymer Bound HALS. In this case a functionalized HALS molecule (e.g. HALS-4) is reactively processed with polymer, grafting the HALS to the polymer chain. This technology has been proposed for a long time but the practical realization of the idea on the industrial scale is not simple. Recently reported results with this technology showed that after optimization of the preparation process, the polymer bound HALS can significantly outperform commercial stabilizers.

Another promising approach is the recently introduced concept of photoreactive HALS molecules. It is based on a capability of a specially designed low molecular weight stabilizer (HALS-5) to become grafted to the polymer by a photochemical reaction. Test results confirmed that this novel technology offers several benefits over conventional stabilization techniques.

REFERENCES

1. Pospíšil, J. (1995) Aromatic and Heterocyclic Amines in Polymer Stabilisation. *Advances in Polymer Science*, 124, 87–189.
2. Gugumus F. (1990) Photooxidation of Polymers and Its Inhibition, in *Oxidation Inhibition in Organic Materials, volume II* (eds J. Pospíšil and P.P. Klemchuk), CRC Press, Boca Raton, Florida, pp. 29–162.
3. Gächter R. and Müller H. (1990) *Plastics Additives*, Hanser Publishers, München.

Keywords: stabilization, mechanism, hydrocarbon degradation, nitroxyl radical, Denisov cycle.

See also: Hindered amine light stabilizers: recent developments; Light and UV stabilization of polymers.

Hindered amine light stabilizers: recent developments

Robert L. Gray

INTRODUCTION

In the absence of stabilizers, many polymer systems such as polypropylene have relatively poor UV stability. Other polymers such as polyethylene, styrenics, polyamides and polyurethanes also benefit from UV stabilization. The practical consequences of unchecked exposure to UV radiation are: discoloration, surface crazing (formation of surface microcracks), Embrittlement and loss of mechanical properties (elongation, impact strength, and tensile strength). The effect of UV exposure can be significantly inhibited through proper selection of UV stabilizers. Light stabilizers can be categorized into four general classifications: screening agents, UV absorbers (UVA), UV quenchers, and Hindered Amine Light Stabilizers (HALS).

The most recent class of light stabilizer is the Hindered Amine Light Stabilizer (HALS). These materials have been shown to function as radical traps, thus interrupting the radical chain degradation mechanism. The cyclic stabilization mechanism proposed for HALS involves multiple regeneration of the active nitroxyl stabilizer. The surprising performance of HALS at relatively low concentrations supports this non-sacrificial mechanism.

While HALS have been demonstrated to provide super light stabilization, co-additive interactions must also be carefully considered when formulating stabilization packages. Resins, phenolic antioxidants, thioethers, pigments, flame retardants, fillers and external pollutants all are known to interact with HALS under certain conditions. Here some

Plastics Additives: An A–Z Reference
Edited by G. Pritchard
Published in 1998 by Chapman & Hall, London. ISBN 0 412 72720 X

of the more common HALS interactions are reviewed and potential remedies offered.

POLYMER EFFECTS

Compatibility and mobility within the polymer matrix can have a significant effect on HALS performance. In the case of thick sections (i.e. garden furniture, bumpers) where a 'suitable' HALS mobility toward the surface layers is necessary, it is a general rule to use relatively low molecular weight monomeric types, either alone or in combination with oligomeric HALS.

The use of monomeric HALS alone generally offers good performance in accelerated aging. However, in really long-term outdoor use, the UV activity is often not completely satisfactory owing to the loss of monomeric HALS from the surface by evaporation or by extraction with contact fluid (rain, detergent solution etc.). In the case of filled compositions (talc, calcium carbonate), they are partially absorbed by the inorganic filler with considerable loss of the UV stabilizing activity.

Blends of monomeric and oligomeric HALS can partially offset the tendency to lose UV stabilizing activity by combining two different migration rates. Although these blends show improvement over the performance of the two single products, they certainly represent a compromise due to the discontinuity in MW distribution and physical characteristics between the two HALS.

More recently, a high performance, oligomeric HALS with a poly(methylsiloxane) backbone (HALS-5) was introduced. The soft polysiloxane backbone is quite flexible compared with traditional carbon-based products. This results in a tighter interaction with the polypropylene chain. Indeed, molecular modeling experiments have shown that HALS-5 can adopt a helical structure along the Si—O—Si axis, with methyl piperidine groups spiraling along this axis. It would appear that this should have good mutual co-penetration possibilities with the helix of polypropylene. This is consistent with the unusually high compatibility observed with polypropylene. Miscibilies in excess of 1:1 (w/w) have been observed.

Figure 1 is an example of UV stabilizing activity, measured as total color variation (delta E), of HALS in polypropylene copolymer plaques, blue color, aged in a weatherometer. HALS-5LM (number-average molecular weight, $Mn = 1100$) completely achieves the performance of the low molecular weight HALS-1. This indicates that the two HALS are characterized by a similar mobility but HALS-5LM offers the advantage over HALS-1 of a much lower volatility and extractability. A further advantage of HALS-5LM (liquid HALS) is its inability to aggregate, that is to precipitate or crystallize out of the polymer matrix, during aging,

**EXPOSURE: WOM Ci65, BPT 63°C, RH 50%
CYCLE 102'/18', Lab (44.4, 4.9, 49.0)**

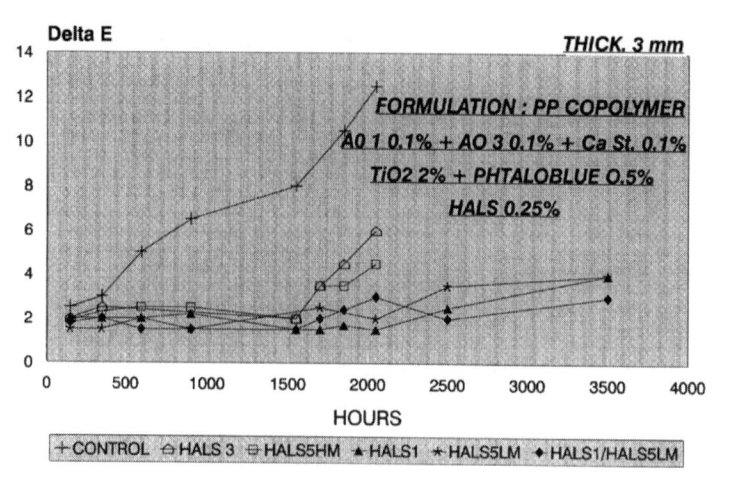

Figure 1 UV stability of polypropylene plaques. (BPT = back panel temperature. Delta E refers to the change in color as measured by movement within the LAB color space co-ordinates).

leaving behind a thin layer of additive. A slightly higher UV effectiveness has been shown in unfilled plaques by the 1:1 blend HALS-5LM/HALS-1. The unsatisfactory results of HALS-5HM (twice the molecular weight of HALS-5LM) and HALS-3 have to be attributed to their reduced mobility.

Figure 2 shows the higher effectiveness of HALS-5LM in comparison with the monomeric HALS-1 in natural weathering in a car bumper composition based on a reactor blend. The higher the ethylene fraction in the polymeric matrix, the higher the superiority of HALS-5 over HALS-1.

In mineral filled items (Figure 3) the monomeric HALS-1 is less competitive in comparison with HALS-5LM alone or in a 1:1 blend with the monomeric one. This is mainly due to the absorption of the lower molecular weight stabilizer by the porous filler surface. The molecular size of HALS-5LM is large enough to allow a coating formulation on the filler surface, reducing the dangerous effect on polymer stability of the metallic impurities always present in natural fillers.

POLYPROPYLENE FIBRE

In fibre applications, the high volatility of monomeric HALS such as HALS-1 made it susceptible to loss during post extrusion heat treatments such as tentering. This is presumably related to the relatively high volatility of HALS-1 which is a result of its low molecular weight and

**NATURAL OUTDOOR AGING SOUTH FRANCE
GREY PP PLAQUES Lab (25.07,-0.09,-0.41)**

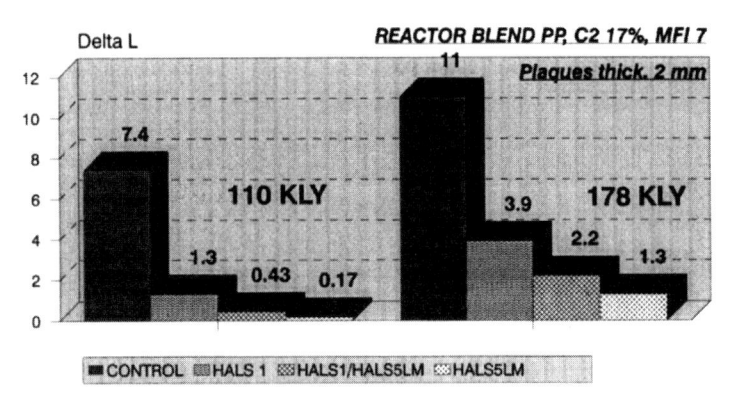

Figure 2 UV stability of reactor blend polypropylene copolymer plaques. (Delta L refers to the change in the L components of the LAB color space co-ordinates. KLY = 1 kcal/cm^2, 4.18 cal = 1 J).

**GREY PLAQUES (L 26.2, a 0.89, b -0.40)
XENOTEST 1200, BPT 83°C, RH 50%,**

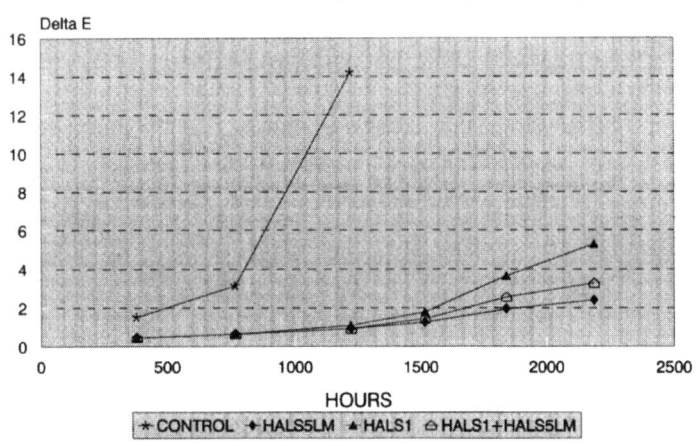

FORMULATION : AO1 0.1% + AO3 0.1% + TALC 25% + HALS 0.25%

Figure 3 Effect of talc filler on UV stability. (See Figure 1 caption for glossary of terms.)

its propensity to migrate quickly to the polypropylene surface. This problem was addressed with the introduction of high molecular weight oligomeric HALS.

In the first set of experiments, polypropylene fibers (18 deniers per fiber, dpf) were exposed in a weatherometer (bpt 63°C, RH 50%) and evaluated for retention of tensile strength after periodic exposure. The UV stabilization activity of several high molecular weight HALS is compared in Figure 4(a). At a 0.2% concentration, all HALS showed a significant improvement over the unstabilized control. Both HALS-5 and HALS-3 showed a substantial performance advantage over HALS-2. HALS-5 maintained a consistently higher level of activity than the other HALS evaluated. The time to retention of 50% of the tensile strength (TS50) for HALS-2, HALS-3 and HALS-5 was 431 hrs, 717 hrs and 799 hrs, respectively.

In the previous data set, the fibers were exposed to artificial weathering in a weatherometer. In an effort to identify any differences between accelerated aging and actual outdoor weathering, fiber samples produced for the previous experiment were sent to Florida for exposure. The results of this outdoor weathering are shown in Figure 4(b). In actual Florida exposure, a much larger differentiation between the HALS is observed. HALS-5 shows a dramatically better performance than HALS-2 and HALS-3. This larger differentiation between HALS-5 and the other HALS in the Florida results is likely due to high compatibility and resistance to extraction (and loss) of HALS-5.

GAS FADE

Recent moves into the residential carpet market required the maintenance of strict control over undesirable color development through processing, weathering and storage (gas fade). In certain systems, HALS have been shown to contribute to color development through an interaction with phenolic antioxidants. This discoloration can be accelerated by exposure of the fiber to NO_x type gases which can be generated by warehouse equipment. This color development can be minimized by replacement of the secondary HALS with a less interacting tertiary HALS. Alternatively, the phenolic antioxidant can be removed from the stabilization system. In this case the phosphite is used as the process stabilizer and the HALS provides both UV and long term thermal stability. Table 1 demonstrates that in the absence of phenolic antioxidants, secondary HALS (HALS-5) provide gas fade results comparable with tertiary HALS (HALS-4).

Figure 4 (a) UV stability of polypropylene fiber exposed in a weatherometer (BPT refers to back panel temperature). (b) UV stability of polypropylene fiber exposed outdoors in Florida (dpf refers to denier per filament).

PP-FIBERS exposed in WOM
Sample : PP-fibre (18 dpf)

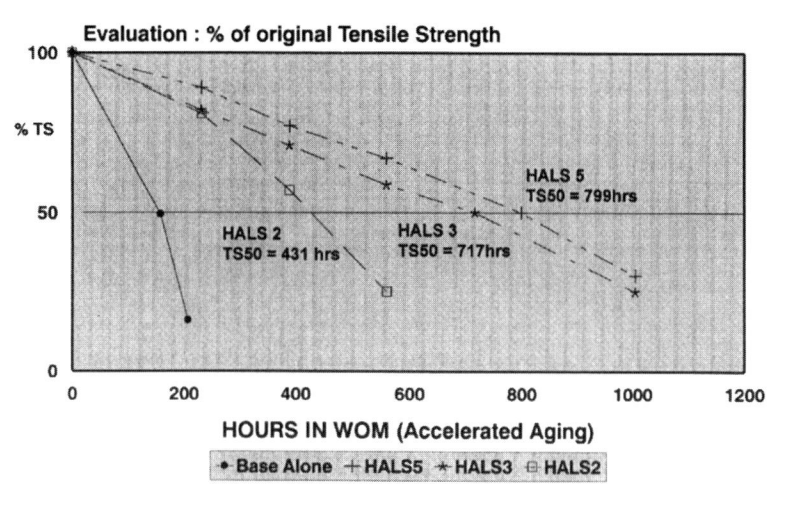

Evaluation : % of original Tensile Strength

HALS 5 TS50 = 799hrs
HALS 3 TS50 = 717hrs
HALS 2 TS50 = 431 hrs

HOURS IN WOM (Accelerated Aging)

• Base Alone + HALS5 ★ HALS3 ▫ HALS2

Aging : WOM, BPT 63C, RH 50%   Additivation level of HALS 0.2%

(a)

NATURAL OUTDOOR AGING -- FLORIDA
Sample : PP-fibre (18 dpf)

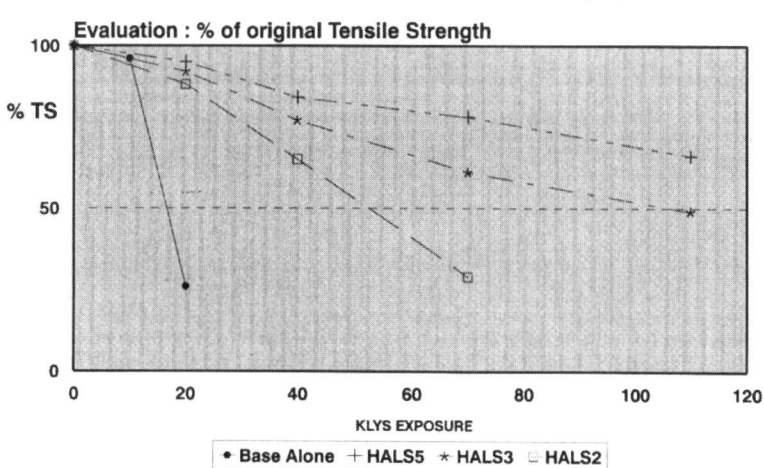

Evaluation : % of original Tensile Strength

KLYS EXPOSURE

• Base Alone + HALS5 ★ HALS3 ▫ HALS2

Aging : Outdoor, 45 South, PMMA   Additivation level of HALS

(b)

Table 1 Gas fading in natural fibers

HALS type	1st cycle	5th cycle
HALS-5	4–5	4
HALS-4	4	4

Grey scale: (1 = Worst, 5 = Best color).
Base: PP + 0.05% Phosphite.

PIGMENT INTERACTIONS

The addition of high molecular weight HALS has been shown to affect color yield or color strength of certain pigment systems. The extent of this effect appears to be related to the type of amine. Secondary amines such as HALS-3 tend to have a greater propensity to affect color yield than tertiary amines such as HALS-2 and HALS-4. Increased polymer compatibility (HALS-5) also can decrease the extent of negative interaction.

Pigments can also decrease the stabilizing efficiency of HALS. Strong interactions between the pigment surface and HALS can result in the immobilization of HALS within the polymer matrix. This effect is most often observed with monomeric, secondary HALS such as HALS-1.

It is important to note that pigment/HALS interactions are quite complex systems. Today, reliable prediction of pigment/HALS performance is not typically possible.

CHEMICAL RESISTANCE

HALS are known to undergo a reaction with mineral acids such as HBr causing an inhibition of activity. For this reason, success in using HALS in combination with co-additives capable of producing acids during processing or exposure has been quite limited. Examples of co-additives of this type are thiosynergists (distearyl-3,3′-thio-dipropionate, DSTDP) and brominated flame retardants. Unexpectedly, HALS-5 has shown an unusual chemical resistance when compared to traditional HALS.

DSTDP

Thioethers such as DSTDP have been used in combination with primary antioxidants to provide extended lifetimes to polyolefins exposed to elevated temperatures. Unfortunately, the mechanism by which these thiosynergists function has been shown to produce sulfenic and sulfonic acids which are capable of further reacting with HALS. Figure 5 shows the effect of DSTDP on HALS performance. While both HALS-1 and HALS-3 show a strong negative interaction with DSTDP, HALS-5 is relatively unaffected.

STRETCHED PP FILMS (40 μm) -- Dry Std Xenon

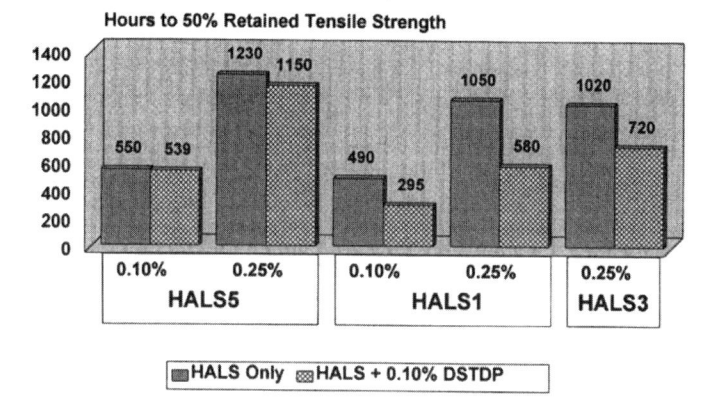

Hours to 50% Retained Tensile Strength

	0.10%	0.25%	0.10%	0.25%	0.25%
HALS Only	550	1230	490	1050	1020
HALS + 0.10% DSTDP	539	1150	295	580	720
	HALS5	HALS5	HALS1	HALS1	HALS3

Base: Process Stabilized PP + 0.1% DSTDP + 0.10-0.25% HALS

Figure 5 Effect of thioether on the UV stabilization performance of HALS.

FLAME RETARDANTS

UV stabilization of polypropylene containing brominated flame retardants has been the focus of intense technical efforts, with only limited success. Acid generated by the flame retardants deactivates HALS, thus severely reducing the HALS' effectiveness. Owing to this interaction, flame retarded polypropylene has been generally restricted to applications requiring little or no UV stability. While some progress has been made, this application continues to be a true technical challenge.

As previously discussed, the key to HALS–flame retardant incompatibility is acid generation by the flame retardant, which in turn deactivates the HALS. The mechanism for bromine radical generation by flame retardants is quite structure dependent. Aliphatic brominated flame retardants are primarily decomposed thermally, which may occur during the extrusion process. Aromatic brominated flame retardants are relatively stable through the processing step but may generate bromine radicals during UV exposure.

Aromatic flame retardants are generally less prone to thermally induced degradation but instead generate bromine radicals through a photo-activated process. Combinations of UV absorbers (UVA) and HALS have been shown to provide outstanding UV stability to polypropylene fiber containing aromatic flame retardants. It appears that the primary benefit of the UVA is its role in providing a UV screen for the flame retardant, thus inhibiting the generation of bromine radicals and hydrobromic acid. With the level of acid minimized,

XENON @ 55°C (ASTM D-4459)

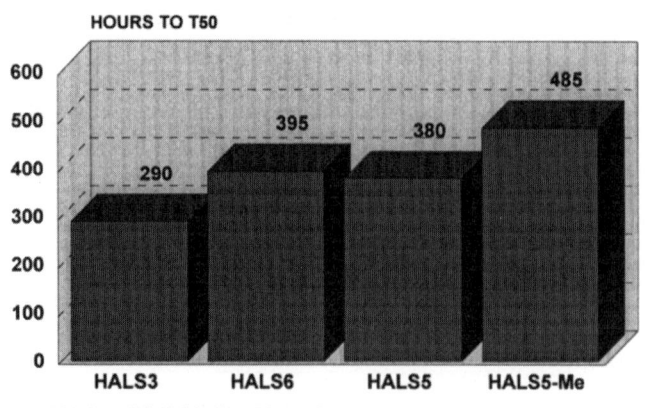

HOURS TO T50

- HALS3: 290
- HALS6: 395
- HALS5: 380
- HALS5-Me: 485

ALL FORMULATIONS CONTAIN FR (6% OBr) + 0.5% HALS + 1.5% UVA

Figure 6 Effect of flame retardants on the UV stabilization performance of HALS.

even relatively basic HALS can be successfully incorporated into these formulations.

The influence of HALS structure on stabilization in these systems is examined in Figure 6. The non-basic NOR HALS (HALS-6 – pKa = 4.2) has a large performance advantage over the basic secondary HALS-3. (The prefix NOR implies a HALS with an active group such as NOC_8H_{17} instead of NH.) Surprisingly, HALS-5 achieved comparable stability to the NOR HALS (HALS-6) despite its basic nature (pKa = 9.8). Activity of this siloxane-based HALS-5 can be further enhanced using a methylated analog of the product. HALS-5–Me demonstrates superior performance over that achieved by the less basic HALS-2.

This performance attribute may be related to the unusual compatibility of HALS-5 in polypropylene. A simplified rationalization of the results can be proposed. The HALS-5 molecules remain well dispersed throughout the bulk of the amorphous polypropylene matrix. Conversely, HALS-3 tends to migrate toward the polar regions of the sample which contain the brominated flame retardant. This unfortunately places the HALS in direct proximity to the site of bromine radical generation.

CONCLUSIONS

Hindered Amine Light Stabilizers have undergone significant advances since their initial introduction. Molecular weight has increased to meet volatility concerns. Tertiary HALS have been developed to address discoloration and negative pigment interaction issues. New low basicity NOR HALS have expanded capabilities in the area of chemical resistance. The latest innovation is the highly compatible siloxane-based HALS. This new generation of HALS appears to provide a significant level of improvement in all of the above mentioned areas. High compatibility not only extends performance but greatly reduces extractibility. In today's era of environmental concerns, polymer permanence will continue to be an increasingly important issue.

BIBLIOGRAPHY

Gächter, R. and Müller, H. (eds) (1990) *Plastic Additives*, Hanser Publishers, Munchen.
Clough, R.L., Billingham, N.C. and Gillen, K.T. (eds) (1996) *Polymer Durability*, American Chemical Society, Washington D.C.
Son, P-N. and Smith, P. (1992) Environmental Protective Agents, in *Plastic Additives and Modifiers Handbook*, 1st edn (ed. J. Edenbaum), Chapman & Hall, London, pp. 208–271.

APPENDIX A

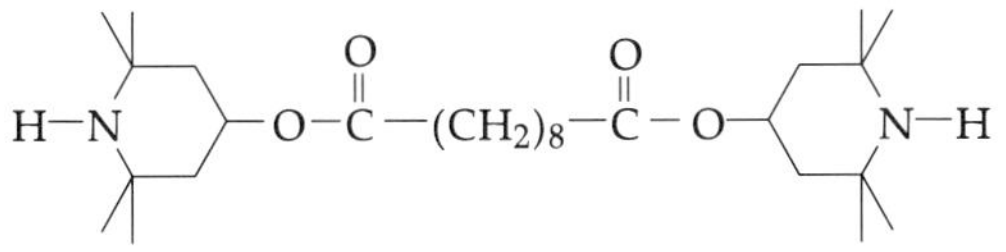

HALS-1

Lowilite 77,
Sanol LS 770,
Tinuvin 770

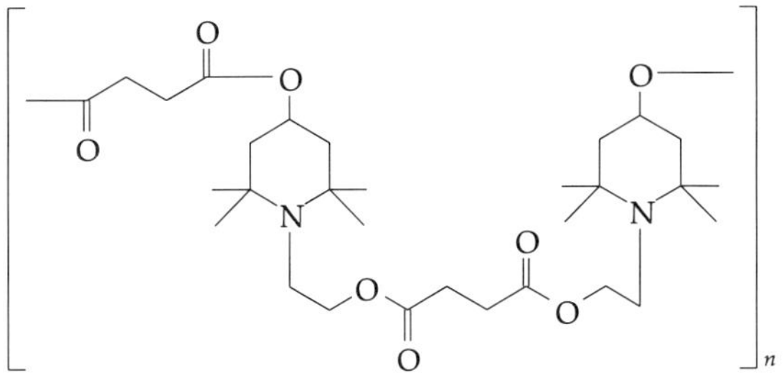

HALS-2

Lowilite 62,
Tinuvin 622

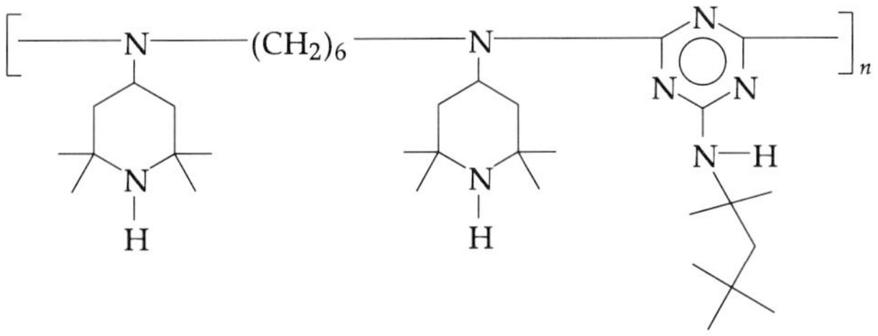

HALS-3

Chimassorb 944

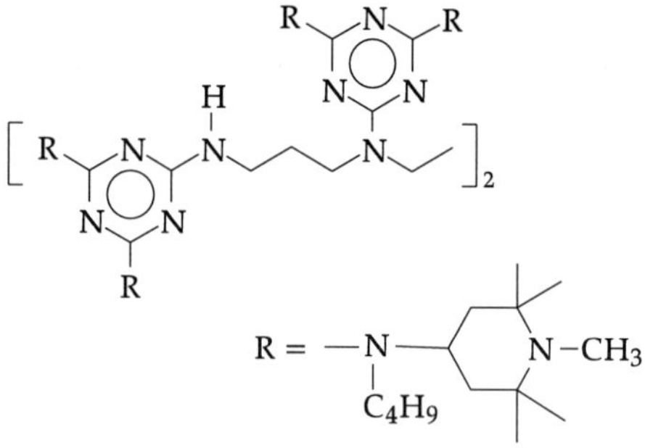

HALS-4
Chimassorb 119

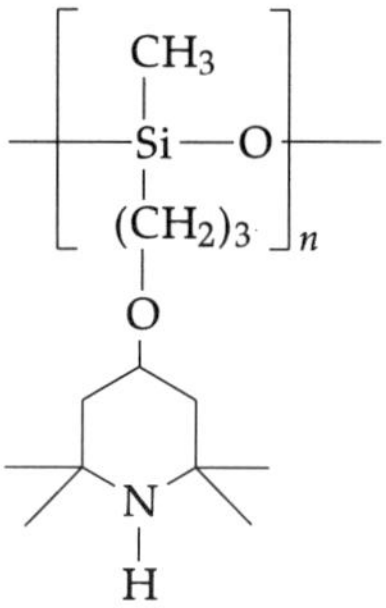

HALS-5
Uvasil 299

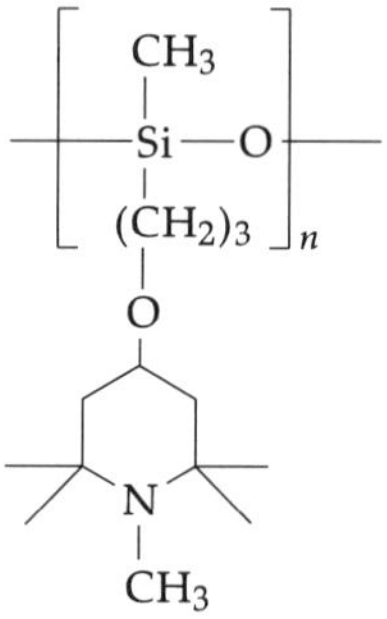

HALS-5–Me
Uvasil 816

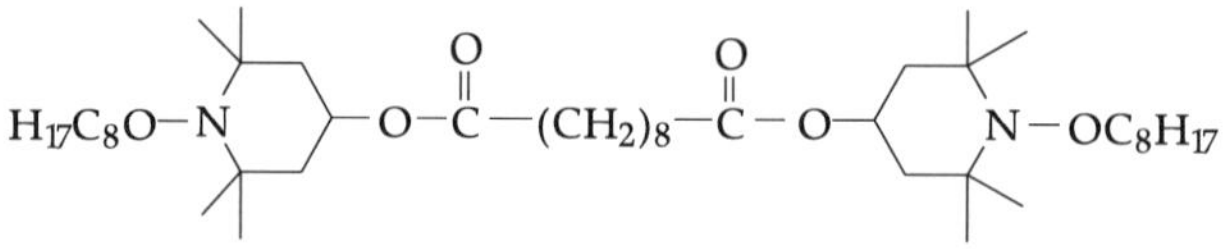

HALS-6
Tinuvin 123

SUPPLIERS OF HALS

Great Lakes Chemical Corp., West Lafayette, IN, USA.
Ciba-Geigy AG, Basel, Switzerland.
Cytec Industries, CT, USA.
Clariant Corp., Basel, Switzerland.

Keywords: UV stabilization, monomeric HALS, oligomeric HALS, pigment interactions, thioethers, chemical resistance, flame retardancy, siloxanes.

See also: Hindered amine light stabilizers: introduction;
Light and UV stabilization of polymers.

Hollow microspheres

Geoffrey Pritchard

CONCEPT

Solid particulate fillers usually increase both the Young's modulus and the density of polymers. The incorporation of hollow microspheres into polymers does not necessarily reduce the density – that depends on the material from which the microspheres are made, and on their wall thickness and volume – but it can enable a better balance between mechanical properties and density to be achieved. There can also be some incidental benefits in terms of processing and surface quality. The microspheres are commonly made of glass, but various thermoplastics, phenolic resin, carbon, and ceramics have also been used. They are capable of improving the flow characteristics of a polymer in the mould, and resin viscosity increases only slightly with increasing filler content. They also reduce mould shrinkage.

The glass grades now commercially available vary greatly in their wall thickness, mean size and size distribution. As a result, a wide range of properties and processing characteristics can be achieved with a given volume fraction of filler. Table 1 shows the effect of hollow glass microspheres on the tensile properties of various plastics. Recently, microspheres made of gas-resistant polymer and containing liquid hydrocarbons have been produced. When the microspheres are added to a resin and heated, the microspheres expand, giving a very low density.

There are difficulties in the use of hollow filler particles. The very thin-walled grades which would impact the lowest density and product weight are easily crushed during injection moulding. When they break, the fragments not only increase the density, they also abrade the equipment and act as stress concentrations, with adverse effects on

Plastics Additives: An A–Z Reference
Edited by G. Pritchard
Published in 1998 by Chapman & Hall, London. ISBN 0 412 72720 X

Table 1 Tensile properties of plastics filled with 5% and 15% by volume of 8 µm, 1.1 g/cm³ hollow glass microspheres.*

Polymer	Tensile strength (MPa)			Elongation (%)		
	5%	15%	0%	5%	15%	0%
Nylon 6,6	69	68	81	15.8	3.2	68.0
Polystyrene, high impact	35	15	38	13.6	16.0	24.0
ABS	35	26	39	8.5	13.9	7.5
Acetal	–	44	57	20.9	10.2	17.4

Source: A.E. Fuchs and B.J. Sutker, Paper 22-E, 44th Annual Conference, Composites Institute, SPI, Inc., Feb. 6–9, 1989, Dallas, TX, USA.

crack initiation in the filled plastics. The wall thicknesses of hollow microspheres are typically in the range 0.4 to 1.5 µm.

When glass bubbles are incorporated in liquid resins of low viscosity, they have a tendency to migrate towards the surface, because they are less dense than the resin, leaving inadequate filler content in the rest of the material.

PRODUCTION

The microspheres typically appear like a very fine flowing white powder. The glass ones are generally made from soda lime borosilicate glass cullet, or high silica glass, which in one process is heated by passing through a flame, along with a chemical blowing agent to generate gas, and the resulting spheres have mean diameters in the range 5 to 90 µm. Another production method involves spray-drying solutions of sodium silicate.

Ceramic microspheres can be made as a by-product of coal combustion. Carbon microspheres are prepared by heating suitable organic materials with a blowing agent. Glass bubbles are vulnerable to strong acids and alkalis, but otherwise their chemical resistance is good. A silane or titanate surface treatment can be applied to improve adhesion to a polymer (see 'Coupling agents').

Table 2 shows some density values achievable with 8 µm glass microspheres.

Table 2 Density of plastics filled with hollow glass microspheres.*

Polymer	Original density (g/cm³)	Wt % filler	Final density (g/cm³)
Polypropylene	0.9	22.4	0.94
Acetal	1.4	22.4	1.25
Nylon 6,6	1.14	22.4	1.13

*Source: Technical data, Sphericel, Potters Industries Inc.

APPLICATIONS

Applications of hollow microsphere-filled polymers include syntactic foam for use where high stiffness is needed along with low weight. Syntactic foam cores can be incorporated in conventional glass fibre laminates to produce lightweight components in transport and in the marine sector. The density of syntactic foam can be adjusted over a wide range. Products requiring this material for underwater buoyancy include riser pipes for oil drilling rigs, cable floats and submersibles. Other uses of hollow microspheres include thermosetting resin based sheet and bulk moulding compounds, sports equipment (notably golf balls for use on miniature courses), microwave cookware and cable insulation. Glass microspheres are frequently used as fillers in resins reinforced with glass fibres, e.g. chopped strand mat. They can improve the processability, stiffness and surface quality of laminates.

In a completely different area, silver coated hollow glass spheres with particle densities of 1.3 to 1.6 g/cm^3 are available for excellent electrical and thermal conductivity, e.g. in special adhesives.

BIBLIOGRAPHY

Guiot, P. and Couvreur, P., 1986, *Polymeric nanoparticles and microspheres*, CRC Press, Boca Raton, Fla., USA.

Fuchs, A.E. and Sutker, B.J., 1989, Lightweight engineering thermoplastics containing hollow spheres, Paper 22-E, 44th Annual Conference, Composites Institute, SPI, Inc., Dallas, TX, USA.

Keywords: density, viscosity, glass, mechanical, syntactic, blowing agent, diameter.

See also: Fillers.

Impact modifiers: (1) mechanisms and applications in thermoplastics

Roberto Greco

INTRODUCTION

It is desirable for plastics articles to be able to withstand cracking when subject to minor impact. Failure to resist cracking is a common feature of plastics and this entry discusses the way in which polymers can be used as toughening additives to overcome the problem.

Several thermoplastics, both of the commodities kind [polystyrene (PS), polyacrylonitrile (PAN), polymethylmethacrylate (PMMA), polypropylene (PP), polyvinylchloride (PVC) etc.] and engineering polymers [polyamides (PA), polyesters (PE), polycarbonates (PC), polyimides (PI), polysulfones (PSF), polyoxymethylene (POM), polyphenylene oxide (PPO) etc.] exhibit glass transition temperatures (T_g) higher than or close to room temperature (R.T.). As a consequence they show, at R.T. or below it, the shortcoming of brittle impact behaviour, which limits their commercial end-uses.

Three main classes of polymer matrices can be identified with respect to their failure characteristics:

I. brittle amorphous polymers, such as PS and SAN, with low unnotched and notched impact strengths, for which crack initiation and propagation stresses are lower than the yield stress;
II. pseudo-ductile engineering polymers, such as PC, PA, PI, PE and PSF, with high unnotched and low notched impact strengths; they exhibit high crack initiation and low crack propagation energies and a brittle-to-ductile transition temperatures as well;

Plastics Additives: An A–Z Reference
Edited by G. Pritchard
Published in 1998 by Chapman & Hall, London. ISBN 0 412 72720 X

III. polymers, such as PMMA, POM, and PVC, exhibiting comparable values of crack initiation and yield stresses with a fracture behaviour intermediate between types I and II.

In all these cases the problem can be solved by adding suitable rubber modifiers to the homopolymer matrices. These will induce, under impact loading, locally diffused microscopic mechanisms of deformation, making the matrix capable of dissipating large impact energies, avoiding catastrophic failure.

The general characteristics of such additives can be briefly summarized as follows:

1. a sufficiently low T_g, typical of elastomeric behaviour at R.T.;
2. effectiveness with minimum amount;
3. optimum particle size;
4. a suitable particle size distribution;
5. a homogeneous dispersion;
6. a good adhesion to the thermoplastic matrix.

Other variables influencing the impact behaviour are the molecular characteristics of the matrix and the type of processing.

TOUGHENING MECHANISMS

General

Well dispersed rubber particles are able to induce in the thermoplastic matrix different mechanisms of toughening:

1. crazing;
2. shear yielding and rubber particle cavitation;
3. combined crazing and shearing yielding.

In a very general sense the mechanisms acting in toughened polymers are essentially the same as those present in their parent homopolymers (types I, II and III). The rubbers operate somewhat like a catalyst, just altering the stress distribution within the matrix.

Crazing

Rubber particles, with an optimum size (ranging from 1 to 5 µm), homogeneously dispersed in a very rigid matrix (type I, such as PS and SAN) can induce a multicraze initiation and termination mechanism, capable of dissipating large impact energies. Crazes initiate within the matrix at the rubber particle equators where a very high stress concentration is built up through the application of an impulsive external load; they undergo termination during their propagation when they impinge upon neighbouring rubber particles. The microcrazes, diffused throughout the deformation zones of the body, are perpendicular to the direction of the applied stress and are accompanied by a marked stress whitening effect. A craze is similar to a microcrack but differs in being bridged across by several microfibrils, made by oriented polymer chains, which can sustain a certain load. The molecular weight (M.W.) of the matrix is an important parameter for crazing: below a critical M.W., in fact, no chain entanglements and therefore no stable crazes can be formed. The fracture under a sufficiently high load is determined by the rupture of the craze fibrils, once a crack of critical size has been developed.

Shear yielding

In homogeneous polymers shear deformation consists of a distortion of the body shape without significant volume variation. In semicrystalline materials shear yielding is very localized and occurs by slip on particular planes of maximum shear stress. In non-crystalline materials the shear yielding is much more diffuse than in the previous case, requiring large co-operative chain movements.

In toughened materials a diffused shear yielding is the main energy dissipation phenomenon, preceded or followed by rubber particle cavitation (the particle must be very small, of sub-micron sizes). Particle cavitation induces a stress whitening effect, visible along the largest deformation patterns of the body. Semiductile polymer matrices of type II, such as PVC, ABS, PC, PA, PE, PI and PSF undergo diffuse shear yielding.

Crazing and shear yielding

Crazing (Figure 1) and shear yielding occur in most cases simultaneously and the differences in behaviour, like those between HIPS and ABS or PVC, can be ascribed to the relative contribution of these two mechanisms to the overall deformation. In HIPS, crazing prevails over shear yielding, whereas in ABS or PVC both phenomena coexist. Moreover their relative importance depends on variables such as: temperature, strain rate, matrix nature and composition, aging, molecular orientation and others. An effective interaction exists between the two mechanisms: the molecular orientation within a shear band, roughly parallel to the applied stress, is perpendicular to the plane of the crazes. This induces a synergistic effect by which shear bands can stop the craze propagation, increasing the material toughness.

The quantitative contribution of these mechanisms to the macroscopic deformation can be measured by volumetric strain tests. Such a technique exploits the fact that crazes contain about 50% by volume of voids:

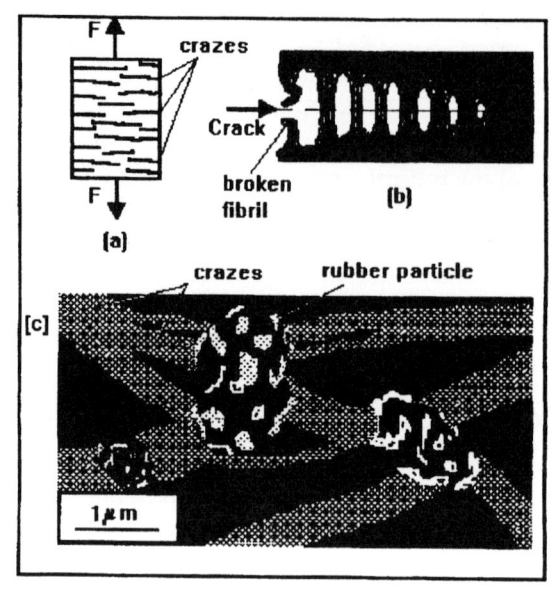

Figure 1 Schematic representation of the crazing phenomenon. (a) Crazed specimen subjected to a tensile force F. The crazes are represented by lines roughly perpendicular to the direction of the applied force. (b) Section of a craze with fibrils, strained by the tensile force F. A crack opening is starting to occur, as evidenced by the breaking of the first fibril on the left-hand side. (c) Multicraze mechanism induced by the presence of rubber particles in a rigid matrix.

therefore multiple crazing determines a large volume increase proportional to the crazed portion of the sample, leaving the cross-section almost unaltered. Shear yielding, in contrast, shows only a slight volume increase due to rubber particle cavitation. For isotropic specimens it is sufficient to measure the longitudinal strain (ε_3) in tensile tests and only one of the lateral strains, since the latter (ε_1 and ε_2) are equal. The volume strain can be expressed in this case as:

$$\Delta V = V/V_0 - 1 = \lambda_1 \lambda_2 \lambda_3 - 1 = (1 + \varepsilon_3)(1 + \varepsilon_1)^2 - 1 \tag{1}$$

where $\Delta V, V, V_0$ are the volume increase, the final and the initial volumes respectively, the λ values are the elongation ratios and the ε terms the strains, along the principal axes 1, 2 and 3.

A tensile stress is the sum of a deviatoric stress plus a hydrostatic tension, which produces, on application of a stress σ, a sudden volume increase $\Delta V(0)$, given by:

$$\Delta V(0) = \sigma/3K \tag{2}$$

where K is the bulk modulus, whose time-dependence can be neglected. Hence the total volume change of a toughened polymer can be analyzed

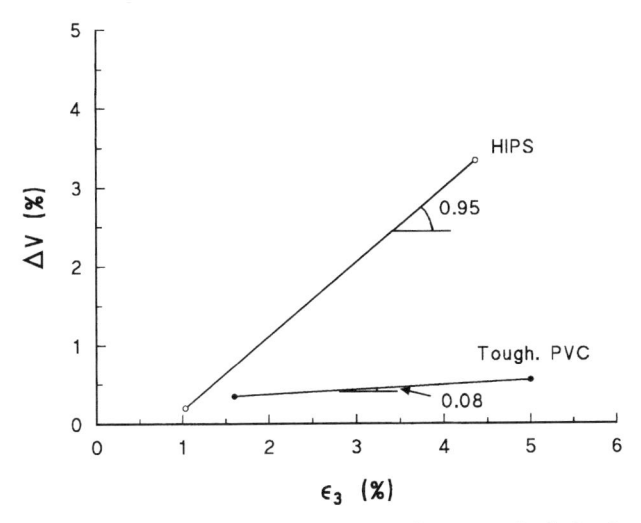

Figure 2 Relative volumetric increment, ΔV, as a function of relative length increment of longitudinal strain, ε_3, in tensile creep tests for HIPS under a stress of 22.7 MN/m^2 and toughened PVC under a stress of 36.0 MN/m^2, at 20°C. (The figure has been drawn by the program FP60. This figure has been replotted from Figure 7.12 of Bucknall (1977).

on the assumption that:

$$\Delta V(t) = \Delta V(0) + \Delta V(\text{crazing}) \quad (3)$$

By derivation of equation 1, one obtains:

$$(\delta \Delta V / \delta e_3)\varepsilon_1 = (1 + \varepsilon_1)^2 \simeq 1 \quad (4)$$

The approximation is valid if no lateral contraction is present, that is, when ε_1 is very small and constant. Thus in a creep tensile test, from the slope of a ΔV versus ε_3 plot, it is possible to determine the relative quantitative contributions of the two mechanisms.

An example is shown in Figure 2, where curves for HIPS and toughened PVC are reported. The HIPS slope is 0.95, indicating that crazing is the main factor responsible (95%) for the time dependent portion of the creep in this toughened polymer. The PVC slope is 0.08, indicating, in contrast, that in this case crazing contributes very little (8%) and the prevailing mechanism is shear yielding.

TOUGHENING ADDITIVES

Butadiene-based graft copolymers

Butadiene (B)-based elastomers constitute one of the most used families of impact modifiers. Their success in the market is due to their sufficiently low T_g (about $-80°C$). However, the presence of double bonds in diene polymers can induce thermal and oxidative degradation at fabrication temperatures and under UV and oxygen exposure. Therefore these effects must be minimized by the use of suitable antioxidants, particularly when added to engineering polymer matrices. In addition to B-based rubbers, other polydienes can be utilized, such as isoprene, chloroprene and several copolymers with S, AN and SAN.

A few examples of thermoplastics toughened by such rubbers are briefly described here:

HIPS (high impact polystyrene)

This was the first example of a toughened polymer (discovered in 1927 and marketed in 1948 by Dow Chem. Co.). It is usually obtained by the following procedure: butadiene (B) or styrene-butadiene rubber (SBR) (introduced during the early 1960s), is first dissolved in styrene (S), then prepolymerized with stirring and finally polymerized to completion. During the stirring a phase inversion occurs: PS becomes the matrix and the rubber the dispersed phase, which contains PS sub-inclusions (often including in turn PB sub-inclusions). The processing is very important in determining the final complex morphology of HIPS, for which the prevailing fracture mechanism is crazing. Effective particles sizes are of the order of a few microns.

ABS (acrylonitrile-butadiene-styrene) and related toughened materials

SAN was the second toughened polymer to be launched on the market (1952). Also in this case B or a B-based copolymer is the rubbery additive. ABS was obtained in the beginning by the mechanical blending of SAN with NBR (styrene-acrylonitrile-butadiene copolymer rubber). ABS can be manufactured by a number of polymerization techniques as well: emulsion, suspension and bulk grafting processes, yielding materials with somewhat different characteristics. ABS polymers are engineering thermoplastics exhibiting good processability, excellent toughness and sufficient thermal stability. They have found applications in many fields, such as appliances, building and construction, business machines, telephone, transportation, automotive industries, recreation, electronics and others. Shear yielding is the main toughening mechanism for these polymers. They can be used in turn as toughening agents for other thermoplastics, such as PC, as explained below.

Butadiene-based block copolymers

Styrene based block copolymers (with a soft block consisting typically of butadiene or isoprene), made by anionic polymerization, are often utilized as toughening additives. A large variety of different structures and properties can be obtained by changing initiator concentration, and the amount and sequence of monomer addition. Other components different from styrene, such as α-methylstyrene, can be used in order to increase the T_g of the hard segment.

Hydrogenation of these blocks improves the thermal, oxidative and UV stability, giving rise to a variety of SBS or of styrene-ethylene-1-butene-styrene (SEBS) block copolymers. The central block is substantially amorphous with a T_g of about $-60°C$, higher than that of PBD ($-80°C$). The hydrogenation can be partial (only the rubber is involved) or total (all the copolymer is hydrogenated). Of course the chemical stability is lower in the former case. These block copolymers have been utilized for the toughening of a number of polymers, such as polyamides, polyesters, PPO and PC.

Ethylene-based rubbers

Ethylene-propylene random copolymers

These rubbers (often referred to as EPM or EPDM, a terpolymer containing, in addition to the EP copolymer, a small percentage of a diene monomer) possess a T_g which is sufficiently low (about $-50°C$), but higher than that of B-based rubbers. They are particularly suitable as impact modifiers for engineering polymers, since in contrast to B-based materials, they exhibit excellent UV, thermal and oxidative stability at the high processing temperatures of these materials. A few examples of polymers toughened by these rubbers are given below.

Toughened PP
Ethylene-propylene rubbers are suitable additives for isotactic polypropylene. They can be added to the matrix by melt-mixing or by block copolymerization, using Ziegler–Natta catalysts (the copolymer is made by adding ethylene, as a second monomer, during the final stages of PP polymerization).

Polyamides and polyesters
In general the non-polar EPM and EPDM cannot be used as such for polyamide and polyester toughening, due to their poor adhesion to these polar engineering polymers. Therefore, as a first step, the rubber must be functionalized before its final mixing with the matrix. Maleic anhydride (MHA) or some other monomer, such as glycidyl methacrylate, is inserted for this purpose onto the rubber backbone by the following series of free

Impact modifiers:

radical reactions:

$$PH + R^\bullet \rightarrow P^\bullet + RH \quad (5)$$

The radical (R$^\bullet$), produced by the thermal degradation of a peroxide (R), extracts a hydrogen atom from the EPR backbone (PH), creating an EPR macroradical (P$^\bullet$), which in turn reacts with the MHA (M):

$$P^\bullet + M \rightarrow PM^\bullet \quad (6)$$

The functionalization (neglecting further M additions to P$^\bullet$) can propagate, transferring the radicals from one site on a PM$^\bullet$ chain to another on the same or a neighbouring chain:

$$PM^\bullet + P \rightarrow PM + P^\bullet \quad (7)$$

Termination by coupling reactions (increasing the EPR molecular weight), and disproportionation (leaving unaltered the EPR molecular weight) will end the complex chemical process. This often involves β-scission degradative reactions, which tend to decrease the EPR molecular weight.

The second and final step is the 'in situ' reaction of the rubber functional sites with amide or hydroxyl groups, forming graft (EPM-g-PA or EPM-g-PBT) copolymers (see equations 8 and 9 below), acting as interfacial agents between the matrix and the rubber:

$$\underset{\text{EPM-g-SA}}{\begin{array}{c}\text{—CH—C}\diagdown \\ | \qquad\quad\;\; \text{O} \\ \text{CH}_2\text{—C}\diagup \\ \phantom{\text{CH}_2\text{—}}\|\\ \phantom{\text{CH}_2\text{—}}\text{O}\end{array}} + \underset{\text{PA6}}{H_2N-(CH_2)_6-CO\sim} \longrightarrow$$

$$\underset{\text{(EPM-g-SA)-g-PA6}}{\begin{array}{c}\text{—CH—C}\diagdown \\ | \qquad\qquad\;\; \text{N—(CH}_2)_6\text{—CO}\sim \\ \text{CH}_2\text{—C}\diagup \\ \phantom{\text{CH}_2\text{—}}\|\\ \phantom{\text{CH}_2\text{—}}\text{O}\end{array}} \quad (8)$$

$$\begin{array}{c}\left\{\begin{array}{c}-\text{CH}-\text{C}\diagdown\\ |\qquad\quad\text{O}\\ \text{CH}_2-\text{C}\diagup\\ \|\\ \text{O}\end{array}\right. + \text{HO}-(\text{CH}_2)_4-\text{O}-\overset{\text{O}}{\underset{\|}{\text{C}}}-\!\!\left\langle\bigcirc\right\rangle\!\!-\overset{\text{O}}{\underset{\|}{\text{C}}}\sim \longrightarrow\end{array}$$

EPM-5-SA   PBT

$$\left\{-\text{CH}-\overset{\text{O}}{\underset{\|}{\text{C}}}-\text{O}-(\text{CH}_2)_4-\text{O}-\overset{\text{O}}{\underset{\|}{\text{C}}}-\!\!\left\langle\bigcirc\right\rangle\!\!-\overset{\text{O}}{\underset{\|}{\text{C}}}\sim\atop |\atop \text{CH}_2-\text{C}-\text{OH}\atop \|\atop \text{O}\right.$$

(9)

(EPM-g-SA)-g-PBT

In this way by the creation of a graft copolymer, acting as an interfacial agent, it is possible to control the rubber adhesion, the particle size, the uniformity of dispersion and the morphological stability of the blends.

Polar or functional comonomers different from propylene

Acrylic or methacrylic acids, esters, anhydrides, carbon monoxide, and sulfonates are used with ethylene in order to provide sites for successive chemical reactions or for adhesive bonding. In the latter case an ionomer is used, which is partially neutralized with zinc or sodium cations, and whose properties depend on several factors:

1. the nature of the polymer backbone;
2. the ionic content;
3. the cation type.

Most of these additives are suitable for polyamide and polyester toughening.

Polyurethanes

TPU elastomers are made by soft segments (polyester or polyether diols) connected by chain extenders to hard segments (aromatic diisocyanates). The monomer choice, the monomer ratios, the addition sequences, and the polymerization procedure are very important in determining the overall rubber properties. They are used in blends with engineering polymers for their polar nature and a large number of variables can be exploited to affect the final material performance: (1) interactions with the matrix; (2) rubber particle size and size distribution; (3) morphology of the particles; (4) thermal history.

Satisfactory impact performances of POM, acetal resins and PC are effectively achieved by the addition of TPU rubbers.

ABS, NBR core–shell graft copolymers

Emulsion graft polymerization processes can be used to obtain core–shell modifiers, in which the core consists of crosslinked poly B ($T_g = -6$ to -85) or NBR ($T_g \simeq -45$). The latex particle size is very crucial for effective toughening of the diverse matrices: styrenic polymers require larger particles ($d > 0.2\,\mu m$) than engineering polymers ($d < 0.2\,\mu m$), due to their different toughening mechanisms. The ABS composition and properties are affected by the following polymerization variables: (1) monomer amount; (2) initiator type and amount; (3) emulsifier concentration; (4) reaction temperature and time; (5) chain transfer agents.

Core–shell modifiers often have an outer shell of PMMA copolymer, which is miscible with, or can be wetted by, several engineering polymers. The result is improved adhesion to the matrices and a better dispersion of the rubbery component. SAN, PVC, PBT, PET, PET–PBT blends, and even blends containing immiscible components (such as ABS, HIPS and PC), can be toughened by adding about 20% of these modifiers.

In other cases it is not sufficient to have physical interactions, but it is necessary to create chemical bonds with the matrix to achieve effective toughening. This is the case with polyamides, and blends of PA with PPO. To this purpose reactive sites must be present on the outer shell of the grafted rubber particles, giving rise to 'in situ' reactions of the kind illustrated above.

Polysiloxanes

A series of polysiloxanes, containing a small amount of crosslinkable vinyl units (and sometimes 3% of silica, as inorganic filler) have been used for the toughening of polysulphones. The blend is obtained by simple melt mixing, since a reduction in viscosity of the matrix is accomplished by the modifier addition as well.

Polyacrylates

10% of polyethylacrylates or polybutylacrylates added to polysulfones, containing up to 4% of silica, gave excellent impact performances, with resilience values about 15 times higher than the homopolymer itself.

BIBLIOGRAPHY

Bucknall, C.B. (1977) *Toughened Plastics*, Applied Science, London.
Paul, D.R. and Newman, S. (eds) (1978) *Polymer Blends*, Vols I and II, Academic Press, New York.
Riew, C.K. (ed.) (1989) *Rubber-Toughened Plastics*, Advanced Chemistry Series, **222**, American Chemical Society, Washington, DC.
Riew, C.K. and Kinloch, A.J. (eds) (1993) *Toughened Plastics I*, Advanced Chemistry Series, **232**, American Chemical Society, Washington, DC.
Collyer, A.A. (1994) *Rubber Toughened Engineering Plastics*, Chapman & Hall, London.

Keywords: brittleness, butadiene-based rubbers, cavitation, crack initiation, crack propagation, crazing, ethylene-propylene random copolymers, high impact polystyrene, impact modifiers, maleic anhydride, particle cavitation, shear yielding, engineering polymers, thermoplastics, toughening additives, toughening mechanisms.

See also: Impact modifiers: (2)–(5).

Impact modifiers: (2) Modifiers for engineering thermoplastics

C.A. Cruz, Jr

INTRODUCTION

Engineering plastics are broadly defined as melt-processable thermoplastics capable of maintaining their dimensional stability and most of their mechanical properties above 100°C and below 0°C [1]. They are generally regarded as light weight substitutes for metals and for the more common types of construction materials used in structural and high performance applications. Growth in engineering plastics has come from the following widely used commercial resins:

- bisphenol-A polycarbonate, a clear and tough plastic;
- polyacetal (polyoxymethylene), a hard crystalline polymer with fatigue resistance and a low friction coefficient;
- polyamides (nylon 6 and nylon 6,6), both with good chemical and abrasion resistance, used in a variety of structural and automotive applications;
- polyesters – poly(butylene terephthalate), PBT, used in automotive and electrical and electronic applications; and poly(ethylene terephthalate), PET, the latter especially interesting due to its recyclability.

Polyethers, polysulfones, polyterephthalamides and polysulfide polymers, and several other high temperature polymers are also part of this group, but they are still produced in low volumes.

On the other hand, the concept of polymer alloying has been amply utilized to produce blends of engineering thermoplastics that combine the properties of their components and form materials with distinct

Plastics Additives: An A–Z Reference
Edited by G. Pritchard
Published in 1998 by Chapman & Hall, London. ISBN 0 412 72720 X

features, thereby avoiding the high cost of developing new polymers. PBT/polycarbonate blends, incorporating the chemical resistance of PBT and the ductility of polycarbonate are an excellent example of this concept. A specialized review of engineering resins with examples of commercial and developmental materials can be found in [1].

A common thread for engineering plastics in terms of their utilization in structural applications is their mechanical strength. Failures in thermoplastics may occur due to large stresses applied at low rates, fatigue or impact. Toughness is a major requirement in end use applications of engineering resins. The purpose of this entry is to provide an overview of the impact modification of these materials and to direct the reader to specialized publications for more in-depth discussions on this topic.

IMPACT PERFORMANCE IN ENGINEERING PLASTICS

The high rates of stress loading associated with an impact process take on a special relevance in polymers, because of their time-dependent mechanical properties. The material behaves in a more ductile fashion when deformed at low stresses or strain rates, and generally becomes harder and more brittle as the strain rate is increased. Ductile deformation processes are preferred when fracture occurs, because they absorb larger levels of mechanical energy. Temperature has a similar effect to strain rate: low temperatures induce brittleness, whereas high ones allow for ductile behavior of the material.

Even though engineering plastics are inherently impact resistant, several factors can significantly affect their performance. Their sensitivity to flaws or notches, however, is particularly notorious. In a qualitative fashion, a notch in the material has the same effect as deforming the material at a low temperature or at a high strain rate. Notch sensitivity has been studied and amply documented for various engineering plastics [2]. The sharper the notch, the higher the propensity of the plastic to undergo brittle failure. In addition, geometry and environmental factors such as temperature, moisture and exposure to weather elements, all affect impact behavior.

TOUGHENING: MEASUREMENT AND ANALYSIS

The measurement of impact resistance can be a complex undertaking. Because of the notch sensitivity effect, analysis of fracture in polymers relies heavily on measuring the energy required to propagate cracks in a structure. Of particular interest are cracks subjected to stresses perpendicular to the faces of the crack (also known as Mode I deformation). Cracks in Mode I deformation generally tend to grow perpendicularly

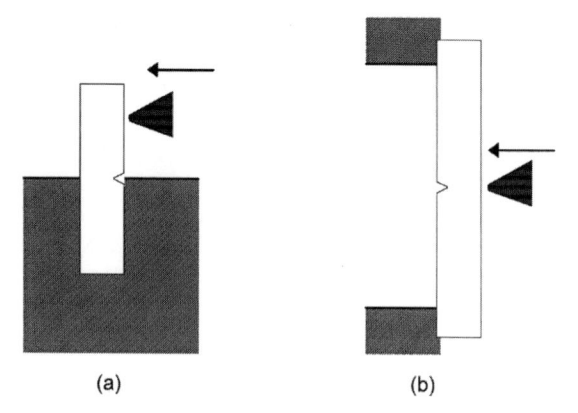

Figure 1 Schematics of notched Izod (a) and Charpy (b) protocols for impact testing.

to the plane in which the maximum tensile stress is applied and, if not arrested, will lead to catastrophic failure of a part.

Several schemes have been developed to measure the energy dissipated during this type of process; however, the most accessible measurements to daily practice are the notched Izod and Charpy protocols. In both cases, a specimen of prescribed dimensions is fixed in a cantilever (Izod) or three-point bending (Charpy) position and struck with a hammer, as described schematically in Figure 1. The hammer is attached to a pendulum, held and let go from a fixed height to strike the sample. The potential energy before and after the pendulum strikes the specimen can be obtained from the initial and final pendulum positions. The difference in potential energy, absorbed by the impact process, divided by the specimen thickness (J/m or ft lb/in, for the Izod test) or area (kJ/m^2, for the Charpy test) provides a normalized value of the 'impact strength' of the material.

These tests offer a rapid and relatively simple method of assessing toughness. Specifics for both protocols and their interpretation are contained in both the American Society for Testing and Materials (ASTM D 256) and the British Standards (BS 2782).

It is important to remark that the impact energy determined in this manner is not a material property. Alterations in the geometry of the specimen and, especially, in its notch radius and thickness, can completely change the results obtained. Therefore, when part design is the objective, it is necessary to consider not only modifications to the standard specimen dimensions but to obtain data from alternative methods. Test protocols that involve unnotched specimens, tensile impact, high speed tensile testing, falling weight and other methodologies are available. An excellent review of the field is provided in [3] and the accompanying chapter.

The following discussion is based on results obtained from the notched Izod or Charpy methods, as these are the most widely used tests to determine impact behavior in plastics. Even with the above-stated limitations, these protocols provide a useful method for discriminating between materials with different degrees of toughness.

CRITERIA FOR TOUGHENING

C.B. Bucknall's book [4], a classic in the field of polymer toughening, provides an excellent introduction to the general ideas for understanding the fracture process in plastics, including engineering resins, and the use of tougheners or impact modifiers.

High impact polystyrene (HIPS), where the presence of elastomeric polybutadiene domains provides toughening to polystyrene, is a material concept that has been successfully extended to engineering resins and other plastics. In this case, the addition of a toughening agent or 'impact modifier' to the engineering resin is usually accomplished by melt blending.

Several factors must be considered when toughening a plastic with a modifier. First, the toughening agent must form a separate or 'second' phase to function effectively. This separate phase, however, should be dispersed in the form of microscopic domains that have specific size requirements for each resin. Adhesion between the impact modifier and the matrix must have an optimum value to ascertain proper mechanical interaction. Since most polymers are incompatible with one another, obtaining proper adhesion requires proper physical and even chemical compatibilization. A low glass transition temperature of the elastomeric portion of the modifier is, as a general rule, required. Because of the melt processing step required for blending, the modifier must be capable of withstanding the relatively high temperatures used during compounding.

It is important, nevertheless, to understand the function of the impact modifier and the mode of deformation of the engineering matrix during fracture.

TOUGHENING MECHANISMS

Early ideas on the toughening process assumed that a significant amount of the energy dissipated during fracture was absorbed by the deformation and rupture of the rubber toughening particles. It is now generally accepted that the role of the toughening agent is not as an energy absorber but, rather, as an agent that enhances energy-dissipating mechanisms inherent to the matrix. Two major kinds of deformation processes have been identified in plastics, namely, shear yielding and crazing. Kinloch and Young [5] elaborate extensively on both processes. When shear yielding occurs, the matrix deforms plastically without a significant change in volume. Crazing, on the other hand, involves the nucleation of microvoids that stretch in a plane perpendicular to the maximum principal stress, but do not develop into a macroscopic crack, because they are stabilized by fibrils of plastically deformed material (see Figure 1 of the entry 'Impact modifiers: (1) Mechanisms and applications for thermoplastics' in this book). Crazing, because of microvoid creation, is a dilatational process, that is, it is accompanied by a volume increase in the matrix. Bucknall [4] has developed methodologies based on volume changes to study and discriminate between crazing and shear yielding processes. Shear yielding leads to crack stabilization and blunting and, eventually, to ductile fracture behavior. Crazing is often a precursor of brittle failure.

The role of elastomeric tougheners is intimately related to the fracture mechanism in operation. In matrices that are inherently shear yielding, impact modifiers act as stress concentrators where shear bands are initiated. When the polymeric matrix tends to craze, the toughening particles induce multiple craze formation and their elastomeric nature prevents the growth of large crazes that could develop into unstable cracks.

IMPACT MODIFIERS: ARCHITECTURE AND MORPHOLOGICAL FEATURES

Bulk or pelletized elastomeric compounds, as well as 'fixed size' core/shell particles, are used to toughen engineering plastics. As stated above, it is especially important to develop the proper particle size for effective toughening as well as the right amount of adhesion.

Compounding conditions regulate, to a large extent, the final particle size to be obtained with bulk rubbers when these are blended into engineering plastics. Core/shell particles, on the other hand, have cores consisting of crosslinked rubbers microencapsulated inside a 'hard' polymeric shell. Their architecture preserves their shape intact through the compounding process. Figure 2 shows the typical architecture of a core/shell particle. The transmission electron micrographs in Figure 3 illustrate the typical particle size distribution of a bulk functionalized rubber (a) and core/shell particles (b), compounded in a nylon matrix at the same weight percentage.

In between these two extreme types of architectures, several other kinds of polymers are used. Specific examples of all these toughening agents are as follows.

1. Ethylene-propylene (EP) and ethylene-propylene diene monomer (EPDM) rubbers are both prototypical elastomeric materials for

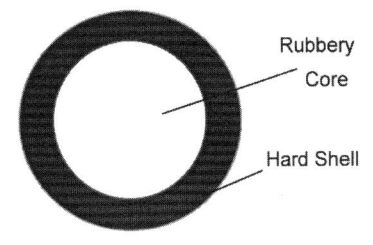

Impact modifiers: architecture and morphological features 391

Figure 2 Schematic architecture of a core/shell modifier.

toughening several engineering resins. Small amounts of reactive monomers such as maleic anhydride are commonly required for proper dispersion and adhesion of these rubbers, which are used particularly in nylon and polyesters.

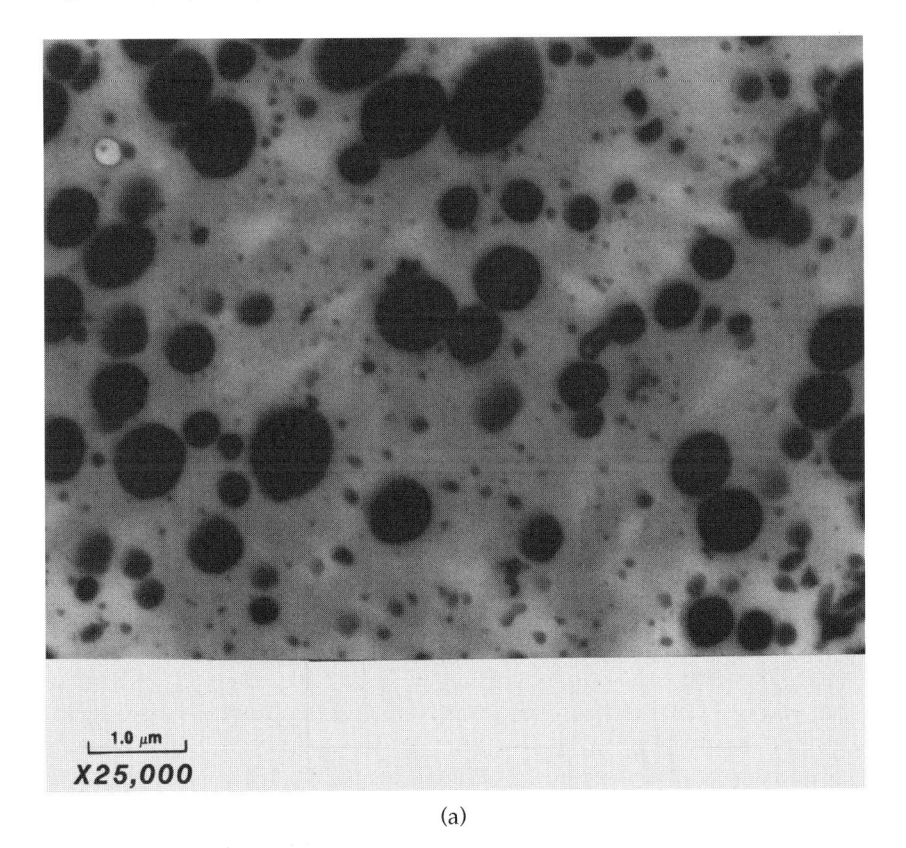

(a)

Figure 3 Transmission electron micrographs of (a) a functionalized bulk rubber and (b) a functionalized core/shell modifier, dispersed in nylon 6. (20 wt% loading in both cases.)

(b)

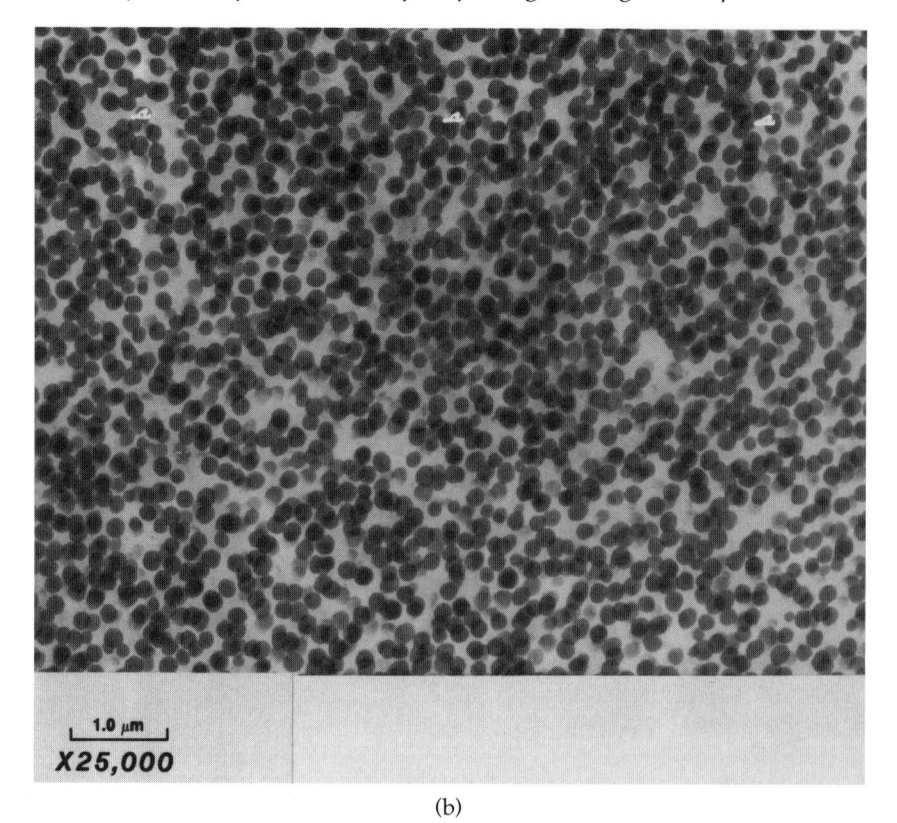

Figure 3 Continued.

2. Styrene-hydrogenated butadiene-styrene (styrene-ethylene-butadiene-styrene) terpolymers offer a further level of sophistication in terms of bulk compounds added to engineering plastics. Commercially available functionalized versions of these can be used either pure or in combination with their precursors for effective toughening of several engineering resins.
3. Acrylonitrile-butadiene-styrene (ABS) polymers, made by a process similar to that for manufacturing high impact polystyrene, provide an architecture intermediate between that of the previous two rubbers and that of core/shells. A crosslinked butadiene-styrene rubber is grafted with a styrene-acrylonitrile copolymer, which functions as a hard stage. This outer stage also provides the necessary means for adhesion to several matrices, such as PBT and polycarbonate.
4. In core/shell polymers, the internal rubbery phase is sometimes made up of an acrylate-based crosslinked polymer, usually poly(butyl acrylate), and the hard shell is based on a poly(methyl methacrylate)

Table 1 Suppliers of impact modifiers

Elf Atochem
Exxon Chemicals
Hoechst AG
GE Specialty Chemicals
Kanegafuchi
Mitsubishi-Rayon
Rohm and Haas
Shell Chemicals
Uniroyal Chemicals

homopolymer or copolymer, grafted to the elastomeric core. Alternatively, a crosslinked butadiene-styrene copolymer core is used instead of the acrylic-based one, to obtain low temperature performance. This type of structure is commonly known as 'MBS' (methacrylate-butadiene-styrene). Modifiers that contain silicon-based rubbers in the core have been used in an experimental fashion.

Functionalization of the shell polymer with special high polarity monomers, such as acid-type ones, is sometimes required for better dispersion and adhesion in very polar polymers, such as nylon.

Core/shell particles can be less efficient than bulk rubbers as toughening agents, because they contain less rubbery material. On the other hand, the hard shell has been well documented to have a less deleterious effect on matrix stiffness than that observed with bulk rubbers.

Table 1 provides a list of suppliers of impact modifiers for engineering resins.

IMPACT PROPERTIES OF ENGINEERING PLASTICS

Impact modifying an engineering polymer with an elastomeric additive has the objective of raising the toughness of the matrix to a specified level with a minimum amount of additive. It is therefore crucial to know the differences that result from modifier efficiency (loading) and effectiveness (absolute value). The following discussion uses nylon 6 and 6,6 as examples, two of the most thoroughly studied systems from a basic viewpoint. Figure 4 shows the effect of adding an elastomeric ethylene-propylene rubber to a nylon 6,6 matrix, as measured by notched Izod testing. The sigmoidal shape of the impact curve is typical of most impact-modified engineering plastics. One must note that the specimens on the high end of the curve deform in a ductile manner upon impact, whereas those in the low end fracture in a brittle manner. The effect of the modifier is to enhance the ductile deformation of the matrix under high deformation rate conditions.

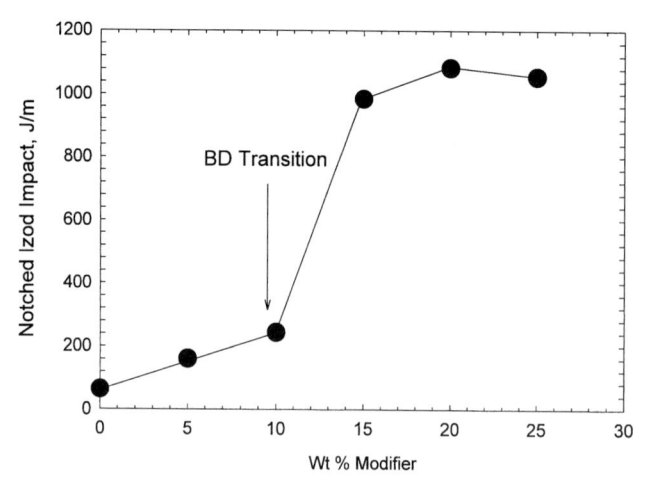

Figure 4 Impact behavior of nylon 66 toughened with a functionalized ethylene-propylene rubber at different loadings (BD = brittle to ductile transition).

Another valued result in determining performance is obtained by studying impact behavior as a function of temperature. Materials that behave in a ductile fashion at room temperature become brittle at a low temperature. This transition in mechanical behavior is known as the Brittle to Ductile Temperature. Figure 5 gives an example of the temperature dependence of the toughness of nylon 6 modified with core/shell particles.

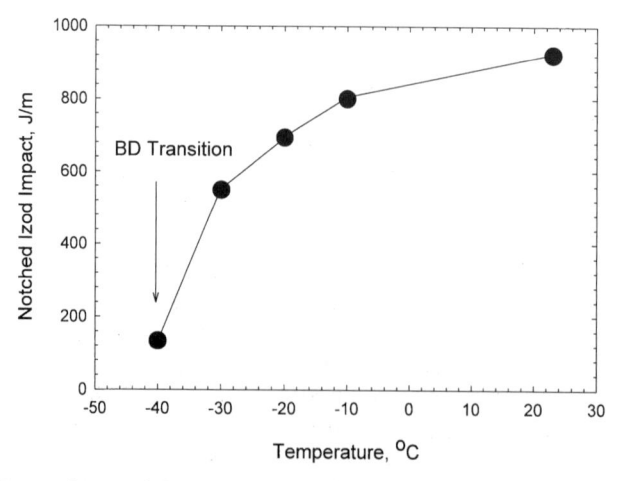

Figure 5 Dependence of the impact strength on temperature for a nylon 6 matrix toughened with 20 wt% of a core/shell modifier.

Toughened engineering plastics 395

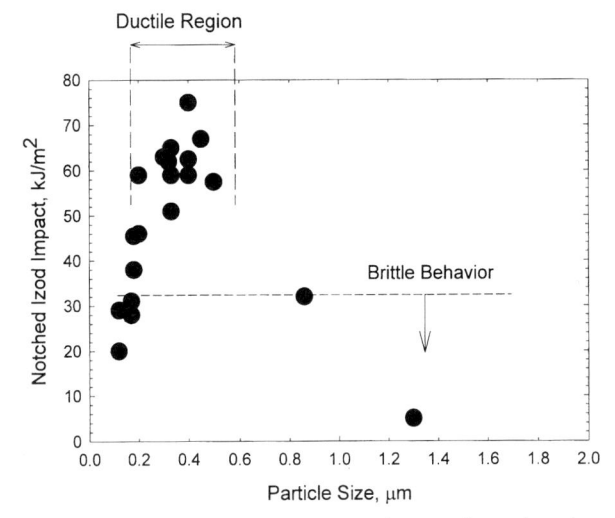

Figure 6 Particle size dependence of the toughness of a nylon 6 matrix at room temperature [6]. (Two types of functionalized ethylene-propylene rubbers were used as impact modifiers.)

Finally, another type of transition is provided by changing particle size at a fixed volume fraction of modifier. Figure 6 [6] illustrates these issues. Only particles with diameters between approximately 0.2 μm and 0.6 μm enhance toughness sufficiently for the matrix to fracture in a ductile manner. Therefore, an optimum particle size range is required for effective toughening. This type of behavior is expected to occur in other engineering plastics, although it has not been as systematically studied in matrices other than nylon.

TOUGHENED ENGINEERING PLASTICS

Several engineering resins, most notably nylons and polycarbonate/PBT blends are sold commercially in toughened grades; however, many applications have specific impact resistance demands which could be either lower or higher than can be met by commercially available materials. Knowledge of the impact performance as a function of modifier loading and temperature allows proper definition of type and loading of modifier to meet the specified impact properties. An important aspect of engineering polymers is crystallinity, which affects the toughness.

Frequently, engineering plastics are reinforced with glass fibers to enhance their stiffness and creep resistance. The presence of glass fibers in the matrix significantly decreases notched Izod impact values. No glass reinforced sample behaves in a 'ductile' fashion. The addition

Impact modifiers: (2) Modifiers for engineering thermoplastics

Table 2 Impact properties of representative engineering plastics and alloys ([3] and *Modern Plastics Encyclopedia*, McGraw-Hill Publications, 1994)

Material	Notched Izod impact, J/m at 22°C			
	Unmodified	Toughened	Glass fiber reinforced	Toughened and glass fiber reinforced
Polyacetal	53	107–800		
Nylon 6*	53–81	961	112–181[a]	187[b]
Nylon 6,6*	27–64	907–1174	51–240[c]	171–267[d]
Polycarbonate	641–854[e]		91–140[f]	
PBT	91–54	801–854	48–85[g]	
PET	13–37		80–117[h]	235[i]
Polytetramethylene adipamide	96	907	107[j]	
Polycarbonate/PBT		No break	160–220	
PPO/polystyrene	160–320	363	91–123	

Note: '%' refers to weight % glass fibers
[a] 30–35% [b] 33% [c] 13–33% [d] 15–33% [e] 3.175 mm thickness. For 6.35 mm thickness: 123 J/m
[f] 30% [g] 30% [h] 30% [i] 35% (supertough grade) [j] 30%
* Dry as molded

of impact modifiers restores only part of the toughness, as seen in Table 2.

The properties presented in the table show average impact resistance values for engineering plastics and alloys. Final fabrication conditions can significantly affect their impact resistance, particularly where crystallinity is present, as is the case in most engineering polymers. Also noteworthy in this series is the impact behavior of polycarbonate, which responds in a ductile fashion in thin samples, even under notched conditions. Thicker specimens are brittle. Tougheners have to be incorporated to re-establish the original ductility. A similar behavior is observed for the other systems, although not to the same extent as in polycarbonate.

High performance polymers, such as poly(ether-ether ketone), PEEK, as well as new resins, still coming on stream, such as poly(ethylene naphthalate), syndiotactic polystyrene, and cyclo-olefinic copolymers, are offering further challenges to impact modification.

REFERENCES

1. Clagett, D.C. (1986) Engineering Plastics, in *Encyclopaedia of Polymer Science and Engineering* (eds H.F. Mark, N.M. Bikales, C.G. Overberger, G. Menges and J.I. Kroschwitz) J. Wiley & Sons, New York, Vol. 6, p. 94.
2. Mills, N.J. (1993) *Plastics. Microstructure, Properties and Applications*, 2nd edn, Halsted Press, New York, Ch. 8.

3. Yee, A.F. (1986) Impact Resistance, in *Encyclopedia of Polymer Science and Engineering* (eds H.F. Mark, N.M. Bikales, C.G. Overberger, G. Menges and J.I. Kroschwitz) J. Wiley & Sons, New York, Vol. 8, p. 36.
4. Bucknall, C.B. (1977) *Toughened Plastics*, Applied Science Publishers, London.
5. Kinloch, A.J. and Young, R.J. (1988) *Fracture Behaviour of Polymers*, Elsevier Applied Science, London.
6. Borgrevve, R.J.M., Gaymans, R.J. and Eichenwald, H.M. (1989) *Polymer*, **30**, 78.

Keywords: crazing, engineering resins, impact modifiers, impact strength, nylon 6, nylon 6,6, polycarbonate, poly(butylene terephthalate), poly(ethylene terephthalate), polyester, shear yielding, toughening, Izod, Charpy, alloys.

See also: Impact modifiers: (1), (3)–(5).

Impact modifiers: (3) Their incorporation in epoxy resins

E.M. Woo

INTRODUCTION

As thermosetting resins, epoxies possess many desirable properties such as high tensile strength/modulus, excellent adhesion, chemical and solvent resistance, dimensional and thermal stability, good creep resistance, and fatigue properties. Two major categories of epoxy resins are generally used: difunctional and tetrafunctional types. The tetrafunctional type is typified by tetraglycidyl-4,4'-diaminodiphenylmethane (TGDDM) (Ciba-Geigy MY-720), and the difunctional type by the diglycidylether of bisphenol-A (DGEBA) family, typified by Epon-828 (Shell), Epi-Rez 510 (Celanese), DER-331 (Dow Chem.), and Araldite-6010 (Ciba-Geigy) etc. Novolac-type epoxy resins are produced by reacting epichlorohydrin with a phenolic novolac resin. The novolac epoxy resins contain epoxide groups and a phenolic backbone with an average of three or more epoxide groups per molecule. Some specialty epoxy resins have also been synthesized to achieve higher T_g, or lower viscosities, or to impart some special functions. Epoxy cure can be achieved over a wide temperature range at various rates by selection of special curing agents. Epoxy can be cured with various aliphatic primary and secondary amines; at elevated temperatures, aromatic amines are used. Anhydrides are also often utilized for curing. For more detailed discussion on the types of epoxy resin and curing agents, readers are advised to consult textbooks or handbooks on epoxies. Major characteristics of epoxy resins includes: (1) excellent adhesion to almost any surfaces, (2) no volatiles upon cure, (3) thermal and mechanical stability over wide temperature ranges, (4) extremely

Plastics Additives: An A–Z Reference
Edited by G. Pritchard
Published in 1998 by Chapman & Hall, London. ISBN 0 412 72720 X

low shrinkage, and (5) easy modification to suit various purposes. These versatile characteristics have helped to gain acceptance in widespread applications, such as adhesives in joining and fastening technology and as encapsulating agents for electronics or microelectronics parts. Over the past few years, increasing applications of epoxy resins have also been found in composites for aircraft, transportation vehicles, sports goods etc. However, epoxy resins are generally brittle due to high crosslink densities. Improvement of impact properties of epoxy matrix composites is generally needed for applications in structural parts. Furthermore, applications in coatings, adhesives, or microelectronics encapsulation etc., also require that the applied epoxy materials exhibit long-term mechanical stability under thermal/stress cycling environments. In such cases, impact improvement of the epoxy resins can usually resist microcracking.

Fracture toughness is a measure of the energy required for propagation of a pre-existing crack in a material under mechanical stress loading (I – tensile, II – shear or III – tearing loading mode). Toughness is usually represented by the stress intensity factor K_{IC} (unit: Pa m$^{1/2}$, Mode-I loading) or by the strain energy release rate, G_{IC} (Joules/m^2). They are related by:

$$K_{IC} = [E \cdot G_{IC}/(1 - \nu^2)]^{1/2} \qquad (1)$$

where E is the Young's modulus and ν is the Poisson's ratio. Linear elastic fracture mechanics (LEFM) proposes that loading of a cracked body is accompanied by inelastic deformation (crazing, shear yielding etc.) near the crack tip (Figure 1(a) and (b)). Thus toughness improvement of epoxy materials usually involves suppressing crack formation or enhancing the energy absorbing capability of the cracked body near crack tips. Figure 1(a) shows that crack-blunting reduces stress concentrations and can usually be achieved if the crack is arrested by a rubbery or glassy thermoplastic polymer particle of appropriate size.

METHODS OF MEASURING FRACTURE TOUGHNESS

Two common methods of determining the stress intensity factor are used: compact tension (CT) (ASTM E399-81) and notched three-point bending. The geometry of the specimens is represented in Figure 2. K_{IC} can be calculated as follows:

CT $\qquad K_{IC} = (P_c/BW^{1/2})f(a/W) \qquad (2)$

Three-point $\qquad K_{IC} = (P_c S/BW^{3/2})f(a/W) \qquad (3)$

where P_c is the load at crack propagation. Other symbols are as indicated in the figure. $f(a/W)$ is a geometry factor and varies with specimen

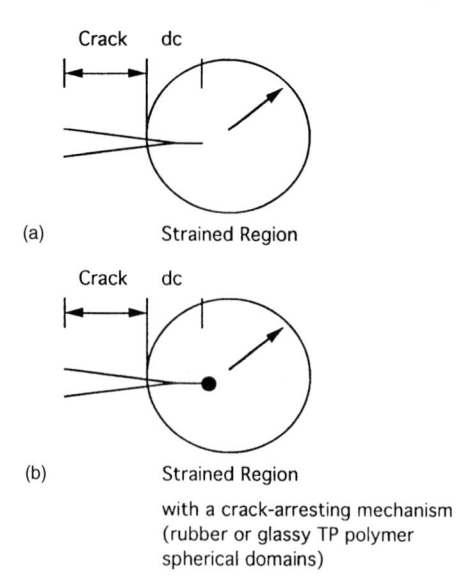

Figure 1 Propagation and deformation near the crack tip: (a) in an untoughened epoxy matrix; (b) in a toughened epoxy matrix with proper crack arresting mechanisms.

geometry. The expressions for $f(a/W)$ are usually listed in most standard textbooks on fracture mechanics.

METHODS OF MODIFICATIONS

Various types of polymeric modifiers have been researched as possible candidates to impart impact resistance of epoxy resins. Sometimes, properties other than impact resistance are to be modified. For example, for adhesive applications, the main focus is usually on improving shear and peel strength. In these cases, epoxies blended with elastomeric nitrile rubbers (Hycar CTBN, B.F. Goodrich), phenolics, nylons (soluble types such as DuPont Zytel-61), and polyurethanes are commonly used. For impact modification, there are several approaches, as discussed below.

Incorporation of liquid reactive rubbers or elastomers

Toughening epoxy matrices using liquid reactive rubbers (such as carboxyl-terminated butadiene acrylonitrile, CTBN, or the amine-terminated equivalent, ATBN) has been widely reported in the literature. Spherical rubber particles of a proper size distribution (usually 1–5 µm) can effectively enhance the toughness through crack blunting or cavitation mechanisms. However, rubber modification of epoxies becomes

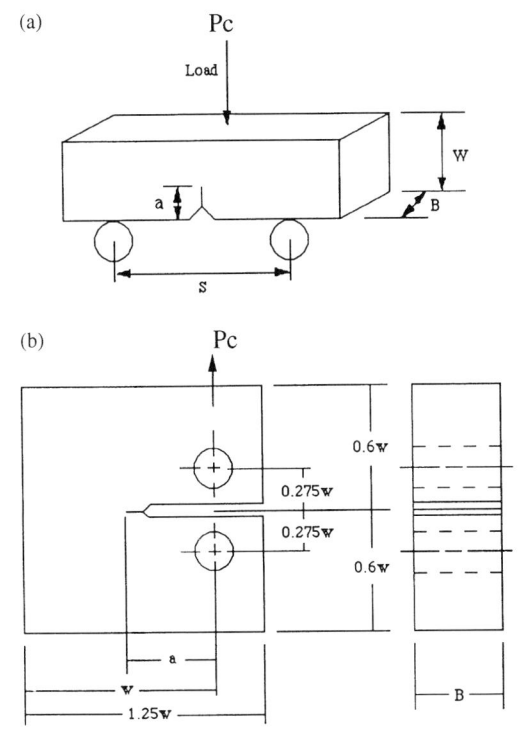

Figure 2 (a) Notched three-point bending (ASTM E399) and (b) compact tension (ASTM E399-81).

ineffective as the crosslink density of the epoxy matrix increases, which may result in inability of the rubber particles to undergo shear yielding or cavitation when surrounded by a highly crosslinked network. Furthermore, toughness improvements in most rubber-modified thermosetting systems usually result in a significant decrease in modulus and T_g of the cured rubber/epoxy resins.

Incorporation of tough thermoplastic polymers

Engineering thermoplastics (amorphous or semicrystalline) are commonly used as modifiers for epoxy resins in order to maintain a delicate balance between the impact properties and the thermal or mechanical performance of the modified epoxy materials. Most thermoplastic polymer additives are non-reactive with the host epoxy/hardener molecules, but may possess functional groups that are responsible for forming a homogeneous mixture by dissolving in the epoxies to be modified. Impact modification may come with a trade-off in the solvent resistance of the modified epoxies, depending on the network structure developed or the phase morphology after curing.

To impart chemical interactions between the polymer modifiers and the epoxy, some thermoplastics are functionalized before use, while a few thermoplastics are capable of reacting with the host epoxy. Alloying with polysulfone or polyethersulfone with reactive functional groups has also been the focus of recent attention in toughness improvement for epoxy resins. Recent studies have demonstrated that polymeric thermoplastics, such as polyethersulfones (PES), polyetherimides (PEI), polycarbonate, poly(phenylene oxide) or a combination of rubbers and thermoplastics etc. can enhance the fracture toughness without seriously sacrificing T_g, strength, stiffness, or other desirable properties of the resin systems. Semicrystalline polymers, such as poly(ether ether ketone) or poly(butylene terephthalate), have also been attempted. However, the difficulty is usually with the proper dispersion of the semicrystalline polymers into the epoxy resins. For that reason, if the semicrystalline polymer can be pre-formed into particulate shapes with functional groups capable of grafting with the epoxy/hardener matrix, a successful toughening method can be developed. Sometimes, the symmetry structure in the semicrystalline polymer is intentionally altered to reduce the crystallinity for greater ease of polymer dissolution in the epoxy.

Thermoplastics addition to epoxy resins is thus expected to alter not only the morphology, but also the cure kinetics. Phase separation takes place and a heterogeneous morphology usually develops if no specific interactions exist between the components. Eventually, the added thermoplastic polymer is excluded from the crosslinking epoxy network. A typical evolution of morphology of a TP/epoxy network during cure is given in Figure 3. Numerous examples of related studies are given in Table 1.

Epoxy modification by combination of thermoplastics and rubbers

More effective modification of the epoxies by use of a combinaton of thermoplastic and rubber has been demonstrated. A complex phase-in-phase morphology after cure can exist, whereby a precipitated thermoplastic-epoxy composite phase is surrounded by a continuous epoxy phase. By incorporating a small quantity (5 phr of CTBN and 20 phr thermoplastic PSu, the fracture toughness of a cured epoxy resin (DGEBA/DDS) can show an impressive increase of 300% in G_{IC} over the base unmodified epoxy resin. In such cases, incorporation of 5 phr of CTBN in the formulation probably makes the epoxy network tougher, by improving the interfacial characteristics of the epoxy and polysulfone (PSu) phase domains. The liquid reactive rubber acts as a coupling between the epoxy and PSu phase domains in much the same way as silane coupling agents enhance the interfacial strength of glass fiber/matrix composites. Because the liquid reactive rubber is used at relatively low loadings, any compromise in hot–wet properties of the

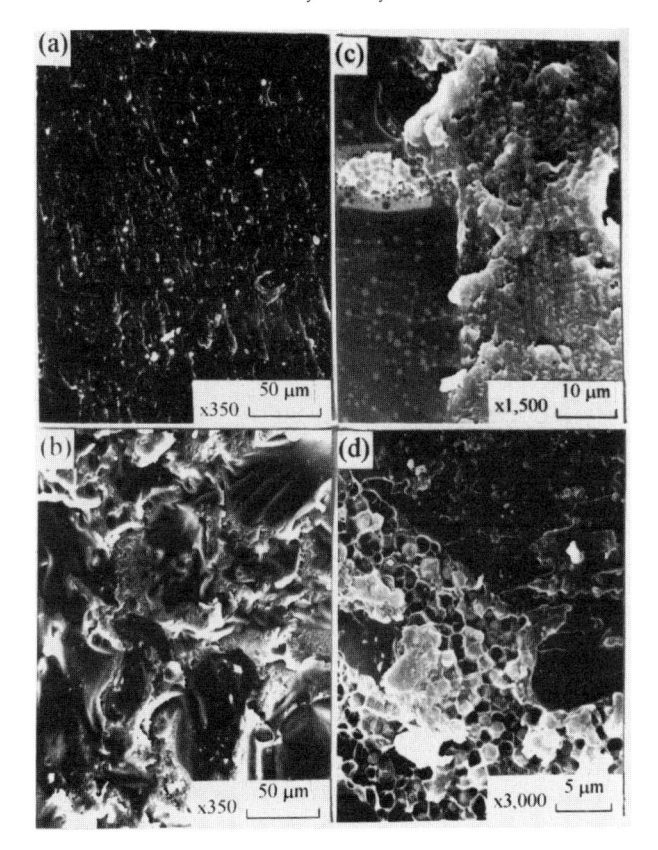

Figure 3 SEM images of 30 phr PEI/epoxy blends cured for different time at 177°C. (a) $t = 10$ min; (b) $t = 15$ min; (c) $t = 22$ min and (d) $t = 120$ min. (Ref: C.C. Su and E.M. Woo, *Polymer*, **36**, 2883 (1995), Courtesy of Elsevier Co.)

modified epoxies is usually kept to a minimum. There are a few outstanding examples of improvement in Mode I or Mode II toughness of epoxies using dual modifiers. For example, phenoxy and CTBN liquid rubber are used to modify epoxies and excellent shear properties are obtained. Polysulfones and CTBN are introduced into DGEBA/DDS epoxy systems. Polyetherimide (PEI) and CTBN are used to modify TGDDM/DDS epoxy systems. In both cases, impressive improvements in the fracture toughness have been reported.

Localized interlaminar modification

Although it has long been recognized that the creation of a phase-separated morphology is an essential means of achieving fracture toughness improvement, the improvement of fracture toughness in most epoxy

Table 1 Polymers as additives for improving the toughness of epoxy resins*

Polymeric additives	Comments	Toughness
Polysulfones (PSu)	Phase separation	+
Functionalized PSu	Phase separation	+
Polyethersulfones (PES)	Phase separation	+
Poly(ether imide) (PEI)	Phase separation	+
Poly(2,6-dimethyl-1,4-phenylene oxide) (PPO)	Phase separation	+
Poly(methyl methacrylate) (PMMA)	Phase separation	NT
Bisphenol-A polycarbonate (PC)	Homogeneous	+
Poly(ethylene oxide) (PEO)	Homogeneous	NT
Poly(butylene terephthalate) (PBT)	Phase separation	NT
Core–shell particles (core: poly(butyl acrylate), shell: crosslinked PMMA)	Multiphase	+
Siloxanes	Phase separation	NT
Elastomers	Phase separation	+
Poly(aryl ether ketone)	Phase separation	+
Poly(alkylene phthalate)	Phase separation	+
Functionalized thermoplastics	Phase separation	NT

* Epoxy resins in most cases are either TGDDM or DGEBA with various amine or anhydride hardeners.
+ Improvement in toughness is reported.
NT Toughness not tested.

composite systems is difficult. Incorporation of high $\bar{M}_w$ thermoplastic polymers into epoxy resins significantly increases the viscosity after blending, and can decrease the solvent resistance of the modified epoxy networks after cure. At low loading levels, improvement in the fracture toughness is usually moderate; increased loading results in processing and impregnation difficulties. Successful formulations are usually hard to devise, due to the extreme difficulty of compromising processing requirements with those of toughness. Whenever toughening of laminated composites is required, it is usually more effective to introduce the polymer modifiers directly to the localized regions where modification is truly needed, for example, in the interlaminar regions between plies.

A schematic is shown in Figure 4. A number of documents have recently proposed interlaminar toughening as a novel methodology for improving the impact resistance of epoxy–matrix laminated composites. Such practices usually require that an epoxy-based prepreg be prepared in two steps. The first step involves conventional impregnation of dry fiber with neat epoxy resins. This avoids the problem of high viscosities and ensures good wetting between fiber and resin. Subsequently, the

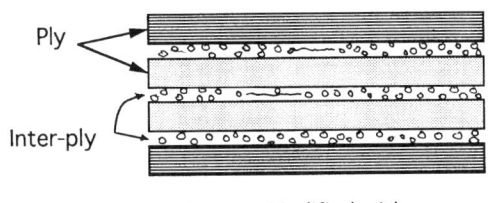

Interlaminar-Modified with
TP/Elastomer Composite Particles

Figure 4 Schematics for interlaminar toughening approach.

polymer modifiers, loaded in a carrier resin, are introduced in the second step on top of the prepreg. Substantial enhancement of interlaminar fracture toughness can be achieved.

BIBLIOGRAPHY

C.C. Su and E.M. Woo (1995) *Macromolecules*, **28**, 6779.
I. Skeist (1977) *Handbook of Adhesives*, 2nd Edn., Chap. 26, Van Nostrand Reinhold Co., New York.
E.M. Woo and K.L. Mao (1996) *Composites, Part A*, accepted.
J.C. Hedrick, N.M. Patel and J.E. McGrath (1993) in ACS Adv. in Chem. Ser. No. 233, *Toughened Plastics I: Science and Engineering*, Ed. by C.K. Riew and A.J. Kinloch.
C.C. Su and E.M. Woo (1995) *Polymer*, **36**, 2883.

Keywords: fracture toughness, crack, impact resistance, reactive rubbers, thermoplastics, phase separation, interlaminar modification.

See also: Impact modifiers: (1), (2), (4), (5).

Impact modifiers: (4) Organic toughening agents for epoxy resins

Raymond A. Pearson

INTRODUCTION

Thermosetting polymers based on epoxy resins often display superior tensile strength and chemical resistance when compared with their thermoplastic counterparts. Such attributes make epoxy polymers ideal matrices for adhesives and composites. However the applicability of epoxy polymers as matrices for adhesives and composites are often limited by an inherent weakness – low flaw tolerance. Ironically, the same crosslinked chemical structure that imparts high strength and superior chemical resistance also promotes brittle behavior.

Brittle behavior can be defined in a number of ways. For example, low elongation to failure in a tensile test, or low impact strength in a notched Izod test are quite common. Unfortunately, neither of the aforementioned tests measures an inherent material property. The values obtained using such tests are a strong function of the size and type of specimen used. Moreover, both elongation to failure values and notched Izod impact values are influenced by defects in machining or casting. To overcome these deficiencies, it is common practice to report fracture toughness values, which are based on the theory of linear elastic fracture mechanics (LEFM).

Although it is not the intention to provide a comprehensive thesis on the origins of LEFM, it is useful to review the test method that is so often used to characterize the fracture toughness of epoxy polymers. The most common geometry to characterize the fracture toughness of epoxy polymers is the single edge notched, three-point bend test

Plastics Additives: An A–Z Reference
Edited by G. Pritchard
Published in 1998 by Chapman & Hall, London. ISBN 0 412 72720 X

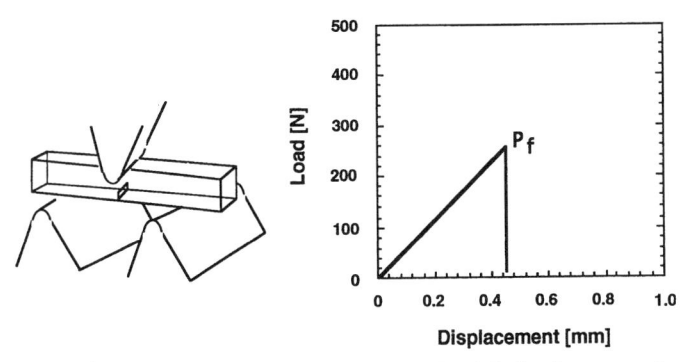

Figure 1 (a) A typical SEN-3PB test. (b) A typical load deflection curve for an SEN-3PB test.

(SEN-3PB) as described by the ASTM D5045 guideline. Such specimens are often tested on a screw-driven materials testing machine capable of recording both the load and the cross-head displacement. A typical specimen and load deflection curve is shown in Figure 1.

LEFM uses the load at fracture and the starting crack length to define a critical stress intensity factor, K_{IC}. The critical stress intensity factor is an inherent material property and does not suffer from the geometry dependence of other mechanical characterizations, provided that plane strain conditions are met. It is common practice to use the term 'fracture toughness' in place of the critical stress intensity factor, which is rather cumbersome. The fracture toughness of a material can be calculated from the load displacement diagram of a 3PB-SEN specimen as follows:

$$K_{IC} = \frac{10^{3/2} PS}{tw^{3/2}} f(X) \tag{1}$$

and

$$f(X) = 3^{1/2} \frac{[1.99 - X(1-X)(2.15 - 3.93X + 2.7X^2)]}{2(1+2X)(1-X)^{3/2}} \tag{2}$$

where P is the critical load for crack propagation in kN; S is the length of the loading span in mm; t is the specimen thickness in mm; w is the specimen width in mm; $f(X)$ is a shape factor function; and X is a ratio of the crack length, a, to the width.

It is common to use a 65 mm × 12.7 mm × 6.4 mm specimen supported by a 50.8 mm span to perform these tests. The fracture toughness can also be converted to the critical strain energy release rate G_{IC} using equation 3:

$$G_{IC} = \frac{K_{IC}^2}{E}(1 - \nu^2) \tag{3}$$

where E is the Young's Modulus in MPa; and ν is Poisson's ratio.

The critical strain energy release rate, G_{IC}, is often described as the 'fracture energy', which may be a misnomer because it can be confused with the energy to fracture. The energy to fracture a SEN-3PB specimen would clearly be a function of the original crack length, the longer the crack length the smaller the energy required. In contrast, the critical strain energy release rate is independent of crack length and is an inherent material parameter. Therefore, in this paper we will clearly define the term 'fracture energy' as another name for critical strain energy release rate in order to avoid any confusion.

As mentioned previously, epoxy polymers possess low flaw tolerance. Typically the fracture toughness values are below $1\,MPa\,m^{1/2}$ or, in terms of fracture energy, below $200\,J/m^2$. These values are extremely low when one compares them with a ductile thermoplastic polymer such as polycarbonate, which has a $K_{IC} = 2.4\,MPa\,m^{1/2}$ and a $G_{IC} = 2400\,J/m^2$. However, there are organic based materials that can be added to epoxy polymers that increase their fracture toughness such that it even exceeds the fracture toughness of common thermoplastic resins!

DESCRIPTION OF THE ADDITIVES

The common organic additives used to toughen epoxy resins can be grouped into several categories. The first distinction between the various categories is the use of elastomeric (soft) versus glassy (hard) additives. The second distinction is between additives that are initially miscible and phase separate upon curing from those which are polymerized separately from the epoxy polymer and are added as preformed particles. A brief description of these four types of organic toughening agents is described below.

Precipitating elastomers

A wide variety of soft rubbery polymers have been evaluated as toughening agents for epoxy resins. In this category, the elastomer is often completely miscible with the liquid epoxy resin and precipitates out during the cure process. Examples of precipitating elastomers evaluated as toughening agents for epoxies include reactive butadiene-acrylonitrile monomers, flexible polyurethanes, liquid chloroprene rubber, naturally occurring vernonia oil, and much recently reactive, hyperbranched molecules.

Precipitating elastomers based on reactive liquid copolymers of butadiene-acrylonitrile are perhaps the oldest and best understood of the precipitating elastomers. Such reactive polymers are produced by the B.F. Goodrich Company in a variety of acrylonitrile contents and are often terminated with reactive endgroups such as carboxylic acid or

$$HO-\overset{O}{\underset{\|}{C}}-R-\left[-(CH_2\cdot CH=CH-CH_2)_x-(CH_2-\underset{\underset{CH_2}{\overset{\|}{CH}}}{CH})_y-(CH_2-\underset{C\equiv N}{CH})_z-\right]_n-R-\overset{O}{\underset{\|}{C}}-OH$$

Figure 2 The chemical structure of a carboxyl-terminated copolymer of butadiene-acrylonitrile (CTBN).

amine. Figure 2 is a schematic diagram of the chemical structure of a CTBN elastomer (CTBN denotes carboxyl terminated copolymer of butadiene acrylonitrile). The advantage of the CTBN oligomers is that the polarity of the reactive liquid elastomer can be changed to assure rubber precipitation and to control rubber particle size for each particular epoxy/hardener system. The use of CTBN oligomers to toughen epoxy polymers has been practised for over 25 years.

An emerging technology is the use of hyperbranched molecules as toughening agents for epoxy polymers. Hyperbranched molecules consist of a multifunctional core, one of several layers of repeating units building up the bulk structure. See Figure 3. A number of reactive functional groups can be applied to the outer shell. As for the CTBN elastomer described above, the hyperbranched molecules are miscible with the epoxy resin and can phase separate upon curing. A potential advantage of these new additives is the ability to obtain low melt viscosities with high molecular weight additives. Unfortunately, at the time of writing a clear product line has yet to be developed.

Preformed rubber particles

This technology is relatively new when compared with CTBN toughened epoxies. The use of preformed rubber particles appears to be gaining in

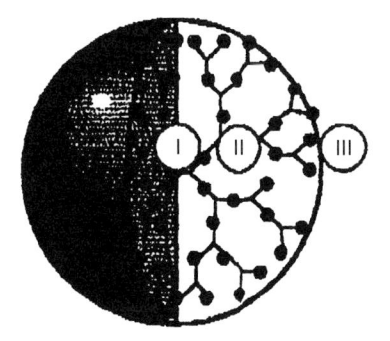

Figure 3 The chemical structure of a hyperbranched molecule. The dendritic molecular structure contains a core (I), the bulk (II) and the shell (III).

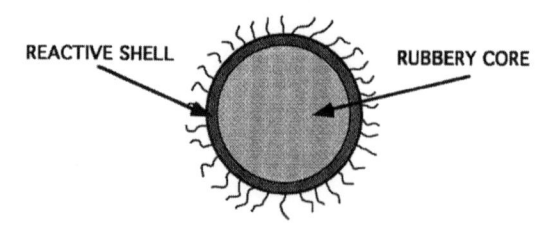

Figure 4 A schematic diagram of a structured, core–shell latex particle.

popularity, as evidenced by the more than 20 papers published in the past 5 years. The advantages of using preformed rubber particles are two-fold. First, the use of preformed rubber particles eliminates the particle size dependence on the resin chemistry and process temperature. Secondly, the particles are inherently immiscible with the epoxy matrix such that none of the elastomer remains in the matrix to lower the glass transition temperatures. Despite these advantages, the use of preformed rubber particles as toughening agents for epoxy resins is not as common as the use of reactive liquid elastomers such as CTBN.

The most common preformed rubber particles used as a toughening agent for epoxy polymers are the so-called structured, core-shell latex particles (Figure 4). These particles typically have a polybutadiene-based core and an acrylate-based shell. Such additives can be purchased as powders from Rohm and Haas or Elf-Atochem and can be purchased as epoxy concentrates from the Dow Chemical Company. The key parameter for these modifiers is the composition of the shell polymer, since the shell chemistry plays a crucial role in the overall blend morphology. It should be noted that it is possible to obtain commercial core–shell latex particles with reactive groups in the shell for improved dispersion of the rubber particles.

An emerging technology is the use of treated rubber particles from ground tires (Figure 5). The particles tend to be cheaper and the surface treatment improves dispersion and adhesion to the matrix. See 'Surface-modified rubber particles for polyurethanes' in this book. Such particles can be purchased from Composite Particles, Inc. in Allentown, Pennsylvania. Unfortunately, the current technology for cryogenic grinding of the rubber produces rather large rubber particles which are not effective in toughening epoxies. However there may be some cost advantages of mixing these rubber particles with the more effective MBS or CTBN types.

Precipitating thermoplastic particles

The technology of thermoplastic-toughened epoxies evolved because rubber particles are incapable of toughening highly crosslinked

Polar surface

- Low surface energy
- Boundary layer
- Inert

Oxidative atmosphere →

- Enhanced wettability/ compatibility
- Cross-linked boundary layer
- Reactive graft sites

Double bond surface

- Inert
- Free radical inhibition

Proprietary reactive gas process →

- Reactive graft sites
- Noninhibiting

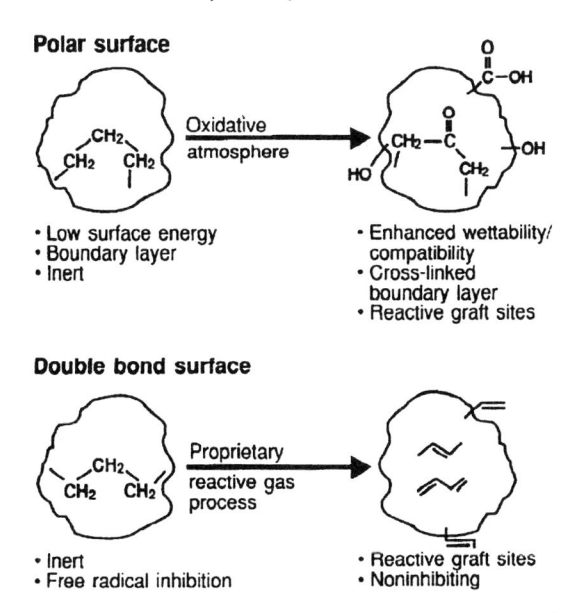

Figure 5 Schematic diagram of treated rubber particles.

matrices. Glassy polymers such as poly(ether sulfone), polyether ketone, poly(etherimide), polyimide) and poly(phenylene ether) have been evaluated as toughening additives for highly crosslinked epoxy polymers. In some cases the change in fracture toughness can be as high as $1.4\,\mathrm{MPa\,m^{1/2}}$! In spite of these successes, the technology of thermoplastic toughened epoxies is not as developed as that for rubber-toughened systems. Perhaps this is merely a consequence of the fact that thermoplastic toughened epoxies are a relatively recent discovery.

The most common thermoplastic modifiers for epoxy resins appear to be polyethersulfone and polyetherimide (Figure 6). Both polymers have been shown to be miscible in bisphenol-A based epoxies. Polyetherimide is often used as received, whereas polyether sulfones often contain reactive end-groups to facilitate miscibility and control the morphology. End-groups such as hydroxyls and amines have been successfully attached to polyether sulfones and some additional increases in toughness have been observed. In general, the use of high molecular weight resins provides greater toughness. However, the use of high molecular weight resins will reduce the miscibility of these polymers in epoxy resins. Parenthetically speaking, the miscibility of these additives in epoxies is much more difficult to predict than for the low molecular weight CTBN modifiers.

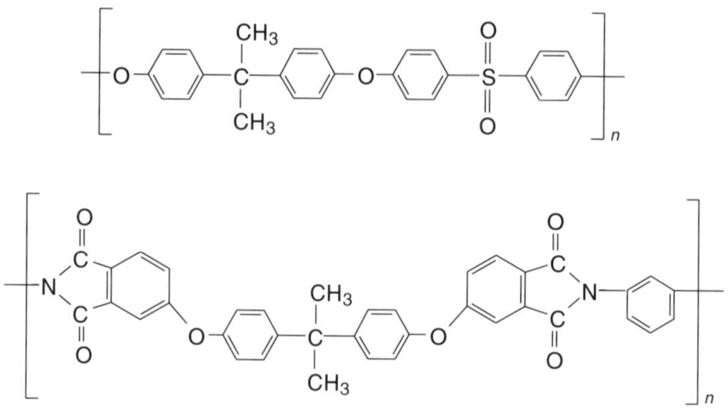

Figure 6 The chemical structure of polysulfone and polyetherimide.

Preformed thermoplastic particles

Nylon 6, 6,12, and 12 polymers (Figure 7) are available as preformed particles from Elf Atochem in a range of particle sizes from 5 to 50 μm in diameter. Toray Pearl particles based on nylon 12 with an average diameter of 6 μm have been found to be very effective toughening agents. Unfortunately, the Toray Pearl particles are presently not available commercially. Improvements in toughness of $1.0\,\text{MPa}\,\text{m}^{1/2}$ have been obtained yet this technology is considered to be in its infancy.

DESCRIPTION OF HOW THE VARIOUS MODIFIERS INCREASE TOUGHNESS

Increases in toughness are achieved by shielding the crack tips from the applied stress. There are two broad types of shielding or toughening mechanisms: crack-tip process zone and crack-wake zone mechanisms. See Figure 8. Crack-tip process zones by definition occur at the crack tip. Examples of crack-tip process zone mechanisms are: (1) rubber particle cavitation and concomitant matrix shear banding; (2) microcracking induced by thermoplastic rubber particles. Crack-wake zones occur by definition behind the crack tip. Particle bridging and stretching is an example of a crack-wake mechanism. It is useful to understand these mechanisms before choosing the appropriate modifier for your specific epoxy.

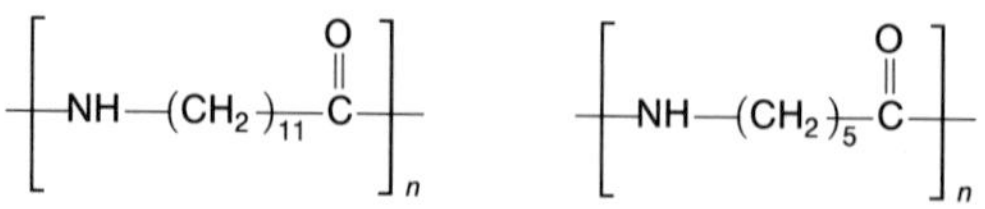

Figure 7 The chemical structures of nylon 12 and nylon 6,12.

Description of how the various modifiers increase toughness 413

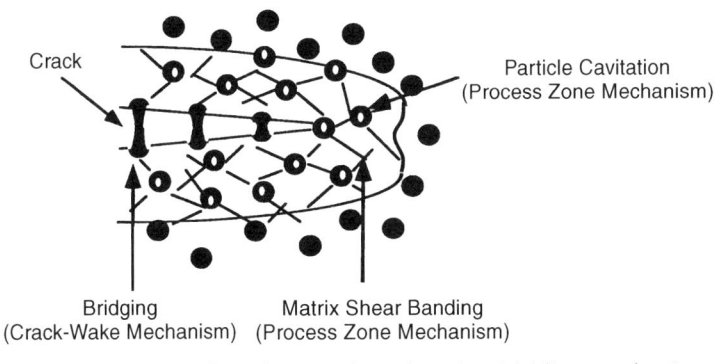

Figure 8 Examples of crack-tip and crack-wake shielding mechanisms.

The most potent toughening mechanism in rubber-modified epoxies is a crack-tip process zone mechanism involving rubber particle cavitation and concomitant matrix shear banding. Such a mechanism requires that the rubber particles be small so that they interact with the process zone stress field at the crack tip and it requires the matrix to shear yield. Highly crosslinked epoxies are extremely resistant to yielding, and so are not likely candidates for the rubber toughening approach. Such an effect is clearly demonstrated in a DDS-cured DGEBA epoxy system as shown in Figure 9.

It should be mentioned that even when rubber toughening is successful there are compromises which must be made, such as reduced modulus, yield strength, creep resistance and, in some cases, glass transition

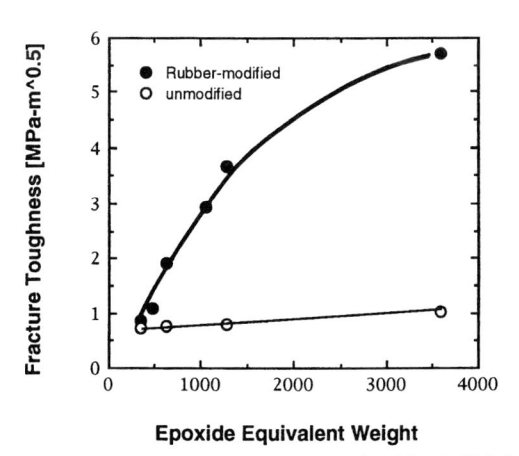

Figure 9 Fracture toughness values for a series of DGEBA/DDS epoxies. Highly crosslinked epoxies (low epoxy equivalent weight, EEW), are difficult to toughen with CTBN rubber.

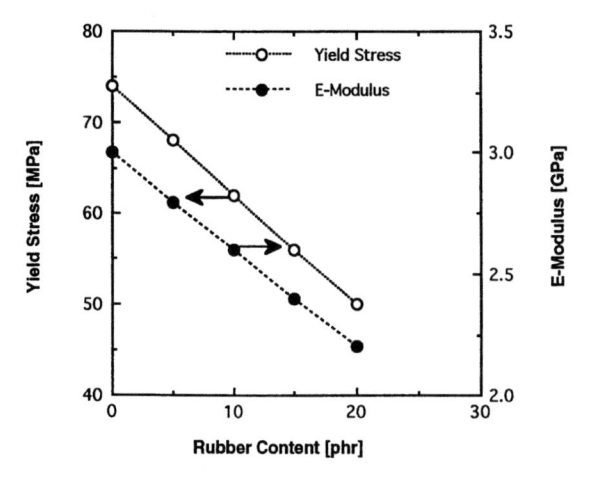

Figure 10 Mechanical properties of a CTBN-modified epoxy (DGEBA/piperidine). Rubber modification has its drawbacks, both yield strength and modulus are reduced.

temperatures. Figure 10 illustrates the reduction in both yield strength and modulus when CTBN additives are used.

Fortunately, crack-wake shielding mechanisms are not significantly influenced by the crosslink density of the epoxy resin and can be used

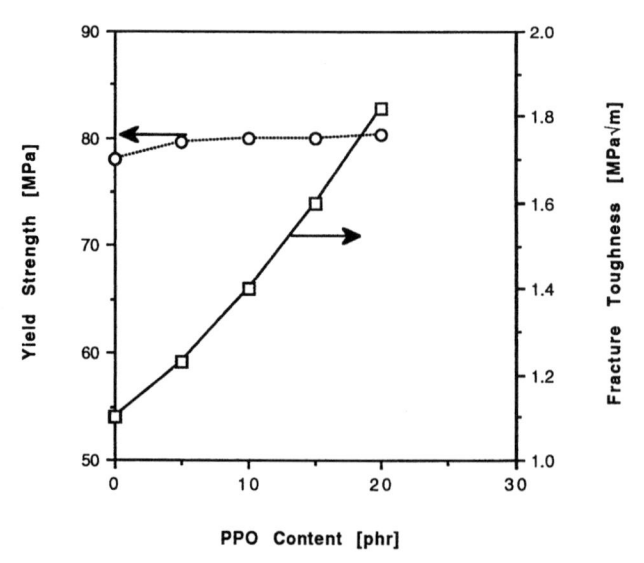

Figure 11 Mechanical properties of a thermoplastic-modified epoxy (DGEBA/piperidine). Note that the toughness increases with thermoplastic content while the yield strength remains unchanged.

rather effectively to toughen highly crosslinked epoxies. Moreover, properties such as modulus and yield strength are not adversely affected by the incorporation of thermoplastic particles. See Figure 11. However, the morphology is difficult to control for the precipitating types and a co-continuous morphology can lead to poor solvent resistance.

SUMMARY

The materials specialist has a wide variety of organic toughening agents from which to choose. The best choice will be determined by the chemistry of the epoxy polymer of interest. If the crosslink density is relatively low then a rubbery toughening agent will work best. A glassy polymer can be used when the crosslink density is high. If low viscosity is required then precipitating toughening agents may be useful and the solubility parameter of the epoxy polymer will determine the chemical structure of the modifier used. Unfortunately the optimization of the toughening agent for a particular epoxy polymer is beyond the scope of this simple introductory chapter but the curious reader should consult the references given below.

REFERENCES

Richard W. Hertzberg, *Deformation and Fracture Mechanics of Engineering Materials*, 4th edn (John Wiley and Sons: New York, 1996) pp. 315–369, 445–451, 644–675.

B. Ellis (ed.) *Chemistry and Technology of Epoxy Resins* (Blackie Academic & Professional: New York, 1993) pp. 117–173.

C. Keith Riew (ed.) *Rubber-Toughened Plastics*, Advances in Chemistry Series vol. 222 (American Chemical Society: Washington, DC, 1989).

A.J. Kinloch and R.J. Young, *Fracture Behaviour of Polymers* (Applied Science, London, 1983).

Keywords: linear elastic fracture mechanics, critical strain energy release rate, precipitating elastomers, hyperbranched molecules, preformed rubber particles, core–shell latex particles, treated rubber, precipitating thermoplastic particles, preformed thermoplastic particles, crack bridging, shear banding, cavitation.

See also: Impact modifiers: (1)–(3), (5).

Impact modifiers: (5) Modifiers for unsaturated polyester and vinyl ester resins

J.S. Ullett and R.P. Chartoff

INTRODUCTION

Unsaturated polyester (UP) and vinyl ester (VE) thermosetting resins are widely used in many commercial applications including protective coatings, architectural materials, bathroom fixtures and automobile body panels. Like other thermosetting resins such as epoxies, UP and VE resins are brittle upon cure and have low resistance to crack propagation, i.e., low fracture toughness. For example, fracture toughness values for UP resins may vary from 0.4 to 0.6 $MPa\,m^{1/2}$ depending on chemical structure. In comparison, polycarbonate has a fracture toughness of the order of 3 $MPa\,m^{1/2}$ and aluminum alloys have fracture toughness values ranging from 16 to 45 $MPa\,m^{1/2}$.

Low fracture toughness values may lead to component failure from low energy impact. In some large sheet products cracking may result from handling. Consequently, improving the fracture toughness of UP and VE resins is of significant commercial importance. A number of techniques are possible. Because UP resins can be made from a variety of components (diacids combined with diols), components can be chosen to add flexibility to the polyester network. For example, long chain aliphatic acids such as adipic acid can be substituted for all or part of the phthalic acid used in a general purpose UP resin formula. Flexibilized resins will have lower heat deflection temperatures in proportion to the amount of flexibilizer used. Flexibilized resins may

Plastics Additives: An A–Z Reference
Edited by G. Pritchard
Published in 1998 by Chapman & Hall, London. ISBN 0 412 72720 X

also have reduced hydrolytic stability because of the increased number of ether linkages.

Another approach for toughening UP and VE resins which has had success with epoxy resins is the use of liquid rubber (or elastomer) additives. The chief benefit of employing liquid rubbers (LR) versus a flexibilized resin is that decreases in hardness, stiffness and heat-deflection properties can be minimized. During cure, the liquid rubber phase separates from the resin and is concentrated in a particulate phase. Very little of the rubber remains in solution with the cured resin so the resin's heat deflection temperature is for the most part unaffected. The toughness of the two-phase, or composite, material will be a function of the microstructure, which in turn will depend on processing and cure conditions. The subject of epoxy resin toughening is covered in separate articles in this book.

MICROSTRUCTURAL FEATURES REQUIRED FOR TOUGHNESS

In order to improve a resin's resistance to crack propagation energy, dissipating mechanisms must be introduced. The amount or volume of material involved in energy dissipation needs to be maximized for toughness to be maximized. Three principal energy dissipating mechanisms have been identified in impact modified thermosets containing a rubbery particulate phase. These are: particle deformation (stretching followed by tearing), dilation of particles and the surrounding matrix, and particle cavitation followed by localized shear yielding in the matrix. Other mechanisms which may contribute to energy dissipation are: microcracking, crack bifurcation, crack deflection and crack pinning. Particle cavitation followed by shear yielding may be the most important mechanism because a large volume of material surrounding the crack tip can be involved. With some of the other mechanisms, such as particle deformation, only particles in the path of the crack are involved.

Figure 1 shows a compact tension specimen in which a starter crack propagated at a controlled rate such that the test could be stopped prior to specimen failure. Whitened areas above and below the fracture surfaces are a result of particle cavitation. This particular material was a flexibilized vinyl ester containing a rubber particulate phase. The large volume of material involved in the fracture process resulted in high toughness and a ductile tearing mode of fracture as opposed to the brittle fracture mode of the base resin.

Microstructural features that are required to improve fracture toughness in thermosets have been identified; however, exact requirements may be material dependent. In general, improvements in fracture toughness are dependent on the volume fraction of the rubbery particulate phase, good dispersion of the particles within the matrix, proper particle

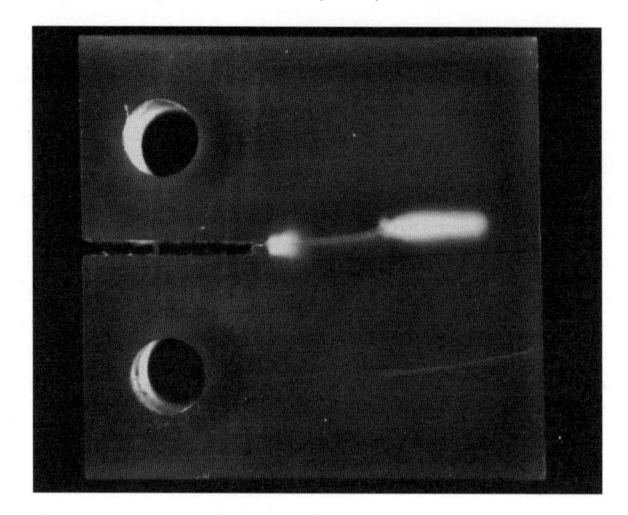

Figure 1 Side view of a compact tension specimen showing extensive stress whitening due to particle cavitation above and below the fracture plane. (Reproduced with permission from Ullett and Chartoff (1995). Copyright 1995 Society of Plastics Engineers.)

size (and distribution of sizes), and the strength of the particle matrix interface. Of these features, the volume fraction of rubbery particles is of central importance. This is also true of other rubber modified materials such as high-impact polystyrene. The toughening mechanism may depend on rubber particle size and the microstructure of the particles. For a good review of the topic of rubber modified plastics see Riew (1989).

LIQUID RUBBER/RESIN COMPATIBILITY

Microstructural features such as volume fraction of rubbery particles and the particle size distribution will depend on the initial resin/modifier compatibility and the cure scheme. From the processor's standpoint, the blend needs to be stable and produce the desired morphology upon cure consistently. Ideally, one would like the liquid rubber modifier and the unreacted resin to form a miscible blend at room temperature. Cure conditions will then control rubbery particle formation. Factors that affect phase separation include cure temperature and time to gelation, which is a function of temperature and catalyst system.

In the case of UP and VE resins complete miscibility with liquid rubbers or elastomers is unusual. Initial compatibility of two polymer components, i.e., the resin and the liquid rubber, will depend on a number of factors such as the molecular weight of each component, the molecular weight distribution of each component, the chemistry of each component, and the presence of a solvent or additional component. In order to obtain compatible resin/rubber blends researchers have tried varying the resin chemistry, including end and pendant groups, and have synthesized block copolymers to be used as toughening additives or as compatibilizing agents for commercially available liquid rubbers.

As described in the literature, many types of liquid elastomers have been used to modify various thermosetting resins. Poly(butadiene-acrylonitrile), NBR, is one of the most often used rubbers and has been modified in a variety of ways to improve its compatibility and reactivity with various resins. Focusing on polyester resins, blends containing NBR with carboxyl end groups (Crosbie and Phillips, 1985b; Ullett and Chartoff, 1995), with vinyl end groups (Crosbie and Phillips, 1985b; Ullett and Chartoff, 1995), and with amine end groups (Subramanium and McGarry, 1994) have been reported.

Other elastomers which have been used to modify polyesters include EPDM (ethylene-propylene-diene-methylene), poly(epichlorohydrin) with hydroxyl end groups (Crosbie and Phillips, 1985b; Ullett and Chartoff, 1995), organic siloxane elastomers, polyacrylates (Ullett and Chartoff, 1995), and polyurethane elastomers (Kim and Chan-Park, 1994).

Cure conditions (temperature, time and catalyst system) can be used to control particle development during cure if the resin/LR blend is miscible prior to cure. However, very few miscible systems have been developed. Figure 2 shows the fracture surface of a tough blend of vinyl ester with poly(epichlorohydrin) elastomer. This system is single phase before cure. Under the proper cure conditions a fine particulate phase develops as shown in Figure 3.

Poor compatibility prior to cure results in large particle sizes and size distributions and limits the amount of liquid rubber toughener which can be added. Figure 4 shows the fracture surface of an unsaturated polyester modified with vinyl-terminated poly(butadiene-acrylonitrile), VTBN. The rubbery particles are much larger than those visible in Figure 3. Methods to improve compatibility and thus reduce particle size have included using copolymers as compatibilizing agents for commercial LR additives (Ullett and Chartoff, 1995) and using a rubber-epoxy copolymer as a toughening agent (Subramanium and McGarry, 1994).

Subramanium and McGarry (1994) pursued a novel approach to toughening a UP resin by reacting it with a rubber toughened epoxy resin in an attempt to form an interpenetrating polymer network. The liquid rubber used to modify the epoxy was an amine terminated butadiene acrylonitrile copolymer. The toughened epoxy was added to the UP resin in various concentrations. The mixtures were cured at elevated temperature forming clear castings. The researchers observed that at rubber contents less than about 8%, the cured material consisted of a discontinuous

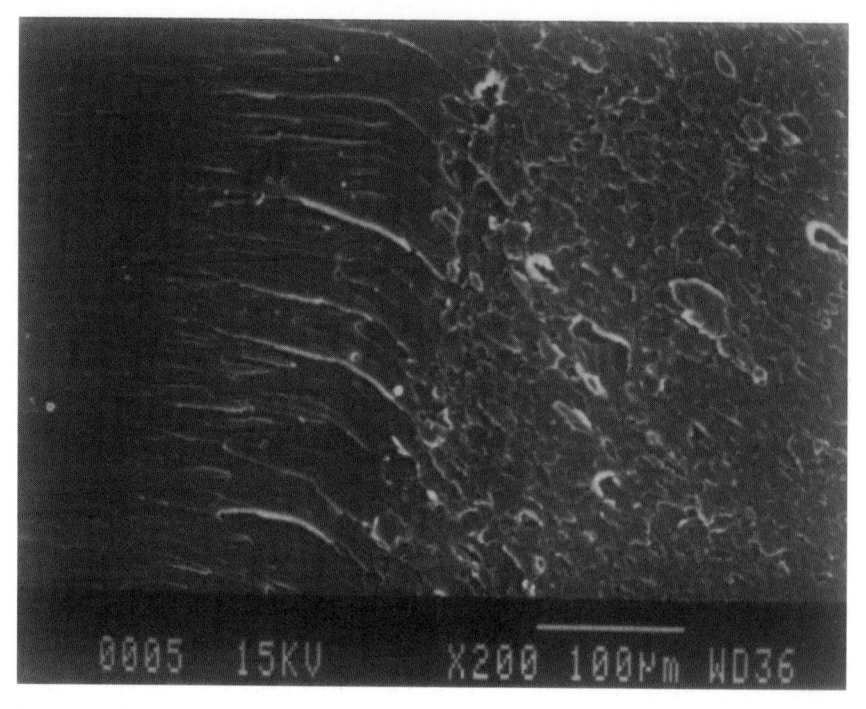

Figure 2 Fracture surface of a vinyl ester resin modified with a poly(epichlorohydrin) elastomer. At low magnification (200×) the second phase particles are not visible. Crack propagation was from left to right.

rubber containing phase in a continuous glassy matrix. The rubber domains were measured to be in the range of a few hundred Ångströms in size. Although a true interpenetrating network may not have been achieved, this type of approach may be valuable in producing large volume fractions of small particles in unsaturated polyester (UP) resins.

EFFECT OF IMPACT MODIFIER ADDITIONS ON PROCESSING

The addition of impact modifiers may affect the viscosity of the resin, its shelf life, and its rate of reaction. Liquid rubbers are low-molecular weight polymers and as a result have viscosities higher than those of unsaturated polyester resins. Because UP and VE resins are most often crosslinked with styrene monomer, liquid rubber additives are sometimes pre-blended with styrene to lower the viscosity for ease of mixing. The viscosity of the final blend may be higher than the base resin due to either the inherently high viscosity of the LR used or the formation of a prepolymer upon mixing, by the reaction of the LR with the base resin. High resin viscosities may lead to a decrease in the time to gelation.

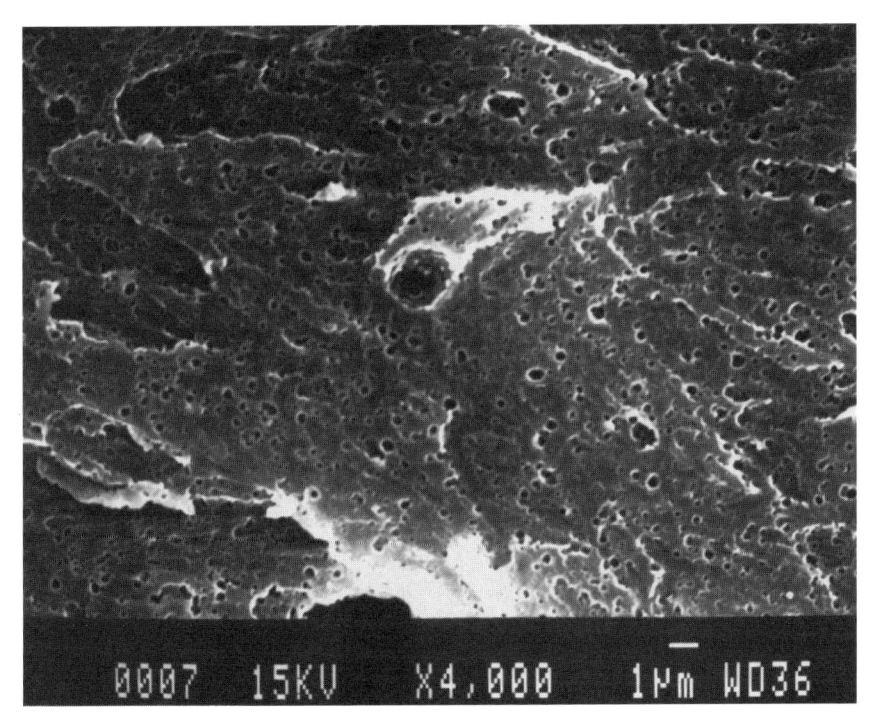

Figure 3 A highly magnified view (4000×) of the fracture surface shown in Figure 2 lightly etched with chromic acid. The fine particulate phase is visible. (Reproduced with permission from Ullett and Chartoff (1995). Copyright 1995 Society of Plastics Engineers.)

Blends that are two-phase before cure may have a limited shelf-life if gross phase separation occurs over time.

The same safety procedures used for handling UP and VE resins should be used with the handling of LR additives. The polymer additives may contain residual monomer and some may come preblended with styrene for ease of mixing. Therefore, gloves and protective clothing should be worn and work done in well ventilated areas.

METHODS FOR TOUGHNESS TESTING

Two quantities can be measured to describe a material's resistance to crack propagation: fracture toughness, K_c, and strain energy release rate, G_c. Different stress states can be applied, with the most common being the opening mode (Mode I). Materials can also be tested in in-plane shear (Mode II) and in torsional shear (Mode III). Testing in these modes typically involves strain rates much lower than those associated with impact tests such as the Charpy and Izod methods. Other methods

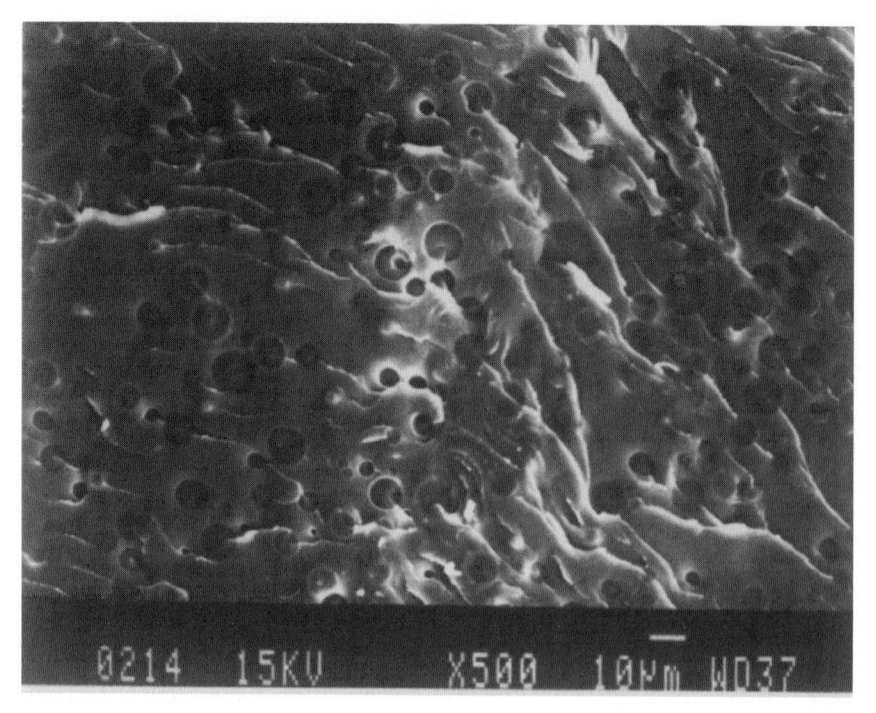

Figure 4 Fracture surface of a rubber modified unsaturated polyester showing relatively large particles which appear to have failed by stretching and tearing. Crack propagation was from left to right.

that have been used to measure the effectiveness of impact modifiers include the Gardner impact test, the Rheometrics impact test, and acoustic emission. The appropriateness of one test compared with another depends on the end-use application. When comparing reported improvements of various impact modifiers, one should compare results from the same type of test (e.g. notched Izod) generated using the same test conditions (e.g. temperature, impact speed).

TOUGHENING EFFECTIVENESS OF LIQUID RUBBERS

The toughening effectiveness of impact modifiers of any kind in unsaturated polyester and vinyl ester resins will depend on the rate of impact or applied strain, atmospheric conditions (temperature, presence of solvents), and the toughness of the base resin. As the applied strain rate increases or the test temperature decreases the effectiveness of the rubber additive may decrease. This is because the rubbery particles and the matrix resin become more brittle under these conditions. As a result, when comparing toughness data from different sources, testing conditions must be considered. The presence of fillers, fibers, and other additives may also alter the effectiveness of impact modifiers (see the section on 'Reinforced systems' below).

There are still many unanswered questions as to why one LR additive more effectively toughens a particular resin than another LR additive. Microstructural features that may affect fracture behavior such as volume fraction, particle size and size distribution, particle–matrix adhesion, and the structure of dispersed phase are difficult to control independently. As a result, the importance of each has not been quantified. Nevertheless, significant improvements in the fracture toughness of rubber-modified UP and VE resins have been realized.

Ullett and Chartoff (1995) measured improvements in fracture toughness (K_{Ic}) of up to 63% for small additions of liquid rubber to a general purpose unsaturated polyester. Crosbie and Phillips (1985a) measured increases in fracture surface energy (G_i) up to eight fold for the UP resins they studied. A resin with a base fracture energy of 95 J/m^2 modified with 10 pph carboxyl-terminated-butadiene-acrylonitrile (CTBN) had a measured fracture energy of 825 J/m^2 (double torsion test, crosshead speed 1 mm/min). In both studies, it was observed that superior toughness resulted from blends in which the rubbery particles cavitated during the fracture process. Similarly, in their study of polyurethane modified UP resins Kim and Chan-Park (1994) observed that materials containing particles that were well bonded to the matrix and that cavitated during fracture had the greatest fracture toughness values.

EFFECT OF IMPACT MODIFIERS ON OTHER BLEND PROPERTIES

Additions of liquid rubbers in quantities typically used for toughening UP resins generally result in slight to moderate reductions (<5% to 25%) in tensile and flexural stiffness (Crosbie and Phillips, 1985a; Subramanium and McGarry, 1994; Ullett and Chartoff, 1995). Tensile strength may increase or decrease depending on many factors such as rubber concentration, strength of the particle/matrix interface, and second-phase particle size.

Small to moderate additions of LRs to UP resins may result in an increase, a decrease, or in no change of the cured UP resin's upper use temperature. From results reported in the literature it is difficult to predict how a particular rubber addition will affect the blend glass transition temperature, T_g, which is an indicator of upper use temperature. Factors that may affect the T_g of the cured blend include resin chemistry, rubber chemistry, and the distribution of rubber between the phases. As the T_g of a typical LR modifier is in the subambient range an increase in the blend T_g is not expected. However, the liquid rubber molecules may act as plasticizers during cure, allowing a highly unsaturated UP resin to attain a greater crosslink density. No explanations have been provided of the examples presented in the literature of T_g increases with LR additions. Crosbie and Phillips (1985a) reported that low level additions of a butadiene acrylonitrile reactive liquid rubber increased the heat deflection temperature of an epoxy modified UP resin slightly. The same rubber addition slightly decreased (<10°C) the heat deflection temperature of a plasticized isophthalic neopentyl glycol UP resin. Ullett and Chartoff (1995) measured small decreases in T_g (<5°C) in some UP resin/rubber blends and small increases (6–9°C) in others. Subramanium and McGarry (1994) measured increases in T_g as high as 10°C for the resin/rubber system they studied.

REINFORCED SYSTEMS

Bulk molding compounds (BMC) and sheet molding compounds (SMC) have gained increased commercial importance in recent years, particularly in the automotive industry. Every year new applications are introduced. BMC and SMC products consist of a thermosetting resin, such as an unsaturated polyester, containing a filler and reinforced with glass fibers. Often the recipe includes a thermoplastic additive known as a low profile additive (LPA) to eliminate cure shrinkage and improve surface quality. Other additives may also be present to control viscosity, add color, and promote mold release.

One problem with cured SMC and BMC products, as with unreinforced UP products, is their brittleness and susceptibility to microcracking. Parts are subject to failure or visible damage under low impact conditions. In autobody-panel applications surface appearance is key. Matrix microcracking which can affect surface appearance is a problem in these materials. Additions of small amounts (e.g. 10 pph resin) of liquid rubber have been reported (McGarry, Rowe and Riew, 1978) to improve the microcrack resistance of SMC type compounds.

The level of measured improvements in toughness attributed to liquid rubber additives is somewhat dependent on the type of test performed as well as the specific SMC recipe. The toughening mechanisms in effect for rubber modified SMC materials have not been well documented. The presence of a low profile additive (LPA), mineral filler, and glass fibers affects the dispersion of the rubbery additive and its effectiveness in toughening the polyester matrix. Interactions between the rubbery additive and each of the typical SMC components have not been well researched.

The filler present in reinforced UP and VE molding compounds can itself act as an impact modifier. Calcium carbonate is the filler most commonly used in SMC or BMC formulations. Other fillers which have been used with molding compounds include silicas, hydrated aluminas, and kaolin clays. The introduction of a rigid filler to a thermoset matrix has, in general, the following effects on mechanical properties: increased modulus with increasing volume fraction of filler, increased fracture energy with increasing volume fraction of filler (often goes through a maximum), decreased strength with increasing volume fraction of filler. A mechanism proposed for hard-particle toughening is pinning of the crack front by hard particles.

A variety of different thermoplastic additives have been used with UP resins to reduce shrinkage and provide a smoother surface finish. The phase structure and morphology of a cured mixture depend on the specific thermoplastic used. The low shrink mechanism involves extensive microcrack formation. Poly(vinyl acetate) or PVAc is one of the more popular LPAs because it can eliminate shrinkage and even result in slight expansion. What is unique about the morphology of PVAc modified resins is that the PVAc forms a continuous phase which coats the UP network. Other thermoplastics, such as PMMA and PS, form a discrete particulate phase such as is formed by liquid rubber additives. There is no general agreement about the effects of LPA on the fracture toughness of UP resins. The combined effect of LPA and liquid rubber additives on the mechanical properties of unreinforced UP resins is not well documented.

APPLICATIONS

An important application for rubber modified polyester resins is in short-fiber reinforced molding compounds. Specific applications within the automotive industry include grill-opening panels and various car and truck body panels. Rubber modified vinyl ester resins are used as adhesives and as primer coatings for storage tanks.

REFERENCES

Crosbie, G.A. and Phillips, M.G. (1985a) Toughening of polyester resins by rubber modification. Part 1: Mechanical properties. *J. of Mat. Sci.*, **20**, 182–192.

Crosbie, G.A. and Phillips, M.G. (1985b) Toughening of polyester resins by rubber modification. Part 2: Microstructures. *J. of Mat. Sci.*, **20**, 563–577.

Kim, D.S. and Chan-Park, C.E. (1994) Mechanisms of modified unsaturated polyester with novel liquid polyurethane rubber. *Proceedings of the American Chemical Society Division of Polymeric Materials: Science and Engineering.* Vol. 70, American Chemical Society, Washington, DC, pp. 104–105.

McGarry, F.J., Rowe, E.H. and Riew, C.K. (1978) Improving the crack resistance of bulk molding compounds and sheet molding compounds. *Poly. Engr. and Sci.*, **18**, 78–86.

Riew, C.K. (ed.) (1989) *Rubber Toughened Plastics*, Advances in Chemistry Series 222; American Chemical Society, Washington, DC.

Subramanium, R. and McGarry, F.J. (1994) Toughened polyester: A novel system. *Comp. Inst. 49th Annual Tech. Conf. Proceedings*, Society of Plastics Industry, New York, session 16c, 1–7.

Ullett, J.S. and Chartoff, R.P. (1995) Toughening of unsaturated polyester and vinyl ester resins with liquid rubbers. *Poly. Engr. and Sci.*, **35**, 1086–1097.

Keywords: unsaturated polyester, vinyl ester, impact modifiers, liquid rubbers, toughening, microstructure, cure, cavitation.

See also: Impact modifiers: (1)–(4).

Light and UV stabilization of polymers

N.S. Allen

MECHANISMS OF PHOTOSTABILIZATION

Some form of photostabilization is essential for almost all polymers if adequate protection against the destructive effects of solar radiation is to be achieved. The photostabilization of light-sensitive polymers involves the retardation or elimination of the various photophysical and photochemical processes that occur during photo-oxidation and may be achieved in many ways, depending on the type of stabilizer and type of degradation or oxidative mechanism that is operative in the polymer. Complete, effective stabilization is, of course, never achieved in commercial practice. A further complication is the influence of processing history and this is not always fully combatted. In many cases antioxidant and light stabilizer degradation products can control the final stability of the polymer. The effectiveness of a stabilizer is measured in various ways depending upon the end-use application of the polymer. This may involve measurements of physical and mechanical property changes such as tensile strength, or chemical changes involving oxidation products such as carbonyl or hydroxyl groups. Over the years four different classes of stabilizing systems have been developed [1–5]. These rely for their stabilizing action on the presence of:

1. an ultraviolet screener;
2. an ultraviolet absorber;
3. an excited-state quencher;
4. a free-radical scavenger and/or hydroperoxide decomposer.

Plastics Additives: An A–Z Reference
Edited by G. Pritchard
Published in 1998 by Chapman & Hall, London. ISBN 0 412 72720 X

Of these it is generally believed that (3) and (4) are the most effective methods [1]. In the latter category the terms chain-breaking acceptor and chain-breaking donor have become in widespread use [5]. The former is simply a molecule possessing a free radical which is capable of accepting another free radical species while the latter normally possesses a labile atom, usually an H-atom, that will react with or terminate another free-radical species. These are exemplified below for different types of stabilizer. It is now an accepted fact that many light stabilizers are multifunctional in their mode of operation in inhibiting photo-oxidation of a polymer. However, this does not necessarily imply that the more diverse the functionality the more effective will be the stabilizer. Many other factors have to be taken into consideration such as light and heat stability of the stabilizer and its compatibility with the polymer. Current theories surrounding the mode of action of various stabilizer types will be considered.

Pigments

This involves reflecting the damaging light from the surface of the polymer through:

1. coating the surface, e.g. paint or metallizing;
2. incorporation of a pigment with high ultraviolet reflectance.

Inorganic pigments are widely used but not for stabilization [1,2]. They are used more for decorative or colour coding. Types used include the following.

Inorganic

Titanium dioxide (rutile form); the anatase form is photoactive. Zinc oxide, barium sulphate, iron oxide (red), chromium oxide, lead oxide and cadmium sulphide (can be photoactive) [1,2,5].

Organic

Phthalocyanine Blues and Greens, Quinacridone Reds, Carbazole Violet, Ultramarine Blue.

In general white pigments give better reflectance in the 300–400 nm region than coloured pigments. Metal powders are sometimes used but they can be active and act as randomly placed mirrors in the matrix. The most effective screener is Carbon Black and here the efficiency depends on the type, concentration and particle size of the pigment. Various types include Furnace Black (17–70 nm), Lamp Black (50–90 nm), Thermal Black (150–500 nm) and Acetylene Black (35–50 nm).

Many of the inorganic and organic pigments discussed earlier also absorb ultraviolet light strongly. Not all plastics will be protected to the same degree because the absorption spectrum of a particular pigment is unlikely to match every polymer in respect of its photoactivation spectrum for degradation. Some pigments are poor protectors, e.g. Ultramarine Blue. Others sensitize, e.g. cadmium sulphide (Cadmium Yellow) and anatase. In PVC, iron pigments are highly catalytic and must be avoided due to the formation of ferric chloride [1]. The performances of a pigment is often controlled by the presence of other stabilizers/anti-oxidants (see below).

Ortho-hydroxyaromatics

These compounds are often still referred to as the classical 'absorber' system because they were originally designed to absorb the ultraviolet portion of the sunlight spectrum in the range 290–400 nm, i.e. the region which is detrimental to most polymer systems. There are three major types of absorbers, namely the 2-hydroxybenzophenones, 2-hydroxyphenylbenzotriazoles, and 2-hydroxyphenyl-S-triazines (see Structures 1 to 3). The modes of action of these stabilizers are now classified into three distinct types [1–3]:

1. ultraviolet absorbers;
2. excited-state quenchers;
3. chain-breaking donors.

(1) 2-Hydroxy-4-n-octoxybenzophenone

(2) 2(2-Hydroxy-5-methylphenyl)-2H-benzotriazole

(3) 2-(2-Hydroxy-4-methyoxyphenyl)-4,6-dimethyl-S-triazine

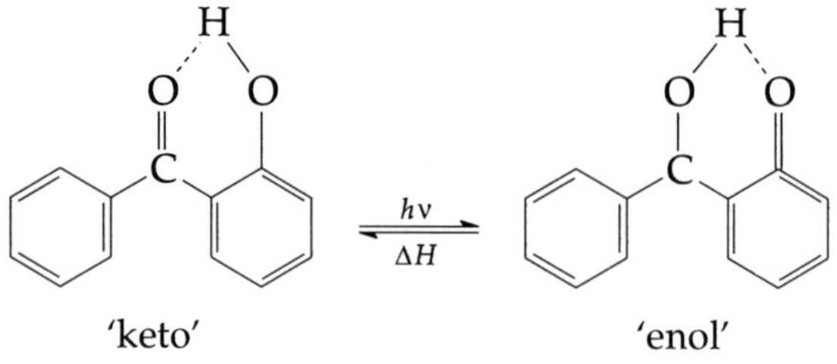

'keto' ⇌ (hν/ΔH) 'enol'

Scheme 1

Ultraviolet absorption is based on the fact that the stabilizer molecule is capable of harmlessly dissipating the absorbed light energy. Taking the 2-hydroxybenzophenones first, they dissipate their absorbed energy by a mechanism that involves the reversible formation of a six-membered hydrogen-bonded ring. The following two tautomeric forms in equilibrium provide a facile pathway for deactivation of the excited state induced by the absorption of light (Scheme 1). The result of this mechanism of light absorption and dissipation thus leaves the stabilizer chemically unchanged and still able to undergo a large number of these activation–deactivation cycles, provided, of course, that no other processes interfere. Until the advent of modern photochemical techniques direct evidence for such a mechanism was difficult to obtain because the role of the stabilizer is essentially that of a 'passive' nature. One feature that did serve as indirect evidence for this mechanism was that the more effective stabilizers exhibited correspondingly stronger intramolecular hydrogen bonding (as measured by NMR spectroscopy) with the carbonyl group and this, of course, is an essential feature of the above reaction [1–3]. However, over the years the concept of photostabilization in commercial polymers by ultraviolet absorption alone has become unacceptable from both a theoretical and practical basis. For example, for effective UV absorption all the light would have to be absorbed at the near surface of the polymer and not through the bulk because it is well-known that polymer photo-oxidation is a surface phenomenon [2]. Many workers maintain that the orthohydroxyaromatic compounds owe their stabilizing efficiency to their ability to quench the excited singlet and triplet states of impurity chromophores in polymers as simply depicted in Scheme 2. In this regard much of the evidence for excited state quenching has originated from the observation of a reduction in the fluorescence and phosphorescence emission intensities and lifetimes of impurity species such as carbonylic groups in polymer systems. However, whilst quenching has been demonstrated on a number of occasions no correlation has been observed between the quenching efficiency and stabilizing activity of a range of stabilizer structures [1–3].

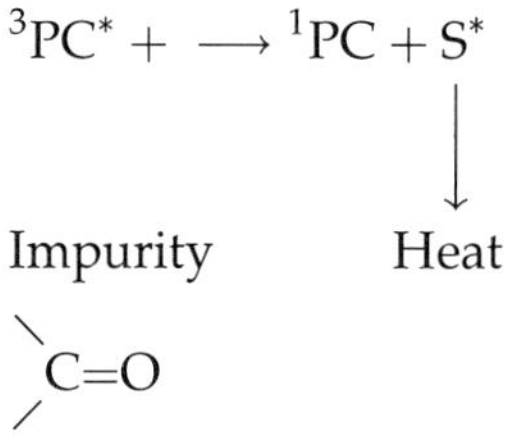

$$^3PC^* + \longrightarrow {}^1PC + S^*$$

Impurity   Heat

$$\underset{/}{\overset{\backslash}{C}}=O$$

Scheme 2

The photostabilizing efficiency of orthohydroxyaromatic compounds has been found to be dependent on processing history and this leads us on to their mode of action as chain-breaking donors. A reduction in photostabilizing activity after processing has been associated with the formation of hydroperoxides which on photolysis give alkoxy and hydroxy radicals that are capable of abstracting the ortho-hydroxyl hydrogen atom by the mechanism shown in Scheme 3. The radical product (A) is no longer capable of undergoing a facile process of intramolecular hydrogen atom transfer [1–3,5]. This mechanism is induced by light as 2-hydroxybenzophenones have no reaction with hydroperoxides in the dark [3].

These stabilizers still have widespread use, but mainly in conjunction with other light stabilizers such as the hindered piperidines. Only in short-term applications do they tend to be used alone in conjunction, of course, with a hindered phenolic antioxidant and/or organic phosphite for process stability [1–3].

Para-hydroxybenzoates

Phenyl-substituted p-hydroxybenzoates (Structure 4) have been used as light stabilizers in polyolefins for some time. They were shown to operate through a photo-Fries rearrangement to give a more effective 2-hydroxybenzophenone product by the following reaction sequence (Scheme 4) to give 2-hydroxybenzophenones as products [1,2]. Obviously for such a reaction to occur, the original stabilizer must absorb appreciably in

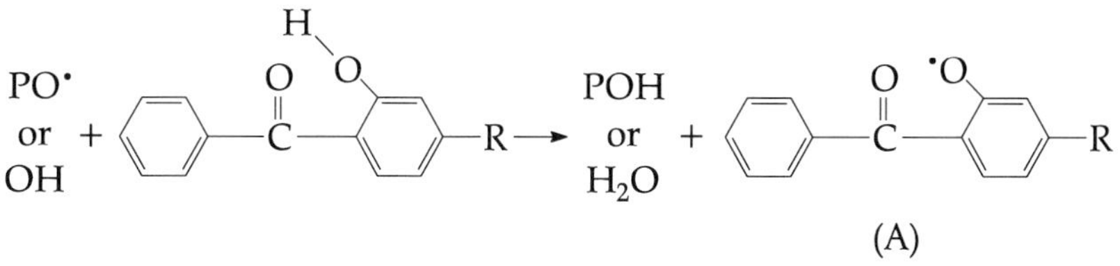

PO•
or +
•OH

POH
or +
H₂O

(A)

Scheme 3

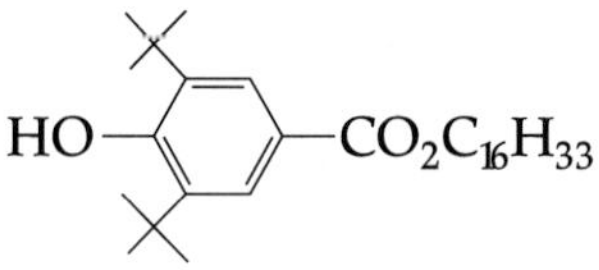

(4) Cetyl ester of 2,6-ditertbutylphenol

the near ultraviolet region and be effectively converted into a product which operates as the light stabilizer. Research studies to date show two mechanisms to be operative [1–3]:

1. a chain-breaking hydrogen donor particularly for alkoxy and hydroxyl radicals;
2. inhibition of auto-oxidation during processing.

The stabilizer is non-absorbing in sunlight and therefore its concentration dependence on simulated sunlight exposure is a function of its reaction with free radicals in the polymer. The advantages of this type of stabilizer in practical use includes its ability to synergize with phosphites and hindered piperidine compounds, especially when used in conjunction with pigments, notably inorganic types [2].

Metal complexes

Despite the advent of the newer and more efficient hindered piperidine stabilizers (see below), metal complexes still receive some attention in the polymer stabilization field, particularly in applications where the former are inefficient, e.g. in agricultural use where pesticides can have a detrimental effect on performance. Metal complexes have probably attracted by far the most controversial arguments in the field

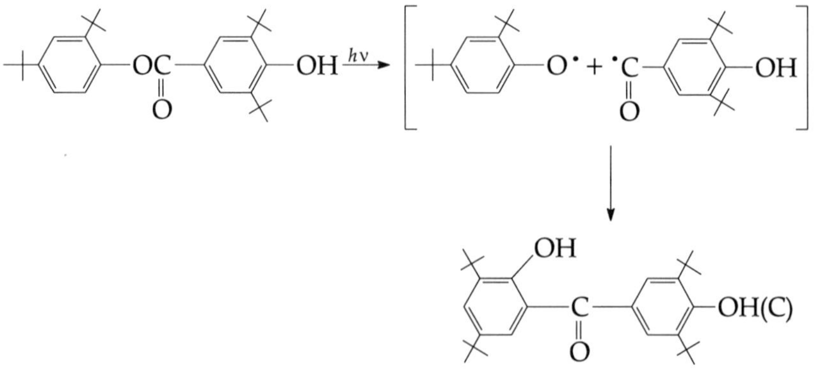

Scheme 4

particularly in the 1970s. However, with the realization of multifunctionality of many types of stabilizers, metal complexes have acquired a list of feasible functions; not all are applicable, of course, in every structural case [1–3].

1. Hydroperoxide decomposition in a dark reaction at ambient temperatures or during processing by:
   (a) the complex itself;
   (b) decomposition products of the metal complex.
2. Radical scavenger during processing/photo-oxidation.
3. Singlet-oxygen quencher.
4. Excited-state quencher.
5. Ultraviolet absorption.

The development of metal complexes, particularly those based on nickel, resulted in compounds which exhibited relatively low extinction coefficients in the near ultraviolet region and yet in many instances were found to be superior in performance to the currently available ortho-hydroxyaromatic compounds [3,5]. This finding resulted in a search for some other feasible mechanistic function. As carbonyl photolysis was considered to be an important initiation process which gives rise to both free radicals and backbone cleavage, then deactivation or quenching of the excited-state precursor should prove an effective means of photoprotection. Effective structures in this respect were tris(dibenzoylmethanato) chelates of Fe and Cr (Structure 5), nickel oxime chelates (Structure 6), and the nickel complex of Structure 7, nickel(II) 2,2′-thiobis(4-t-octylphenolato)-n-butylamine.

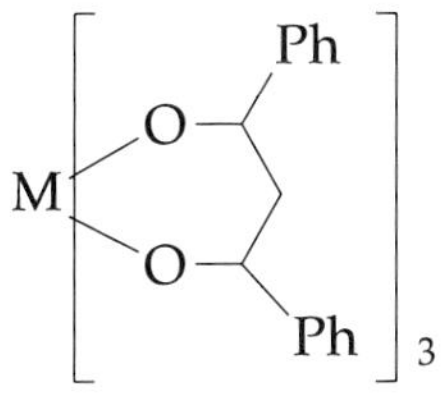

(5) Tris(dibenzoylmethanato)chelate of Cr or Fe

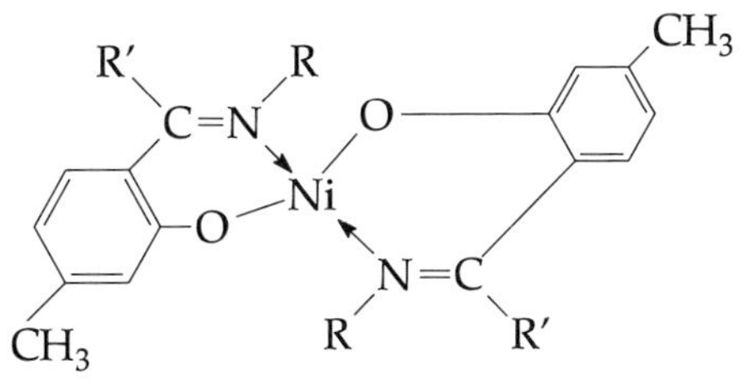

R′ = CH$_3$

(6) Ni oxime chelate

(7) Ni(II) 2,2'-thiobis(4-octylphenolato)-*n*-butylamine

R = −C(CH$_3$)$_2$CH$_2$C(CH$_3$)$_3$

Excited-state quenching has, however, been strongly disputed as an important stabilizing mechanism, particularly on a practical basis where it is calculated that up to 10% w/w of uniformly distributed stabilizer would be required in order to produce effective quenching.

Apart from the ability of metal complexes to quench the excited states of photoactive chromophores they have also been found to be effective quenchers of active excited singlet oxygen, especially those chelates containing sulphur donor ligands such as Structure 7. The interrelationships between the structure of the metal complex and singlet-oxygen quenching efficiency are far from being understood however, and although there is no doubt about the ability of some complexes to quench singlet oxygen the absence of large amounts of unsaturation in polyolefins and many other polymers suggests that this stabilizing function is insignificant in the overall process [3].

Essentially, the mode of action of metal complexes is now associated almost wholly with their ability to prevent the formation of, or catalytically to destroy, hydroperoxides either during processing and/or photo-oxidation. Peroxide decomposers fall into two classes, namely stoichiometric reducing agents and catalytic hydroperoxide decomposers. Catalytic hydroperoxide decomposers apparently destroy hydroperoxides through the formation of an acidic product in a radical generating reaction involving the hydroperoxides. Here the parent compound acts as a reservoir for the 'real' antioxidant. A wide variety of sulphur-containing antioxidants fall into this category, including Structure 7. Zinc and nickel complexes of the dithiocarbamates, dithiophosphates, xanthates, mercaptobenzothiazoles, and diothiolates all destroy hydroperoxides by a peroxidolytic mechanism. In all cases there is a pro-oxidant stage, the contribution of which is a function of the particular anti-oxidant. Scheme 5 is typical for the metal dialkyldithiocarbamates. The sulphur is oxidized to sulphur oxides and eventually sulphur dioxide, which are effective in destroying hydroperoxides [5].

Most of these types of metal complexes owe their light-stabilizing action to their ability to operate as effective thermal antioxidants as well as extending this behaviour to a light-catalysed reaction. Many

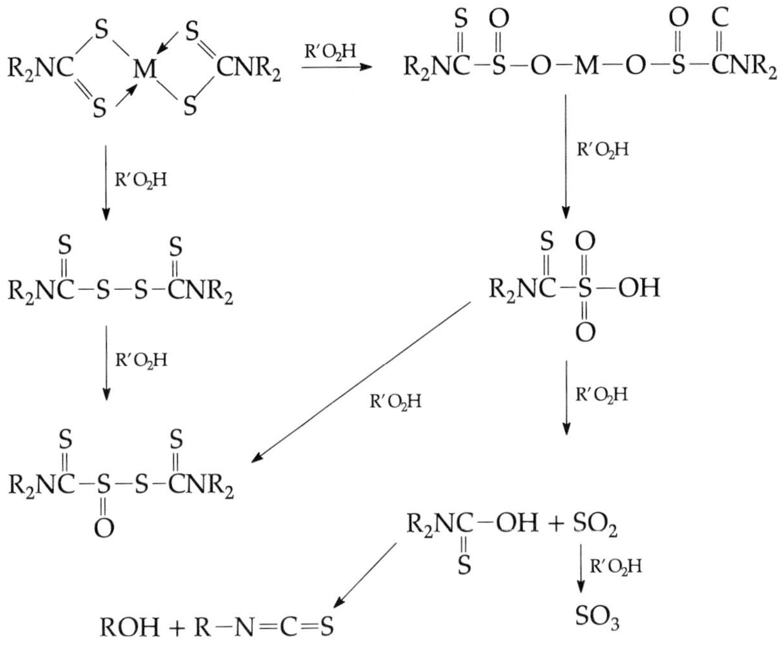

Mechanisms of photostabilization 435

[Scheme 5 - chemical reaction scheme showing photostabilization mechanisms involving R₂NC-S-S-CNR₂ metal complexes and their oxidation products]

Scheme 5

correlations have also been observed between actual processing history and light stabilizing performance of metal complexes.

Hindered piperidine compounds

This is the newest and most efficient class of light stabilizer and, in fact, is now widely established in many polymer applications. A range of different commercially acceptable monomeric and polymeric types for polyolefins are now available and are shown in Structures 8 and 9 as examples of monomeric and polymeric types. These light stabilizers have also attracted much controversy with regard to mode of action, and these mechanisms may be summarized as [1–4]:

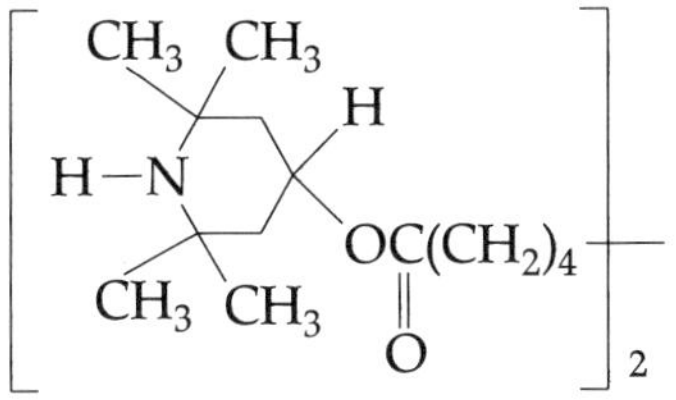

(8) Bis(2,2,6,6-tetramethyl-4-piperidinyl)sebacate

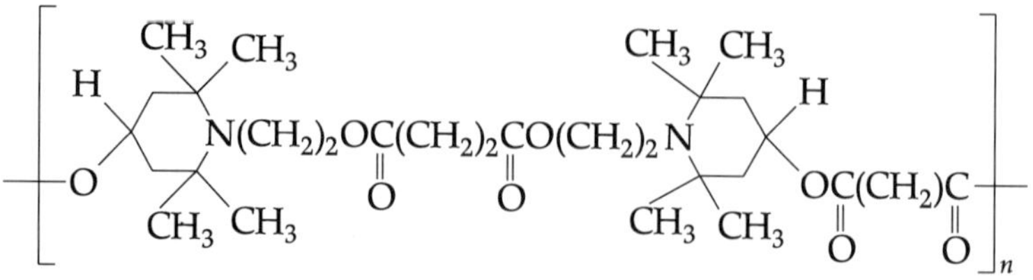

(9) Polyester of succinic acid with N-β-hydroxyethyl-2,2,6,6-tetramethyl-4-hydroxypiperidine

1. chain-breaking donor/acceptor redox mechanism through the nitroxyl/substituted-hydroxylamine intermediates;
2. decomposition of hydroperoxides by the amine during processing;
3. inhibition of photoreaction of α,β-unsaturated carbonyl groups in polyolefins;
4. reduction in quantum yield of hydroperoxide photolysis;
5. singlet-oxygen quenching – polydienes only;
6. complexation with hydroperoxides/oxygen;
7. complexation with transition-metal ions;
8. excited-state quenching by the nitroxyl radical.

These compounds certainly exhibit no absorption in the near ultraviolet region and are ineffective excited-state quenchers, although there is some evidence to suggest that in polybutadiene they quench excited singlet oxygen. The overall stabilizing efficiency of these compounds is associated with the following cyclic (regenerative) mechanism (Scheme 6) [1,3–5].

Here the amine is initially oxidized through to the nitroxyl radical by reaction with hydroperoxides or via the breakdown of an amine oxygen complex. There is much evidence in the literature in favour of this

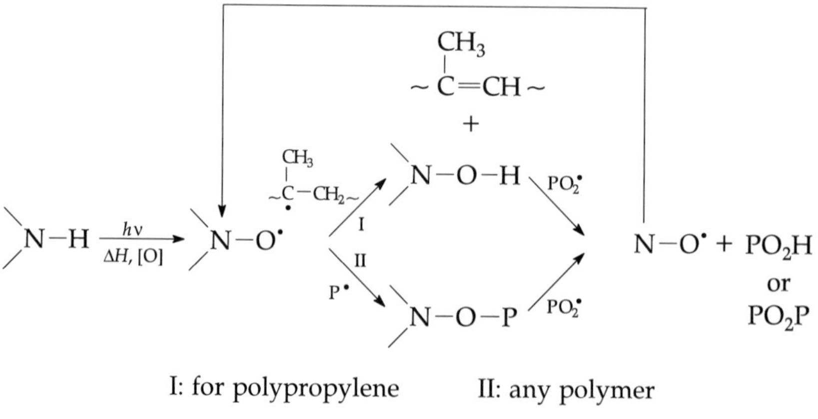

I: for polypropylene   II: any polymer

Scheme 6

$$\diagdown\!\!\!\!\!\diagup\!\!\text{NOR} + \text{R}'\text{OO}^{\bullet} \longrightarrow \diagdown\!\!\!\!\!\diagup\!\!\text{NO}^{\bullet} + \text{R}{=}\text{O} + \text{R}'\text{OH}$$

This basic mechanism is also proposed for acylperoxy radicals

$$\diagdown\!\!\!\!\!\diagup\!\!\text{NOR} + \text{R}'\text{C(O)OO}^{\bullet} \longrightarrow \diagdown\!\!\!\!\!\diagup\!\!\text{NO}^{\bullet} + \text{R}{=}\text{O} + \text{R}'\text{COOH}$$

Scheme 7

mechanism occurring in the polymer at elevated temperatures, despite room-temperature mode-system studies to the contrary.

During irradiation the concentration of nitroxyl radicals grows rapidly but thereafter falls to a very low steady-state value of 10^{-4} M. This low level of nitroxyl radicals is thought to be insufficient to account for the high photoprotective efficiency of hindered piperidines. Much of the nitroxyl is converted into either the hydroxylamine (N–OH) or substituted hydroxylamine (N–O–P) produced by reaction of the nitroxyl radicals with different types of macroalkyl radicals. These reactions are now well established, and, in fact, may occur thermally or photochemically. The hydroxylamine and substituted hydroxylamines are now clearly established as the most effective stabilizing intermediates and act as reservoirs for nitroxyl radicals by reaction with peroxy radicals. This latter reaction is somewhat controversial because for each nitroxyl radical generated one molecule of hydroperoxide is produced. However, more recent studies have now shown that this latter reaction proceeds to produce ketones, alcohols and carboxylic acids as products and not hydroperoxides (Scheme 7). Assuming that the former ketones are inactive then this would account for the efficiency of the hindered piperidine compounds. Low molecular weight ketones in polymers are non-active species because they are not integrated into the chain and hence the Norrish reactions are unimportant [1–3]. The main problem in this area has been identified and provides confirmation of the involvement of the substituted hydroxylamine in stabilization.

The above cyclic mechanism has been carefully scrutinized by many workers in the past few years, and, in fact, it has been concluded that it alone cannot fully account for the high photoprotective efficiency of the parent amine molecule. The nitroxyl radical itself is a radical scavenger but is not as effective as hindered phenols in competing with oxygen for alkyl radicals. To account for this deficiency in the cyclic mechanism it has been suggested, and indeed confirmed, by many workers, that hindered piperidine stabilizers and their derived nitroxyl radicals form weakly bonded localized complexes with hydroperoxides in the polymer (Scheme 8). This mechanism raises the local concentration of nitroxyl radicals in regions where alkyl radicals are generated after the

$$\text{>N-O}^{\bullet} \quad \text{P-H} + \text{PO}_2\text{H} \rightleftharpoons \text{>N-O}^{\bullet} \quad \text{PO}_2\text{H} + \text{P-H}$$

Scheme 8

photocleavage of hydroperoxide groups, followed by, of course, hydrogen atom abstraction from the polymer by the derived hydroxyl/alkoxy free radicals. Under these conditions the nitroxyl radicals would then effectively compete with oxygen for the alkyl radicals. Evidence for the association mechanism originates from infrared solution studies, and from the observation of an increase in stabilizer absorption from solution by oxidized polypropylene containing higher concentrations of hydroperoxide groups. One recently observed effect of this mechanism is to reduce the quantum yield of hydroperoxide photolysis.

Reaction of the parent amine with transition-metal ions is also a possibility. The hindered piperidines have also been found to inhibit the photolysis of luminescent α,β-unsaturated carbonyl groups (impurities) in polypropylene.

SYNERGISM AND ANTAGONISM

Synergism

These are important factors which need to be considered in the overall final package. Many light stabilizers and anti-oxidants give synergism. There are two complementary effects here [1–5] as follows.

1. Absorbers prevent the antioxidant from undergoing photolysis.
2. Antioxidants inhibit/destroy hydroperoxides (POOH) which can attack the absorber via PO$^{\bullet}$ and $^{\bullet}$OH radicals.

Another example of synergism is the interaction between metal complex light stabilizers and secondary phosphite antioxidants. Here the phosphite will destroy the hydroperoxide during processing and protect the metal complex. There is also evidence of a ligand exchange where some nickel phosphite is produced. Pigments synergize with antioxidants such as lead chromate and iron oxide. Some pigments have the ability to absorb additives onto the particle surface where it can do the most good or possible harm (antagonism) [1,2,5].

Antagonism

Interaction with hindered piperidine light stabilizers

Despite the high efficiency of these stabilizers they often interact unfavourably with many other additives used in commercial polymers for various purposes, such as antioxidants and fire-retardants. Hindered phenolic antioxidants vary widely in terms of their interactions with hindered piperidine compounds depending on the nature of the polymer and processing history. A recent wide ranging survey of phenolic antioxidants and hindered piperidine compounds and their interactions during thermal and photochemical oxidation in both polypropylene and high density polyethylene has shown that during thermal oxidation (oven ageing) the interactions are seen to be synergistic in most cases, whereas on photo-oxidation the majority of the effects are seen to be antagonistic [1,5]. For high density polyethylene the effects were also highly synergistic during oven ageing, whereas during photo-oxidation the effects were found to be variable, both synergism and antagonism being operative. In many cases the stability of the phenolic antioxidant itself was the determining factor in controlling performance. Synergism thermally and photochemically is probably associated with the fact that the generated nitroxyl and hydroxylamine products from the parent amine are scavenging macroalkyl and macroperoxy radicals and protecting the phenolic antioxidant.

Antagonism on photo-oxidation is associated with any of the following four processes [1–5]:

1. oxidation of the phenol to an active quinone by the nitroxyl radical (Scheme 9);
2. inhibition of hydroperoxide formation by the phenolic antioxidant, thus preventing the cyclic mechanism above;
3. reaction of the nitroxyl radicals with radical intermediates from the phenol (Scheme 10);
4. formation of a salt between the acidic phenolic antioxidant and the basic piperidine stabilizer.

These would effectively remove nitroxyl free radicals from the cyclic mechanism above. With thioesters there is a very strong antagonism, both thermal and photochemical, which is associated with reaction of

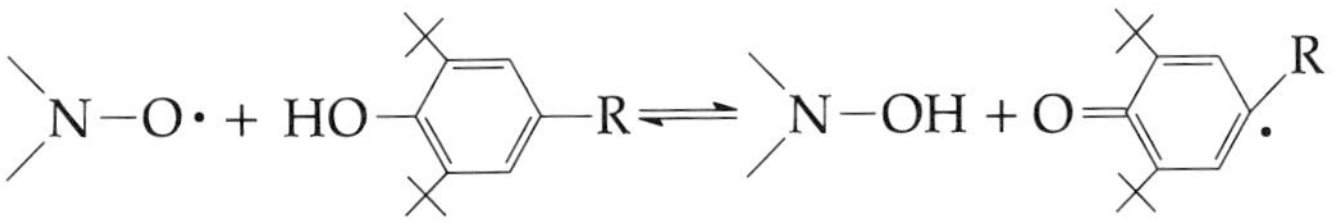

Scheme 9

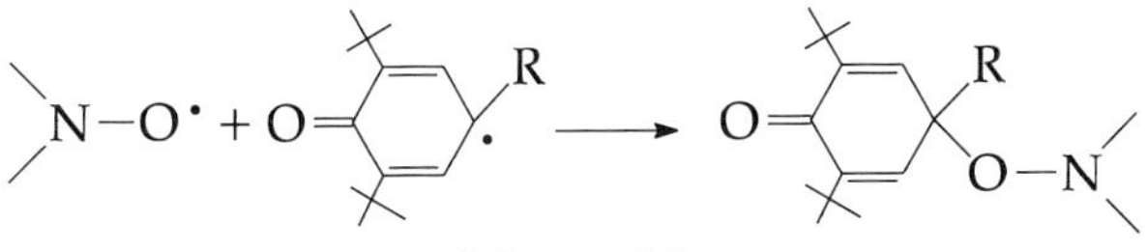

Scheme 10

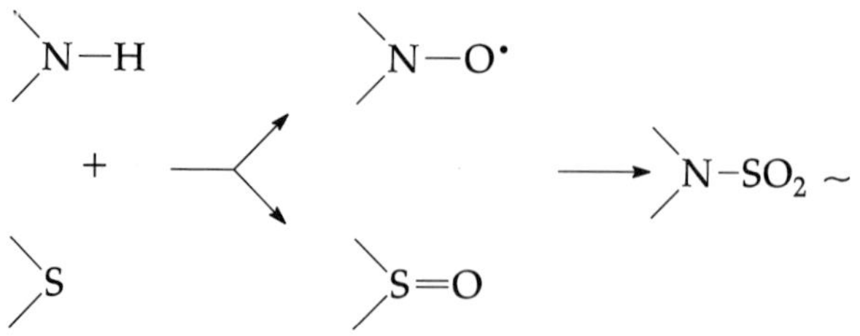

$$\begin{array}{c} \diagup \\ \diagdown \end{array}\!\!N\!-\!H \qquad \begin{array}{c} \diagup \\ \diagdown \end{array}\!\!N\!-\!O^\bullet$$

$$+ \qquad \longrightarrow \qquad \begin{array}{c} \diagup \\ \diagdown \end{array}\!\!N\!-\!SO_2\!\sim$$

$$\begin{array}{c} \diagup \\ \diagdown \end{array}\!\!S \qquad \begin{array}{c} \diagup \\ \diagdown \end{array}\!\!S\!=\!O$$

Scheme 11

nitroxyl radicals with sulphenyl radicals to give inactive sulphonamides (Scheme 11).

Another important interaction is that between absorbers such as 2-hydroxybenzophenones and 2-hydroxybenztriazoles and hindered piperidine compounds. Comparisons of embrittlement data in the literature indicate that antagonism is operative although this tends to conflict with an observed protective effect by the hindered piperidine compound on the stability of the absorber. This protective effect is due to the ability of the hindered piperidine compound to decompose and remove any hydroperoxides which may be detrimental to the absorber through Scheme 3.

FACTORS CONTROLLING CHOICE OF STABILIZER [1–3]

1. Compatibility: Many stabilizers exude from the polymer during fabrication, storage or irradiation. One way of preventing this is to incorporate long alkyl chains into the stabilizer molecule which acts as a link.
   Other ways include:
   (a) grafting onto polymeric chains;
   (b) polymeric stabilizers;
   (c) copolymerizable stabilizers.
2. Stability to light: Many stabilizers are photolysed early in the degradation process and the products are often better light stabilizers.
3. Stability to high temperature processing, e.g. some nickel chelates containing sulphur donor ligands will decompose above 250°C and turn black due to nickel sulphide.
4. The stabilizer must not react in an unfavourable manner with the polymer. Nickel complexes cannot be used in nylon polymers because the nickel complexes with the amide groups.
5. Low toxicity.
6. Colour.
7. Cost should be the lowest possible, consistent with stabilizer performance.

REFERENCES

1. N.S. Allen and M. Edge (1992) 'Fundamentals of Polymer Degradation and Stabilisation', Chapman and Hall, London.
2. J.F. Mckellar and N.S. Allen (1979) 'Photochemistry of Man-Made Polymers', Elsevier Science Publishers Ltd.
3. J.F. Rabek (1990) 'Photostabilisation of Polymers', Chapman and Hall, London.
4. N.S. Allen (Ed.) (1983) 'Degradation and Stabilisation of Polyolefins', Chapman and Hall, London.
5. G. Scott (Ed.) (1990) 'Mechanisms of Polymer Degradation and Stabilisation', Chapman and Hall, London.

Keywords: photostabilization, degradation, ultraviolet, hydroperoxide, free radical, pigment, singlet-oxygen, hindered piperidine, nitroxyl.

See also: Hindered amine light stabilizers: an introduction;
Hindered amine light stabilizers: recent developments;
Antioxidants.

Low profile additives in thermoset composites

Kenneth E. Atkins

INTRODUCTION

The use of fiber reinforced unsaturated polyester resin composites in the form of sheet molding compound (SMC), bulk molding compound (BMC) and preform molding exceeded 475 000 tons (approximately one billion pounds) in North America and Europe in 1995. Many applications were in the transportation industry such as automotive body panels, truck components and semi-structural parts (Table 1). Other applications are in sanitary ware (bathtubs, showers, sinks), applicances, business machines and electrical components.

A key to the success of these composites has been the development of 'low profile additive' technology. These additives are little-known outside the industry but have had a major impact on composite use.

When unsaturated polyester fiber reinforced composites are cured and molded without 'low profile' additives, considerable shrinkage (~7 volume % on organics) results. This yields molded parts with:

- major warpage and the inability to hold tolerances;
- poor mold reproduction (rough surface) and fiber pattern;
- internal cracks and voids;
- a depression ('sink') on the surface opposite reinforcing ribs and bosses.

Such defects made it impossible to consider these materials for their current applications.

'Low profile' additives are specially designed and controlled thermoplastic polymers which provide shrinkage control when introduced into

Plastics Additives: An A–Z Reference
Edited by G. Pritchard
Published in 1998 by Chapman & Hall, London. ISBN 0 412 72720 X

Introduction

Table 1 Examples of low profile transportation applications

North America

Company/Model	Application
Ford Mustang	Hood
Chrysler Sebring JX	Hood
Ford Crown Victoria	Grill Opening Panel
General Motors Chevrolet Camaro	Doors
	Roof
	Hatch
	Spoiler
Ford Taurus/Mercury Sable	Integrated Front End Module
Ford Aeromax Truck	All Cab Panels
Dodge Caravan	Cowl Plenum
General Motors	Head Lamp Reflectors

Europe

Company/Model	Application
Citroen Xantia	Hatchback
Renault Espace	Hood, Hatchback Bumper
IVECO Eurocargo, Eurotech, Eurostar	Hood, Bumper, Fender, Step Panel, etc.
Mercedes 17–35 Tons	Front Grille Front Spoiler
Scania Saab G	Front Grille Air Deflector
Alfa Romeo Spider	Hood
Citroen Xantia, Citroen XM, Peugeot 405, Peugeot 605	Front End Panel

these thermosetting materials. Typical BMC formulations with and without low profile additive are shown in Table 2.

While the unsaturated polyester resins can vary widely in structure, the ones that perform the best with 'low profile' additives contain high levels of unsaturation. One of the most commonly used is based on 1.0 mole of maleic anhydride and 1.0 mole of propylene glycol. The maleate is largely isomerized to fumarate during the condensation reaction. The resultant polymer is dissolved in a crosslinking monomer (normally styrene) at about 60–65% unsaturated polyester and 35–40% monomer. This is the form used commercially.

Low profile additives in thermoset composites

Table 2 BMC formulations

	Parts by weight	
Unsaturated polyester resin*	26.1	15.7
Low profile additive in styrene†	–	10.3
Calcium carbonate	52.6	52.7
Zinc stearate	1.0	1.0
t-butylperbenzoate	0.3	0.3
Glass fibers (12.5 mm chopped)	20	20
Shrinkage control (linear shrinkage) mils/inch	4.0	0.0

Panels molded at 3 mm thickness at 150°C/1000 psi (68 bars) for 2 minutes

* Unsaturated polyester alkyd containing about 30 wt% styrene.
† A 40 wt% solution of poly(vinyl acetate) in styrene.

SHRINKAGE CONTROL

'Low Profile Additives' are additives consisting of thermoplastics. They got their name from a test used early in their development to measure performance. Because these materials decreased shrinkage when added as a styrene solution they allowed for better mold reproduction. A profilometer used to measure steel smoothness was employed on the molded panels to trace the surface. The composites with the best mold reproduction from smooth surfaces gave fewer and lower 'peaks and valleys', hence a 'lower profile'. Examples of profilometer tracings with and without 'low profile additives' are shown in Figures 1 and 2.

The type of thermoplastic used as a 'low profile' additive can greatly influence the shrinkage control within the resultant composite. The earliest polymers identified were non-polar materials such as polystyrene

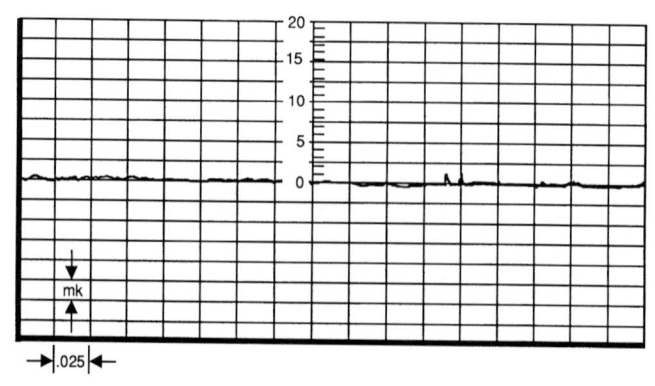

Figure 1 Surface profile of standard molding: with low profile additives (units: µ inch/inch).

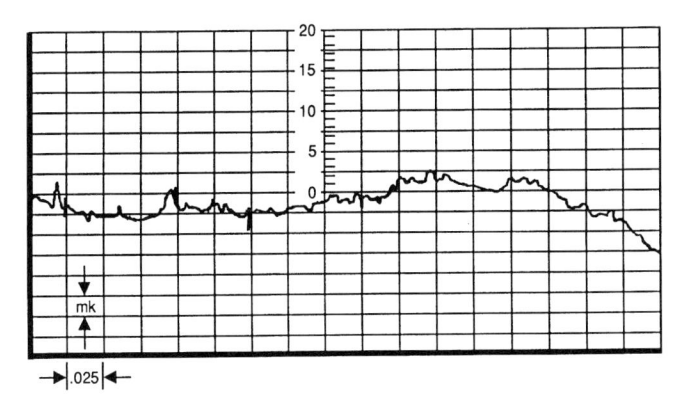

Figure 2 Surface profile of standard molding: without low profile additives (units: μ inch/inch).

and polyethylene. These substances reduced the shrinkage of the composite by about half. However, this was still not satisfactory for many applications. More polar polymers came into use, particularly polymethylmethacrylate and poly(vinyl acetate). These materials allowed essentially 'zero shrinkage' parts, leading to many new applications, particularly with poly(vinyl acetate). Major examples of these applications were the automotive grill opening panels in the USA, truck body parts in the USA and Europe and injection molded rear or hatch back doors in Europe. Figure 3 shows the comparative shrinkage control capability of several thermoplastic materials.

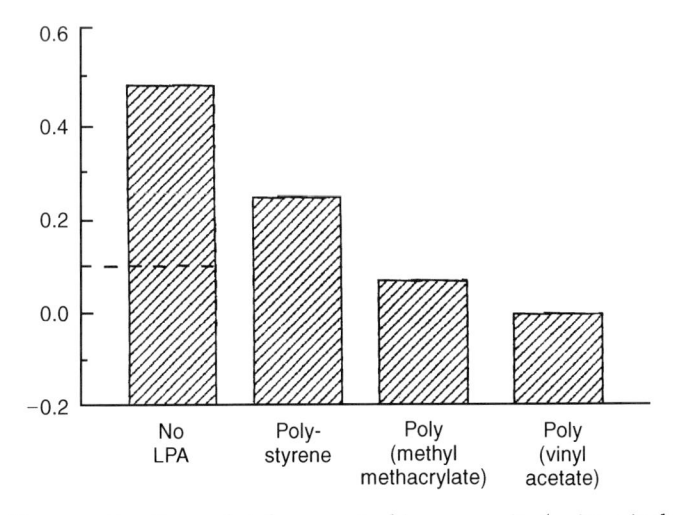

Figure 3 Comparative linear shrinkage control in composite (units: μ inch/inch).

Table 3 SMC formulations

	Parts by weight
Unsaturated polyester (60% in styrene)	60
Low profile additive (40% in styrene)	40
Calcium carbonate (3–5 µm)	200
t-Butylperbenzoate (peroxide initiator)	1.5
Zinc stearate (mold release)	4.0
Magnesium oxide (chemical thickening agent)	As needed (0.5–1.0)

Because of their lesser ability to control shrinkage, the non-polar polymers such as polystyrene and polyethylene are often classified as 'low shrink' rather than 'low profile' additives. Usually, low profile additives are supplied as 30–40% polymer solutions in styrene monomer. Polyester resin manufacturers also package the low profile additives dissolved in their resins. These are referred to as 'one pack' systems. As the industry has expanded, other thermoplastics have been identified which have shrinkage control properties. These are also now used commercially in a variety of applications. Examples of these other polyers are saturated polyesters, polyurethanes, stryene-butadiene copolymers and polycaprolactones. Poly(vinyl acetate) based materials are probably still the most used 'low profile' additives, being useful with the broadest range of unsaturated polyester resin structures. Relative proportions of the organics used in most formulations are 30–50% polyester alkyd, 10–20% thermoplastic and 40–50% styrene.

A modification of the low profile additive became necessary with the advent of chemically thickened sheet molding compound (SMC) materials. This process involves mixing all formulation ingredients (unsaturated polyester resin, low profile additive, styrene, filler, catalyst, mold release agent and thickening agent) except the fiberglass. This paste is then combined with the fiberglass on a continuous machine which makes SMC. Generally the fiberglass content is 20–30 wt% and length about 25 mm. A typical formulation is shown in Table 3.

RESIN THICKENING

A key to this material is the chemical thickening reaction that takes place during a few days' maturation at near room temperature. The paste viscosity at the time of compounding is 15–50 000 cps whereas at the time of molding it is 10 000 000 to 50 000 000 cps. During this increase the compound converts from a sticky, difficult to handle material to a dry, leather-like sheet which can easily be cut into appropriate sizes for molding. Also this increase in viscosity allows for good fiberglass reinforcement distribution (or 'carry') throughout the molded part at typical compound mold coverages of 25–40%.

The chemical thickening reaction that takes place involves reaction of the carboxyl groups on the unsaturated polyester resin with alkaline earth oxides or hydroxides such as magnesium oxide, magnesium hydroxide, calcium hydroxide or calcium oxide as illustrated below.

$$\text{UPE}-\overset{\overset{\displaystyle O}{\|}}{C}-O-Mg-O-\overset{\overset{\displaystyle O}{\|}}{C}-\text{UPE}$$

UPE = unsaturated polyester

However, when low profile additives were applied to these new chemically thickened formulations rather than the BMC and preform unthickened systems, problems were observed. The reaction of the unsaturated polyesters' carboxyl groups with the magnesium oxide changed the compatibility between the resin and low profile additive. This caused a separation of the components which resulted in an 'exudate' on the surface of the SMC. Not only did this complicate the compound handling but caused molding defects such as scumming, sticking and porosity.

While numerous approaches were tried, the solution to these problems was to modify the low profile additive polymer with carboxyl groups. This allowed these polymers to enter into the chemical thickening reaction as shown below.

$$\text{UPE}-\overset{\overset{\displaystyle O}{\|}}{C}-O-Mg-O-\overset{\overset{\displaystyle O}{\|}}{C}-\text{LPA}$$

It is critical that the level of carboxyl in the low profile additive be carefully controlled. Because of the generally higher molecular weight of the low profile additive compared with the unsaturated polyester resin it has a profound effect on thickening rate and level when it contains carboxyl groups. Too high a level can give unacceptable processing properties and can significantly reduce shrinkage control (see below). Obviously, too low a level does not produce the desired effects.

Levels of carboxyl content must be worked out with each thermoplastic structure and molecular weight. With a poly(vinyl acetate) of about 100 000 molecular weight a copolymer of about 99.0% vinyl acetate and 1.0% acrylic acid is quite effective in chemically thickened applications.

MECHANISM

The mechanism of shrinkage control in these systems is complex. Studies generally agree on the following points.

1. The thermoplastic and crosslinked unsaturated polyester materials must form separate phases.

2. Microvoids are formed in the phases due to the differential in thermal shrinkage between the phases as the molded parts cool. There is some evidence that the number and size of these microvoids influences shrinkage control.
3. Consideration of the polarity and thermal coefficient of expansion of various thermoplastics can explain differences in shrinkage control performance. For example, poly(vinyl acetate) occupies a greater specific volume than poly(methyl methacrylate), corresponding to better shrinkage control with poly(vinyl acetate). Polystyrene being much less polar than these materials does not separate as efficiently from the relatively non-polar crosslinked polyester matrix, hence it is not as good a shrinkage control agent.
4. Too high a level of carboxyl in the low profile additive in chemically thickened systems can result in too much compatibility through these bonds with the unsaturated polyester resin. This results in less two-phase formation and hence poorer shrinkage control.
5. The phase separation and hence shrinkage control is much more efficient at elevated temperatures such as 130–180°C compared with 25–80°C. However, progress has recently been made in the lower temperature areas.

The phase separation necessary for efficient shrinkage control causes problems when there is a desire to pigment these composites internally. At the best shrinkage control, a non-uniformity of color (mottling), can result, as well as a distinct reduction in color depth (hazing). The 'low shrink' additives such as polystyrene and polyethylene have provided a compromise, giving some shrinkage control with reasonable pigmentation. Recently progress has been made towards true 'zero shrinkage' pigmentable systems by special co-extrusion of poly(vinyl acetate) with certain pigments prior to introducing them into the formulation. This provides both the low profile additive and pigment in one package. Work has also been done in pre-pigmenting a filler package to provide better performance.

A list of suppliers of low profile additives is given in Table 4.

Table 4 Low profile additive suppliers

Union Carbide Corporation	Danbury, CT, USA
Union Carbide (Europe) S.A.	Meyrin, Switzerland
Ashland Chemical Co.	Dublin, OH, USA
BASF	Ludwigshafen, Germany
AOC	Colliersville, TN, USA
Cray Valley	Paris, France
Jotun Polymer	Sandefiord, Norway

BIBLIOGRAPHY

1. Kia, H. (1993) *Sheet Molding Compound – Science and Technology*, Hanser Publishers.
2. Paul, D.R. and Newman, S. (1978) *Polymer Blends*, Volume 2, Academic Press.
3. Kikelaar, M., Wang, B. and Lee, L.J. (1994) Shrinkage Behavior of Low Profile Unsaturated Polyester Resins. *Polymer*, Vol. 35, No. 14, pp. 3011–3021.
4. Atkins, K.E. and Rex, G.C. (1994) Internal Pigmentation of Low Profile Composites, Part III, *Proceedings of the Society of Plastics Industry Composites Institute 49th Annual Conference*, Section 13-D.

Keywords: low profile additive, unsaturated polyester resin, sheet molding compound, bulk molding compound, shrinkage control, fiber reinforcement, composite.

Lubricating systems for rigid PVC

Joseph B. Williams, Julia A. Falter and Kenneth S. Geick

INTRODUCTION TO RIGID PVC

Overview

More than 30 billion pounds of polyvinyl chloride (PVC) were sold worldwide in 1995. Almost 60% of PVC is chlorine, which basically comes from sodium chloride (table salt). This results in a polymer that is inherently low cost, and one that is affected much less than most plastics by the cost of petroleum. However, the density of rigid PVC ranges from 1.30 to 1.58 g/cc, which is relatively high for a polymer. Because most plastic applications are volume dependent, PVC's high density can result in more expensive finished items than those made with a lower density polymer.

PVC polymer has the unusual property that its thermal decomposition temperature is lower than its melting point. Therefore, to process it one must add additives to retard the thermal decomposition mechanism and to lower its melting point. Those that retard thermal decomposition are known as heat stabilizers.

Those additives that lower the melting point and T_g are plasticizers. PVC formulations containing plasticizers are known as flexible or semi-flexible (depending upon the amount of plasticizer) PVC. However, well over half of PVC usage is in rigid PVC, rPVC, applications. These are made from formulations that do not contain plasticizers. For rPVC, the additives that make processing possible are lubricants.

Rigid PVC has good chemical resistance, and may be either opaque or transparent depending upon the formulation.

Plastics Additives: An A–Z Reference
Edited by G. Pritchard
Published in 1998 by Chapman & Hall, London. ISBN 0 412 72720 X

Formulations

Large volume opaque rPVC applications include pipe, siding, lineals and pipe fittings. In general the cost of the formulation increases in that order. The first three are extruded applications, while pipe fittings are injection molded. A typical siding formulation consists of:

- PVC suspension resin (100 parts);
- heat stabilizer system (0.75 to 1.50 parts);
- calcium stearate (0.75 to 1.50 parts);
- acrylic processing aids (1.0 to 1.5 parts);
- lubricant system (1.0 to 1.5 parts);
- acrylic or chlorinated polyethylene impact modifier (4.0 to 6.0 parts);
- calcium carbonate filler (3.0 to 8.0 parts);
- titanium dioxide (7.0 to 12.0 parts);
- color pigments (0.1 to 2.0 parts).

Large volume clear rPVC applications include thin calendered sheet and blown bottles. A typical blown bottle formulation consists of:

- PVC suspension resin (100 parts);
- heat stabilizer system (1.5 to 2.0 parts);
- acrylic processing aids (1.5 to 2.5 parts);
- lubricant system (1.0 to 1.5 parts);
- MBS impact modifier (10 parts);
- blue toner (<0.001 parts).

Manufacture

The first step in the manufacture of rPVC products consists of mixing PVC polymer powder with all the appropriate additives in a high intensity solids mixer. The mixer must be such that the ingredients are heated through shear during the mixing. Normally the mixer is loaded with the PVC resin, and the other additives are added at specific temperatures. After all the ingredients are added, the hot (about 120°C) mixture is dropped to a cooling, low shear mixer.

The powder mixture is then used directly in an extruder to produce finished products (siding, pipe etc.) or pellets that are used in other plastic processing equipment.

FUNCTION AND CLASSIFICATION OF LUBRICANTS

Purpose of lubricants

Like all polymers, PVC is made up of long chain molecules. These are highly viscous in the melt phase, and tend to stick to the metal components of processing equipment. These properties can be overcome by lubricants. The major function of lubricants in rPVC (and other polymers) is to decrease internal and external friction. This results in:

- reduced shear, which results in lower temperature and reduced tendency for polymer degradation;
- reduced equipment wear;
- increased production rates;
- reduced energy consumption.

Lubricant characteristics

Lubricants are normally classified as external or internal. Although both types of lubricants are mixed with the PVC polymer as described above, they function differently. External lubricants are largely insoluble in PVC. They work 'externally' by migrating to the surface of the polymer melt during processing and lubricating the melt from the metal of the processing equipment. Internal lubricants are mostly soluble in PVC. They work 'internally' by 'lubricating' the movement of PVC molecular chains past one another during processing.

However, most lubricants have some combination of both external and internal characteristics. The lubricant's solubility is determined by molecular structure and its polarity in relation to the polymer. Total solubility, which indicates high bond strength between lubricant and polymer is undesirable and can actually embrittle the polymer. The desirable effect for an 'internal' lubricant is for a slightly weaker attraction between lubricant and polymer in which the molecules arrange themselves in the direction of flow and slide by each other creating a ball bearing effect. This reduces shear stress between polymer molecules, thus lowering melt viscosity and temperature buildup.

Lubricants which are less soluble or incompatible with the base polymer provide external lubrication in the molten phase. In this case the bond strength between polymer and lubricant is weak and under high shear the lubricant is, in effect, squeezed out of the compound. Thus, the lubricant provides external lubrication by covering the surfaces of the processing equipment, reducing friction at the interface of the polymer and the metal surfaces of the machinery.

The fact that essentially all lubricants have both external and internal characteristics make it impractical to describe lubricants by external versus internal classification. Instead, PVC lubricants will be grouped by chemical class, and the effect of each class on rPVC properties will be covered.

CHEMICAL CLASSES OF LUBRICANTS

Although there is a fairly wide range of chemical classes of lubricants for rPVC, most are thought of as waxes or soaps. The five major chemical classes to be covered are:

- amides;
- hydrocarbon waxes;
- fatty acid esters;
- fatty acids;
- metallic soaps.

Amides

Although the term 'amides' is commonly used as a class of lubricants for rPVC, in fact only one amide, ethylenebisstearamide (EBS), is normally used. EBS is often referred to as 'amide wax'. It is made from largely renewable sources, being the reaction product of about 90% stearic acid from animal or vegetable fats or oils and about 10% ethylene diamine as shown in the following reaction.

$$2C_{17}H_{35}\overset{O}{\overset{\|}{C}}OH + H\overset{H}{\overset{|}{N}}CH_2CH_2\overset{H}{\overset{|}{N}}H \rightarrow C_{17}H_{35}\overset{O}{\overset{\|}{C}}\overset{}{\underset{H}{\overset{|}{N}}}CH_2CH_2\overset{}{\underset{H}{\overset{|}{N}}}\overset{O}{\overset{\|}{C}}C_{17}H_{35} + 2H_2O$$

EBS has a good balance of internal and external lubricating properties. It was the first major lubricant used in rPVC. It has a broad processing latitude and is quite forgiving of formulation errors. Until the late 1980s, EBS was the lubricant of choice for siding and pipe in the United States. Since then the switch from single screw to twin screw extruders has required more external lubrication, and most producers now use a paraffin/oxidized polyolefin lubrication system.

Hydrocarbons

This class of lubricants is composed of several subclasses, which are often referred to as separate classes. The various subclasses include paraffin waxes, microcrystalline waxes, polyethylene waxes, and oxidized polyethylene waxes. All are based on $-(CH_2)_n-$ species where n varies from about 20 to over 80.

The paraffin and microcrystalline waxes are mixtures of solid, saturated hydrocarbons that are produced by distillation and solvent refining from crude oil. The paraffin waxes are mostly linear alkanes with chain lengths varying from about 20 to about 50 carbons. The microcrystalline waxes contain many more branched chains and/or cyclic chain molecules. The number of carbon atoms in a microcrystalline wax can vary from about 30 to over 80. Consequently their average molecular weight is much higher than that of paraffin waxes.

The polyethylenes are produced by directly polymerizing ethylene at relatively low pressures and temperatures. The molecular weights range from about 500 to 1000, which corresponds to 35 to 70 carbon atoms. The molecular chains can be either linear or branched depending upon the catalyst and conditions used to make the product. They differ from the paraffin and microcrystalline waxes in end groups and catalyst residues.

Oxidized polyethylene waxes are produced by melting polyethylene waxes in a column, and forcing air up through the column. This oxidizes the homopolymer at branch points, breaking the molecular chains and forming acid groups. Other oxidized species such as esters, aldehydes, ketones, hydroxides and peroxides are also formed. This results in products with a broad molecular weight distribution with an average molecular weight of about half that of the starting polyethylene wax.

All of the hydrocarbon waxes are very external in rPVC. One would expect the oxidized polyethylene waxes, because of their polarity, to have more internal characteristics than the other hydrocarbon waxes. However, they seem to be the most external in processing equipment, probably because of their excellent metal release characteristics. The lower molecular weight paraffin waxes do have some internal character, but, in general, rPVC lubricated with any of the hydrocarbon waxes requires additional internal lubricant(s).

Esters

There are many types of ester lubricants that are used in various rPVC formulations:

- simple esters;
- glycerol esters;
- polyglycerol esters;
- montan esters;
- partial esters of polyfunctional alcohols;
- fully esterified esters of polyfunctional alcohols.

Except for the montan esters, these esters are made from various alcohols and from fatty acids; these lubricants are often referred to as fatty acid esters. Esters, in general, are extremely versatile and can range from internal to external lubricants with increasing carbon chain length and degree of esterification. Therefore, an ester's lubrication characteristics can be tailored for specific applications.

The monoesters of the higher saturated fatty acids are colorless, odorless, crystalline solids. These are the most internal functioning of the esters. The most common is glycerol monostearate (GMS). GMS is a mostly internal lubricant with some external properties. Increasing the degree of esterification increases the external lubrication characteristics.

Polyglyceryl esters are made from polyglycerol and fatty acids. They are more external than GMS, and their external character increases as the degree of polymerization and esterification increases. An increasingly important characteristic of both GMS and polyglyceryl esters is their low toxicity. Most are actually approved in the USA as food additives.

Montan wax is naturally occurring fossil vegetable wax. It is found in almost all lignite deposits in the world. It consists of esters of montanic acid with long chain aliphatic alcohols. The chain length of both the montanic acid and the alcohols is about 28 to 32 carbons. Montan waxes have a broad range of lubricant functionality with both internal and external properties. The use of montan waxes in rPVC formulations is limited due largely to the cost associated with purifying and upgrading crude montan wax to plastics additive quality. However, montan waxes are utilized in rPVC formulations for the blow molding of water bottles, where good internal lubrication and high melt strength are required.

Fatty acids

Although a large number of fatty acids are commercially available, the only one that is used to any extent as a lubricant for rPVC is stearic acid. The most common stearic acid used is the so-called 'triple pressed' grade, which is a blend of palmitic (carbon chain length of 16) and stearic (chain length of 18) acids. It is a good external lubricant with a balance of some internal lubricating properties. Its major drawback is its volatility.

Metallic soaps

The largest volume class of lubricants for plastics, in general, is metallic soaps. The largest volume metallic soap is calcium stearate. More pounds of calcium stearate are used in rPVC than any of the other materials covered in this section, but its classification as a lubricant in rPVC is not straightforward. For example, calcium stearate in rPVC

- improves flow (internal lubrication),
- increases internal shear (opposite of most internal lubricants), which leads to improved physical properties of finished parts,

- acts as an acid scavenger,
- increases mold release (external lubrication).

Calcium stearate is more of a colubricant, both internal and external, than an independent lubricant. However, because of its capacity for co-stabilizing, co-lubricating, and physical property improvement, it is used in essentially every pound of opaque rPVC produced in the world. Its weaknesses are that it cannot be used in clear formulations due to haze formation, and it tends to react with external lubricants to increase die buildup and plateout.

Other smaller volume metallic stearates used as lubricants include those based on aluminum, lead, sodium, tin and zinc.

Mixed

As rPVC processing becomes more sophisticated, rates increase, and parts become more complicated, it becomes more critical to have well-balanced lubrication systems with good internal and external characteristics. This usually requires two or more lubricants – a mixture of those with mostly internal characteristics and those with mostly external characteristics.

Examples include paraffin/oxidized polyethylene/calcium stearate, oxidized polyethylene/fatty acid ester, mixtures of fatty acid esters, paraffin/amide etc. (A conceptual graph of the external/internal characteristics of the major rPVC lubricants is shown in Figure 1.)

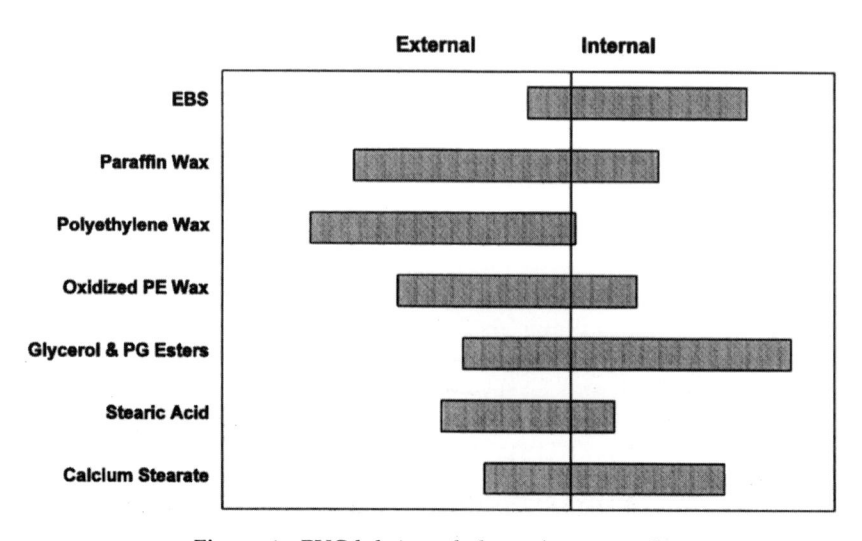

Figure 1 rPVC lubricant balance (conceptual).

TESTING OF LUBRICANTS

Lubricating effects

As previously stated, internal lubricants are soluble in the resin system and act by reducing friction between polymer molecules, leading to lower melt viscosity and lower energy requirements needed for processing. External lubricants are generally incompatible with the polymer and act to reduce the friction at the interface between the polymer and the surface of the processing equipment.

As it is very costly to run commercial scale equipment, initially the effects of a newly developed lubricant system on the properties of a rPVC compound are measured using laboratory scale equipment such as rheometers, dynamic two-roll mills and small scale molders and extruders. An example of the effect of various lubricants on output in a laboratory extruder is shown in Figure 2.

Determining whether a material is an internal or external lubricant can be defined by its effects on the fusion and mill stick times and melt viscosity of the rPVC compound. Internal lubricants will not significantly affect fusion or mill stick times of the compound but will lower melt viscosity. External lubricants will increase both the fusion and mill stick times but will not significantly lower the compound melt viscosity. These properties can be measured on a torque rheometer and a dynamic two-roll mill.

Processing properties

The effect of lubricant systems on processing can be measured on laboratory sized extruders and injection molders; however, reformulation or

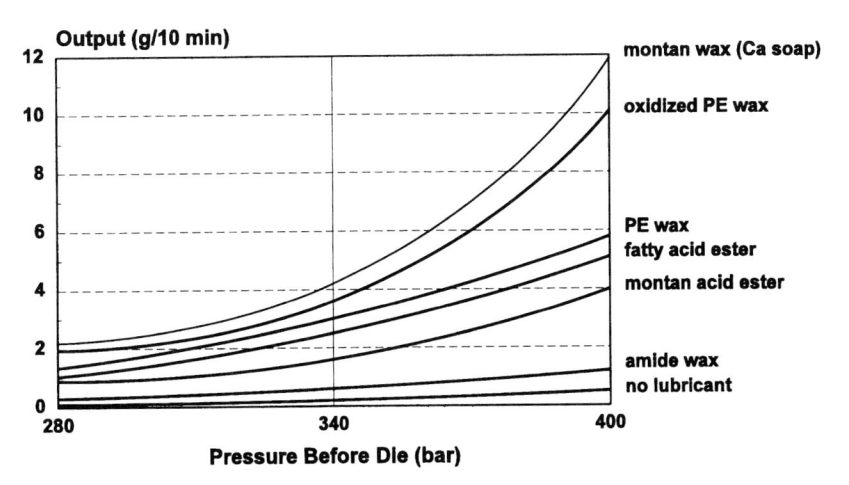

Figure 2 Rheometer flow properties of rPVC.

tweaking of the lubricant loading or type may be necessary once the compound is tested in commercial trials.

Rigid PVC can be processed on a wide variety of equipment such as extruders (both single and counter-rotating twin screw), calenders and injection molders. The lubricant system has to be balanced for the process and end use property requirements.

Single screw extruders require a balance of internal and external lubrication while twin screw extruders, calenders and injection molders require more external lubrication. Not only does the system need to be balanced for the processing equipment, the lubricants cannot adversely affect required end use properties.

A fully optimized formulation provides high outputs, low scrap rates, high-quality finished products and the required compound physical properties. Deficient or excessive amounts of lubricant lead to reduced processing efficiencies or can even shutdown the operation. A balanced lubricant system (right amounts of both internal and external lubrication) provides control over the compound fusion and thermal stability times, output rate, blooming, compound clarity and physical properties.

BIBLIOGRAPHY

Nass, L.I. and Heiberger, C.A. (eds) (1988) *Encyclopedia of PVC*, Vol. 2, 2nd edn, Marcel Dekker, Inc, New York and Basle, pp. 31–43, 263–390.
Riedel, T. (1990) Lubricants and Related Additives in *Plastics Additives*, 3rd edn (eds R. Gächter and H. Müller), Hanser Publishers, Munich, Vienna, New York, pp. 423–470.
Edenbaum, J. (ed.) (1992) *Plastics Additives and Modifiers Handbook*, Van Nostrand Reinhold, New York, pp. 41–55, 773–842, 858–867.
Štěpek, J. and Daoust, H. (1983) *Additives for Plastics*, Springer-Verlag, New York, Heidelberg, Berlin, pp. 34–42.

Keywords: lubricants, solubility, compatibility, amides, waxes, fatty acids, esters, metallic soaps, melt viscosity, extruders, rheometer.

See also: Release agents.

Mica

C.C. Briggs

INTRODUCTION

The key characteristic of mica as a filler for plastics is its ability to be processed to give tough thin plate shaped particles, with aspect ratios (diameter divided by thickness) higher than for any other mineral. Special processing, not used for plastics applications, can achieve aspect ratios as high as 1000.

COMPOSITION AND PROPERTIES

Mica is the name given to a group of alumino silicate minerals characterized by having a layered structure which is easily cleaved to give thin, flexible sheets.

The main commercial source of mica is pegmatite rock, where large 'books' of mica can be hand separated from feldspar and quartz. Finer flake size mica is mechanically separated from deposits of schist where quartz is the main co-mineral. Large tonnages of mica are generated as a by-product of kaolin production and some of this fine particle size mica is separated by froth flotation. Another commercially important source is as a by-product from apatite (calcium phosphate) mining.

The characteristics of different mica deposits are summarized in Table 1. Muscovite is a potassium aluminium silicate which is transparent and almost colourless. It is chemically unreactive and is stable to about 600°C when dehydroxylation takes place. Phlogopite, sometimes known as magnesium mica, is generally coloured brown (but can also be green depending upon oxidation state) due to the iron which is also present

Plastics Additives: An A–Z Reference
Edited by G. Pritchard
Published in 1998 by Chapman & Hall, London. ISBN 0 412 72720 X

Table 1 Characteristics of mica mineral deposits

Deposit type	Pegmatite	Schist	Kaolin	Other
Mica flake size	Very large	Medium	Small	Medium
Types*	Muscovite 1 Phlogopite 5	Muscovite 2 Biotite 5	Muscovite 2	Phlogopite 3
Co-minerals	Feldspar Quartz	Quartz Garnet	Kaolin Silica	Apatite/Limestone Pyroxene
Flake aspect ratio	High	Medium	Low	Medium/High

*Commercial importance (1 = High, 5 = Low)

in the structure. When the ratio of iron to magnesium is high, the colour becomes black and this type of mica is known as biotite.

PROCESSING

Following the removal of co-minerals, either by hand or mechanically, mica is ground by various methods to give particles with top sizes ranging from 5 mm to 10 μm. In order of decreasing particle size, the methods of comminution used are: hammer mills, pin mills, wet rod mills, wet pan mills and jet mills. Mechanical sieving is used to control particle size down to about 100 μm top size. Air classification is more effective for finer particle sizes.

The aspect ratio of the final flake products depends upon the origin of the mica and the processing method. Large flake pegmatite mica and wet milled mica can have aspect ratios in excess of 100. By contrast, processed mica flakes from schist and kaolin deposits tend to have aspect ratios in the area of 30 to 60.

Mica processing methods are designed to achieve a high aspect ratio and give undamaged flat flakes. This ideal becomes increasingly difficult with decreasing particle size.

Below 100 μm diameter, wet grinding is necessary if a high aspect ratio is to be achieved. This slow and expensive process increases the chance of flake delamination in preference to breakage across the flakes and gives flat, smooth edged platelets.

SURFACE TREATMENT

Silane coating is the only widely used surface treatment for mica. Hydroxyl groups on the mica surface are well suited to reaction with the silanol groups of hydrolysed silanes.

Silane coating can be effectively carried out using an organic solvent but this is an expensive route. The most common method is high shear 'dry' mixing in which temperature control can be critical depending upon the type of silane used. Where mica is being wet milled, silane coating can be efficiently carried out at the wet grinding stage. It should also be pointed out that silanes can be added during compounding, rather than by pre-coating the mica.

Many types of silane are used on mica depending upon the polymer resin involved. The functional group on the silane is chosen to enhance bonding between the mica particle and the polymer matrix. Achievement of strong bonding with inert resins such as polypropylene requires a highly reactive group such as azido to give chain scission and create active sites for bonding. An alternative and more favoured route is to incorporate acrylate or other reactive groups into the matrix by blending standard PP with a minor proportion of chemically modified PP or PE (e.g. maleic anhydride modified polymers) and to use less exotic silane such as amino or vinyl to interact with the active sites on the modified polymer component.

It should be noted here that uncoated mica gives adequate interaction with polar resins such as thermoplastic polyesters and nylons, so that the additional benefits of silane coating are rarely worth the extra cost, and even in non-polar resins, most applications do not justify the additional expense of silane surface treatment.

MICA FILLED PLASTICS

After taking account of compounding costs, mica filled compounds are generally more expensive on a volume basis than the unfilled resin. Consequently, mica is not viewed as a cost saving filler, but rather as a means to modify mechanical properties. The key properties which mica affects are:

(a) *Rigidity*
 Only glass fibre rivals high aspect ratio mica in rigidizing polymer compounds.
(b) *Heat deflection temperature* (HDT)
 High aspect ratio mica approaches the performance of glass fibre in raising the working temperature as measured by the heat deflection temperature test. When softening point is measured, mica exceeds the performance of glassfibre. This is because softening point measures surface hardness and mica flakes, parallel to the surface, spread the applied force over a larger area.
(c) *Dimensional stability*
 Mica filled compounds give low shrinkage and warpage after moulding and the final compounds have a low coefficient of expansion. Mica is unsurpassed in its ability to reduce warpage induced by

other factors (e.g. in injection moulded glassfibre filled PP). This is because warpage almost always has more than one dimensional component and mica flakes strongly resist being simultaneously bent in more than one direction.

(d) *Flexural and tensile strength*

Uncoated mica does not have an important effect upon these properties, but appropriate silane coated grades can give marked improvements in strength due to improved bonding of the mica flakes with the polymer matrix.

(e) *Impact strength*

Incorporation of mica normally reduces impact strength. A simple improvement in bonding via silanes normally reduces impact strength still further. This is because the increased rigidity allows less energy to be absorbed prior to break.

This problem can, in principle, be overcome by bonding a rubbery layer to the mica which also interacts with the matrix. However, such solutions are not straighforward and detract from some of the other beneficial effects of using mica, especially rigidity.

APPLICATIONS

Mica was first used in thermosetting resins in the 1920s when the main application was in phenolic resins for electrical use. The name 'Formica' also alludes to an early use of mica in thermosets. Mica is still used in phenolics, epoxies and polyester resins though thermosetting applications are now small compared to usage in thermoplastics. The only area showing growth is as a cheaper substitute for glassfibre and glass flakes in polyester applications.

Mica-filled polypropylene is now the largest volume market for mica. Some typical applications and benefits are:

(a) loudspeaker cones and cabinets (rigidity and vibration control)
(b) underbonnet car parts (high HDT, dimensional control)
(c) fan cowlings (rigidity and high HDT)
(d) automative cabin mouldings (reduced scratch visibility)

A typical application in nylon is automotive headlamp housings where dimensional control is critical. Substantial quantities of mica are also used in thermoplastic polyesters.

A developing, and potentially large, market for mica is as a partial replacement for glassfibre in polypropylene and nylon. In injection moulded parts, where glassfibre can give severe warpage problems, partial replacement of the glassfibre by mica can markedly reduce this warpage. Some reduction in impact strength occurs but the final balance of properties is often quite adequate for the application concerned. An extra benefit is that mica is less costly than glassfibre. A typical application in PP is dashboard armatures and in nylon is automotive rocker box covers.

Finally, mention should be made of the increasing use of mica as a decorative filler in both thermosetting and thermoplastic mouldings. Pearlescent micas in which a thin layer of titanium dioxide is deposited on the mica surface have been available for many years. These fine flakes are normally silver but a wide range of colours is available. Pearlescent mica is best established in thin film applications.

More recently coarser mica flakes in a wide range of sizes and colours have been introduced to create decorative speckled effects. The high aspect ratio of mica leads to a strong visual effect at loading levels as low as 0.25 phr.

CONCLUSION

Mica is a specialist filler for plastics, used to achieve mechanical or visual properties not available via other routes.

BIBLIOGRAPHY

S.J. Lefond (ed.) *Industrial Minerals and Rocks*, AIME, 1983.
H.S. Katz and John V. Milewski, *Handbook of Fillers for Plastics*, Chapman & Hall, London, 1987.

Keywords: aluminosilicate, aspect ratio, muscovite, phlogopite, biotite, polypropylene, rigidity, heat distortion, warpage, impact strength, hardness, particle size, hammer mill, surface treatment.

See also: Fillers.

Nucleating agents for thermoplastics

Robert A. Shanks and Bill E. Tiganis

INTRODUCTION

The morphology of semi-crystalline polymers is crucial to their application. Homogeneous nucleation generally does not provide consistent properties, as nuclei appear throughout the crystallization, giving a broad distribution of sizes, and hence crystallization will be controlled too much by the processing conditions. Heterogeneous nucleation gives crystals of consistent size, as all nuclei are present at the start of crystallization and the nucleant controls crystallization. Heterogeneous nucleation may be caused by initiator residues, impurities etc., or preferably by addition of a specialized nucleating agent. The latter is preferred as control of the system is deliberate.

Nucleation can occur on the surface of a filler or reinforcement. Indeed the action of fillers is, in part, due to the changed crystal structure in the environment of the filler. Nucleating agents must form very small particles within the polymer melt. They tend to have rigid planar structures which are expected to provide the site for the first few polymer segments to adsorb and provide stable embryo nuclei [1].

PRIMARY NUCLEATION

Crystallization occurs in two stages – nucleation and growth. The free energy of crystallization involves the sum of (1) the free energy for formation on a stable nucleus embryo (ΔG^*) and (2) the free energy for

Plastics Additives: An A–Z Reference
Edited by G. Pritchard
Published in 1998 by Chapman & Hall, London. ISBN 0 412 72720 X

diffusion of molecular segments to join the growing crystal ($\Delta G\eta$). ΔG^* decreases with temperature below the melting temperature, while $\Delta G\eta$ increases, creating a maximum in the nucleation rate. Homogeneous nucleation may occur through a fringed micelle, a bundle of polymer chains with long sections remaining random, or by chain folding to reach the critical nucleus dimensions. Chain folded nuclei are more probable than fringed micelle nuclei. Homogeneous nucleation is rarely reached, and most polymers crystallize from heterogeneous nuclei [1].

$$S = S_o \exp[-\Delta G\eta/R(T_c - T_\infty)] \exp[-Kg/T_c(\Delta T)f] \qquad (1)$$

where $S =$ radial growth rate, $T_c =$ crystallization temperature, $T_\infty =$ temperature where crystallization ceases, $T - T_g$, $f =$ a factor accounting for the temperature dependence of the heat of fusion, Kg is dependent on the crystalline regime [2]. Heterogeneous nucleation involves the creation of a nucleus on an existing foreign surface, which greatly decreases ΔG^*. This will decrease the critical nuclei size so that nuclei can form at lower undercooling (ΔT). This process is termed primary nucleation, and a constant nucleation rate is expected under isothermal conditions.

SECONDARY NUCLEATION

When a folded layer on the surface of a crystal has finished growing, a new nucleus needs to form on the surface for continued growth of the crystal. Secondary nucleation requires a high undercooling because it has a low temperature dependence. The secondary crystallization layers are completed by an attachment–detachment mechanism [1].

NUCLEATION ON FILLER SURFACES

Particulate and fibrous fillers can provide oriented growth from their surfaces. Transcrystalline structures are formed, because the closely packed surface nuclei inhibit lateral growth, leaving perpendicular growth to predominate near surfaces. Nucleation activity of the surface may be because of specific sites on the filler or polymer adsorption giving a greater melt density near the surface than in the bulk. Transcrystallinity of polypropylene has been observed after shearing of fibres through a melt.

POLYMORPHISM IN NUCLEATED POLYMERS

When a polymer can exist in more than one crystalline form, the time–temperature conditions and type of nucleant can determine the form

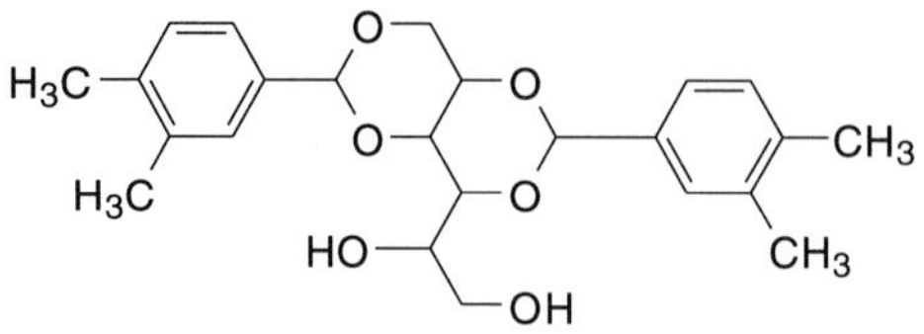

Figure 1 Structure of bis-(3,4-dimethylbenzylidene sorbitol diacetal).

which predominates. The most common example is provided by the monoclinic (α) and hexagonal (β) forms of polypropylene. Epitaxial growth is the parallel growth of one crystal onto another, involving a parallelism of two crystal lattice planes that have nearly identical arrangement of atoms. The polypropylene can crystallize to give one of its crystalline forms, depending on the nucleating agent which is used. Homoepitaxy is where the polypropylene crystallizes on a polypropylene crystal, often forming lamella branching. Heteroepitaxy is where the polymer crystallizes on an inorganic or organic compound, such as the salts of benzoic acid. Lattice matching may be important between the salts and the oriented grown polymers [3].

NUCLEATING AGENTS

Nucleants may (1) be melt sensitive, i.e. they melt below or near the processing temperature, or (2) melt insensitive, i.e. they do not melt below normal processing temperatures. The melt sensitive nucleants form a physical gelation network within the polymer ([4] and Millad technical datasheet) while the melt insensitive nucleants provide single nucleation sites within the polymer.

Typical nucleating agents include:

- lithium, sodium, potassium benzoate;
- sodium salts of organophosphates;
- finely divided (<40 nm) clays, silica flour;
- Millad, bis-(3,4-dimethylbenzylidene sorbitol diacetal) (Figure 1);
- 4-chloro-, 4-methyl- and 4-ethyl- substituted forms of dimethylbenzylidene sorbitol
- diacetal [4].

EXPERIMENTAL METHODS FOR STUDYING NUCLEATION [5]

Differential scanning calorimetry (DSC) can be used to study the onset of crystallization on cooling from the melt where nucleated polymers have higher onset temperatures (Figures 2 and 3). Isothermal studies provide kinetic data where Avrami analysis (equation 2, where x = reduced

Experimental methods for studying nucleation [5]

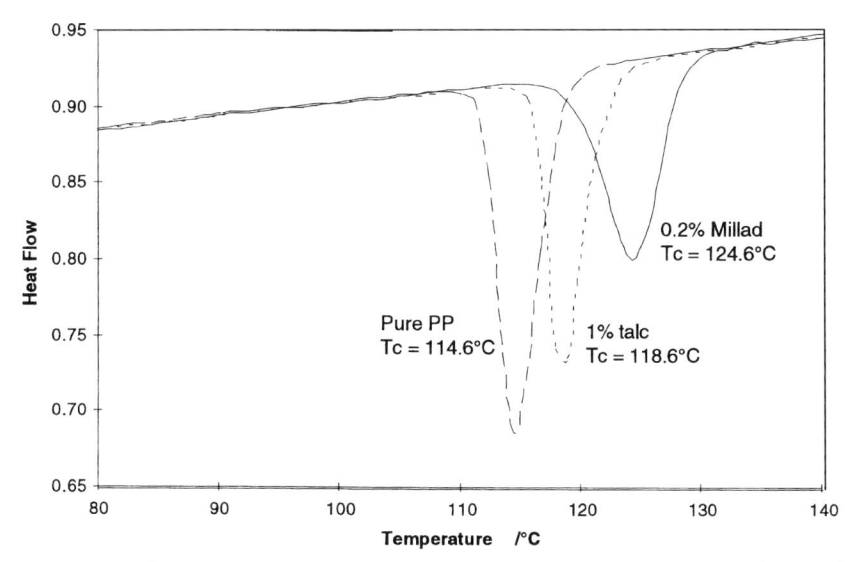

Figure 2 DSC of polypropylene crystallization – pure (114.6°C), 1% talc (118.6°C) and 0.2% Millad (124.6°C).

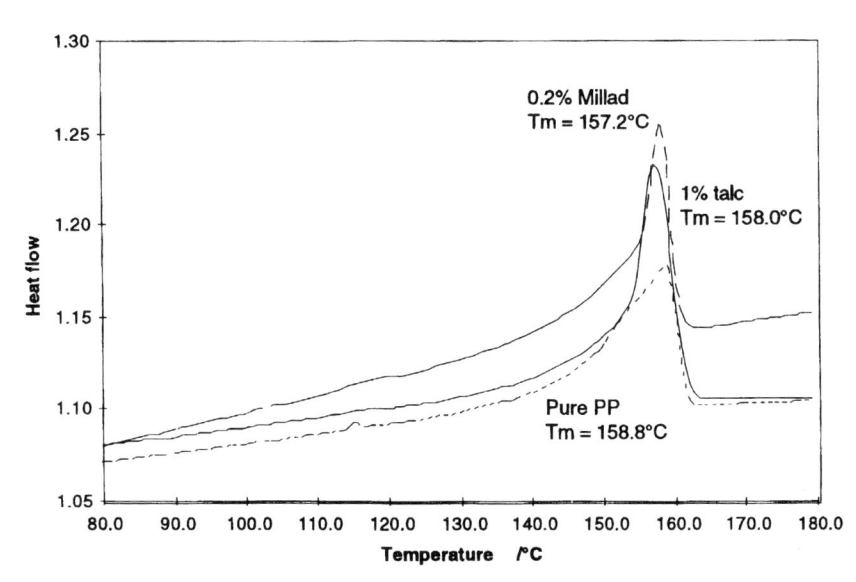

Figure 3 DSC of polypropylene melting – pure (158.8°C), 1% talc (158.0°C) and 0.2% Millad (157.2°C).

Table 1 Crystallization times for polypropylene and nucleated polypropylene

Isothermal crystallization temperature (°C)	120	122	124	126	128	130
Crystallization time (s) for PP	71	86	139	179	264	380
Crystallization time (s) for PP–0.2% Millad	42	42	53	106	136	181

crystallinity, k = rate coefficient, t = time) will yield a lower exponent (n) for heterogeneously nucleated polymers.

$$(1 - x) = \exp(-kt^n) \qquad (2)$$

Table 1 shows crystallization times for polypropylene, pure and with 0.2% Millad, at various crystallization temperatures. The Millad nucleant can be seen to have decreased the crystallization times. Table 2 shows crystallization and melting parameters for polypropylene, pure and with various nucleants with Millad at different concentrations. The crystallization temperatures are increased by nucleants while the melting temperatures are not significantly changed. Crystallinity was increased by Millad but decreased by the other materials, which were not specifically intended for nucleation.

Optical microscopy with polarized light, especially in conjunction with a hot stage, can be used to observe nucleation density and crystal size directly. Figures 4 and 5 show the change in crystal size between pure polypropylene and polypropylene nucleated with 0.2% Millad. Quantitative analyses can be performed using digital image analysis. These observations can be extended with the higher resolution and depth of field of scanning electron microscopy and transmission electron microscopy.

Wide angle X-ray scattering is used to measure crystallinity, distinguish crystalline morphologies and obtain crystal dimensions and

Table 2 Crystallization and melting of nucleated polypropylene

Polypropylene composition	Onset T_m (°C)	T_m (°C)	Onset T_c (°C)	T_c (°C)	ΔHm (J/g)
PP no nucleant	151.2	155.2	118.5	115.0	81.2
PP–Millad 0.1%	153.0	157.4	127.5	120.9	83.1
PP–Millad 0.15%	150.5	158.7	127.2	118.6	83.8
PP–Millad 0.2%	147.7	158.7	128.4	124.3	84.8
PP–Millad 0.3%	145.2	158.1	128.9	125.4	84.6
PP–sodium benzoate 0.25%	150.0	157.1	121.3	117.9	81.5
PP–boron nitride 0.25%	152.5	156.7	118.4	115.1	72.1
PP–talc 1.2%	154.4	157.9	121.7	118.4	80.1
PP–calcium carbonate (1) 2%	151.2	155.9	117.1	113.2	74.4
PP–calcium carbonate (2) 2%	151.8	156.3	116.0	120.1	62.7

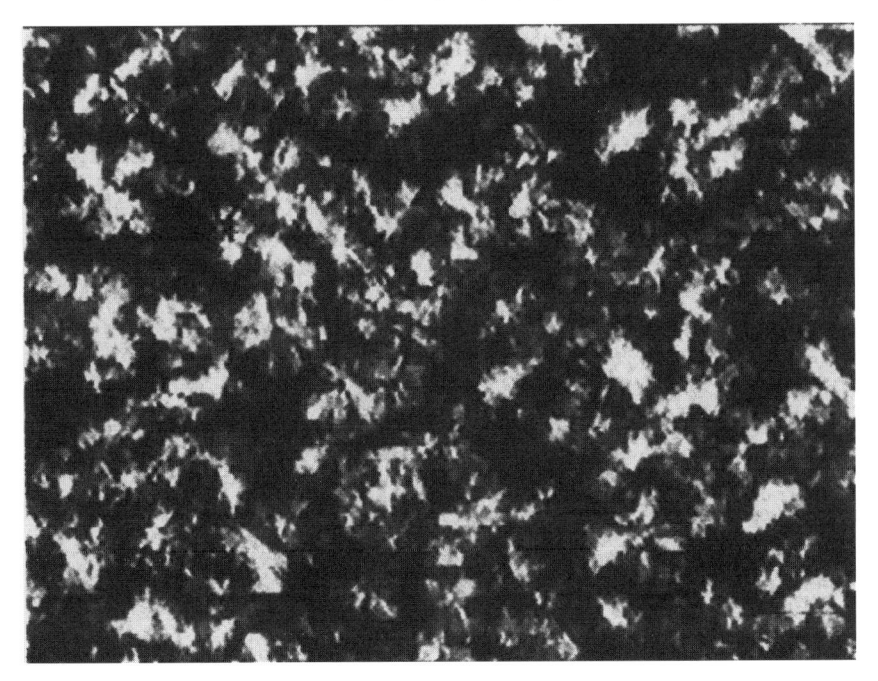

Figure 4 Polarized optical microscope image of unmodified polypropylene after crystallization at 125°C, ×200.

atomic spacings. Small angle X-ray scattering is used to measure larger crystal dimensions.

NUCLEATION FOR SPECIFIC POLYMERS

Poly(ethylene terephthalate) has slow nucleation and crystallization rates. Typical nucleating agents are: minerals such as chalk, gypsum, clay, kaolin, talc, mica and silicates; pyrophyllite, pigments such as cadmium red, cobalt yellow, chromium oxide; titanium dioxide, magnesium oxide, antimony trioxide, carbonates, sulfates, boron nitride, sodium fluoride and carbon black, salts of carboxylic acids, montan wax, halogenated alkanes, benzophenone, several polymers, and many other organic liquids [3]. The organic liquids may increase the mobility of the polymer when present in small proportions.

Polyamides-6,6, -6,10 and -6, can be nucleated with 0.1% highly disperse silica. Polyamide-6,6 powder is used as a nucleant for lower melting polyamides. Other nucleants such as molybdenum disulfide, iron sulfide, titanium dioxide, talc and sodium phenylphosphinate are used [4].

Polypropylene usually crystallizes in the α form, however a β form also occurs. The β forms are easily identified by X-ray diffraction, or polarized

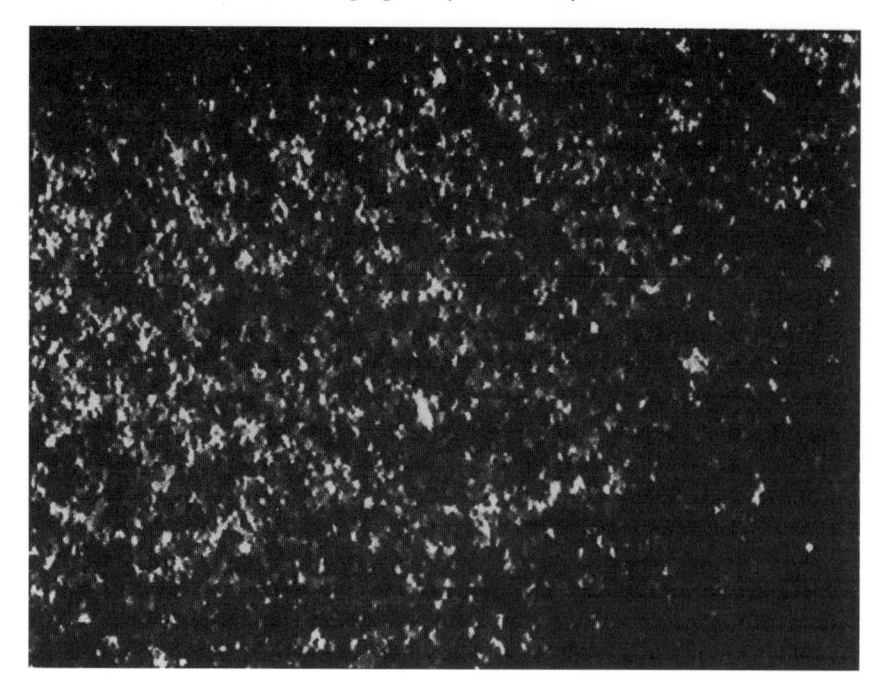

Figure 5 Polarized optical microscope image of polypropylene with 0.2% Millad nucleant after crystallization at 125°C, ×200.

optical microscopy, due to its high birefringence. The β form has been obtained by crystallization at higher temperatures ($T_c = 120$–$140°C$) and can only be studied if the sample temperature is maintained above 110°C [3]. β-nucleating agents can provide pure β crystals and some of these are: γ-quinacridone, triphenol ditriazine, aluminium quinizarin-sulfonic acid, disodium phthalate, calcium phthalate, and wollastonite (calcium silicate). β crystallization is reduced in copolymers of propylene with ethylene. Talc is a nucleating agent for the α form [5,6].

Polyethylene is a rapidly crystallizing polymer and is rarely used with nucleating agents. Potassium stearate has a large effect with high density polyethylene. Some organic pigments can cause nucleation, giving high internal stresses and severe distortions in mouldings. Polybutene has been nucleated with adipic and p-aminobenzoic acids [4].

NUCLEATION OF POLYMER BLENDS

Polymer blends usually consist of immiscible components and if miscible in the melt the polymers crystallize in mutually separate crystalline phases. The presence of a second phase can provide nucleation sites regardless of whether it is liquid or solid. Heterogeneous nuclei may migrate preferentially into one of the phases, thereby removing nuclei from the other phase; this will retard nucleation in one phase [7].

CONCLUSION

Nucleation of crystallization of polymers can be heterogeneous or homogeneous. Heterogeneous crystallization predominates in most important thermoplastics as nucleation must be controlled. Even when nucleating agents are not added, initiator residues, impurities from reactors or processing, other additives such as fillers or blended polymers and processing aids can provide the fluctuations in the melt necessary for nucleation. Efficient nucleation may be used to enhance mechanical properties or to provide consistent optical properties.

REFERENCES

1. Galeski, A., Nucleation of Polypropylene, in Karger-Kocsis, J. (1994), Polypropylene Structure, Blends and Composites, Vol. 1 Structure and Morphology, Chapman & Hall, London, pp. 119–139.
2. Phillips, R.A. and Wolkowski, M.D. (1996), Structure and Morphology, pp. 113–176, and Becker, R.F., Burton, L.P.J. and Amos, S.E. (1996), Additives, pp. 177–210, in Moore, E.P. (Ed.), Polypropylene Handbook, Hanser Publishers, Munich.
3. Petermann, J., Epitaxial Growth on and with Polypropylene, in Karger-Kocsis, J. (1994), Polypropylene Structure, Blends and Composites, Vol. 1 Structure and Morphology, Chapman & Hall, London, pp. 119–139.
4. Jansen, J. (1985), Nucleating Agents for Partly Crystalline Polymers, in Gachter, R. and Muller, H. (Eds), Plastics Additives Handbook, Hanser Publishers, Munich, pp. 671–683.
5. Cheng, S.Z.D., Janimak, J.J. and Rodriguez, J., Crystalline Structures of Polypropylene Homo- and Copolymers, in Karger-Kocsis, J. (1994), Polypropylene Structure, Blends and Composites, Vol. 1 Structure and Morphology, Chapman & Hall, London, pp. 119–139.
6. Varga, J., Crystallisation, Melting and Supramolecular Structure of Isotactic Polypropylene, in Karger-Kocsis, J. (1994), Polypropylene Structure, Blends and Composites, Vol. 1 Structure and Morphology, Chapman & Hall, London, pp. 119–139.
7. Long, Y., Shanks, R.A. and Stachurski, Z.H. (1995), A Practical Guide to Study Kinetics of Polymer Crystallisation, Progress in Polymer Science, 20, 651–701.

Keywords: nucleation, crystallization, semi-crystalline polymers, spherulite, morphology, polarized optical microscopy.

Optical brighteners

Geoffrey Pritchard

Optical brighteners are organic substances which are used in plastics to correct discoloration or to increase whiteness and brightness. They absorb ultraviolet radiation below 300 nm and emit it as visible radiation below about 550 nm. The usage of optical brighteners in plastics is very small compared with their applications in other markets. This explains why one of the references cited below is about detergents.

The main requirements for optical brightening agents are that: they must not be toxic; they must be fast to light; they must not migrate or leach out of the plastic; they must not adversely affect or be affected by other additives present; and they must be soluble in the polymer matrix, although the concentrations used are typically no more than 10 ppm. Masterbatches are used in order to control the small quantities. This means that a batch of polymer containing a large quantity of the additive is diluted with a larger quantity of polymer containing no additive, thus avoiding the need to meter very small quantities of anything at the main mixing stage. Polystyrene, polyolefins, ABS and PET are typical materials modified by optical brighteners.

Examples of optical brightening agents are the hindered amines, and derivatives of benzophenone and benzotriazole. Most are heterocyclic, often with sulphur present. A typical structure is given in Figure 1.

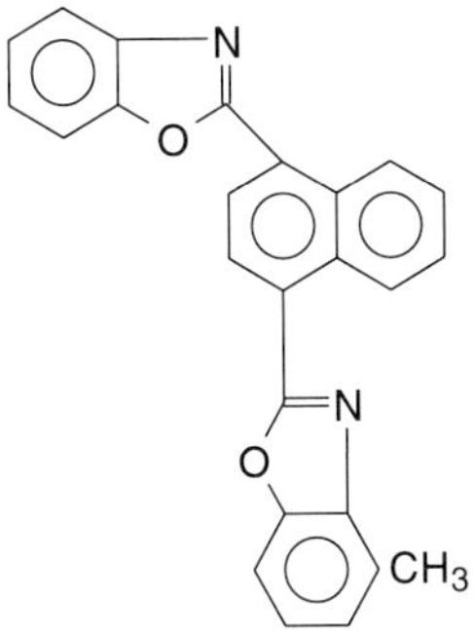

Figure 1 An optical brightener.

BIBLIOGRAPHY

Rubel, T. (1972). *Optical brighteners, technology and applications*, Noyes, New York.
Lange, K. Robert (ed.) (1994). *Detergents and cleaners, a handbook for formulators*, Hanser, Munich.

Keywords: brightness, white, hindered amine, benzophenone, benzotriazole, heterocyclics, masterbatch.

Paper for resin bonded paper laminates

R.J. Porter

WHAT ARE LAMINATES?

This book is about additives for plastics; in this context, we shall treat paper and certain related materials as additives for resins. As will become apparent, there are also additives (such as pigments) in the paper used to make laminates, so this topic is relevant to the book's central theme in more than one way.

Paper laminates are structures consisting of either thermosetting (the usual type), or thermoplastic resins, reinforced by a variety of fibrous materials. Such laminates are produced in a multilayer construction, either having several layers of the resin and fibrous material, or consisting of a single sheet of the reinforced resin, on top of a substrate such as particleboard or fibreboard. The end uses of the finished article can be divided into decorative applications and so-called 'industrial' ones.

The decorative laminates are made from paper and resin combinations, using a variety of processes which have evolved over half a century in response to changing fashions and end uses. The original process for the industry was the one which produced what are now referred to as 'high pressure laminates' or 'HPL'. In all laminate processes the combined resin and fibrous material are brought together under elevated temperature and pressure, to create conditions in which the resin flows and forms a homogeneous mass, and which upon cooling produces an apparently continuous sheet of a plastic-like material. The name 'high' pressure needed to be used to differentiate the procedure from later processes,

Plastics Additives: An A–Z Reference
Edited by G. Pritchard
Published in 1998 by Chapman & Hall, London. ISBN 0 412 72720 X

which evolved to meet different needs. High pressure laminates are made today almost exclusively from paper, and they are composites consisting of a resin-impregnated decorative surface sheet on top of a number of sheets of similarly impregnated structural papers which form the core of the laminate, contributing much to the high mechanical strength of this type of material.

'Low pressure' laminates began to appear in the 1960s, when cheap wood-based substrates such as particle board and fibre board began to find application in furniture making. Initially these substrates were given an attractive appearance by bonding sheets of high pressure laminate to their outer surfaces. However, this was expensive, and ways were sought to adapt the laminate process to produce a decorative particle board directly. To avoid unacceptable deformation of the substrate in the laminate press, lower pressures were essential; hence the new generation of laminates known as 'low pressure laminates' or 'LPL'. The lower pressures brought problems relating to resin flow and resin distribution, and resulted in new resin chemistry, together with new demands on the paper maker.

More recent developments have used the technology from both sectors to create a continuous process for laminate manufacture. Whilst HPL and LPL are both semi-batch processes (resin impregnation is continuous, but laminates are produced in batches of up to a few hundred in HPL and singly in LPL), rolls of impregnated decorative paper and impregnated core paper may now be fed continuously into a dynamic press. This development entailed considerable engineering ingenuity, together with yet more changes to resin technology and paper specifications. These so-called 'continuous HPL' laminates now command a significant part of the overall laminate market.

Whilst strictly speaking it is not a laminate, a related product made from paper and resin is the artificial veneer of 'foil'. These veneers are of single sheet construction, and are made by impregnation with very durable and flexible resin mixtures such as modified acrylics, polyurethanes, and mixtures with urea formaldehyde. The product is fully cured, and the base paper is usually a printed woodgrain design, so the finished article resembles a piece of wood veneer. It is applied by glue bonding to a suitable substrate, for furniture making. Figure 1 indicates the structure of various decorative laminates.

The remaining sector of the laminate industry is the so-called 'industrial laminates'. These materials have a diverse range of applications in such fields as electrical/electronics, construction, leisure, aircraft, boats and others. Sometimes paper is used as the fibre component, but usually more exotic materials such as glass fibre, polyester, cotton, carbon fibre and Kevlar are found. Fibre choice is dictated by the very special properties required of the laminate such as (a variety of) electrical characteristics,

HPL

Overlay (optional)
Decorative sheet
Several core sheets

Balancer sheet

Typical thickness 1 - 2 mm

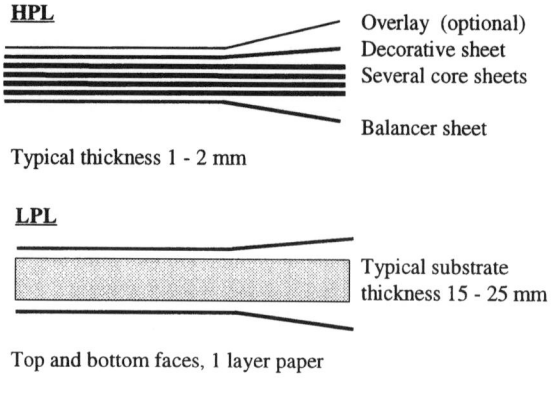

LPL

Typical substrate thickness 15 - 25 mm

Top and bottom faces, 1 layer paper

CONTINUOUS

Similar to HPL, usually fewer core sheets.
Typical thickness 0.8 - 1.2 mm.

Figure 1 Types of laminate construction.

flame retardancy, mechanical strength, flexural strength, water resistance, chemical resistance etc.

Probably the most prolific application is in the electrical industry, where laminates are used as insulating formers in transformers, switchgear and printed circuit boards.

MANUFACTURING PROCESS

All the laminate types mentioned are manufactured by a process which involves first applying resin to the fibrous material, and secondly heating and pressing the product.

The aim of the impregnation process is to apply a defined quantity of resin to the fibres, to distribute the resin uniformly across the sheet, and to produce the desired distribution of resin from one surface to the other. The latter may or may not be homogeneous – many applications require asymmetry of application.

The drying of the treated sheets needs to be such that any pre-cure of the resin meets the requirements of the subsequent pressing process. Figure 2 shows a typical impregnation process.

Pressing is similar in principle for all types. A heated platen or belt softens the resin, and flow occurs, together with a pre-determined degree of cure. After a defined heating cycle the laminate is cooled, either naturally in ambient conditions, or as part of the pre-determined press cycle with the aid of a coolant. Figure 3 shows typical pressing configurations. Table 1 indicates typical pressures, temperatures and press cycle times.

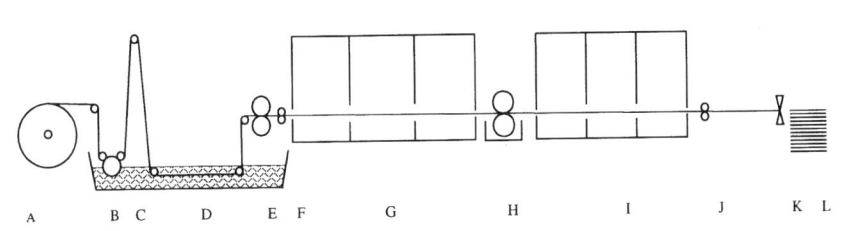

Figure 2 Typical impregnation process. A, paper reel; B, pre-wet station; C, sky roll; D, saturation; E, metering rolls; F, smoothing rolls; G, graduated temperature airfloatation dryers; H, optional mid-process treatment (additional resin/pigmentation); I, graduated temperature airfloatation dryers; J, tension control; K, guillotine; L, cut-to-size sheets. Dimensions: length about 5 m A–F; 50 m F–L; height at C = 2 m (variable); speed = 10–40 m/min.

Laminates may be of a rigid, flat type, i.e. truly of a thermosetting species, or of the type referred to as 'postforming'. This type is made with a combined resin and paper chemistry which results in laminates from the press being only partially cured. They lend themselves to a subsequent forming process in which heat and pressure are applied, to induce the laminate to curve around a contoured substrate, whilst being bonded to it. To ensure that there is no cracking during this deformation, modified resins are employed, and papers with special formulations are used; sometimes these formulations include cotton as one of the fibres, but not of necessity. The paper chemistry has to be engineered to suit that of the resin system, and it is usually the subject of much secrecy.

Patterns or designs are sometimes pressed into the surface of the laminate, either by etched effects on the metal press plate, or from an interleaving textured foil of thin aluminium or special paper.

Resin types are determined by the end-use of the laminate. Such types as melamine formaldehyde, urea formaldehyde, polyester, polystyrene, epoxies etc., are all common-place, either alone or in mixtures. Some laminates, notably decorative HPL, employ two types in the same construction. The decorative surface is obtained with melamine formaldehyde, which is desirable for its colourless, tasteless, non-toxic characteristics, but cheaper and stronger (although brittle) phenol formaldehydes are used in the core of the laminate.

PAPERS FOR HPL

The use of paper as the reinforcing/decorative fibre is found predominantly in the decorative field. Several types are used, each defined by the specific role to be played in the laminate. Table 2 compares the physical characteristics of the various types.

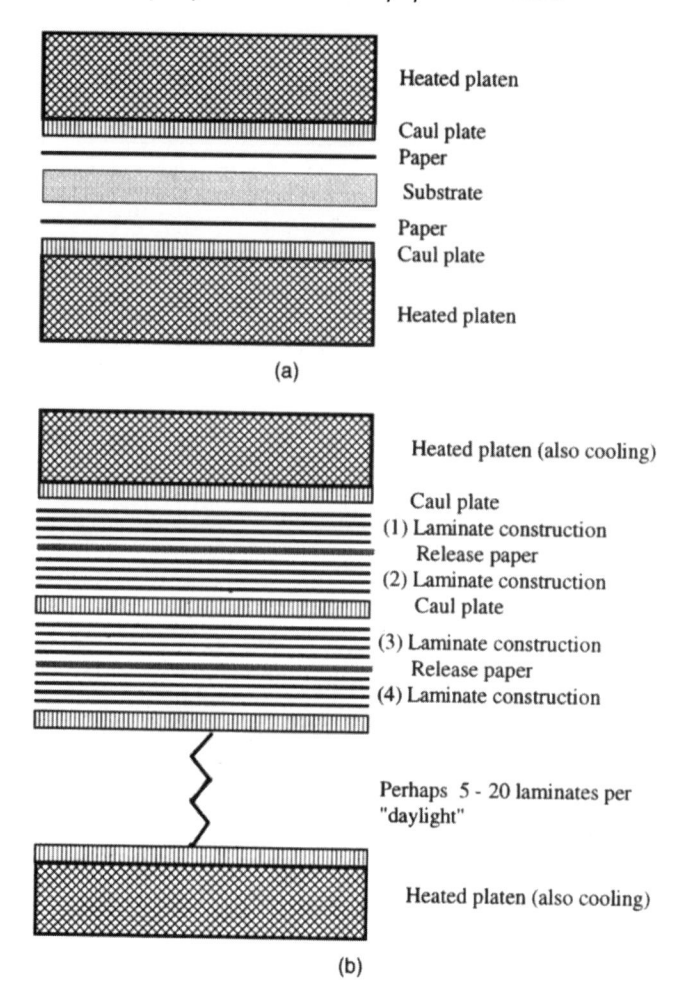

Paper for resin bonded paper laminates

(a)

- Heated platen
- Caul plate
- Paper
- Substrate
- Paper
- Caul plate
- Heated platen

(b)

- Heated platen (also cooling)
- Caul plate
- (1) Laminate construction
- Release paper
- (2) Laminate construction
- Caul plate
- (3) Laminate construction
- Release paper
- (4) Laminate construction

Perhaps 5 - 20 laminates per "daylight"

Heated platen (also cooling)

(c)

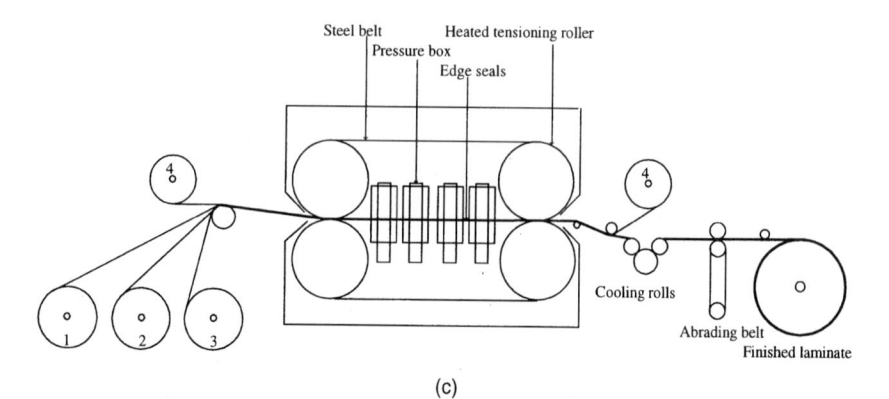

Steel belt, Heated tensioning roller, Pressure box, Edge seals, Cooling rolls, Abrading belt, Finished laminate

Table 1 Typical press conditions for various laminate types

Process parameters	Type of laminate		
	HPL	LPL	Continuous
Typical pressure (kg/cm^2)	70–105	15–35	18–22
Typical temperature (°C)	140–150	150–175	150–175
Typical cycle time: heating,	50–60 min	60 sec	12–40 sec[†]
cooling	10–30 min	*	5–15 m/min
Throughput of press			
Number of boards	>100/cycle	1/cycle	–

[*] Cooled naturally at ambient conditions.
[†] Cooled continuously (chilled rolls).

Overlay

This is a very light weight paper – in the range 16–40 g/m^2 – found as the uppermost layer on some high pressure laminates. Its function is to carry a very heavy surface layer of resin, to afford additional protection to the decorative surface below. Normally it is to be found on printed designs, where even a relatively light scratch or abraded area would penetrate through the print to the underlying base colour. In the case of especially demanding applications such as laminate flooring, then hard particulate material such as alumina or silica can be introduced as a filler in the overlay paper. The type of filler is chosen with its refractive index in mind, to ensure that, in the right proportion, it becomes transparent in the laminate. The fibre furnish for making overlay is chosen for its very high absorbency and is usually of a very high alpha cellulose content. The latter ensures that the refractive index of the paper is very close to that of the melamine resin, ensuring transparency upon lamination. The paper maker needs to ensure the highest possible standards of cleanliness. It must produce a paper which is very absorbent, and strong enough to process through the impregnator without breaking. Typically an overlay paper might have a Klemm absorbency up to 70 mm/10 min and a wet tensile strength around 6 N/15 mm (i.e. the force required to break a strip of paper 15 mm wide is 6 N). It is designed to pick up well in excess of double its own mass of dried resin. The species of wood fibre used for overlays is predominantly of the softwood sulphate type

---

Figure 3 (opposite) (a) Typical single daylight press. (b) One daylight of a multi-daylight press. (There may be perhaps 20 daylights in one press.) (c) Typical continuous press configuration. 1, decorative paper; 2, core paper; 3, vegetable parchment; 4, textured/embossed release paper unwind and rewind.

Table 2 Typical physical characteristics of paper types

Physical characteristic	Overlay	Decorative sheet	Core paper
Grammage (g/m^2)	16–40	75–200	80–105
Thickness (μm)	30–100	115–300	120–160
Klemm absorbency (mm/10 min)	55–80	25–60	60–100
Gurley proposity (s/100 ml)	1–5	3–30	5–15
Tensile strength			
Dry (N/15 mm)*	20–30	30–40	Over 100
Wet (N/15 mm)*	4–7	5–8	8–10
Smoothness (Bekk sec)	Below 10	15–150	Below 10
pH of extract	5.5–8.0	5.5–8.0	6.5–7.5

* Force in Newtons required to break a strip of paper 15 mm wide.
Note: As a general comment: LPL types tend to be more dense, less absorbent and less porous than HPL types.

such as Scandinavian Pine, supplemented by relatively small proportions of short-fibred hardwoods such as eucalyptus. The final choice is dictated by the combination of physical specification for the paper together with optimum transparency.

Decorative surface sheet

This may be a printed design or a so-called 'plain colour', or 'solid colour'. In terms of fibre furnish most decorative papers, of either type, have a very high proportion of short fibred hardwood such as eucalyptus. This species has excellent characteristics in respect of producing a sheet of paper of good 'look-through' or 'formation' to use the more technical term. This is essential for uniformity of resin distribution. Short fibres also exhibit excellent printing characteristics. In order to achieve adequate strength the fibre furnish includes a proportion of longer fibred softwood pulps such as Scandinavian Pine. One should expect to find proportions of hardwood to softwood in excess of 80/20.

Printing papers are usually relatively lightweight – say 60–80 g/m^2. They have to be made with a very smooth surface in order to accept the ink from the high definition gravure printing process used to impart the design. Various types of paper machine processes are available to produce the smooth surface, from traditional steel calender rolls to MG cylinders and so-called 'soft calenders'. Printed designs are often produced on a multi-station printing machine, applying perhaps six separate colours. Good dimensional stability is an essential ingredient of these papers, to ensure precise registration of each colour component of the final design. This has become particularly important as the printing industry moves from solvent based inks to aqueous systems. Apart from this interaction with printing inks and, of course, appropriate strength requirements for the printing machine, the paper must be designed such that after printing, it lends itself to trouble free impregnation and pressing.

Plain colours cover a wide range of effects, from high opacity whites, through just about every possible variant of the colour spectrum, to black.

White papers are obtained by using very large proportions of rutile titanium dioxide (up to 40% by weight retained in the paper). Suitable techniques have been developed which ensure optimum retention of the titanium dioxide in the paper, whilst endeavouring to keep the particles of pigment in a highly dispersed state to maximize light scattering, for good opacity. Inevitably there is compromise between seeking the presence of discrete particles in the sheet (the best for opacity) and having large flocs or agglomerates (the best for retention). Much research and development involving either polymeric retention aids or other mechanical or chemical means is held very secretively by the companies competing in this field.

Coloured papers are especially complex in their retention chemistry. They need also to be of certain levels of opacity, so they include titanium dioxide as well. The colour formulations in modern decorative laminates tend now to be very complex. Historically one of the main sources of coloured pigmentation was from materials derived from 'heavy metals' such as lead, cadmium and chromium. However, more environmentally acceptable alternatives were sought in the 1980s, and all the major decorative laminate paper producers are following the lead taken by a British papermaker, who had completely replaced these pigments with organic alternatives by about 1985. (See the entries on 'Dyes for the mass coloration of plastics', and 'Pigments for plastics', for further information on related issues.)

These newer materials necessitated very complex mixtures to achieve matches to the older pigments, which would be sufficiently light-fast and non-metameric. This complexity made for very difficult paper making chemistry. It is not possible to use dyes for decorative laminates. They do not have the required resistance to fading upon exposure to light (especially direct sunlight), and currently there is increasing interest in the use of laminates in exterior applications, so weather-resistance places even greater demands on the colour medium, which dyes could not fulfil. The paper, apart from being the medium which imparts the colour and opacity to the laminate, contributes significantly to the chemistry of the laminate, through interface with the resin chemistry. It also contributes to the mechanical properties of the laminate. There needs to be close cooperation between the laminate scientist and the paper technologist to optimize their respective processes. Typical specifications for these papers would include some measure of tensile strength (wet and dry), absorbency and porosity, and some chemical parameter which relates to resin cure.

Core papers

The unseen central core of high pressure decorative laminates is made from a number of layers (dictated by the required thickness of the laminate) of an unbleached Kraft paper. That is to say, papers made from softwood pulps extracted by the Kraft process, and with no bleaching. Such papers have high fibre length, resulting in a web of very high strength properties such as tensile strength and tear resistance. They are designed, by virtue of their absorbency and porosity, to be very easy to saturate with (usually) phenol formaldehyde resins, and they are able to absorb a large volume of resin. In many cases they are produced at mills having integrated pulp and paper manufacture, with the consequent efficiency and economy of such a process.

Backing paper or balance sheet

The bottom layer of the laminate is often a single sheet of a Kraft paper, similar to the core paper. However, it may be pigmented, and may be saturated either with phenolic or melamine based resins. Its function is to give a suitably keyed surface for bonding to a substrate, following a sanding process. It also plays a role in ensuring that the laminate does not warp.

PAPER FOR LPL

In these laminates a single layer of paper is applied to the two surfaces of a substrate, which is usually particle board. This sector is very sensitive indeed to resin economics; also, the distribution of resin in the finished laminate plays a very important role in the performance of the laminate. These papers are therefore designed to have very fast resin penetration, and to have a controlled distribution of resin within the sheet. Ideally one needs all the voids in the sheet to be filled, and to have a relatively rich deposit of resin at the decorative surface. These objectives are met by the correct choice of fibre mixture in making the paper, together with appropriate manipulation of the paper making process, such as refining, wet pressing and some form of calendering. The aim is to optimize the combination of void volume, density, smoothness, absorbency and porosity; and of course to ensure adequate strength for the impregnation plant.

PAPERS FOR DECORATIVE FOILS

These tend to be quite low grammage papers, in the range 40–60 g/m², and are designed to be either saturated or coated with resins which are usually mixtures of acrylics and urea-formaldehyde. They follow the principles of the other decorative papers, in that they need to be reasonably strong, and they must have the appropriate receptivity to their particular resin process. There are at least two further key requirements: (1) they are usually printed to very high quality standards, so the paper maker must ensure that this requirement is met; and (2) since the end product is glued to the substrate, it is essential that the foil is free of pinholes, to avoid glue penetration on to the decorative surface.

PAPERS FOR INDUSTRIAL LAMINATES

These laminates use a wide range of fibrous materials, among which paper has its part to play. Paper for industrial grades is usually highly specified in respect of electrical properties, notably its conductivity. This aside, the key requirements relate to impregnation and pressing performance. They include absorbency and porosity controls to ensure the desired resin pick-up, and strength properties to ensure both impregnator runnability and the mechanical properties of the laminate. Mostly such papers are 'unloaded' – that is to say free of any pigmentation. However, they are occasionally found with fillers which are present for a variety of reasons, such as augmentation of resin saturation, flame retardency, or other special needs.

ENVIRONMENTAL CONSIDERATIONS

Legislation and social attitude require increasing attention from the paper maker and the laminator. Coloured papers especially have seen radical changes in their chemistry with the advent of organic pigments to replace heavy metals. Chlorine remains an issue, because decorative papers need to be bleached, and some organic coloured pigments contain chlorine. The bleaching issue is being largely addressed by the pulp producers; there has been a major switch towards bleaching processes which do not use elemental chlorine (so-called 'ECF' pulps) and many are developing viable processes which are totally chlorine free ('TCF' pulps). The pigment problem is not currently thought to be addressable; brightly coloured pigments of certain shades can be achieved with halogenated organic compounds or with lead and cadmium compounds, but as already mentioned, current thinking excludes the heavy metals. The paper makers have become very skilled at recycling their own waste. However, there are severe restraints arising from the complex pigment chemistry, together with very exacting cleanliness standards, which effectively preclude the possibility of using general (i.e. post consumer) waste in the manufacture of these papers.

For the laminator there are increasing concerns about free formaldehyde emissions along with other resin derived chemicals, and all producers have programmes to reduce or eliminate such emissions. The question of disposal of the laminate at the end of its life cycle receives much attention. They are not biodegradable; incineration raises objections. Much work is being done to find other useful materials that might be made from reconstituted laminate. Some producers have processes which allow a proportion of waste laminate to be re-cycled in new production, as part of the core-material.

BIBLIOGRAPHY

Norman E. Beach, (1967) *Plastic Laminate Materials – Their Properties and Usage*, Fosters.

Keywords: high pressure/low pressure laminates, overlay, alpha cellulose, plain colours, opacity, pigment retention chemistry, titanium dioxide, chlorine free pulps.

Pigments for plastics

Robert M. Christie

INTRODUCTION

Pigments and dyes are distinctly different types of colourants. A pigment is a finely-divided solid which is essentially insoluble in its polymeric application medium. Pigments are incorporated by a dispersion process into the polymer while it is in a liquid phase and, after the polymer solidifies, the dispersed pigment particles are retained physically within the solid polymer matrix. In contrast, a dye dissolves in the polymeric application medium and is usually retained as a result of an affinity between individual dye molecules and molecules of the polymer. Pigments are generally preferred to dyes for the coloration of plastics mainly because of their superior fastness properties, especially migration resistance.

The main reason for incorporating pigments into plastics is to introduce colour (including black and white), either for aesthetic reasons and market appeal or because of functional demands. However, the optical role of a pigment can extend wider than simply providing colour, because it plays a decisive part in determining whether the medium is opaque or transparent.

Pigments may often perform useful functions that are more wide-ranging than their optical role, for example mechanical reinforcement or the inhibition of polymer degradation. On occasions, the incorporation of pigments can produce problems in plastics, such as the warping of polyolefins as a result of uncontrolled nucleation.

Pigments may be introduced into plastics by a variety of methods. Direct dry colouring, in which the pigment is incorporated into the molten polymer often along with other additives using high-shear dispersing

Plastics Additives: An A–Z Reference
Edited by G. Pritchard
Published in 1998 by Chapman & Hall, London. ISBN 0 412 72720 X

equipment, may be used. However, many manufacturers of plastic articles find it more convenient to make use of pre-dispersed concentrates or masterbatches of pigment in a liquid additive such as a plasticizer or in a compatible resin. Such concentrates are then easily incorporated by mixing into the final polymer composition at an appropriate stage of the processing sequence.

Pigments are conveniently classified as either inorganic or organic types. The properties of a pigment are primarily dependent on its chemical constitution. However, other factors influence the properties as a result of the fact that pigments are used as solid crystalline particles. One of these is the crystal structure, i.e., the way in which the molecules pack in their crystal lattice. Certain pigments, notably titanium dioxide and copper phthalocyanine, exist in different polymorphic forms with significantly different optical and stability properties. Further important factors, especially in influencing the strength or intensity of colour of pigments, are particle size and shape. Organic pigments generally show an increase in colour strength as the particle size is reduced, while with many inorganic pigments there is an optimum particle size at which the colour strength reaches a maximum. Other important factors which influence the dispersion properties in particular are the degree of aggregation of pigment particles and the nature of the particle surfaces.

REQUIREMENTS OF PIGMENTS FOR PLASTICS APPLICATION

The ability to produce the desired optical effect in the plastic product is obviously a prime requirement. However, the pigments must also be capable of withstanding the effects of the environment in which they are placed, both in processing and in their anticipated useful lifetimes. A pigment will be selected for a particular application on the basis of its technical performance but with due regard also to toxicological considerations and, inevitably, cost.

Optical properties: colour and opacity

The optical properties of materials are a result of a combination of two effects arising from the way they interact with visible light: absorption and scattering. An object appears coloured when it selectively absorbs certain wavelengths of visible light. The brightest, most intense colours are in general provided by the use of organic pigments. The colours of inorganic pigments are as a rule weaker and duller. High transparency in a plastic material requires the absence of light scattering centres either within the structure of the polymer itself or as a result of additives present. To produce a coloured transparent article, an inherently transparent polymer is coloured either with dyes which dissolve in the polymer or with organic pigments. Organic pigments are generally low refractive index materials manufactured in a fine particle size form, much smaller than the wavelength of visible light, ensuring that they cause minimal light scattering and thus are highly transparent. When there is considerable light scattering the plastic article will appear opaque. Inorganic pigments, a prime example being titanium dioxide, are usually high refractive index materials and therefore highly scattering and so when incorporated into plastics they provide opacity.

Fastness properties

The heat stability of a pigment refers to its resistance towards changing colour at high processing temperatures, and this is clearly an important factor for many plastics applications. Colour changes resulting from inadequate heat stability, which can lead to off-shades or a failure to match shades, may be due to thermal decomposition of the pigment, to increasing solubility of the pigment at elevated temperatures or to crystal phase changes. In the case of thermoplastics, heat stability is generally a critical feature in pigment selection. The degree of heat stability required will depend not only on the processing temperature for the polymer in question, which can range from 150 to 350°C, but also on the time of exposure. Heat stability requirements in the coloration of thermosets tend to be less severe than is the case with thermoplastics.

The properties of lightfastness and weatherfastness are clearly related although, on occasions, pigments which show good resistance to fading when exposed to light can perform less well under the combined attack of sunlight and moisture. These properties are determined principally by the chemical structure of the pigment but they may also depend to an extent on its concentration in the polymer and on the nature of the polymer. Lighter shades, especially in combination with white pigments, tend to show poorer lightfastness than deeper shades. Resistance towards changing colour when exposed to acids, alkalis or other chemicals is occasionally important for plastics applications, e.g. for thermosets cured in the presence of acid catalysts, for acrylic polymers incorporating peroxide curing agents or in containers for reactive liquids. The coloration of PVC generally requires the use of acid-resistant pigments.

Migration resistance

Four aspects of migration of colour from plastic materials may be identified: contact bleed, bloom, plate-out and solvent bleed, all of which are associated to a certain extent with solubility of the pigment in the polymer. Contact bleed occurs when a coloured plastic material causes staining of a dissimilarly coloured material with which it is in contact. Pigments which show some solubility in the polymer or in a plasticizer present in the formulation are liable to give contact bleed problems as a result of molecular diffusion. Bloom is observed when a powdery deposit of pigment appears on the surface of a plastic product giving rise to poor rub-fastness. This occurs when a pigment of limited solubility dissolves in the hot polymer, producing a supersaturated solution when cooled. In time, pigment molecules diffuse to the surface of the polymer and crystallization occurs. Plate-out refers to an accumulation of additive on the metal surfaces of plastics processing equipment. The reasons for plate-out are not fully established, but it appears to be a migration phenomenon associated with the presence in the polymer of a number of additives of different types, e.g. pigments, lubricants and stabilizers. Solvent bleed occurs when colour is leached from a plastics article when immersed in a solvent. Absence of solvent bleed is essential to ensure, for example, that a plastic bottle does not contaminate its contents.

Dispersion properties

Pigments are finely divided solid materials consisting of clusters of particles in varying degrees of aggregation. Two different types of particle cluster can be considered: aggregates and agglomerates. Aggregates are groups of primary particles (the smallest particles not normally further subdivided) joined at their faces. Agglomerates are considered to be groups of aggregates or primary particles joined at edges and corners. Agglomerates are much less tightly bound together than aggregates and are therefore easier to disrupt. During dispersion, agglomerates, and to a certain extent aggregates, are mechanically reduced to smaller aggregates and primary particles. It is essential that a pigment is well-dispersed in a plastic material in order to achieve an even colour distribution with the absence of coarse, undispersed particles ('specks'). In addition, control of the degree of subdivision may be used to optimize optical performance. For example, with organic pigments colour strength increases as particle size is reduced so that efficient dispersion is essential.

The term 'dispersibility' refers to the ease with which the desired degree of dispersion is achieved. Over the years, pigment manufacturers have dramatically improved the dispersibility of the products produced commercially, thus minimizing the energy requirements of the dispersion process. This is generally achieved by a surface treatment process in which the pigment particles are coated with surface active agents or organic resins. These treatments reduce the degree of particle aggregation and, in addition, the aggregates which are formed are of lower mechanical strength and hence easier to disrupt.

Electrical properties

Pigments which reduce the insulating properties of PVC are, for example, unsuitable for electrical cable, although such a defect is more often due to residual electrolyte on the surface of the pigment than to the pigment itself.

Toxicological and environmental considerations

Plastics are generally perceived as relatively 'safe' materials so that it is essential to assess whether the incorporation of pigments introduces any hazards, either in the workplace or by exposure of the general public to the product. It is therefore reassuring that most pigments may be considered as relatively non-toxic, inert materials. It may be argued that the insolubility of pigments means that they pass through the digestive system without absorption into the bloodstream and so present little hazard. However, toxicological considerations are of critical importance for applications where ingestion is a possibility, such as plastics in contact with food and in toys and graphic instruments and there is long-standing legislation limiting the use of certain pigments, notably those containing lead, hexavalent chromium and cadmium, in such applications.

As a result of growing concern on the more general impact of these pigments on the environment, recent years have seen the extension and growth in complexity of legislation limiting the use of these pigments further and this process seems set to continue into the future. The issues which are becoming of increasing concern to pigment manufacturers and users include environmental problems during manufacture, such as the management of waste water and other residues, safety in production and hygienic materials handling, toxicological and ecotoxicological safety of the final products, traces of potentially harmful impurities such as PCBs, dioxins, free aromatic amines and non-bound heavy metals, and environmental fate at the disposal stage. As this scrutiny intensifies, the use of some other pigments, such as those containing other heavy metals and organic halogens, may in due course be questioned. The development of alternative pigments which offer equivalent technical performance but which are more acceptable environmentally is probably the most important challenge currently facing pigment manufacturers.

INORGANIC PIGMENTS

Natural inorganic pigments derived from mineral sources have been used as colourants since prehistoric times and a few, notably some iron oxides, remain of some significance today. However, they have been largely superseded by a range of technically superior synthetic inorganic pigments including white, black and a variety of coloured products. Inorganic pigments are generally well-suited to plastics applications because of their ability to provide excellent resistance to heat, light, weathering, migration and chemicals at relatively low cost, and in those respects they can offer technical advantage over many organic pigments. On the other hand, they frequently suffer from the disadvantage of inferior intensity and brightness of colour and organic pigments are used in preference where good colour properties are of prime importance. Inorganic pigments, because of their high refractive index, are generally used where opacity is required.

White pigments

A wide range of insoluble white powders find extensive use in plastics. These products may be classified as either non-hiding or hiding white pigments. Non-hiding white pigments, more commonly referred to as extenders or fillers, are inexpensive white powders used in large quantities by the plastics industry.

They are products of relatively low refractive index and thus they are capable of playing only a minor role in providing opacity. They are used in plastics mainly to reduce cost. Commonly-used non-hiding white pigments include calcium carbonate, barium sulphate, talc (hydrated magnesium silicate), china clay (hydrated aluminium silicate) and silica.

Hiding white pigments, as a result of their high refractive index, are capable of providing high opacity in plastics applications. Titanium dioxide (TiO_2) is by far the most important hiding white pigment used in plastics. It owes its dominant position to its ability to provide a high degree of opacity and whiteness (maximum light scattering with minimum light absorption) and to its excellent durability and non-toxicity. The pigment is manufactured in two polymorphic forms: rutile and anatase. The rutile form with its higher refractive index and better weathering properties is much more important commercially than the anatase form. However, the anatase form exhibits lower absorption in the blue-violet region of visible light below 420 nm and is frequently used, often in conjunction with fluorescent brightening agents, when a distinct blue-whiteness is desired. Titanium dioxide pigments produce maximum opacity at a particle size of around 0.25 µm. However, finer particle size grades (around 0.18 µm) which have a cleaner blue tone because of higher scattering of the lower blue-violet wavelengths are generally preferred for plastics, especially for those polymers with an inherent yellowish colour.

As a polymer containing a white pigment degrades on exposure to natural weathering and its surface erodes, the release of pigment particles may produce a white powdery deposit ('chalking'). Titanium dioxide pigments, as a result of their strong absorption of UV radiation, provide the polymer with a degree of protection against this degradation. The rutile form provides inherently greater protection although in a few applications, such as in white sidewall car tyres, the controlled chalking resulting from the use of the anatase form is utilized to advantage. Surface treatment of the TiO_2 particles with silica, alumina or other inorganic or organic materials is used to enhance weathering properties, to improve dispersibility and to minimize the irreversible yellowing which can result from the reactivity to titanium dioxide towards phenolic anti-oxidants used in polyalkenes, polysytrene and ABS.

There are other white hiding pigments produced commercially, including zinc oxide, zinc sulphide, lithopone (a mixture of ZnS and $BaSO_4$) and antimony oxide, but they are of only minor importance in plastics.

Black pigments

Carbon black is by far the most important black pigment and is the second most important in terms of volume of all pigments used by the plastics industry, ranking behind only titanium dioxide. Although there is a convincing argument that carbon black ought to be regarded chemically as an organic pigment, it is more commonly classified amongst the inorganics. In applications where their role is to provide a black colour, carbon black pigments exhibit high tinctorial strength and an outstanding range of fastness properties at relatively low cost. However, carbon blacks can adopt a number of other important functions when incorporated into polymers. The pigments excel in their ability to protect polymers against weathering as a result of a combination of UV absorption and their ability to function at the particle surfaces as traps for radicals formed in photodecomposition. Carbon blacks can also function as thermal antioxidants, particularly in LDPE. Some grades of the pigments, which exhibit a high degree of structure involving chain-like clusters of particles, exhibit reinforcing characteristics which are of considerable commercial importance especially in rubber applications. Properly selected grades may be used to improve either the electrical conductivity or insulating properties of a polymer.

Coloured inorganic pigments

The most important coloured inorganic pigments are iron oxides, which provide shades ranging from yellow and red to brown and black. These pigments are manufactured either from natural deposits or by synthetic routes. However, the synthetic products, which offer the advantages over their natural counterparts of higher chemical purity and better control of physical form, are generally preferred by the plastics industry. Red iron oxide pigments consist principally of anhydrous Fe_2O_3 in its α-crystal modification. Yellow iron oxides, although often described as hydrated oxides, are more correctly formulated as the oxidehydroxide, FeO(OH). Brown iron oxides are somewhat variable in composition, consisting principally of non-stoichiometric mixed oxidation state [Fe(II)/Fe(III)] oxides. Iron oxide pigments are characterized in general by excellent durability, high inherent opacity, good UV-screening properties, low toxicity and low cost, but rather low intensity and brightness of colour. The yellow pigments at temperatures above 175°C undergo dehydration to Fe_2O_3 with a resulting shade change, clearly a critical limitation to their use in plastics. The only other 'simple' oxide pigment of major significance is the chromium oxide, Cr_2O_3, a tinctorially weak dull green pigment which is used in a range of plastics applications where outstanding durability, including exceptional thermal stability, is of greater importance than brightness of colour.

The mixed phase oxides are a group of coloured inorganic pigments which were developed originally for use in ceramics but which have subsequently found widespread application in plastics because of their outstanding heat stability and weathering characteristics combined with moderate colour strength and brightness. Important commercial examples include cobalt aluminate blue and nickel antimony titanium yellow. Structurally, these pigments may be considered to consist of stable colourless host lattices such as rutile (TiO_2) and spinel ($MgAl_2O_4$) into which are incorporated varying quantities of transition metal ions, giving rise to a variety of coloured products which retain the excellent durability characteristics of the host lattice.

Cadmium sulphide and sulphoselenide pigments provide a range of colours from yellows to reds and maroons, rather brighter and stronger than other classes of inorganic pigments. Cadmium sulphide itself (α-CdS) is yellow. Partial replacement of Cd^{2+} by Zn^{2+} ions in the lattice produces progressively greener shades of yellow, while replacement of sulphur by selenium gives rise to the orange, red and maroon sulphoselenides. The use of cadmium sulphide pigments is limited severely on the grounds of their potential toxicity. In the European Union there is a Directive which restricts their use in applications where they are not seen to be essential. Accordingly their use is currently based on necessity, related to their excellent technical performance, in particular thermal stability to 600°C, unmatched in their shade areas by any organic pigments. Their principal use tends to be restricted to engineering polymers requiring high processing temperatures such as polyamides, acetals and PTFE. It seems certain that in years to come the decline in the use of cadmium pigments will continue as more environmentally-acceptable alternatives are developed.

Lead chromate pigments provide a range of hues from greenish-yellow to yellowish-red. The mid-shade yellows are essentially pure $PbCrO_4$. The greener lemon chromes are solid solutions containing $PbSO_4$, while incorporation of molybdate ions into the lattice gives the orange and light red molybdate chromes. These pigments, which provide bright colours with good fastness properties, find extensive application in paints but are used much less in plastics. Like cadmium sulphides, their use is limited by the potential toxicity hazards associated in this case with the presence of both lead and hexavalent chromium.

Ultramarine blue exhibits excellent fastness to light and heat, although rather poor resistance towards acids. It has a complex zeolytic sodium aluminosilicate structure containing trapped S_3^- radical anions which are responsible for its colour. The pigment is widely used to provide bright clean reddish-blue shades in plastics, although it has only a fraction of the tinctorial strength of copper phthalocyanine blue. Prussian blue, or iron blue, $Fe_4(Fe(CN)_6)_3 \cdot nH_2O$, is the longest established of all synthetic colouring materials currently in use. It is a low cost blue pigment with good light fastness, but relatively poor resistance to heat and alkalis and its use in plastics is restricted largely to LDPE.

ORGANIC PIGMENTS

Organic pigments are characterized in general by high colour strength and brightness and good transparency, although they are somewhat variable in the range of fastness properties which they offer. The most important commodity yellow, orange and red organic pigments currently in use in plastics are a long-established series of azo pigments. A critical event in the development of the organic pigment industry was the discovery in the 1920s of copper phthalocyanine blue, the first product to offer outstanding colouristics combined with a range of excellent fastness properties. Since then a wide range of structural types of organic pigments emulating the properties of copper phthalocyanines in the yellow, orange, red and violet shade areas has been introduced to meet the high performance demands of many plastics, synthetic fibres and paint applications. They tend, however, to be rather expensive.

Azo pigments

Azo pigments, both numerically and in terms of tonnage produced, dominate the yellow, orange and red shade areas in the range of commercial organic pigments. Azo colourants are generally described as structures containing one or more azo (−N=N−) groups, even though many have been shown to exist preferentially in the alternative structural hydrazone (−NH−N=) form.

(1) Monoazoacetoacetanilide, C.I. Pigment Yellow 97

(2) Disazoacetoacetanilide, C.I. Pigment Yellow 83

Simple classical monoazo pigments such as the Hansa Yellows and Toluidine Red find little general use in plastics as they are too soluble and hence liable to migration. Migration resistance of azo pigments may be enhanced by increasing the molecular size, by incorporating metal ions or by introducing functionality, such as the amide group (–NHCO–), which leads to strong intermolecular association in the crystal lattice and hence low solubility. Compound (1) is an example of a highly-substituted, higher molecular weight monoazo pigment recommended for use in polystyrene, polypropylene and rigid PVC. The most important yellow azo pigments, used in polyethylene, polypropylene and flexible PVC, are the disazoacetoacetanilides (Diarylide or Benzidine Yellows), exemplified by compound (2), which exhibit good colour strength and migration resistance although somewhat inferior lightfastness. The disazopyrazolones are a small group of orange pigments which are structurally related to the Diarylide Yellows and are similar in their properties and applications. The most important classical red azo pigments are metal salts, such as the calcium 2B toner (3). These are products of good colour strength, brightness and migration resistance which are used in many thermoplastics applications.

There are two main types of high performance azo pigments. Disazo condensation pigments are a range of durable, if rather expensive, yellow, red, violet and brown products with complex high molecular

(3) Calcium 2B toner, C.I. Pigment Red 48: 2

weight structures. The second type are products containing a benzimidazolone group which range in shade from yellow to bluish-red and browns and exhibit excellent fastness properties, including thermal stability to 330°C. Their good stability to light and heat and their insolubility is attributed to strong intermolecular hydrogen-bonding and dipolar forces involving the benzimidazolone group in the crystal lattice.

Copper phthalocyanines

(4) Copper phthalocyanine, C.I. Pigment Blue 15

Copper phthalocyanine (4) is arguably the single most important organic pigment. It finds widespread use in most plastics applications because of its brilliant blue colour and its excellent resistance to light, heat, migration, acids and alkalis. In addition, in spite of its structural complexity, copper phthalocyanine is a relatively inexpensive pigment as it is manufactured in high yield from low cost starting materials. There are two commercially important crystal modifications of the pigment, the reddish-blue α-form and the more stable greenish-blue β-form. The α-form of copper phthalocyanine may be stabilized by the presence of a single ring chlorine substituent, and for plastics these are preferred to the unstabilized pigments, which tend to convert in application to the β-form with a corresponding shade change. Copper phthalocyanine green pigments are products in which most of the sixteen outer ring hydrogen atoms are replaced by halogens, and include polychloro, polybromo and polybromochloro derivatives, the shade of the pigments becoming progressively yellower with increasing bromine content. The copper phthalocyanine greens, which exhibit properties comparable to the blues, are also excellent pigments for plastics.

Miscellaneous high-performance organic pigments

(5) Quinacridone, C.I. Pigment Violet 19

(6) Diketopyrrolopyrrole, C.I. Pigment Red 254

Many of this group of products owe their good stability to light and heat and their extreme insolubility to extensive intermolecular association as a result of hydrogen-bonding and dipolar forces in the crystal lattice. A notable example is provided by quinacridones, such as compound (5), which are used widely in plastics offering red and violet colours and heat stability to 400°C. A number of vat dyes developed originally for textile applications are suitable, after conversion to an appropriate pigmentary physical form, for use in many plastics applications. Examples of these so-called vat pigments include the anthraquinones, Indanthrone

Blue and Flavanthrone Yellow, red to violet thioindigo derivatives and Perinone Orange. Other high performance pigments used in plastics include the red and maroon perylenes, the yellow to red tetrachloroisoindolinones and Dioxazine Violet. Arguably the most significant development in organic pigments in recent years has been the discovery and development of diketopyrrolopyrrole (DPP) pigments, such as compound (6). These pigments are capable of providing brilliant saturated red shades with outstanding durability, including excellent thermal stability. As such they are of considerable interest for the pigmentation of plastics, and their use seems certain to increase in years to come.

PIGMENTS FOR SPECIAL EFFECTS

Fluorescent pigments

The striking brilliance of a fluorescent colour results when a molecule absorbs visible radiation and re-emits an intense narrow band of visible light at somewhat higher wavelengths, reinforcing the colour already present due to normal visible light absorption. Fluorescent pigments consist of low concentration solid solutions of fluorescent dyes in a transparent resin ground to a fine particle size. Fluorescent pigments tend to show inferior lightfastness and thermal stability compared with normal coloured pigments, factors which restrict their use to a certain extent in plastics. Their main uses in plastics are for visual impact in toys, packaging and safety applications.

Pearlescent pigments

Pearlescent pigments give rise to a white pearl effect accompanied by a coloured irridescence. The most important pearlescent pigments are thin platelets of mica coated with titanium dioxide which partly reflect and partly transmit incident light. Simultaneous reflection from many layers of oriented platelets creates the sense of depth characteristic of pearlescent lustre and, where the particles are of an appropriate thickness, colours are produced by interference phenomena. For maximum pearl effect in a plastic material, sufficient shear to produce a good dispersion of the pigment but avoid disintegration of the platelets is desirable.

Metallic pigments

The most important metallic pigments are those based on finely-divided aluminium flakes. They may be used in plastics to simulate the colour of the metal or, in combination with transparent organic pigments, to produce novel metallic colouristic effects. Pre-dispersed preparations in liquid plasticizer or resin granules are invariably used because dry aluminium flakes are potentially explosive. The shear used in dispersing aluminium pigments into polymeric materials should be minimized in order to prevent disruption or deformation of the flakes which would lead to a reduction in the metallic effect. Bronze pigments are used similarly to give a gold or copper effect.

BIBLIOGRAPHY

Pigment Handbook (1973) 1st edition, ed. by T.C. Patton, Vols I–III, John Wiley and Sons, New York; (1988) 2nd edition, ed. by P.A. Lewis, John Wiley and Sons, New York.

Christie, R.M. (1993) Pigments: Structures and Synthetic Procedures, Oil and Colour Chemists Association, London.

Kaul, B.L. (1993) 'Coloration of Plastics using Organic Pigments', Rev. Prog. Coloration, **23**, 19.

Christie, R.M. (1994) 'Pigments, Dyes and Fluorescent Brightening Agents for Plastics: an Overview', Polymer International, **34**, 351.

Herbst, W. and Hunger, K. (1993) Industrial Organic Pigments, VCH Verlagsgesellschaft, Weinheim.

Keywords: pigment, colour, opacity, transparency, dispersion, dispersibility, titanium dioxide, cadmium, carbon black, azo, phthalocyanine.

See also: Dyes for the mass coloration of plastics;
Fillers;
Mica;
Conducting fillers for plastics: (1) Flakes and fibers.

Plasticizers

C.J. Howick

PLASTICIZERS FOR POLY(VINYL CHLORIDE) POLYMERS

Poly(vinyl chloride) (PVC) is the second largest selling pure polymer in Western Europe, second only to polyethylene in terms of tonnage sales of pure resin. However, in terms of tonnage sales of finished goods, the order is changed, because of the more widespread use of additives in PVC applications. This entry concerns one of the more important additives used.

PVC as a pure resin has very poor properties and requires the use of additives to produce products of acceptable quality. The need for the use of PVC additives can be thought of in two ways: negatively, it could be thought that the use of such additives introduces unwanted complexity and additional price. More positively, the use of additives gives the end product producer an additional ability to tailor the properties of the product to the market without the need for a change in polymerization conditions. Additionally, many PVC additives are available at prices below that of the resin so that, especially if volume cost savings are taken into account, the judicious use of additives frequently results in cost savings. These observations are particularly relevant when discussing plasticizers.

Plasticizers are typically organic liquids which can be added to PVC to obtain a product with flexibility. If PVC is processed with thermal stabilizers (necessary for all PVC applications) but without plasticizer, a rigid product is obtained, the material being termed unplasticized PVC or PVC-U (this includes the well known application of window profiles). The successful addition of a plasticizer will result in the formation of a product with a degree of flexibility, such as a cable insulation or sheathing, a floorcovering or flexible profile. The clear differentiation

Plastics Additives: An A–Z Reference
Edited by G. Pritchard
Published in 1998 by Chapman & Hall, London. ISBN 0 412 72720 X

Table 1 Usage of plasticizers in PVC applications

Product	PVC type	Typical plasticizer level* (phr)	Typical plasticizer type	Comments
Cables	Suspension	50–60	DOP/DIDP Speciality phthalates Trimellitates	Specialities for high temperature applications
Calendered sheets (including calendered flooring)	Suspension	50–70	DOP/DIDP	Often high mineral filler loadings
Spread flooring	Plastisol	40–50	DOP/DIDP Phosphates Adipates Speciality phthalates	BBP use for foam layers. Low stain resistance needed
Synthetic leather	Plastisol	50–80	DOP 911P for automotive applications	Low fogging requirement for automotive
Wallcovering	Plastisol	40—60	DOP/DINP	Low viscosity requirement
Food film	Suspension	40—50	DOA/polymeric	Food contact legislation

* phr = Parts per hundred of resin, by weight.

and diversity of properties brought about by the addition of a plasticizer indicates the importance of plasticizers in PVC technology (see Table 1).

PLASTICIZER ACTION

Plasticizers act to reduce the glass transition temperature (T_g) of PVC. Under this definition, any substance capable of producing a reduction can be classed as a plasticizer, but in common usage other properties (see below) will place restrictions on the choice of material to be used as a plasticizer. For many applications, a reduction of T_g to ambient temperature is required, whereas in certain technical applications the temperature range for material flexibility is larger.

In rigid, unplasticized PVC (PVC-U), strong interactions exist between neighbouring polymeric chains, resulting in a rigid network which gives the polymer its well characterized properties of rigidity, tensile and impact strength (see Figure 1). When PVC is heated, these interactions are lost, resulting in a flowable, formable melt which can be satisfactorily processed. For PVC-U however, the attractions are re-formed as the processed material is cooled, as there are no substances competing for

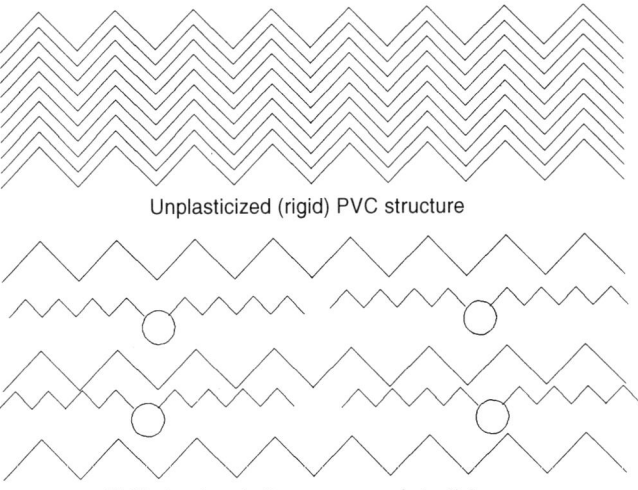

Figure 1 PVC structures with and without plasticizer.

inter-chain attractions. When PVC is heated in the presence of a plasticizer, however, an intimate mix of polymer and plasticizer is formed in the melt. When the melt is cooled, plasticizer molecules can set up their own interactions with the PVC chain (see Figure 1). As plasticizers usually possess relatively long alkyl chains, they have the effect of screening the polymer chains from each other, thereby preventing them from re-forming the chain–chain interactions which give the unplasticized polymer its rigidity. As long as there is sufficient compatibility between polymer and plasticizer, the plasticizer will be retained in the flexible product that is formed.

PLASTICIZER TYPES

Introduction

The vast majority of plasticizers are organic esters. As they are used in thermal processing with PVC resins, they usually have relatively high boiling points (>300°C). The choice of plasticizer will depend on the properties required of the final product, the applications technology used to make it and the price boundaries present for the process and product. Many types of plasticizers are available.

General purpose plasticizer esters

The most widely used are the family of phthalate esters. These are produced by the esterification of phthalic anhydride with various alcohols.

These plasticizers can be further grouped in a number of ways; one convenient distinction is into general purpose and speciality plasticizers. General purpose types include the three phthalates: di-2-ethylhexyl phthalate (DEHP or DOP), di-isononyl phthalate (DINP) and di-isodecyl phthalates (DIDP). These are the esters of phthalic anhydride with the alcohols 2-ethylhexanol, iso-nonanol and isodeconal respectively, and they account for about 75% of plasticizer usage in Western Europe. Other phthalates are also manufactured, but find use more in applications where specific properties are required.

Properties required of a plasticizer

Naturally, the properties required of a plasticizer for a flexible PVC application will depend on the properties required of the final product as well as the method of application involved. It is important here to state that many different types of PVC resin exist. Suspension PVC resins for flexible applications possess relatively large and porous particles which will absorb the liquid plasticizer when added. The resulting blend, generally termed a dry-blend or powder blend, can then be extruded, injection moulded, blow-moulded or calendered. Paste or plastisol-forming PVC polymers have relatively small particles of limited porosity which, when mixed with plasticizer, form a plastisol or paste. This resembles a paint and can be processed by spreading, casting, spraying and rotational moulding.

Speciality plasticizer esters

Speciality phthalates

In addition to the general purpose phthalates described above, other phthalates also find use in PVC applications, although they are usually confined to applications for which the general purpose phthalates do not give the performance required of either the end product or the processing technology. This includes the relatively fast fusing di-isoheptyl phthalate (DIHP) and fast fusing butylbenzyl phthalate (BBP) which, owing to their above average polarity, give higher than average fusion rates with PVC, making them suitable for low temperature applications. Also, and especially in the case of BBP, they give very good foaming properties with plastisol grade PVC resins. At the other extreme of the activity scale, di-isoundecyl phthalate (DIUP), di-undecylphthalate (DUP) and di-isotridecyl phthalate (DTDP) find usage in high temperature cable applications in which exceedingly low volatile loss figures are required.

Phthalate esters based on linear and semi-linear alcohols in the C9 to C12 range are now a very popular material in the automotive synthetic leather market because they combine the properties of low plastisol viscosity with little tendency to volatilize and cause windscreen fogging.

Adipates

These materials show similar activity to the phthalate analogues but possess the added benefits of very low viscosity, thus making them particularly attractive to the plastisol sector, and giving excellent low temperature flexibility properties. Additionally, di-octyladipate (DOA) has been used in food contact applications where it combines excellent technical properties (cling) with legislative approval. In plastisol applications the low viscosity property of the plastisol produced with an adipate has various processing advantages. In floorings the use of adipate can extend the lower end of the temperature flexibility range. This latter property often also makes adipate esters the material of choice for underground cables in arctic environments.

Trimellitate esters

These esters are structurally similar to phthalates but the presence of a third carbonyl functionality and third alcohol chain greatly enhances the high temperature properties, making them the material of choice for applications in which high temperature thermal stability is required. Additionally, if the alcohol used in the trimellitic anhydride esterification process is predominantly linear, a material possessing good low temperature flexibility properties results, thus making the material a highly versatile one. For the end user, the only drawback is that, owing to the extensive additional chemistry involved in producing these properties, these materials tend to attract price premiums over other types of plasticizer.

Other speciality plasticizers

Other plasticizers are used in various applications where a specific property or properties are required of the end product. These include phosphate esters used in fire retardant applications such as cables and spread contract flooring; polymeric plasticizers (usually a polymeric adipate, used in low migration food contact applications); benzoates (fast fusion), and sebacates and azelates (which are similar to adipates but give enhanced properties). For further information the reader is directed to the recent definitive text by A.S. Wilson.

BIBLIOGRAPHY

Sears J.K. and Darby J.R. (1982) The Technology of Plasticisers, Wiley and Sons, New York.

Wilson, Alan S. (1995) Plasticisers, Principles and Practice, The Institute of Materials, ISBN 0 901716766.
Howick C.J. (1995) Plasticisers for Poly(Vinyl Chloride) RAPRA Reviews, Vol. 11, No. 4, 239.

Keywords: glass transition temperature, PVC, phthalates, adipates, trimellitates, polymeric plasticizers.

See also: Plasticizers: health aspects.

Plasticizers: health aspects

D.F. Cadogan

INTRODUCTION

About 1 million tons of plasticizers are used annually in Western Europe, mainly in the plasticization of PVC. The vast majority of plasticizers (around 90%) are esters of phthalic acid (phthalates) with a wide variety of long chain alcohols containing up to 13 carbon atoms. The remainder are also esters or polyesters and include those based on adipic, trimellitic, phosphoric, sebacic or azelaic acids. When considering the possible health aspects of plasticizers it is logical therefore to concentrate on phthalates.

The widespread use of phthalates over the last 40 years has led to their toxicology being extensively researched and understood. Many different phthalates have been studied but the particular phthalate which has been most thoroughly investigated has been di-(2-ethylhexyl) phthalate (DEHP). This is because it is the most widely used plasticizer and, being a well-defined single substance, has often been considered as a model for the other phthalates.

Plasticizers possess an extremely low order of acute toxicity; LD_{50} values are mostly in excess of 20 000 mg/kg body weight for oral, dermal or intraperitoneal routes of exposures.

In addition to their low acute toxicity many years of practical use coupled with animal tests show that plasticizers do not irritate the skin or mucous membranes and do not cause sensitization.

The effects of repeated oral exposure to plasticizers for periods ranging from a few days to two years have been studied in a number of animal species including rats, mice, hamsters, guinea-pigs, marmosets and monkeys [1]. These studies have shown that some plasticizers may cause adverse effects in the liver and reproductive systems of certain species.

Plastics Additives: An A–Z Reference
Edited by G. Pritchard
Published in 1998 by Chapman & Hall, London. ISBN 0 412 72720 X

LIVER EFFECTS

In 1980 a two year feeding study carried out as part of the NTP/NCI Bioassay Programme in the United States indicated that DEHP causes increased incidence of liver tumours in rats and mice and that di-(2-ethylhexyl) adipate (DEHA) had a similar effect in mice but not rats. In these studies the levels of plasticizers fed were very high, this being possible because of their low acute toxicity.

These findings led to a large number of more detailed investigations being carried out on a variety of plasticizers and different animal species. These have revealed the following.

- Plasticizers are not genotoxic.
- Oral administration of plasticizers, fats and other chemicals including hypolipidaemic drugs to rodents causes a proliferation of microbodies in the liver called peroxisomes which are considered by some authors to be linked to the formation of liver tumours.
- Administration of plasticizers, fats and hypolipidaemic drugs to non-rodent species such as marmosets [2] and monkeys [3] (primates considered to be metabolically closer to humans) does not lead to peroxisome proliferation and liver damage. In addition some hypolipidaemic drugs which cause peroxisome proliferation in rodents have been used by humans for many years with no ill effects.
- These species differences have also been observed in in-vitro studies. Phthalates, their metabolites and a variety of other peroxisome proliferators caused peroxisome proliferation in rate and mouse liver cells but not in those of humans, marmosets or guinea pigs.

On the basis of these differences in species response and taking into consideration the extremely high exposure in the animal tests, it was concluded some years ago that phthalates do not pose a significant hazard to man. This view is borne out by the EU Commission decision of 25 July 1990 which states that DEHP shall not be classified or labelled as a carcinogenic or an irritant substance [4].

This has recently been reaffirmed in a comprehensive review [5] which concludes that:

> 'peroxisome proliferators constitute a discrete class of non-genotoxic rodent hepatocarcinogens and that the relevance of their hepatocarcinogenic effects for human hazard assessment is considered to be negligible'.

The International Agency for Research on Cancer (IARC) has classified DEHP [6] as 'an agent possibly carcinogenic to humans'. However this is based only on the rodent studies and does not take into account the more recent understanding of the underlying mechanisms. It is possible that this classification will be open to change in the future following

REPRODUCTIVE EFFECTS

Some phthalates, at high dose levels, have been shown to cause reproductive effects in rats and mice. However the relevance of these findings to humans is limited due to the dose level being far in excess of human exposure and there being significant species differences in response.

Dose level relative to human exposure

The reproductive toxicity of some phthalate esters has recently been reviewed by the Commission of the European Communities [8]. This review concludes that testicular atrophy is the most sensitive indicator of reproductive impairment and that the rat is the most sensitive species. The no effect level for body, testes, epididymis and prostate weights and for endocrine and gonadal effects of DEHP in male animals after oral administration was considered to be 69 mg/kg body weight per day. This conclusion was endorsed by the Danish Institute of Toxicology [9].

This no effect level is orders of magnitude higher than estimated human or environmental exposure. There are a number of recent estimates of the average daily lifetime exposure to DEHP which have been produced by a variety of authoritative bodies. These include:

- Europe [10]   2.3–2.8 µg/kg/day
- USA [11]   4 µg/kg/day
- Canada [12]   6 µg/kg/day.

These estimates are in line with the most recent UK Ministry of Agriculture, Fisheries and Food (MAFF) investigations [13] which give the DEHP dietary intake as 2.5 µg/kg body weight (mean) and 5.0 µg/kg body weight (max).

A reasonable estimate of the average daily lifetime exposure to DEHP would therefore be 5 µg/kg body weight/day.

Comparing the estimated exposure to DEHP with the level at which no effects are observed in rats indicates that the margin of safety for the general public is around 14 000. Taking into consideration the significant difference in species sensitivity between rodents and primates, the safety factor is in fact even greater.

If we assume that DEHP accounts for around 50% of the total phthalate consumption then the total exposure to phthalates would be approximately 10 µg/kg/day and the margin of safety 7000.

Species differences

Primates appear resistant to the reproductive effects seen in rodents.

A recent study with DEHP in marmosets which has been carried out in Japan shows that with repeated dosing at up to 2500 mg/kg for 13 weeks there is:

- no difference in testicular or prostate weight;
- no histopathological changes in the testes, epididymis, seminal vesicle or prostate;
- no morphological changes in the various cells of the testis including Leydig's cells, spermatogonia, spermatocytes and spermatids;
- no differences in blood testosterone or oestradiol levels.

The lack of effect on the testes of primates is in agreement with the findings of Rhodes et al. [2]. They found no reduction in testicular weight in the marmoset after dosing for 14 days at 2000 mg/kg/day.

This is very different from the response in the rat and is most likely due to the marked differences in metabolism between rats and primates (marmosets, cynomolgus monkeys and humans).

These differences in metabolism are shown in a number of studies [2–5] and may be summarized [6] as follows:

- In rats DEHP is metabolized via a greater number of oxidative steps than in man or monkeys.
- The main metabolites excreted in man and monkeys are MEHP (25%) (no oxidation) and MEHP which has undergone one oxidative step to give a hydroxyl group at the $\omega - 1$ position (65%).
- In rats virtually no MEHP is excreted and the major metabolite (75%) is a dicarboxylic acid formed by two oxidation steps at the ω position.
- In man and monkeys about 80% of the metabolites are excreted conjugated with glucuronic acid whereas in rats the level of conjugation is virtually zero.

ENDOCRINE MODULATION

Recently there have been reports of reduced sperm count in human males. It has been hypothesized that this may be due to exposure to chemicals which mimic natural oestrogens. It must be emphasized that this remains a hypothesis. It is not yet clear that there is a general problem in humans and no evidence that chemicals in general or any chemicals specifically are the causative agents. However, among other things, the publication of this hypothesis has sparked interest in the development of screening tests which could be used to identify oestrogenic substances.

As a consequence of their widespread usage, phthalates have been one of the groups of chemicals which have been tested for endocrine modulating effects using the recently developed in-vitro tests and the more well established in-vivo ones.

In-vitro screening tests

These are only intended to indicate which substances should be investigated further via in-vivo studies. Thus the relevance of these in-vitro tests to real life is not known and in addition they suffer from considerable inter-laboratory variation.

The only phthalates which have given positive results have been butylbenzyl phthalate (BBP) and dibutyl phthalate (DBP). These are only weakly positive [17] and equivocal in that they have proved negative in some studies [18–21].

All the more common, commercially significant phthalates have been tested and found to be negative [19].

The monoesters, mono butyl, monobenzyl and mono 2-ethylhexyl phthalate have all proved negative [20].

In-vivo studies

The most recent studies specifically intended to look for oestrogenic effects are a series of uterotrophic assays [20, 22]. These are tests which measure increase in uterine weight, a process which is under oestrogenic control. They have shown that all the phthalates ranging from DBP to diisodecyl phthalate (DIDP) produce no oestrogenic effects.

In addition numerous multigeneration fertility studies have been carried out on many different phthalates. The most recent of these are two-generation studies which demonstrate that exposure of rats to diisononyl phthalate (DINP) [23] and DIDP [24] in utero, during lactation, puberty and adulthood does not affect testicular size, sperm count, morphology or motility or produce any reproductive or fertility effects. No outcome which might be anticipated from hormone modulation was observed. The maximum level dosed was around 600 mg/kg bw/day.

In a recent study Sharpe et al. [25] fed low levels of BBP, octyl phenol, octyl phenol polyethoxylate and diethyl stilbestrol to pregnant rats. All the test chemicals caused a small reduction in testicular size of the offspring when they reached maturity; however, effects on the ventral prostate were not observed with BBP. In addition, in unpublished studies, the same workers have seen no effects on the testes with BBP at a ten-fold increase in dose level or when administered subcutaneously. These findings have not been confirmed yet by other laboratories or in different animal species and strains and are contradicted by the recent uterotrophic assays [20, 22] outlined above. In addition Richard Sharpe has emphasized, both in the press and on television, the limitations of his study and has warned against overinterpretation of his data.

CONCLUSIONS

Plasticizers have a very low acute toxicity and do not cause irritation or sensitization.

Repeated oral exposure to high levels of some plasticizers has caused adverse effects in the liver and reproductive systems of certain species. However, as a consequence of large species differences in metabolism and response it may be concluded that they pose no significant carcinogenicity or reproductive threat to humans.

The vast majority of in-vitro screening tests and, more importantly, all in-vivo oestrogenicity and two-generation studies indicate that none of the phthalates nor any other plasticizers possess oestrogenic activity.

REFERENCES

1. International Programme on Chemical Safety (1992) *Environmental Health Criteria 131, Diethylhexyl Phthalate*, World Health Organisation, Geneva.
2. Rhodes, C., Orton, T.C., Pratt, I.S., Batten, P.L., Bratt, H., Jackson, S.J. and Elcombe, C.R. (1986) Comparative pharmacokinetics and subacute toxicity of di(2-ethylhexyl)phthalate (DEHP) in rats and marmosets: extrapolation of effects in rodents to man. *Environmental Health Perspectives*, **65**, 299–308.
3. Short, R.D., Robinson, E.C., Lington, A.W. and Chin, A.E. (1987) Metabolic and peroxisome proliferation studies with di(2-ethylhexyl)phthalate in rats and monkeys. *Toxicology and Industrial Health*, **3**(1), 185–194.
4. Official Journal of the European Communities, 17/8/1990, L 222/49.
5. Ashby, J., Brady, A., Elcombe, C.R., Elliott, B.M., Ishmael, J., Odum, J., Tugwood, J.D., Kettle, S. and Purchase, I.F.H. (1994) Mechanistically-based human hazard assessment of peroxisome proliferator-induced hepatocarcinogenesis. *Human and Experimental Toxicology*, **13**, Supplement 2.
6. International Agency for Research on Cancer (IARC) (1982) IARC Monographs on the evaluation of the carcinogenic risk of chemicals to humans, **29**, 269–294.
7. International Agency for Research on Cancer (IARC) (1995) *Peroxisome Proliferation and its role in Carcinogenesis, Technical Report No 24*, Lyon.
8. Sullivan, F.M., Watkins, W.J. and van der Venne, M.Th. (eds) (1993) *The Toxicology of Chemicals, Series Two: Reproductive Toxicity, Volume 1, Summary Reviews of the Scientific Evidence*, Commission of the European Communities.
9. Neilsen, E. (1994) *Evaluation of Health Hazards by Exposure to Phthalates and Estimation of Quality Criteria in Soil and Drinking Water*, The Institute of Toxicology/National Food Agency, Denmark.
10. European Centre for Ecotoxicology and Toxicology of Chemicals (ECETOC) (1994) *Assessment of Non-occupational Exposure to Chemicals, Technical Report No. 58*, Brussels.
11. US Agency for Toxic Substances and Disease Registry (1992) *Toxicological Profile for DEHP*.

12. Health and Environment Canada (1994) *Priority Substances List Assessment Report, Bis(2-ethylhexyl)phthalate.*
13. UK Ministry of Agriculture, Fisheries and Food (MAFF) (March 1996) Food Surveillance Information Sheet No 82.
14. Astill, B.D. (1989) Metabolism of DEHP: Effects of prefeeding and dose variation, and comparative studies in rodents and the cynomolgus monkey. *Drug Metabolism Reviews*, **21**(1), 35–53.
15. Albro, P.W., Corbett, J.T., Schroeder, J.L., Jordon, S. and Mathews, H.B. (1982) Pharmocokinetics, interactions with macromolecules, and species differences in metabolism of DEHP. *Environmental Health Perspectives*, **45**, 19–25.
16. Huber, W.W., Grasi-Kraupp, B. and Schulte-Hermann, R. (1996) Hepatocarcinogenic potential of di(2-ethylhexyl)phthalate in rodents and its implications on human risk. *Critical Reviews in Toxicology*, **26**(4), 365–481.
17. Jobling, S., Reynolds, T., White, R., Parker, M.G. and Sumpter, J.P. (1995) A variety of environmentally persistent chemicals, including some phthalate plasticisers, are weakly oestrogenic. *Environmental Health Perspectives*, **103**(6), 582–587.
18. Soto, A.M., Sonnenschein, C., Chung, K.L., Fernandez, M.F., Olea, N. and Serrano, F.O. (1995) The E-screen assay as a tool to identify estrogens: an update on estrogenic environmental pollutants. *Environmental Health Perspectives*, **103**(7), 113–122.
19. Balaguer, P., Gillesby, B.E., Wu, Z.F., Meek, M.D., Annick, J. and Zacharewski, T.R. (1996) Assessment of chemicals alleged to possess oestrogen receptor mediated activities using in-vitro recombinant receptor/reporter gene assays, *SOT 1996 Annual Meeting*, Abstract 728 cited in *Fundamental and Applied Toxicology, Supplement, The Toxicologist*, **30**(1), Part 2, March 1996.
20. Meek, M.D., Clemons, J., Wu, Z.F. and Zacharewski, T.R. (1996) Assessment of the alleged oestrogen receptor-mediated activity of phthalate esters, Presented at the *17th Annual SETAC Meeting*, Washington, 18–21 November 1996.
21. Gaido, K.W., Barlow, K.D. and Leonard, L. Use of a yeast-based oestrogen receptor assay to assess chemical interactions with the oestrogen receptor, *SOT 1996 Annual Meeting*, Abstract 731 cited in *Fundamental and Applied Toxicology, Supplement, The Toxicologist*, **30**(1), Part 2, March 1996.
22. Uterotrophic assays in immature rats, Zeneca Central Toxicology Laboratory, Report Nos CTL/R/1278–1281, 1996.
23. Nikiforov, A.I., Keller, L.H. and Harris, S.B. Lack of transgenerational reproductive effects following treatment with diisononyl phthalate (DINP), *SOT 1996 Annual Meeting*, Abstract 608 cited in *Fundamental and Applied Toxicology, Supplement, The Toxicologist*, **30**(1), Part 2, March 1996.
24. Nikiforov, A.I., Trimmer, G.W., Keller, L.H. and Harris, S.B. Two-generation reproduction study in rats with diisodecyl phthalate (DIDP), Presented at *Eurotox '96*, September 22–26, 1996.
25. Sharpe, R.M., Fisher, J.S., Millar, M.M., Jobling, S. and Sumpter, J.P. (1995) Gestational and lactational exposure of rats to xenoestrogens results in reduced testicular size and sperm productions. *Environmental Health Perspectives*, **103**(12), 1136–1143.

Keywords: toxicity, phthalates, genotoxic, metabolism, oestrogenic effects, sperm count, fertility

Editor's note. This contribution was included in the book because the topic was being widely discussed at the time of writing. From time to time, other additives have also received considerable attention but space is not available to summarize the health aspects of all additives for plastics. It should not be assumed that plasticizers are the only additives suspected of presenting health problems, nor that they are necessarily more hazardous than other additives.

Polymer additives: the miscibility of blends

Nadka Avramova

INTRODUCTION

Polymer blends are physical mixtures of at least two structurally different polymers, which adhere together with no covalent bonding between them. Each constituent can be a polymer or a copolymer, with a linear, branched or crosslinked structure. When one of the mixed polymers is the minor component, it can be considered simply as an additive. In the search for new polymeric materials, the blending of polymers is a promising method for obtaining desirable properties using already known polymers. Such systems should provide a relatively simple solution to complex economic and technological problems. Polymer blends are now of great scientific and industrial interest. Several blends of commercial importance already exist. The commercial polymer blends can be plastic–plastic, plastic–rubber or rubber–rubber. Their application leads to a reduction in the amount of the more expensive material necessary, and/or to an improvement in the properties. In this way high performance materials can be developed from synergistically interacting polymers. Another possible application of polymer blending is recycling industrial plastics waste.

One of the major factors affecting the structure and properties of polymer blends is their miscibility. Most polymer pairs exhibit pronounced immiscibility, and only certain pairs are thermodynamically miscible. Generally, polymer blends are at least two-phase non-homogeneous systems.

Plastics Additives: An A–Z Reference
Edited by G. Pritchard
Published in 1998 by Chapman & Hall, London. ISBN 0 412 72720 X

THERMODYNAMICS OF POLYMER MIXING

Homogeneity in mixtures depends on the change in thermodynamical potential, G, as a result of mixing:

$$\Delta G = \Delta H - T\Delta S$$

where ΔH and ΔS are the changes in enthalpy and entropy of the system on mixing, respectively. T is the absolute temperature. At equilibrium, homogeneous dissolving of one polymer into another is a spontaneous process if $\Delta G < 0$. Therefore miscibility depends on which of the two factors (ΔH and ΔS) is predominant. Mixing always leads to an increase in entropy, but for long polymer molecules the change in entropy is small. As mixing is usually an endothermic process, blends are normally heterogeneous. Interactions between macromolecules of the mixed components, for instance hydrogen bonding, are favourable for their association and miscibility of blended polymers.

Observed miscibility of a given polymer pair is related to a set of physical conditions (temperature, pressure, molecular weight, chain structure, etc.). $\Delta G < 0$ when the binary polymer–polymer interaction coefficient, x_{12}, is negative. The contributions to x_{12} are dispersion forces, free volume and the specific interactions. The tendency of blends to mix or to separate over a long period of time determines their true thermodynamic miscibility. For example, depending on the conditions, some polyolefin blends are thermodynamically miscible. Most polyolefin blends are compatible, with an enhancement of their physical performance. Therefore, blending is widely used in the polyethylene industry.

In some cases, when equilibrium cannot be attained, the systems appear homogeneous in spite of their endothermic mixing. Such blends involve polymeric plasticizers, graft copolymers, or block copolymers. Polymeric plasticizers which are dispersed to a molecular level in the melt remain trapped in a rigid matrix after cooling. Small amounts of block or graft copolymers can be dispersed on a molecular scale in a homopolymer corresponding to the terminal or pendant polymer chains of the block or graft. In these cases the important criterion for homogeneity is the size of the domains and not the heat of mixing, i.e. the criterion is kinetic and not thermodynamic, since equilibrium cannot be attained.

Crystallization of blended polymers leads to their phase separation, because the formation of isomorphic crystals occurs very rarely. Blends of poly(ethylene terephthalate) and poly(butylene terephthalate) remain miscible in the amorphous phase after crystallization of both components. Unlimited mutual solubility of polymers is exceptional. It can be achieved under certain conditions, for example, in blends of poly(vinyl chloride) and butadiene-nitrile rubber or poly(vinyl acetate) and cellulose nitrate.

Usually the solubility of one polymer in another is very low. Solubility of polystyrene in poly(methyl methacrylate) is 0.9 wt% and in polyisoprene it is 0.4 wt%. As a function of molecular weight, the compatibility of polymers increases significantly when the molecular weight corresponds to oligomers. Miscible polymers become immiscible as a result of the rise in molecular weight.

The miscibility of polymers can be qualitatively predicted by the stability of their solution in a common solvent. Theoretically, miscibility is considered at a molecular level. From the practical point of view, miscible polymers are those which answer certain requirements, for example they possess transparency, adhesion between components, etc. In fact, most such materials are microheterogeneous.

Two polymers can be immiscible, partially miscible or completely miscible, depending on the phase structure, i.e. on whether the polymer blend consists of two separate immiscible phases, or whether there is partial mixing at the molecular level, or the two polymers form only one thermodynamically stable phase.

The ideal mutual solubility of polymers is rarely achieved because of the small effect of the entropy of mixing. Miscible blends are characterized by optical transparency, homogeneity at the level of 50–100 Ångström units (5–10 nm) and a single glass transition region.

METHODS OF PREPARATION

The principal commercial methods for preparing polymer blends are melt mixing, solution blending, and latex mixing. The mixing process is carried out at high temperature or in diluents or in a suspending medium. Polymer–polymer chemical bonds can form as a result of chemical reactions during mixing. These reactions can usually be avoided or limited if necessary.

A strong bond between free polymer and block or graft copolymer interface is formed. Therefore, block and graft copolymers are used to achieve strong interface adhesion. For example, graft polymers are used in ABS materials (acrylonitrile-butadiene-styrene resins) so as to have a strong bond between the rubber and plastic phases.

Mixing of polymers in the molten state assures a system which is stable in moulding operations, avoiding the need for removal of diluents. Of course, degradation of either or both polymers can occur during heating. As a result of thermal degradation, the product of mixing can consist of a complex mixture of block, graft and crosslinked copolymers. Preparation of a simple blend by melt mixing is used for systems in which thermal degradation does not ordinarily occur. Otherwise, a material with high get content can be prepared, and primary bonding between the two polymers can be achieved.

Solution blending is satisfactory for mixing at low temperatures without degradation. However, the removal of diluent usually leads to changes in the domain sizes and even to complete polymer separation from the blend. In this way macroheterogeneous systems are obtained.

CHARACTERIZATION

The methods used to study polymer blends involve characterizing each polymer in its new environment, and determining the phase composition, phase separation, interface interactions between polymers, blend morphology etc. Such information is very often obtained by means of light scattering, light microscopy, electron microscopy, small angle X-ray scattering, small angle neutron scattering, differential scanning calorimetry, nuclear magnetic resonance, and mechanical and dielectric studies.

If the polymer blend is homogeneous, it appears optically clear. The blend has a single refractive index, intermediate between those of the individual components, and the material is transparent. Heterogeneous blend with particles large enough to scatter the visible light appear opaque if the sample is not very thin. An example of a heterogeneous blend is acrylonitrile-butadiene-styrene (ABS). Transparency, however, cannot unambiguously answer the question of whether the polymer blend is miscible.

Differential scanning calorimetry of miscible blends demonstrates the existence of a single broad glass transition region, intermediate between the corresponding glass transitions of the pure components. The glass transition temperature changes regularly with blend composition (Figure 1). Heterogeneous blends exhibit glass transitions for each of the individual components; the values obtained, and the extent of any broadening observed, are only slightly affected by the presence of the other component.

Microscopic studies are suitable for heterogeneous systems. The degree of dispersion can be observed and the final properties of the blends can be predicted.

Characterization of microheterogeneous blends is difficult, due to the small sizes of the particles. Investigations of such multiphase systems can be done by electron microscopy. Besides miscibility, interface interactions can also be estimated by scanning transmission electron microscopy with a resolution of a few nanometres.

Another useful way of blend characterization is low voltage scanning electron microscopy. Even small composition changes can be observed by this method, which also allows the estimation of interphase interactions.

As mentioned above, polymer–polymer chemical interactions can take place during the blending process. As a result, new graph or block

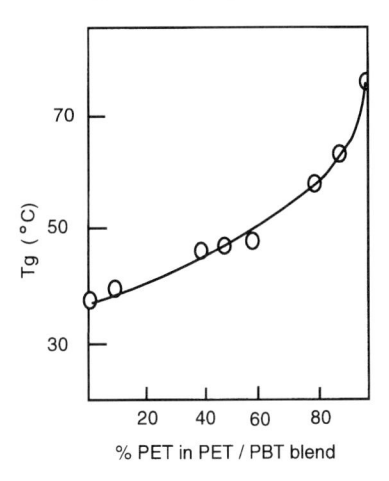

Figure 1 Dependence of the glass transition temperature, T_g, on the composition of PET/PBT blends (Avramova, 1994).

copolymers arise in the mixing stage. The complete characterization of most systems involves a combination of phase separation, glass transition determination, and spectroscopic data.

HOMOGENEOUS POLYMER BLENDS

Owing to the thermal instability of pure poly(vinyl chloride), PVC, its processing is difficult. In order to avoid thermal degradation, and improve the processability, plasticizers are involved. Low molecular weight plasticizers migrate to the polymer surface and give soft blends. Polymeric plasticizers are less mobile and volatile, do not affect the glass transition substantially, and yield blends with appropriate rigidity. However, the tensile properties of the main polymer are lowered by addition of plasticizers. PVC gives homogeneous blends with plasticizers as high molecular weight oils. Other polymer plasticizers are polystyrene for poly(methyl vinyl ethers), cellulose nitrate for poly(oxyethylene) glycols, and vinyl polymers for polyesters. The main advantage of miscible blends is the homogeneous structure and stability of the system.

Poly(vinyl chloride) is blended with other plastics to provide flame retardancy.

HETEROGENEOUS POLYMER BLENDS

From the practical point of view, the hetero-phased systems are advantageous. The dispersed phase improves the toughness of brittle polymers or has a reinforcing effect. The problem with the stability of such systems is solved by compatibilization. For example, in the rubber industry, stabilization of the multiphase systems is achieved by chemical or physical crosslinking. Compatibilization involves addition or generation of an agent modifying the interfacial properties of blended polymers.

Heterogeneous polymer systems yield high-impact plastics. A typical example is polystyrene, which is very brittle on its own, and therefore its application used to be restricted. Styrene polymers are now of great commercial importance because of the use of rubber additives. Rubber–polystyrene blends are tough and stiff materials. Rubber–plastics blends in the form of high impact polystyrenes and ABS resins help to stop crack propagation by means of the dispersed rubber particles. A brittle plastic mixed with dispersed crosslinked rubber yields a touch, stiff blend. The high impact styrene materials are blends of styrene with natural or synthetic rubbers. The rubbers are grafted with polystyrene to achieve a good rubber-to-plastic interaction.

Liquid crystalline polymer additives reinforce the main polymer due to the dispersion of the long rigid macromolecules into the polymer matrix. If molecular orientation of the liquid crystalline polymer is achieved, a high performance material can be obtained.

BIBLIOGRAPHY

Avramova, N. (1994) Amorphous blends of poly(ethylene terephthalate) and poly(butylene terephthalate) through ultraquenching. *Polymers & Polymer Composites*, **2**(5), 277–285.

Manson, J.A. and Sperling, L.H. (1979) *Polymer Blends and Composites*, Plenum Press, New York–London.

Mark, H.F. and Gaylord, N.G. (eds) (1969) *Encyclopedia of Polymer Science and Technology*, John Wiley & Sons.

Paul, D.R. and Newman, S. (eds) (1981) *Polymer Blends*, Academic Press, New York–San Francisco–London.

Utrasski, L.A. (1989) *Polymer Alloys and Blends: Thermodynamics and Rheology*, Hanser. Munich–Vienna–New York.

Keywords: polymer blends, miscibility, thermodynamics of mixing, homogeneous blends, heterogeneous blends.

See also: Compatibilizers for recycled polyethylene.

Processing aids: fluoropolymers to improve the conversion of polyolefins

Koen Focquet and Thomas J. Blong

INTRODUCTION

A link between the relatively good processability of low density polyethylene (LDPE) and the excellent toughness of linear low density polyethylene (LLDPE) has been a key subject for many researchers. Fluoropolymer-based additives provide this link. Today, the use of fluoropolymer processing additives has expanded greatly, from an early way of minimizing melt defects, to improving the throughput rates and properties of LLDPE. These additives now provide benefits in a host of polyolefin extrusion applications.

MATERIALS

Fluoropolymers of the kind used as polymer processing additives (PPAs) are quite impervious to chemical attack and thermal degradation. They are of low surface energy, and are generally incompatible with other polymers.

Microscopic examination of a polyethylene (PE) containing PPA reveals discrete micron-sized, droplet-shaped particles of the fluoropolymer. Figure 1 shows an ideally dispersed PPA in a PE matrix. Typical PPA use levels vary from 200 to 1000 ppm, depending on the application.

The PPAs commercially available from the 3M Company are Dynamar™ FX-9613 and Dynamar™ FX-5920A. They are free-flowing

Plastics Additives: An A–Z Reference
Edited by G. Pritchard
Published in 1998 by Chapman & Hall, London. ISBN 0 412 72720 X

Figure 1 Microscope picture of dispersed PPA particles measuring about 5 μm.

powders and will hereafter be referred to as PPA-1 and PPA-2. The latter was designed for processing improvements in systems containing inorganic fillers, pigments, and anti-blocking agents.

MECHANISM

When the polymer is extruded, the applied shear-field allows the PPA to phase-separate from the PE matrix and to form a thin, persistent coating on the metal surfaces of the extrusion equipment. This is illustrated in Figure 2. Once a dynamic equilibrium is established, the differential between the surface energies of the two polymers allows for reduced friction during extrusion of the PE.

Evaluation of a PPA-modified PE by capillary rheometry results in a decrease in extrusion pressure as this coating is formed. Eventually an equilibrium is reached, and the host polymer appears to have a lower viscosity than expected.

Figure 3 shows that the apparent viscosity of LLDPE (MFI 0.7; density 0.925 g/cm^3) obtained by a capillary die ($l/d = 20/0.5$) can be reduced by

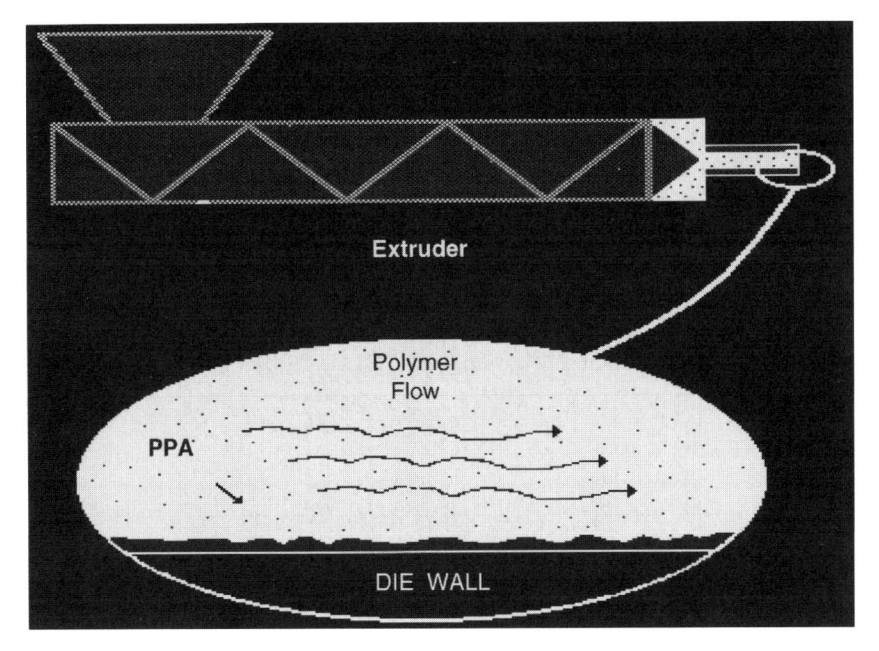

Figure 2 PPA mechanism.

40% or more. Consequently, as indicated in Figure 4, the critical shear stress can be shifted towards higher shear rates.

The formation of the PPA coating is not instantaneous. Figure 5 shows the elimination of melt fracture over a period of time during film fabrication using the same LLDPE. The time required for establishing the dynamic equilibrium depends on the type of host resin and on the final formulation during the conversion step. PPA use levels are often

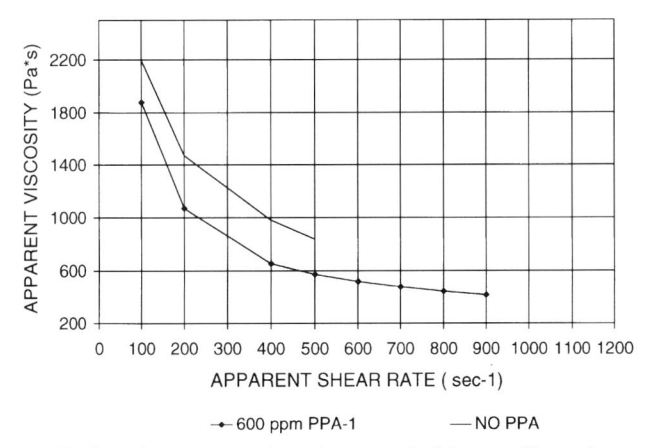

Figure 3 Reduced apparent viscosity recorded by capillary rheometry.

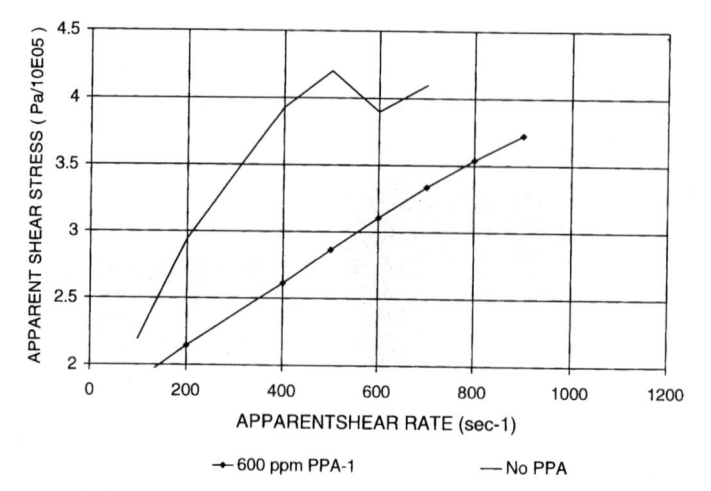

Figure 4 Reduced apparent shear stress recorded by capillary rheometry.

determined by the need to allow complete coating within an acceptable time period.

In Figure 6 we can see how reduced gate pressures are recorded during actual processing. Consequently, lower torque and more efficient power consumption may be expected.

The same PPA coating can provide an additional benefit by reducing or eliminating accumulation of die deposits and internal die build-up. If left unchecked, these unwanted accumulations may eventually degrade and exit the extrusion process as die-lip build-up, and/or visible contamination within the extrudate. The low surface energy coating of the PPA can minimize this problem.

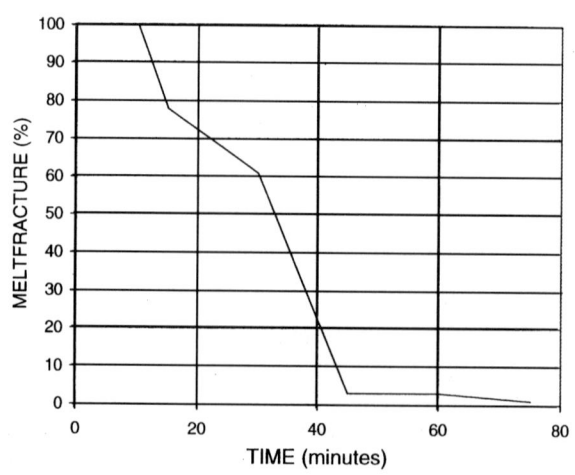

Figure 5 Melt fracture elimination during LLDPE blown film fabrication.

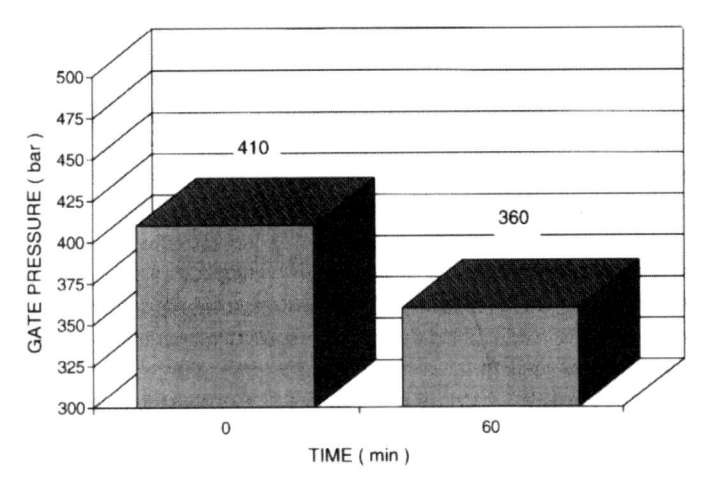

Figure 6 Reduced gate pressure during LLDPE blown film fabrication.

ADDITIVE INTERACTIONS

Although fluoropolymers are virtually inert, and thermally stable (Thermo-Gravimetric Analysis, TGA, indicates that the first 1% weight loss occurs at about 400°C), several ways have been documented for other additives to interact with the PPA. These include chemical reaction, adsorption, abrasion, and competition with the coating mechanism of fluoropolymer based additives. Several tests have been developed to reveal whether the additive is behaving synergistically or antagonistically towards the PPA's ability to coat the capillary die.

Through the study of additive interactions, a new generation of synergistic processing aids has been developed to overcome many of these interferences. For example, PPA-2 gives improved performance in the presence of inorganic additives, compared with processing additives containing only fluoropolymer. PPA-2 can be masterbatched with inorganics, such as anti-blocking agents, with minimal effect on its performance.

THE EFFECT OF PPA ON FINISHED PROPERTIES

After exiting an extrusion process, the inclusion of a PPA has little if any further intended purpose. The effect of its long term presence in finished articles is sometimes questioned.

To evaluate this concern, LLDPE blown films have been fabricated with PPA levels two or four times those typically used. Evaluation of mechanical properties such as tear, tensile and dart impact test indicated no positive or negative effect due to the PPA's presence. Given that the fluoropolymer has a density twice that of polyethylene, its volumetric effect is only half its weight contribution to the PE mixture.

Because the coating mechanism of PPAs can reduce pressure and torque during extrusion, PPAs can also be utilized to extrude under previously unattainable conditions such as higher throughput rates, or through narrower die gaps and at lower temperatures. Changes in these variables allow for greater control of PE molecular orientation, and hence optimization of mechanical properties. Extrusion through narrower die gaps can improve properties such as dart impact behaviour, and the machine and transverse direction balance of tear properties. The effect of allowing increased cooling rates is especially noticeable with products like PPA-2. Visible reductions in haze and improvements in gloss are observed. Gloss improvements have also been documented in applications in HDPE blow-moulded parts.

Finally, the components of PPA-1 and PPA-2 are optimized to prevent their migration from within solidified polyethylene. It is only in the melt phase under shear that the PPAs are allowed to function. To confirm this, the same films that were extruded for mechanical property measurements were evaluated over a year-long period for any surface changes. Surface energy and contact angle measurements, along with ESCA (Electron Spectroscopy for Chemical Analysis) and SSIMS (Static Secondary Ion Mass Spectroscopy) verified that there was no physical or chemical presence of intentionally elevated PPA levels.

SUMMARY

Fluoropolymer-based processing additives provide a range of benefits beyond melt fracture elimination. Among these benefits are improved production capacities, and a better control of molecular orientation and final physical properties. Owing to their design and low volumetric contributions, no adverse effects have been detected which can be attributed to the PPA's presence. Careful selection of other additives used with the PPA can minimize the PPA requirement. Finally, the use of newer products, such as PPA-2, can overcome additive interferences to further enhance polyethylene properties and productivity.

BIBLIOGRAPHY

Blong, T., Focquet, K. (1995) Fluoropolymer-Based Additives Improve Efficiency of Polyolefin Processing and Product Characteristics. *Conference 'AddCon 1995'*, Brussels, RAPRA Technology, Shawbury, UK.

Blong, T., Klein, D., Pocius, A., Strobel, M. (1994) The Influence of Polymer Processing Additives on the Surface, Mechanical, and Optical Properties of LLDPE Blown Film. *SPE ANTEC 94*.

Blong, T., Duchesne, D., Brandon, M. (1993) Processing Additives in High Density Polyethylene Extrusion Blow Molding Applications. *SPE ANTEC 93*.

Johnson, B., Blong, T., Kunde, J., Duchesne, D. (1988) Factors Affecting the Interaction of Polyolefin Additives with Fluorocarbon Elastomer Polymer Processing Aids. *TAPPI 1988 PLC Conference*.

Keywords: fluoropolymer, processing additive, melt fracture, die build-up, PE, LDPE, LLDPE, HDPE.

Processing aids for vinyl foam

John Patterson

VINYL FOAM MARKETS

Foamed PVC has been commercially available for at least 20 years, but is now becoming increasingly important because of recent improvements in technology and a broadening of applications. It has becomes one of the fastest growing markets in the vinyl industry, sparking enthusiasm among processors over new opportunities for vinyl foam, particularly in many wood-replacement applications.

There are currently three major existing markets for PVC foam: (1) sheet, where the major application is signage, (2) profile, which is mainly used for trim and molding, and (3) foam core pipe for drain, waste and vent applications. These markets have grown worldwide at an annual rate of greater than 12% over the last three years. The benefits that vinyl foam offers include: low cost per unit volume, high rigidity, improved thermal insulation, high water resistance, and most importantly, the ease and versitility of being able to work with it. Rigid vinyl foam is, in many ways, just like wood. Impressive growth rates are forecast for vinyl foam, particularly in wood replacement applications in the building and construction industries, where its water resistance and high stiffness per unit weight are important attributes.

Most commercial processing aids have a composition consisting of high molecular weight copolymers of methyl methacrylate and alkyl acrylates, where MMA is the major component. The glass transition temperature of these copolymers is generally greater than that of PVC.

Plastics-Additives: An A–Z Reference
Edited by G. Pritchard
Published in 1998 by Chapman & Hall, London. ISBN 0 412 72720 X

VINYL FOAM PROCESSES

Rigid vinyl foam is usually produced by adding a chemical blowing agent to a PVC formulation that contains, in addition to the resin, a thermal stabilizer, internal and external lubricants, fillers and colorants, and a processing aid which is typically acrylic. During processing, the blowing agent thermally decomposes to produce gases. A small percentage of the blowing agent is left unreacted, and this provides nucleation sites for bubble formation. On exiting the die, the supersaturated gas–resin mixture expands, with the pressure drop forming the foam as discrete bubbles in a continuous PVC matrix. The bubbles continue to grow until equilibrium is reached between the gas pressure and the surface tension on the cell walls. As this occurs, the extrudate is also cooled on its external surfaces, which tends to terminate the bubble growth process producing solid, high-density skin layers and low-density cellular core. These general principles hold true for all the different processing techniques used to produce rigid vinyl foam.

With regard to the processing of rigid vinyl foams, there are three main extrusion processes: free-foaming, inward foaming and coextrusion. In the free-foaming process, the melt (containing a precise amount of blowing agent) is allowed to expand freely on exiting the die. This process gives a density about 0.70–$0.75\,\text{g/cm}^3$ and offers limited surface hardness. In the inward-foaming or Celuka process, the foaming is directed toward the core by the presence of a mandrel in the die. Cooling is applied to the entire surface immediately after exiting the die, resulting in a harder, smoother surface with a higher density at the surface and much lower density at the core. The overall density of the inward-foaming product is typically 0.5–$0.6\,\text{g/cm}^3$. A third option is coextrusion, which places a solid skin over a foam core. It offers greater flexibility in processing, and close control over density, foam structure, and skin/core ratio.

VINYL FOAM FORMULATIONS

Vinyl foam requires several ingredients not found in typical rigid PVC formulations. Typical formulations for both free-foam and inward-foam vinyl extrusions are shown in Table 1. Most PVC foam formulations do not contain an impact modifier, but do incorporate both a blowing agent and an acrylic processing aid. Virtually every ingredient in the PVC formulation influences cell structure, skin development, and ultimately the physical properties of the finished part in some way or other, but probably the one having the most effect is the processing aid.

Table 1 Typical vinyl foam formulations

Free foam formulation		Inward foaming formulation	
PVC resin	100.0	PVC resin	100.0
Thermal stabilizer	1.2	Thermal stabilizer	2.0
Internal lubricants	0.5	Co-stabilizer	0.5
External lubricants	1.0	Internal lubricants	0.6
TiO_2	1.0	External lubricants	0.75
$CaCO_3$	5.0	TiO_2	3.0
Azodicarbonamide	0.6	$CaCO_3$	2.0
Acrylic processing aid	6.0	$NaHCO_3$	1.5
		Azodicarbonamide	0.2
		Acrylic processing aid	7.0

FUNCTION/BENEFITS OF PROCESSING AIDS IN RIGID VINYL FOAM

By definition, the function of a processing aid is to facilitate the processing of vinyl compounds. It has long been considered that better processability meant quicker fusion, and subsequently it was thought that the key role of a processing aid was to promote the fusion rate of PVC. Quick fusion, however, does not always guarantee a good homogeneous melt. The main function of a processing aid is, in fact, to promote the breakdown of PVC particles and ensure that a homogeneous melt is obtained. Besides melt homogeneity, the key functions of a processing aid include (1) increased melt strength, (2) increased melt extensibility, and (3) increased melt elasticity.

Processing aids are used in rigid vinyl foam formulations for much the same reasons that they are used in other PVC formulations. The processing aid provides faster fusion and ensures a more homogeneous melt, but most importantly it helps increase the melt elasticity, which contributes to density reduction [1,2]. The processing aid also imparts melt strength to the PVC foam extrudate to form a smooth skin surface during formation of the cellular structure.

Effect on fusion

In the initial stages of the fusion and melting of PVC, the polymer tends to behave as a polyolefin, exhibiting reduced adhesion to metal processing surfaces. This delays heat transfer and thus fusion. It is of particular concern in extrusion processes where some degree of adhesion is necessary to convey material along the flights of the screw. The fusion-promoting properties of a processing aid can be observed using a torque rheometer where torque as a function of temperature can be measured. The curves

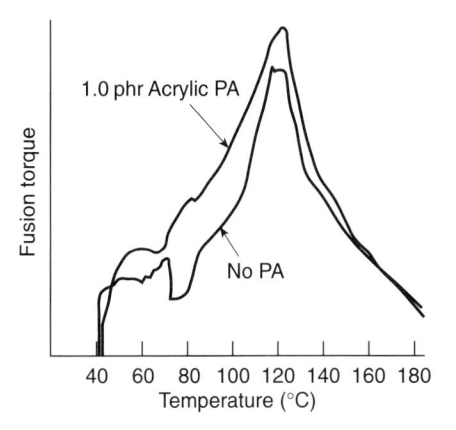

[Figure: Fusion torque vs Temperature (°C) showing curves for "1.0 phr Acrylic PA" and "No PA", x-axis from 40 to 180°C]

Figure 1 The effect of processing aid on the fusion of rigid PVC compound (Haake Rheocord bowl torque versus temperature, heating at 4°C/min).

shown in Figure 1 clearly show the effect of the processing aid on the initial fusion process of the compound [3,4]. The initial torque peaks at the lower temperatures are due to powder compaction (very little particle breakage has occurred) and the grains and subgrains starting to break into the primary particles. The primary particles actually start to fuse at the major large torque peak and the process continues beyond the torque maximum. With the addition of only 1 phr processing aid, this torque peak, where primary particles begin to break down, is shifted to significantly lower temperatures.

Faster fusion at lower temperatures in the vinyl foam extrusion process gives better (more efficient) dispersion of blowing agent, a faster melt seal which prevents gases from escaping and results in lower densities, and a more homogeneous melt in the die.

Effect on melt homogeneity

In the processing of a PVC compound it is important to have a homogeneous melt in order to achieve a high-quality product. A homogeneous melt is one where there is a minimum of primary particles present to minimize melt fracture. Gonze used a capillary rheometer with a short die length to measure the state of fusion of PVC as a function of processing temperature [5]. He processed PVC over a range of temperatures and then measured the extrusion die entry pressure necessary to maintain a constant shear rate in a rheometer. The plot shown in Figure 2 shows the transition from particulate to molecular flow as a function of processing temperature. This die entry pressure technique can also be useful to show the effect of processing aids on the temperature at which this transition occurs. Work done by Rozkuszka and Meyers [6] and presented in

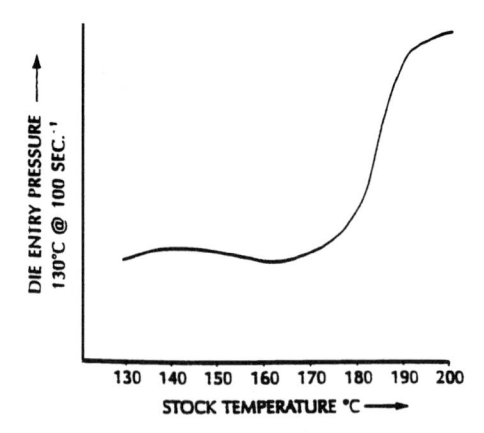

Figure 2 Melt flow versus temperature (transition from particulate to molecular flow).

Figure 3 shows that the transition occurs at a lower temperature when a processing aid is present in the formulation. This results in a more homogeneous melt (as well as a less variable melt morphology) at the normal processing temperatures of PVC. In addition, because this transition in flow occurs at a lower temperature, the compound is subjected to less overall heat history.

A more practical example of the effect of processing aids on achieving a homogeneous melt is shown in Figure 4. In these photographs, a tin-stabilized, high-molecular-weight (K69) PVC is processed on a two roll mill at 250°F (121°C) with and without 2 phr of an acrylic processing aid. The stock without processing aid shows a non-homogeneous melt on the roll and a badly fractured rolling bank. With the processing aid,

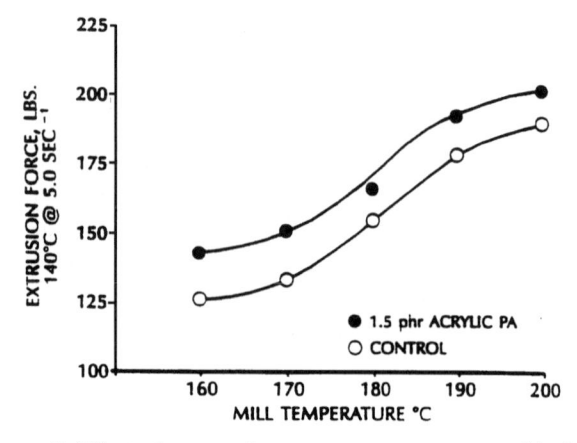

Figure 3 Effect of processing temperature on state of fusion.

Figure 4 The effect of processing aid on melt stock on two roll mill at 250°F: (a) without processing aid; (b) with processing aid.

the stock on the roll is clear, smooth, and homogeneous and the rolling bank is also smooth.

Effect on melt strength and melt elasticity

It is difficult to separate the effects and the importance of these two properties. A combination of tensile strength, elongation and elasticity defines the 'toughness' of a melt. The acrylic polymers that are typically used as processing aids are generally compatible with PVC and, with their long chains, interact to produce a stiffer and more elastic melt. Increased rupture stress and extensibility are provided by the addition of processing aids which renders the PVC far more resistant to rupture-induced defects. In this way they prevent the surface of the melt from tearing as it exits the die and prevents the expanding gases from blowing through cell-wall membranes and surface skin in producing PVC foam products. Although the practical effects of melt strength are abundantly clear to the processor, measuring the melt strength quantitatively is usually difficult. The Gottfert Rheotens is a device that uses a gear-like strain gauge-instrumented 'puller' to draw a fully fused melt from a right-angled (vertical drop) extruder. While the extruder output rate is stabilized, the geared take-off increases in speed until the melt (extrudate) breaks. Typical stress–strain curves from a Rheotens experiment at 190°C are presented in Figure 5. The fused PVC compounds contain 0 and 2 phr acrylic processing aids. The use of 2 phr acrylic processing aid in this experiment resulted in about 30% improvement in ultimate elongation and about 67% increase in tensile strength.

Melt elasticity is an important factor in establishing melt stability as the melt enters and proceeds through the die in extrusion. The higher die

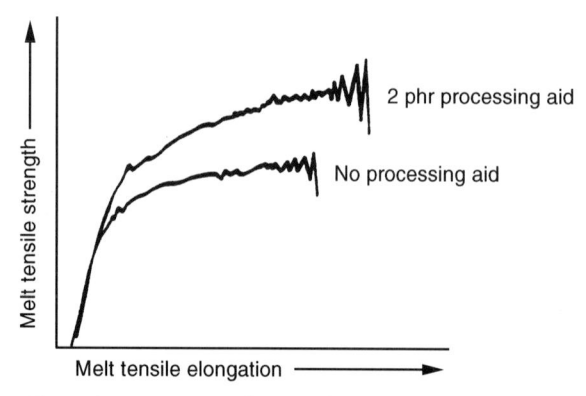

Figure 5 The effect of processing aid on melt strength of rigid PVC. With processing aid, tensile strength is 19.9 kg/cm^3, elongation is 104%. Without processing aid, tensile strength is 11.9 kg/cm^3, elongation is 80%.

entry pressure observed in Figure 3 when processing aid is present, is also a good indicator of higher melt elasticity [7]. Greater melt strength and elasticity gives better gas retention, lower and more uniform density, finer cell structure, better skin formation, more uniform free blowing prior to sizing, and easier sizing operation.

GUIDELINES FOR THE USE OF PROCESSING AIDS IN RIGID VINYL FOAM

The type and level of processing aid in the PVC foam formulation significantly influences foam density by trapping the blowing agent gases and preventing the cell structure from collapsing. The degree of foaming depends on the viscosity and melt elasticity of the PVC formulation, the gas pressure evolved, and the degree of solvation of the gas in the melt. Therefore variations in melt rheology, blowing agent concentration and cooling conditions have a profound influence on the cell morphology (cell size and distribution) and subsequent physical properties.

Processing aids are typically incorporated into a rigid foam formulation at levels up to 10 phr according to the processing technique used. It has been shown that increasing the level of the processing aid in a vinyl foam formulation reduces foam density and provides a smooth, hard surface. Table 2 presents data on the reduction in density that results from increasing the acrylic processing aid level from 2 to 8 phr. Surface quality also improves. In the visual rating of the surface quality of extruded rods, a set of standards is used with a scale from 1 to 5 (1 being a good, smooth surface, and 5 being a very rough surface).

It has been shown that acrylic processing aids have different efficiencies which result in different foam densities. There appear to be two main effects operating to influence the efficiency of the acrylic processing aid: molecular weight and composition. Large changes in molecular weight have a significant influence, and compositional changes also show some

Table 2 The effect of processing aid on foam density and surface quality

	Level (phr)	Density (g/cc)	Surface quality*
Without processing aid	–	0.85	4
With processing aid	2	0.7	2
	4	0.65	2
	6	0.59	1
	8	0.56	1

*1 = good, smooth surface; 5 = very rough surface.

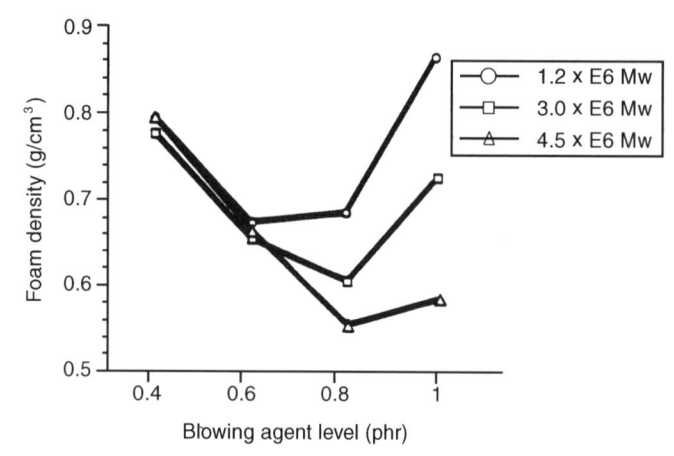

Figure 6 The effect of processing aid molecular weight on the Gas Containment Limit (GCL).

effects. The common denominator between these two parameters may be their ability to influence the melt elasticity of the formulation. As molecular weight rises, the minimum foam density that is achieved drops. So it is important to select the acrylic processing aid with a high molecular weight and a composition that most affects elasticity. An example of the effect of processing aid molecular weight is shown in Figure 6.

REFERENCES

1. E.C. Szamborski and R. Marcelli (1976) *Plast. Eng.*, Nov., Issue 11, p. 49.
2. M.T. Purvis and R.P. Grant (1978) U.S. Patent 4,120,833, Oct..
3. D.L. Dunkelberger (1987) *History of Polymeric Composites*, R.B. Seymour and R.D. Deanin, Eds, VNU Science, p. 177.
4. R.P. Petrich and J.T. Lutz, Jr. (1989) *Thermoplastic Polymer Additives: Theory and Practice*, J.T. Lutz, Jr., Ed., Marcel Dekker, New York, pp. 381–415.
5. A. Gonze (1971) *Plastica*, 24(2), 49.
6. K. Rozkuszka and C. Meyers, private communication.
7. H.Y. Parker (1989) *Encyclopedia of Polymer Science and Engineering*, 2nd end, John Wiley & Sons, New York, p. 307.

Keywords: PVC, vinyl foam, acrylic, blowing agents, torque, melt strength, melt extensibility, melt density, cell morphology.

Recycled plastics: additives and their effects on properties

Francesco Paolo La Mantia

INTRODUCTION

The degradation undergone by plastics products during their processing and subsequent lifetime worsens all the characteristics of the polymers, and recycling operations therefore give rise to secondary materials with poor properties. This holds for all recycled polymers, both homogeneous and heterogeneous, but in the latter case other difficulties arise from recycling [1]. The incompatibility between the different polymer phases, indeed, leads to materials having properties even lower than those of the pure components. Therefore, to obtain recycled polymers with acceptable properties it is necessary to protect the material from degradation phenomena and/or to try to enhance the crucial properties. Finally, for mixed plastics, the use of compatibilizing agents (see entry on 'Compatibilizers for recycled polyethylene') can be very useful. This means that for re-stabilization, the addition of fillers, modifier agents and compatibilizers is, in many cases, a necessary operation to carry out during recycling. The use of similar additives, or at least some of them, is of course usual also with virgin polymers, but it can be particularly important for recycled materials.

In this entry, after a brief look at the concepts of plastics recycling, examples of the effect of some additives on the properties of the recycled materials will be given.

Plastics Additives: An A–Z Reference
Edited by G. Pritchard
Published in 1998 by Chapman & Hall, London. ISBN 0 412 72720 X

BASIC CONCEPTS OF PLASTIC RECYCLING

The main problems in plastics recycling are:
- degradation that has occurred during processing and during the product lifetime;
- incompatibility between polymers.

Of course, this latter point must be considered only when mixed heterogeneous plastics are recycled. During the processing and the lifetime of the plastics products, some external influence such as heat, mechanical stress or ultraviolet radiation, can fundamentally change the structure and the morphology of the original polymer. Although photo-oxidation and thermomechanical degradation can cause a wide variety of changes in the structure of plastics materials, it is possible to summarize the main effects of these degradative processes as:

- variations in molecular weight and molecular weight distribution;
- the formation of chain branching;
- the formation of oxygenated compounds, unsaturation etc.

It is worth noting that the change in molecular weight can also induce changes in the crystallinity. The degree and type of degradation depend on the processing conditions and on the nature of the polymer, but, of course, all the changes undergone by the chemical structure will cause variations in the properties of the material. Severe processing conditions, polymers with high molecular weight, or repeated processing operations can cause significant decreases in the polymer characteristics. The degradation leads to deterioration of the mechanical properties (decrease in elongation at break and impact strength, in particular) and also to discoloration and other surface damage. To protect the materials against thermomechanical degradation, stabilizers must be used, and to enhance the mechanical properties, fillers and impact modifiers can be added.

It is a well-known fact that only a few polymer pairs are miscible; for all the other polymer pairs the concept of compatibility is used. Compatibility is in itself difficult to define and quantify. From a technological point of view, we can define the compatibility on the basis of the variations in a macroscopic property as a function of the blend composition. The polymers are compatible when some property of the blend is higher than those of the components, semi-compatible when the properties are between those of the components, but not additive, and incompatible when some property is lower than those of the components. The synergistic effect (giving a maximum in the graph of property versus composition) is found only in blends with strong interactions between the two phases. Instead, the antagonistic effect is typical of those polymer pairs with strong repulsion (for example, apolar polymers and strongly polar polymers). In most cases, the properties of the blends are intermediate between those of the two components (partially compatible or semicompatible) although most of them show values lower than might be expected on the basis of an additive law. The morphology of the incompatible blends suggests that the poor mechanical properties are due to the very limited adhesion between the two phases. To obtain secondary materials with acceptable properties from mixed plastics it is then necessary to enhance the adhesion between the phases. To do this, i.e. to compatibilize the blends, two main methods can be used:

- physical bonding between the phases;
- chemical bonding between the phases.

Both points can be achieved by adding small amounts of a third component to the blend to act as a bonding agent between the two incompatible phases. Depending on the chemical structure of both compatibilizers and components of the blends, the compatibilization mechanism is different. When the macromolecular compound is a copolymer having monomer units which are identical to, or at least compatible with, each phase, there is physical bonding. The copolymer is miscible with the two phases, creating a bond between the two completely immiscible phases.

In other cases the third component is a polymer miscible with one of the phases, and includes some functional groups grafted on to it. This component is then miscible in one of the polymeric phases while the functional group can react with some functional group of the other polymer, giving rise to chemical bonding between the two phases.

STABILIZING AGENTS

Most polymers are stabilized against the degradation that they are expected to undergo, both during processing and during their useful lifetime. These processes are particularly important for those polymers, like polypropylene (PP) and polyvinyl chloride (PVC), which are very prone to thermal degradation. Before recycling occurs, the stabilizing additives lose their effectiveness and are not able to protect the polymer during recycling operations and during their second lifetime. Moreover, the presence of oxygenated and other functionalized groups, formed during degradation, enhances the degradation kinetics, rapidly reducing the properties of the polymers. Re-stabilization is the only chance to stop or to slow the degradation, and the deterioration of the mechanical properties [2–4].

The melt index, MFI, of an extrusion grade of PP is reported as a function of the extrusion passes for both stabilized and unstabilized samples [4]. MFI increases dramatically (Figure 1) and this means a rapid reduction in the molecular weight of the polymers because of considerable chain scission during processing. The dramatic increase in MFI is drastically reduced by adding the stabilizing agent, a mixture of a phosphite and an antioxidant, before each recycling step. With this kind

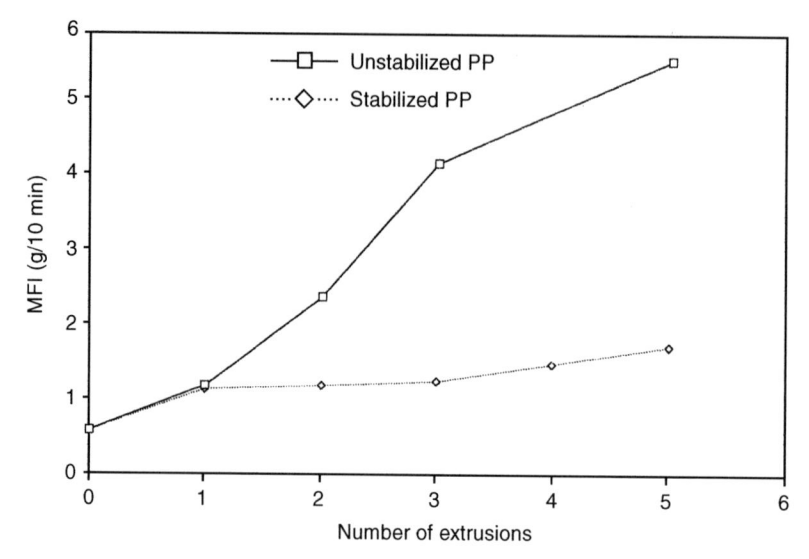

Figure 1 Melt index as a function of the number of extrusions for unstabilized and stabilized polypropylene (from [4] with permission).

of stabilization, MFI increases with each recycling operation, but its value is considerably lower than that of the unstabilized sample. Other phosphite based stabilizers give similar results [2–4].

The stabilized polymers retain most of their properties, as reported in Figure 2, where the elongation at break is reported as a function of the number of extrusions [4]. The elongation at break is highly sensitive to the structural modifications induced by degradation, and indeed a rapid decrease is observed for the unstabilized polymer. The elongation at break, on the contrary, is almost unaffected by reprocessing of the stabilized PP.

By adding a stabilizing system before every recycling step, the structure and the characteristics of the polymers are protected against thermo-mechanical degradation, and the final properties of the secondary materials are only slightly lower than those of the virgin polymers. The re-stabilization also improves the outdoor weathering resistance of the recycled plastics which, because of their previous history, would otherwise be limited [2, 3].

FILLERS AND MODIFIER AGENTS

Inorganic fillers or polymeric impact modifiers can be used to improve the mechanical properties and the thermal resistance of the polymers, both virgin and recycled. Inorganic fillers, like calcium carbonate, glass fibers

Fillers and modifier agents

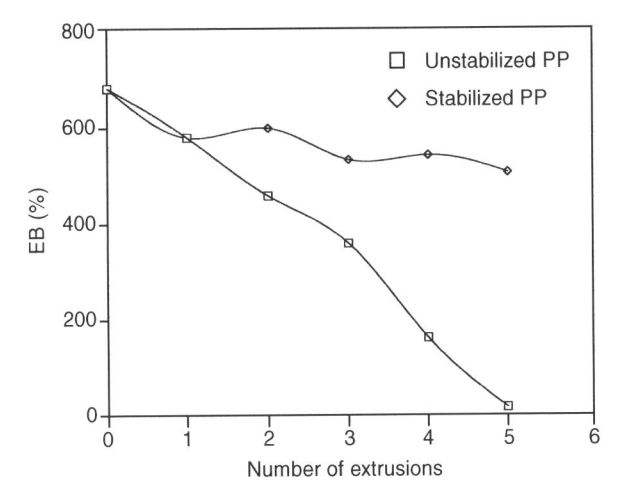

Figure 2 Elongation at break, EB, as a function of the number of extrusions for unstabilized and stabilized polypropylene (from [4] with permission).

etc., enhance the elastic modulus, the dimensional stability and the thermal resistance while a reduction in the elongation at break is in general observed. Impact modifiers, like elastomers, improve elongation at break and impact strength, whereas the elastic modulus is reduced. Glass fibers and wollastonite improve the modulus, tensile strength and heat distortion temperature of recycled polypropylene [5] while the elongation at break is only slightly reduced. Table 1 reports the heat distortion temperature (HDT) of recycled PP alone and in the presence of different contents of glass fibers (GF) and wollastonite (W). HDT, the maximum working temperature that the polymer can sustain without serious loss of stiffness, is greatly enhanced by adding the two fillers, and the short glass fibers are more efficient.

Similar features have been reported for other polymers with many inorganic fillers like calcium carbonate, mica etc. [6].

As already mentioned, impact modifiers are in general elastomeric compounds. Some of them are reported in Table 2 together with the thermoplastic matrix to which they are added.

Table 1 HDT values of recycled PP with glass fibers (GF) and wollastonite (W)

Filler content, %	HDT, GF (°C)	HDT, W (°C)
0	55	55
20	125	95
40	138	108

Table 2 Impact modifiers and relative thermoplastic matrix

Impact modifier	Thermoplastic matrix
EVA, CPE, MBS	PVC
EPM, EPDM, SBS	PA
Acrylic rubbers	Polyesters

Table 3 Mechanical properties of recycled PET/HDPE blends in the presence of impact modifiers

Material	Tensile strength (MPa)	Notched Izod impact strength, (J/m)	Elongation at break (%)
PET/HDPE	28	24	6
PET/HDPE + 20% EPDM	25	47	15
PET/HDPE + 20% SBS	34	30	8
PET/HDPE + 20% SEBS	36	40	16

The influence of some impact modifiers on impact strength and elongation at break is particularly beneficial while the modulus is usually reduced because of the rubbery nature of these compounds. Some mechanical properties of recycled PET/HDPE blends in the presence of elastomers are reported in Table 3.

COMPATIBILIZERS

The rate determining step in the compatibilization process when adding a third component is the synthesis of new copolymers or new functionalized polymers able to induce binding between the incompatible polymers. Some commercially available materials are reported in Table 4.

By adding small amounts of compatibilizing agents, remarkable improvements in the mechanical properties of some incompatible blends have been observed. Some examples are reported in Table 5.

Table 4 Compatibilizers and appropriate polymer pairs

Compatibilizer	Polymer blend
LDPE-g-PS	PE/PS
CPE	PE/PVC
Acrylic-g-PE, -PP, -EPDM and	Polyolefin/PA and
Maleic-g-PE, -PP, -EPDM, -SEBS	Polyolefin/Polyesters

Compatibilizers

Table 5 Mechanical properties of compatibilized and uncompatibilized recycled blends

Property	PP/PET	PP/PET/Comp1	Ny6/PP	Ny6/PP/Comp2
Elongation at break (%)	12	210	105	30
Notched Izod impact strength (J/m)	21	105	37	88

Comp1 is a maleic anhydride grafted SEBS, and Comp2 is maleic anhydride grafted PP. Elongation at break and impact strength are dramatically enhanced by adding these compatibilizing agents. This improvement cannot be attributed to the simple presence of these compounds because of their very low concentrations (about 4% wt/wt). In fact, the presence of the copolymers strongly improves the adhesion between the two phases, and induces a dramatic reduction in the number of particles in the dispersed phase, as is clear from the SEM images of the uncompatibilized PP/PET blend, Figure 3(a), and of the same compatibilized blend, Figure 3(b).

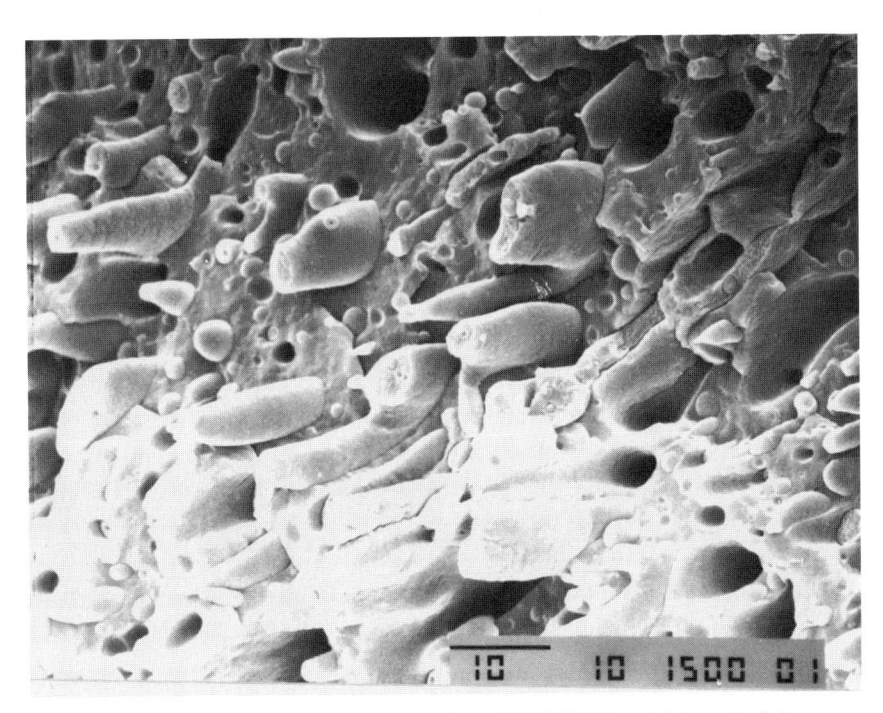

Figure 3 SEM images of the uncompatibilized PP/PET blend, (a), and of the same compatibilized blend, (b).

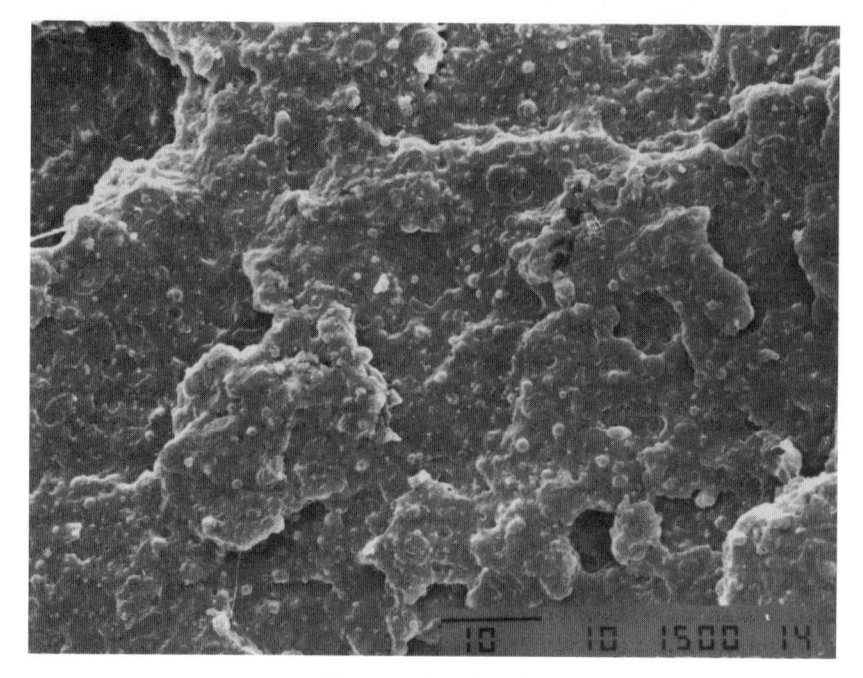

Figure 3 Continued.

NOMENCLATURE

EVA	Ethylene vinylacetate copolymer
CPE	Chlorinated polyethylene
EPM	Ethylene-propylene rubber
MBS	Methyl-butyl-styrene rubber
EPDM	Ethylene-propylene-diene rubber
SBS	Styrene-butylene-styrene rubber
SEBS	Styrene-ethylene-butylene-styrene rubber
LDPE-g-PS	Low density polyethylene-grafted-polystyrene

REFERENCES

1. La Mantia, F.P. (1996) Basic Concepts on the Recycling of Homogeneous and Heterogeneous Plastics, *Recycling of PVC and Mixed Plastics* (ed. F.P. La Mantia), ChemTec, Toronto.
2. Dietz, S. (1992) Phosphite Stabilizers in Postconsumer Recycling, in *Emerging Technologies in Plastics Recycling* (eds Andrews, G.D. and Subramanian, P.M.) ACS Symposium Series 513, Washington, 134.
3. Herbst, H., Hoffmann, K., Pfaendner, R., Sitek, F. (1995) Stabilizers Allow Production of Higher Added-Value Post-Use Plastics, *Polymer Recycling*, 1, 157.
4. Marrone, M., La Mantia, F.P. Re-Stabilization of Recycled Polypropylenes, in press, *Polymer Recycling*.

5. Vinci, M., La Mantia, F.P., Properties of Filled Recycled Polypropylene, in press *J. Polym. Eng.*
6. Stepek, J., Daoust, H. (1983) *Additives for Plastics*, Springer-Verlag, New York.

Keywords: polymer, recycling, stabilizer, compatibilizer, filler.

See also: Compatibilizers for recycled polyethylene.

Reinforcing fibres

J.E. McIntyre

INTRODUCTION

The use of fibres and fabrics as additives to reinforce matrix materials in structures that are often referred to as composites goes back into pre-history, as in the use of straw to reinforce clay bricks. As usual, nature developed such structures first. Examples are wood (cellulosic fibres in a lignin matrix) and bone (collagen fibres in an inorganic matrix). A composite need not be based on fibres – it is a material or product formed by intimate combination of two or more distinct physical phases, so the shells of crustaceans (calcium carbonate in a chitin matrix) are composite structures. However, the word composite now commonly brings to mind structures consisting of fibres embedded in a matrix of some other material, whether plastic, ceramic or metal (i.e. fibrous composites), and even tends to be used particularly for structures in which the fibres are laid out in organized fashion before the matrix material is consolidated around them.

In terms of plastics, there is a clear division between processes in which fibres are incorporated into a polymer before shaping by, for example, extrusion, and processes in which fibres are assembled into a desired shape or arrangement before incorporation into a polymer and retain that organization in the final product. A further division is between processes where the matrix, at the stage where it is combined with the fibres, is (1) already fully polymeric and requiring essentially no removal of other materials, (2) already fully polymeric, but in a dissolved or suspended state and requiring removal of a solvent or dispersing medium, and (3) not already polymeric, or not yet fully polymerized, and requiring a polymerization stage after combination with the fibres in order to develop the desired molecular weight.

Plastics Additives: An A–Z Reference
Edited by G. Pritchard
Published in 1998 by Chapman & Hall, London. ISBN 0 412 72720 X

Fibrous additives are used in polymers to improve the mechanical properties, notably the tensile strength, tensile modulus, toughness, creep and resistance to impact. They may themselves be polymeric, carbonaceous, metallic, glassy or ceramic in nature. Their importance derives particularly from their dimensional anisotropy and in many cases also from the anisotropy of their mechanical properties.

FIBRES AND FIBROUS ASSEMBLIES

Fibres for use as additives are available, and used, in a wide range of physical forms. The main divisions are between continuous filaments, where the axial ratio (length:diameter) is very high and the individual fibres can be regarded for most purposes as endless; staple fibres, where the length is usually from 1 to 100 cm and the aspect ratio is typically in the range 1000 to 10 000; and flock, where the length is at most a few millimetres and the aspect ratio is of the order 100. With flock may be classed pulp fibres, which are very short fibrous materials in which the cross-sectional shape varies widely both within and between fibres. Whiskers are small single crystals in filament form, which are made by crystal growth processes.

Both continuous-filament and staple fibres may be collected together in an approximately unidirectionally oriented form. The continuous-filament fibres may be assembled as a tow, where the number of filaments may be large and the degree of cohesion between them is low, or as a yarn, where the number of filaments is at most a few hundred and there is cohesion between them arising most usually from twist or localized intermingling. The staple fibres may be assembled as a relatively thick sliver or a roving, which represent two consecutive stages in their conversion by textile spinning into yarns, and which exhibit a low degree of cohesion, or in the thinner form of a yarn, which like the continuous-filament yarns has a relatively high lateral cohesion, usually, but not necessarily, arising from twist. The word 'roving', however, when applied to glass fibres, has come to mean also an assembly of continuous filaments (i.e., a tow). The word strand also receives an unusual meaning in glass fibre technology: it is a collection of parallel untwisted continuous filaments held together by a size. A glass roving thus consists of a number of parallel strands.

A still higher degree of complexity is exhibited in the fabrics used in polymer matrices. Woven fabrics, which are formed by interlacing threads or yarns, usually have the yarns oriented octagonally to each other and therefore exhibit maxima in tensile modulus in these directions, but may, as in the case of triaxially and tridimensionally woven fabrics, have more than two directions of orientation. Knitted fabrics are formed by the intermeshing of loops of yarn; they are generally more flexible and extensible than woven fabrics, but may be reinforced by insertion of additional more highly oriented weft yarns. Stitch-bonded fabrics of the type where the structure is held together by stitches running along the fabric are useful when the stitches serve to maintain the relative positions of highly-oriented warp threads during composite manufacture. There is a further wide range of nonwoven fabrics formed from continuous-filament or staple fibres in which the cohesion of the fabric is obtained by such processes as needle-punching or localized thermal fusion.

Two-dimensional and three-dimensional braids are made by intertwining or orthogonal interlacing of two or more yarns to form a complex integral structure. Three-dimensional braiding, in particular, can be used to make high yarn-volume, near-net-shape complex preform structures such as tubes and T-beams, using a variety of conventional and high-modulus continuous filament yarns. Such products have high multidimensional resistance to deformation.

SHORT-FIBRE THERMOPLASTIC COMPOSITES

Processes where the fibre is incorporated into the polymeric matrix material before shaping are in general cheaper and more versatile than those where the fibres are first arranged in a shape approximating to that finally required and the polymer is applied as a further stage, or where there is a further polymerization stage after the shaping. Hence short-fibre thermoplastic composites are particularly suitable for large-scale mass-production techniques.

During screw extrusion of polymers containing fibrous additives, the stresses applied to the fibres are high. The longer the fibres are initially, the higher the stresses that are developed. If, like glass, the fibres are subject to brittle fracture at low extensions, they undergo tensile or flexural failure during the process, so that the average fibre length is reduced and the length distribution, if originally narrow, is widened. This occurs not only during injection moulding and extrusion moulding processes, but also during the prior extrusion process used to form granules from a mixture of short fibres and polymer. It does not occur in, for example, compression moulding, which is however a relatively slow process. The fibre length of glass is typically reduced to 50–500 μm.

During extrusion processes, the shear forces tend to align the fibres parallel to the direction of flow, but relaxation, for example in a mould, tends to restore an isotropic distribution of fibre orientation. However, the shear forces in the shaping process impart further anisotropy, which may not relax out of the structure, particularly if the cooling rate is high. Among the consequences is a tendency for the fibre orientation near surfaces to be preferentially parallel to the surface. The complexity of fibre length distribution and of orientation distribution make mathematical analysis of the mechanical properties of the final products rather difficult.

FIBRE-REINFORCED THERMOSETS

The main techniques for producing fibre- and fabric-reinforced composites from thermosetting resins are:

1. hand lay-up or spray-up: labour-intensive but suitable for complex shapes;
2. compression forming: the shape is formed by placing the fibrous structure and pre-polymer together either between matched die faces or on a single die face to which a pressure difference is applied using a vacuum bag, a pressure bag, or an autoclave;
3. injection forming: the fibrous structure is introduced into a mould which is then closed; the pressure inside is reduced and the pre-polymer introduced by applying external pressure. This technique is also applied to random short-fibre composites by incorporating the fibres into the pre-polymer before injection, and is applied not only to the traditional slow curing resins but also to fast-curing systems, such as polyurethanes and certain epoxies, when it is known as R-RIM (reinforced reaction injection moulding). Because these pre-polymers have much lower viscosity than thermoplastics, the product exhibits more random orientation and less length degradation of the fibres;
4. die forming: pultrusion involves impregnating a tow by pulling it from a collecting device through a resin bath followed by a heated die;
5. mandrel forming: filament winding involves winding filaments or a tow, the latter often in a planar tape-like arrangement, on to a mandrel; the pre-polymer is supplied either before winding (as in pultrusion) or after winding on to the mandrel.

MECHANICAL PROPERTIES OF COMPOSITES (Figures 1–4)

The primary purpose of continuous fibres in composites with the fibres uniaxially aligned is to give the structures tensile properties in that direction approaching those of the fibres. In some cases the matrix may even be regarded as being essentially a protective coating for the fibres, holding them in position and preventing damage. The transverse tensile properties remain low. In order to obtain a better balance of properties, laminae containing unidirectionally oriented fibres are stacked before curing with the fibre directions at right angles or, if additional shear strength is required, at various angles to one another. The contribution of short fibres to the tensile properties, even if they are preferentially aligned parallel to the direction

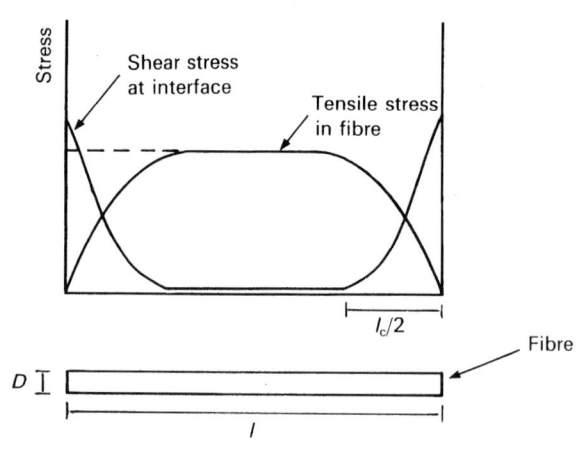

Figure 1 Variation of tensile strength in an axially stressed fibre and shear stress at the interface in a short fibre-reinforced composite; critical length of fibre = l_c (Cox, H.L. (1952) *Brit. J. Appl. Phys.*, **3**, 72).

of test, is lower than that of continuous fibres, and lower the shorter the fibres. Below a certain critical length of fibre, failure of the composite will occur without failure of the fibres.

Similarly, creep behaviour is very different for composites containing uniaxially-oriented continuous filaments and randomly oriented short

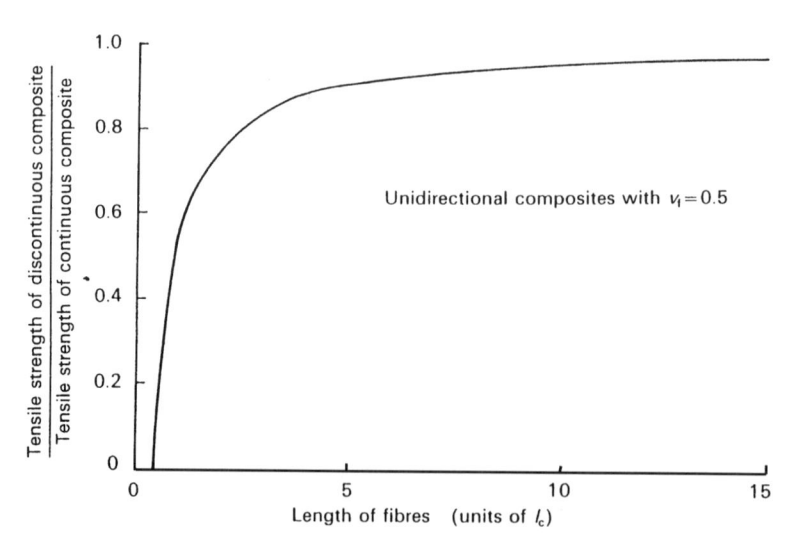

Figure 2 Tensile strength of a composite containing discontinuous fibres as a proportion of the strength using continuous fibres of the same type. Critical length = l_c; volume fraction of fibre = v_f (Matthews and Rawlings, 1994).

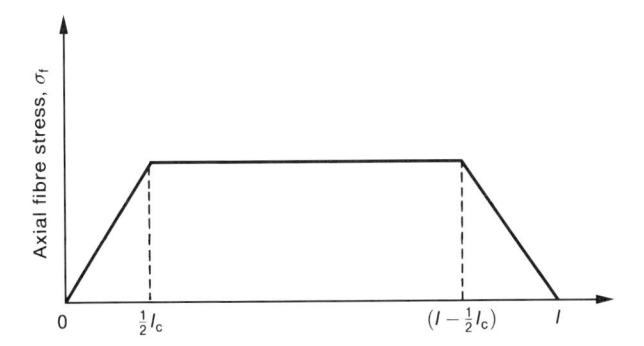

Figure 3 Stress transfer from polymer to a fibre of length, l, greater than the critical length, l_c (Powell, P.C. (1983) *Engineering with Polymers*, Chapman & Hall, London, p. 177).

fibres. In the former, measured parallel to the filaments it is dominated by the creep behaviour of the filaments, which have usually been chosen to minimize creep. In the latter, creep is a complex volume-, length- and orientation-related function of the creep behaviour of the two phases.

If the breaking strain of the fibre is above a critical value, which depends upon the nature of both the fibre and the matrix, the impact strength and toughness of composite structures based on brittle fibres

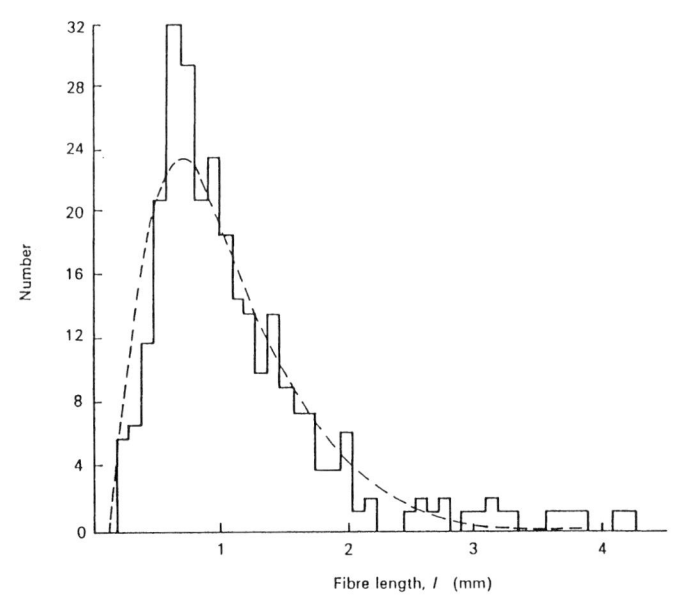

Figure 4 Distribution of fibre lengths from an injection moulding using brittle fibres of initial length 6 mm (Matthews and Rawlings, 1994).

are directly related to the breaking strain of the fibre, and are improved by increased adhesion to the matrix. However, if the breaking strain of the fibre is lower than the critical value, better impact strength is obtained when the adhesion is reduced.

SELECTION OF FIBRES FOR USE AS ADDITIVES

A very wide range of chemical and physical properties is available in fibrous additives. These include such key factors as thermal stability (melting or softening behaviour, shrinkage temperature, thermal decomposition temperature), density, strength and mode of failure under tensile and compressive stress, tensile modulus, and adhesion or lack of adhesion to the matrix. A further key characteristic is cost. Much the most widely used fibre is glass, largely because of its low cost. In the following sections the properties of most of the relevant fibres are reviewed. Table 1 summarizes some properties of a range of fibres.

INORGANIC FIBRES

Asbestos

Asbestos is a naturally occurring mineral fibre. The most important form is chrysotile, a hydrated magnesium silicate belonging to the serpentine

Table 1 Properties of some fibres used to reinforce plastics

Fibre	Density (g cm^{-1})	Tensile strength (GPa)	Tensile strength (Ntex^{-1})	Tensile modulus (GPa)	Tensile modulus (Ntex^{-1})	Extension at break (%)	Mode of failure
HT polyester	1.38	1.3	0.95	15	12	8	D
HT nylon 6,6	1.14	1.0	0.9	6	5.4	15	D
HT acrylic	1.18	0.5	0.45	12	10	20	D
HT polypropylene	0.91	0.55	0.6	5	5.5	15	D
HT vinal	1.30	1.0	0.8	20	16	9	D
Ultra-drawn polyethylene	0.96	0.93	1.0	85	90	5	D
Gel-spun polyethylene	0.97	3.0	3.1	175	180	3	D
HT para-aramid	1.44	3.3	2.4	75	55	3.6	F
HM para-aramid	1.47	2.4	1.7	160	110	1.5	F
HT carbon	1.75	4.0	2.3	240	140	1.6	B
HM carbon	1.8	2.5	1.4	390	220	0.6	B
MP carbon	2.15	2.2	1.0	690	320	0.3	B
E-glass	2.54	3.5	1.4	73	29	3	B
S-glass	2.49	4.9	2.0	87	35	5	B
α-alumina	3.90	1.4	0.35	380	100	0.4	B
Stainless steel	7.86	3.8	0.5	200	25	2	D

HT: high tenacity; HM: high modulus; MP: mesophase pitch; D: ductile; F: fibrillar; B: brittle.

group of minerals. Amphiboles, such as crocidolite, have been used only to a minor extent. Asbestos fibres are of short, irregular length and variable diameter, suitable for incorporation into moulding plastics but little used there on cost grounds. Their main importance has been in friction materials (e.g. clutch and brake linings) and thermally-resistant gland packings and seals. Woven or nonwoven mats of asbestos have often been used in the latter applications. In the case of friction materials, where the matrix is usually a phenolic resin, asbestos raises the coefficient of friction, thus improving torque transmission, and also improves the resistance to wear; in gaskets, where the matrix is usually an elastomer, it acts as a reinforcing filler.

Use of asbestos in general has declined severely due to its bad reputation for producing respirable particles of a size that can cause lung disease on long term exposure. Although the problem is less acute in plastics than in other uses, fibres such as glass, steel and para-aramid have made inroads into some of these markets.

Glass fibres

Several different types of glass fibre have been used to reinforce plastics, but much the most widely used is E-glass, in which the silica network is modified by other oxides, principally those of calcium, aluminium and boron. S-glass, with a higher aluminium content and some magnesium, is designed to give higher tensile strength.

Glass fibres are formed by melt spinning – for E-glass, gravity fed extrusion through fine holes at a temperature typically of about 1300°C. The fibres are attenuated by being collected on a rotating drum at a speed higher than the extrusion velocity to produce a continuous-filament strand, with average diameters in the range 2 to 25 μm, according to the attenuation conditions. Short fibres are produced either by chopping continuous filaments to the required length or by melt blowing, a process in which air directed at the freshly extruded fibres extends them, and which produces short fibres of variable length and diameter.

Glass fibres, although dimensionally anisotropic, are structurally isotropic, apart from minor differences between the surface and the interior. They are very brittle, and it is usual to protect them against surface damage, leading to crack initiation, during production and storage, and also to improve their affinity with and adhesion to polymer matrices, by applying sizes and coupling agents. (See Coupling Agents.) The sizes may be washed off before use and replaced by coupling agents, or a single application may incorporate both functions. Among the most important of these surface treating agents are alkoxysilanes that contain groups, such as vinyl, allyl, amino, glycidyl or mercapto groups, that can react with or are compatible with the polymer matrix. Chrome complexes and titanates are also useful. Different reactive groups give optimum results with different matrix polymers, particularly in the case of thermoset resins.

Chopped fibres of controlled length, often about 3 mm, but sometimes longer, contribute the main fibrous ingredient in filled thermoplastics for injection and extrusion moulding. Particulate mineral reinforcing fillers and flame retardants may also be included in special grades. Usually from 15% to 40% of glass fibre is incorporated in order to give very significantly higher tensile and flexural strength, tensile and flexural modulus, and heat distortion temperature, and in many cases in impact strength. Above about 40%, processability and mechanical properties decline steeply. In the special case of liquid-crystalline polyester matrices, the glass reduces the anisotropy caused in the polymer by elongational flow during moulding, thus improving the abrasion resistance and weld-line strength, but also reducing the impact strength.

Methods of combining the glass with a thermoplastic matrix vary. In some cases the short glass fibre and plastic are fed independently to the feed hopper of the moulding extruder. Better blending is achieved by pre-forming granules, either by impregnating parallel continuous strand bundles with molten thermoplastic and cutting the resulting composite into granules, or by pre-extruding a mix of short glass fibre with the thermoplastic and granulating the extrudate. Polymer producers include granulated glass-reinforced thermoplastics in their range.

Continuous-filament glass is widely used, alone or in hybrid composite structures with other fibres, in epoxy, unsaturated polyester, phenolic and melamine resins. Techniques for composite formation have already been outlined. In undirectional composites requiring high tensile strength and modulus, S-glass may be used because of its additional contribution to these properties.

Ceramic and other refractory fibres

Although the uses of ceramic fibres in composite structures lie mainly in ceramic-matrix and metal-matrix composites, where their outstanding chemical and thermal resistance are important, there are a few applications in organic polymers. Their relevant properties are low thermal expansion, low electrical conductivity, low dielectric constant, high stiffness, good compressive strength, and in most cases complete resistance to combustion. On the other hand they are very brittle, hard to process, and mostly considerably more expensive than carbon and para-aramid fibres. They have, for example, been used in hybrid structures with carbon and para-aramid and in electronic circuit boards. The fibres available or potentially available include alumina, combinations of alumina with other oxides such as silica or zirconia, silicon carbide (sometimes also containing oxygen or titanium), silicon nitride and carbonitride, and boron deposited on tungsten or carbon.

Metal fibres

Although metal fibres can be made from a wide range of metals and alloys, much the most important are those consisting of stainless steels. They are made by multiple-stage die drawing of wires, singly for relatively thick fibres or combined together in bundles for the finer fibres. The main use is that of continuous filaments of diameter about 200 µm as reinforcement for flexible composites, particularly tyre cord, where they hold a substantial share of the market, notably in the belt for radial tyres. Good adhesion to the rubber matrix is essential for this purpose, so the filaments are given a thin brass coating. The key characteristics here are high modulus, toughness, and thermal stability of the mechanical properties. Short steel fibres are also incorporated into the treads of some specialized tyres to improve their resistance to abrasion and cuts. Short, fine fibres with diameters about 8–12 µm, are used in friction materials such as brake linings, where higher thermal conductivity relative to other fibres is advantageous, and in electrically conductive plastics.

Carbon fibres

There are two main sources of carbon fibres. The first source is pyrolysis of continuous-filament fibres made from various regenerated and synthetic fibres, at one time mainly from viscose rayon but now almost entirely from acrylic fibres. In the initial stages, the fibres are subjected to stretching and controlled oxidation; the later stages involve heat treatment at temperatures rising finally to about 1200°C for high-strength products and as high as 3000°C for ultra-high modulus products. The second source is pyrolysis of pitch fibres, melt spun from a highly aromatic low polymer produced by pyrolysis of a purified natural pitch fraction. Both processes are slow and complex, so the product is rather expensive. However, the excellent strength and modulus have made these fibres, particularly those from an acrylic source, of major importance in high-performance reinforced plastics.

The acrylic-based carbon fibres can be engineered to have high modulus (HM) or high tenacity (HT), or a balance between these two properties, and are available in numerous grades. The pitch-based carbon fibres fall into two classes: those made from fibres spun from isotropic pitch, which have relatively low strength and modulus, and those made from mesophase (liquid-crystalline) pitch, which have exceptionally high modulus. A general consequence of very high modulus is a correspondingly low extension to break, which ranges from 2% for some HT fibres down to about 0.2% for some mesophase-pitch fibres.

Carbon fibres exhibit rather poor adhesion to matrix polymers, so it is usual to carry out oxidative treatments of various types to modify the morphology and chemical structure at the surface, in particular to increase its polarity and introduce reactive groups. Such treatments greatly improve the adhesion as measured, for example, by the interlaminar shear stress.

Most carbon fibre is used in oriented continuous-filament form, but randomly dispersed chopped fibre is also widely used to reinforce thermoplastic matrix polymers.

ORGANIC-POLYMER FIBRES

Natural fibres

None of the vegetable and animal fibres are of major importance as additives for polymers, although cotton yarns were extensively used in flexible composites, including tyre cords, before the development of manufactured fibres to replace them. Cellulosic fibres, derived from cotton or cheaper lignocellulosic sources such as jute and hemp, are used to some extent in gaskets and in other thermosets, and despite their low char temperatures are sometimes added to thermoplastic polymers.

Polypropylene fibres

The fibre with the lowest density (*ca.* $0.9 \, \text{g cm}^{-3}$) of all those readily available, polypropylene, is made by melt extrusion of the isotactic polymer, which melts at a temperature of about 170°C. Although widely available in conventional fibre forms, polypropylene is also particularly suitable for the production of split-film and fibrillated-film fibres and network structures based on film fibrillation processes. These are made by extruding and uniaxially orientating films, then subjecting them to processes involving various combinations of slitting, twisting, abrasion, and splitting by passage over rotating needle rolls. These products are inexpensive and may be used in either continuous or chopped staple form in composite structures. The poor chemical adhesion to matrix polymers and the low melting point have restricted their use in plastics but they are used to improve the crack resistance of cement-based structures.

Polyethylene fibres

Although several types of polyethylene polymer, branched and unbranched, are available on a large scale, fibres derived from them are of little importance. Two speciality types of polyethylene fibre are, however, sometimes used in composite structures.

One of these products is based upon a linear polymer of controlled molecular weight and molecular weight distribution, which is melt-spun under conditions that permit orientation of the spun yarn by application of a particularly high draw ratio. The very high degree of orientation this produces leads to a very high modulus and to a moderately high tensile strength. The other product is based on a linear polymer of very high molecular weight, which is dissolved at a low concentration in a high-temperature solvent and extruded through a spinneret into a cold liquid, where it forms a gel fibre. Very high draw ratios can be applied to this gel-spun fibre, and the product exhibits exceptionally high tensile strength and modulus.

The low density of these fibres – about 0.97 g cm^{-3} – means that in terms of specific stress and specific modulus (i.e. on a mass per unit length basis) they rank very highly. However, they are limited in composites by their low melting temperatures (about 140°C), tendency to creep, and the need for special surface-activation processes, such as corona discharge treatment, to develop adhesion to matrix polymers. They are sometimes used alone, but more often in hybrid yarn and fabric structures with glass or carbon fibres in an epoxy or unsaturated polyester resin matrix to improve the impact resistance and energy absorption. Curing temperatures should not exceed 125°C.

Acrylic fibres

Products based on the acrylonitrile copolymers used for standard acrylic textile fibres are of little importance in polymer matrices. Specialized polymers of higher, often 100%, acrylonitrile content are converted into continuous-filament and staple fibres for use in composites such as friction products.

Vinal (vinylon) fibres

Poly(vinyl alcohol) fibres are made by wet- or dry-spinning hydrolysed poly(vinyl acetate) from a solution in water. They exhibit mechanical properties similar to those of nylon and polyester; their outstanding property from the composite point of view is that their highly polar and hydroxyl-rich chemical structure leads to excellent adhesion to most thermosetting resins. They are not widely used outside East Asia.

Nylon (polyamide) fibres

The two most important nylon fibres are nylon 6 (made from ε-caprolactam) and nylon 6,6 (made from hexamethylenediamine and adipic acid). They have moderately high tensile strength, limited modulus, excellent toughness, abrasion resistance, recovery and adhesive properties, and melting temperatures of about 220°C and 250°C respectively. Nylon 6,6 was at one time widely used in tyre cords and other flexible composites, but polyester and steel, on grounds of cost and modulus, have largely displaced it except for certain heavy-duty and aircraft tyres.

Polyester fibres

Much the most important polyester fibres are those made from poly(ethylene terephthalate) (PET), the condensation product of ethylene glycol with either terephthalic acid or dimethyl terephthalate. The polymer, of melting temperature about 250°C, is melt-spun and drawn to fibres of moderately high tensile strength and modulus. These are important in continuous-filament form, particularly as woven fabric, in flexible composites. Their inherently poor adhesive properties are improved by a priming treatment with epoxy- or isocyanate-based materials, usually by the fibre producer. Typical products are tyre cords and conveyor belts. Continuous polyester filaments are also used in hybrid rigid-composite structures with glass or carbon fibres, where they are cheaper than high-modulus polyethylene but make a smaller mechanical contribution.

Aramid fibres

There are two types of aramid (fully aromatic polyamide) fibre: those with high thermal stability and flame resistant properties but mechanical properties similar to those of standard textile fibres such as polyester, sometimes referred to as meta-aramid fibres because the main commercial products are of this chemical structure; the others with additionally exceptionally high tensile strength and modulus, sometimes referred to as para-aramid fibres on similar grounds. The former type is not widely used in composite structures, but some of the most important uses of the latter lie in this area.

Poly(p-phenylene terephthalamide) (PPT) filaments are wet-spun through an air gap from a liquid-crystalline solution in 100% sulphuric acid into water. Unlike conventional polymers, PPT becomes highly oriented during the spinning process, thus developing very high tensile strength and modulus. The product at this stage has many commercial uses, for example in flexible composites such as tyres, but a brief thermal treatment at a very high temperature develops still higher orientation and modulus, although at the expense of reduced extension to break. This further product is competitive with carbon fibres in many uses. An alternative product, based on a copolymer containing 50 moles % of *p*-phenylene terephthalamide repeating units, is spun from an isotropic solution in an organic solvent and subjected to orientational drawing at a very high temperature and draw ratio. This process gives fibres with properties similar to those of the unannealed PPT fibres.

The main uses for the higher modulus product are as continuous-filament tows for impregnation with thermosetting, particularly epoxy, resins. Para-aramids are also available as pulp fibres, of high surface area and nominal length down to 0.8 mm, as staple fibres, cut to lengths from 5 to 100 mm, and as uniformly cut flock fibres of length down to about 1 mm. These products are sufficiently stable to be incorporated into thermoplastics as well as into thermosets, and are available as injection mouldable pellets containing, for example, aramid staple fibre in nylon 6,6 or in a thermoplastic elastomer. The short fibres and pulps are also used in friction products and gaskets, and the pulps as additives providing thixotropic properties to sealants and related products.

Para-aramid fibres inherently have relatively poor surface adhesion properties unless pre-treated, but are available from the producers with enhanced adhesive properties. A particularly important feature of their mechanical behaviour is that beyond a certain flexural couple they undergo failure by fibrillation on the strained outer side and by a crushing mechanism, involving formation of 'kink bands' in the structure, on the compressed inner side. This behaviour does not result in rupture, as observed with a brittle fibre, and therefore permits retention of some mechanical strength in the reinforcing material.

Liquid-crystalline aromatic copolyesters (LCPs)

This class of fibre is melt-spun from a nematic liquid-crystalline phase, which leads to the production of highly-oriented filaments at quite low elongational shear rates. The fibres have high modulus and, if thermally post-treated over an extended period to raise the molecular weight, they also have very high tensile strength. However, so far they are limited as additives by their fusibility and poor adhesive properties.

Other high-strength, high-modulus fibres

Some extremely expensive fibres have been developed for specialized situations, notably for the aerospace market. They exhibit very high strength and modulus combined with unusually high thermal and chemical stability. Of these, the polybenzobisoxazole (PBO) fibres are at present the most readily available.

ELECTRICALLY CONDUCTIVE FIBRES

For some purposes, good electrical conductivity or resistance to development of electrostatic change is needed in a composite structure. A range of conductive fibres is available for incorporation to satisfy these requirements. These include fine metal fibres, already discussed, but also glass, carbon and other fibres coated or impregnated with conductive material, usually a metal such as aluminium.

BIBLIOGRAPHY

Bunsell, A.R. (ed.) (1988) *Fiber Reinforcement for Composite Materials*, Elsevier Science Publishers, Amsterdam.

Kelly, A. (ed.) (1989) *Concised Encyclopedia of Composite Materials*, Pergamon Press, Oxford.

Lee, S.M. (ed.) (1990–91) *International Encyclopedia of Composites, Vols 1–6*, VCH, New York.

Carlsson, L.A. (ed.) (1991) *Thermoplastic Composite Materials*, Elsevier Science Publishers, Amsterdam.

Matthews, F.L. and Rawlings, R.D. (1994) *Composite Materials: Engineering and Science*, Chapman & Hall, London

SOME SUPPLIERS OF REINFORCING FIBRES FOR PLASTICS

Owens-Corning Fiberglas, DSM (Netherlands), DuPont, Vetrotex, Sumitomo Corporation, Tenmat, Nippon, Asahi-Schwebell, PPG Glassfibers.

Keywords: glass, carbon, aramid, nylon, polyester, cellulose, polyethylene, matrix, filaments, roving, fabric, tow, aspect ratio, braid.

See also: Fibres: the effect of short glass fibres on the mechanical properties of thermoplastics.

Release agents

Geoffrey Pritchard

INTRODUCTION

Mould release agents are substances which help to separate the moulding, i.e. the product, from its mould when it has been made. The choice of release agent depends on the size and complexity of the moulding operation, and on the quality of the surface finish required.

INTERNAL RELEASE AGENTS

Internal release agents are used in moulding compounds, including injection moulded plastics. Examples are zinc stearate, which is one of the commonest; calcium stearate, aluminium stearate, and stearic acid for lower temperature moulding. All these are typically added to the polymer during compounding at about the 1 to 3% level. Internal release agents can adversely affect the mechanical properties if used in higher concentrations.

Considerations should be given not only to the effectiveness of a release agent but also its cost, the health and safety aspects, the effects on other properties, and the ease of use. In this last respect, physical form is important. The stearates are dusty powders, but alkyl phosphates are also available, and have the advantage of being liquid.

EXTERNAL RELEASE AGENTS

The remainder of this entry is concerned with external mould release agents for use with thermosetting resins, mainly in open mould and low throughput operations. In the case of large fibre glass polyester or epoxy

Plastics Additives: An A–Z Reference
Edited by G. Pritchard
Published in 1998 by Chapman & Hall, London. ISBN 0 412 72720 X

mouldings, excessive adhesion of the moulding to the mould surface can be disastrous with a large and expensive product. The mould surfaces are coated with a suitable parting or release agent. The reliability of a mould release system is a crucial part of the production process. It is advisable to test a release agent before using it.

Hydrocarbon waxes can be used, subject to temperature limitations. High grade, silicone-free Carnuba wax is effective. Hard, medium and soft waxes are available. Several thin coats (up to about six) are applied and polished. Wax can also be used in an emulsion form.

Poly(vinyl alcohol) is also frequently used. It is available in coloured solutions, and can be sprayed or brushed. One problem with poly(vinyl alcohol) is the low viscosity of the solution. This means that it drains down slopes and collects in recesses or sharp corners. It may then take a long time to dry, and will not offer protection in these local areas. The use of water/methanol solvent mixtures gives a faster evaporation of solvent after application than water alone. A slow rate of evaporation can result in a long delay before it is safe to proceed to the next step, e.g. the application of a gelcoat which must be applied to a dry surface for satisfactory cure.

Another polymer used as a release agent in the glassfibre moulding industry is cellulose acetate, dissolved in acetone. PTFE and related fluorinated polymers can be sprayed on. Silicone release agents are effective with epoxy resins but not usually with unsaturated polyesters, except in hot press moulding. They are also ineffective with certain polyurethanes.

Various types of sheet are used for mouldings which are not too intricate. Examples include polythene for thermosetting sheet moulding compounds (SMC), chlorinated rubber, regenerated cellulose (Cellophane®), cellulose acetate in sheet form, and poly(ethylene terephthalate) (i.e., Melinex®, Mylar® etc.).

It is important to be able to remove release agents from a mould. Poly(vinyl alcohol) can either be stripped off or washed off with warm water. Cellulose acetate requires cellulose thinners. Wax can be removed with a solvent such as xylene. Detergents can sometimes be effective.

BIBLIOGRAPHY

Warring, R.H. (1971) *The New Glassfibre Book*, Argos Books Ltd, Kings Langley, Herts, UK.

Keywords: mould, wax, silicone, poly(ethylene terephthalate), poly(vinyl alcohol), phosphates, PTFE, cellulose, xylene, gelcoat.

See also: Lubricating systems for rigid PVC.

Rice husk ash

M.Y. Ahmad Fuad, Z. Ismail, Z.A. Mohd Ishak and A.K. Mohd Omar

INTRODUCTION

The properties of plastics can be significantly modified by the incorporation of fillers. Rice husk ash (RHA) fillers are derived from rice husks, which are usually regarded as agricultural waste and an environmental hazard. Rice husk, when burnt in open air outside the rice mill, yields two types of ash that can serve as fillers in plastics materials. The upper layer of the RHA mound is subjected to open burning in air and yields black carbonized ash. The inner layer of the mound being subjected to a higher temperature profile results in the oxidation of the carbonized ash to yield white ash that consists predominantly of silica.

Reports on the application of RHA as a filler material in thermoplastics are believed to be quite limited [1]. Perhaps the earliest similar work on such applications was the incorporation of rice husk flour into phenol-formaldehyde resin. The use of RHA as a reinforcing agent for synthetic and natural rubbers and as a replacement for carbon black in epoxidized natural rubber has also been proposed [2]. The use of RHA to synthesize zeolite, a type of catalyst, has been carried out, and the incorporation of zeolites into polypropylene and nylon was patented by Canard *et al.* [3]. Early work on the treatment of rice husk was described in the United States Patent, 3,574,816 [4]. The husk was roasted in a controlled condition for less than 60 seconds and roasting occurred at a specified temperature, i.e. between 600 to 800°C. The husk was roasted until it lost 30 to 70% of its original weight. Severe roasting of the rice husk results in a weight loss of more than 70% and will convert the organic siliceous material of the husk

Plastics Additives: An A–Z Reference
Edited by G. Pritchard
Published in 1998 by Chapman & Hall, London. ISBN 0 412 72720 X

Table 1 The chemical and physical properties of typical RHA fillers

Property	WRHA	BRHA
Chemical composition (%)		
CaO	0.36	0.12
MgO	0.16	0.078
Fe_2O_3	0.041	0.022
K_2O	0.69	0.95
Na_2O	0.034	0.018
Al_2O_3	0.025	0.023
P_2O_5	0.57	0.27
SiO_2 (silica)	96.20	53.88
Loss of ignition (LOI)	1.62	44.48
Physical properties		
Particle size (μm)	6.6	19.5
Surface area (m^2/g)	1.4	26.8
Density (g/cm^3)	2.2	1.8

into an unreactive crystalline silica. On the other hand, moderate roasting to give less than 30% weight loss gives a mixture of a carbonaceous material and amorphous silica. In relation to the present study, white RHA (WRHA) may be related to the former product, i.e. crystalline silica, and black RHA (BRHA) to the latter (amorphous silica).

WRHA has been analysed and found to have about 95% silica content. BRHA on the other hand has typically only about 54% and a substantial carbon content, i.e. about 44% (Table 1).

COMPOUNDING AND COUPLING AGENT TREATMENT

In one study, the outcome of which is discussed later, the RHA fillers were compounded into polypropylene by means of a Brabender twin screw compounder. The compounds were extruded through a twin rod die into a water bath, pulled and pelletized. Prior to compounding, some of the fillers were treated with an organofunctional silane coupling agent. A special coupling agent with built-in peroxide was utilized so that it could generate free radicals to initiate an addition reaction with the inert polypropylene matrix.

MECHANICAL PROPERTIES

It is a well known fact that incorporation of fillers into thermoplastics increases the flexural modulus, i.e. the stiffness of the material [5]. Likewise addition of both the BRHA and WRHA fillers into polypropylene

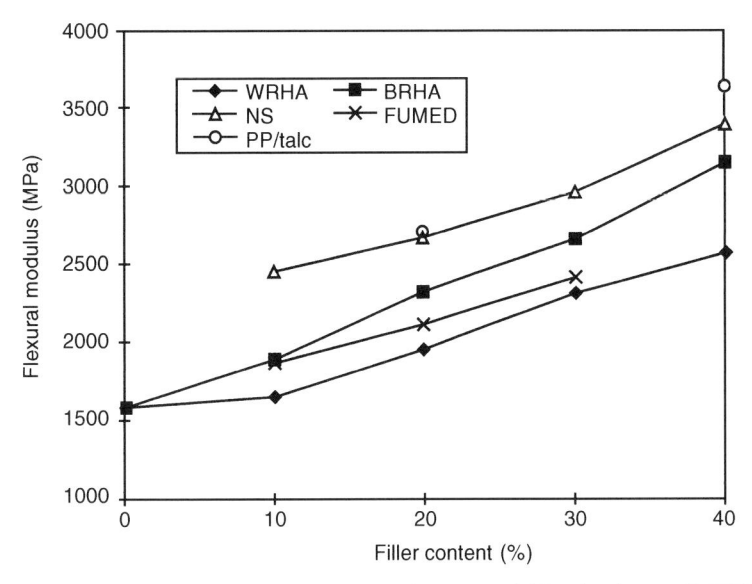

Figure 1 Modulus of RHA composites as a function of filler loading. NS, FUMED and PP/talc represents Neuberg Silica, fumed silica and talc-filled polypropylene composites respectively.

resulted in an increase in the flexural modulus of the resultant composites (Figure 1). The BRHA composites showed a more prominent modulus improvement than the WRHA composites. The flexural moduli of the RHA composites were quite comparable to those of polypropylene/ fumed silica composites. Nevertheless, the stiffness was less than that of polypropylene filled with Neuberg silica and talc fillers. Upon treatment with the coupling agent, the RHA composites did show an improvement in modulus at higher filler loadings, but the stiffness improvements were somewhat less than those in the basic RHA composites (i.e. without any coupling agent treatment).

Incorporation of fillers into a polymer matrix may increase or decrease the tensile strength of the resulting composite. Fibre type fillers normally improve tensile strength as the fibres are able to support stresses transferred from the polymer. For irregularly shaped fillers, the strength of the composites decreases through the inability of the fillers to support stresses transferred from the polymer matrix. In the above mentioned study, the tensile strength of both uncoupled RHA composites decreased steadily with filler loading (Figure 2). RHA fillers are irregular and form agglomerates. These factors (and the absence of coupling agent) result in poor adhesion of the RHA particles to the polypropylene matrix. Hence the drop in the tensile strength is quite an expected trend.

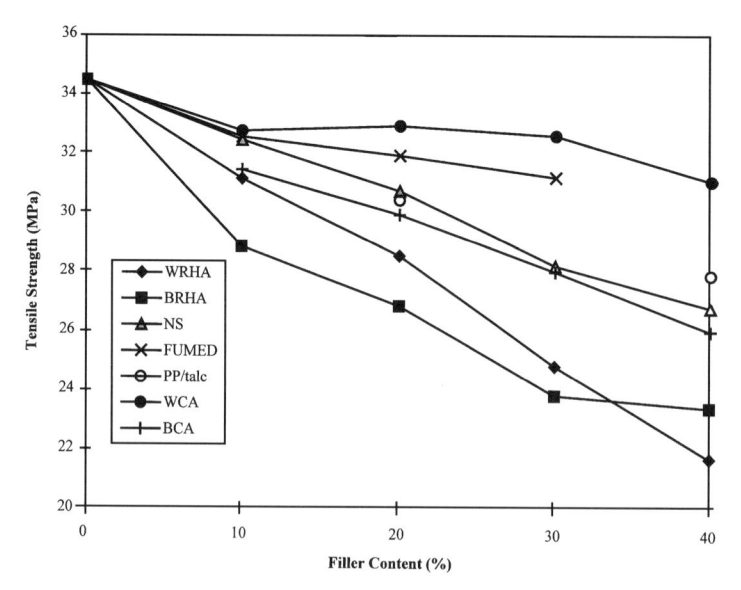

Figure 2 Effect of filler loading on the tensile strength of the RHA composites. WCA and BCA indicate WRHA and BRHA composites respectively, treated with a silane coupling agent.

A significant improvement in the tensile strength of the RHA composites was achieved by treating the RHA fillers with the silane coupling agent. The big drop in tensile strength is checked by the action of the coupling agent, particularly in the case of WRHA. The true coupling capacity of the coupling agent is displayed here where it improves the tensile strength of the RHA composites by a large margin. This improvement can be attributed to the improvement in the filler–matrix adhesion. Treatment of fillers by the coupling agent modifies the nature of the interface and results in a pronounced modification of the WRHA–polypropylene interface.

Morphological studies have indicated that failures of all BRHA composites are brittle in nature while WRHA composites experience ductile failures [2]. It appears that the incorporation of the WRHA does not alter the ductile mode of failure of the polypropylene matrix. On the other hand, slight addition of the BRHA results in a marked transition of the failure mode (to brittle fracture).

The impact strength of the composites falls with increasing filler content for both WRHA and BRHA composites. This is probably due to poor wetting of the particles by the polypropylene matrix. This gives rise to poor interfacial adhesion between the filler and the polymer matrix, resulting in weak interfacial regions. During the impact test, the crack travels through the polymer as well as along the weaker interfacial regions. The latter cannot resist crack propagation as effectively as the polymer matrix, thus reducing the impact strength. Increasing the filler content merely increases the interfacial regions that exaggerate the weakening of the composites towards crack propagation.

The fall in the impact strength is more prominent in BRHA composites. The WRHA composite shows significantly better impact resistance than the black. This can be attributed to the greater agglomeration of the BRHA particles. It is believed that there are several adhesion forces holding the particles together in agglomerates, viz. interlocking, electrostatic, van der Waals, liquid bridge and solid bridge forces. These combined interparticle forces are, however, weaker than the chemical bonding of the polypropylene matrix and are therefore easier to overcome by external mechanical energy (impact energy in this case) during fracture. Consequently BRHA particles are thought to have a higher degree of agglomeration and to yield a more brittle composite that is more susceptible to cracking.

Application of the coupling agent causes considerable reduction in the Izod impact strength of the composites. Improved filler–matrix adhesion does not favour the impact resistance of the RHA–polypropylene composites. The tradeoff in impact strength (lower value) for modulus (higher value) seems to be inevitable. The probable reason is the inability of the resin material to slip over the surface of filler particles when the composite is subjected to the impact force.

ECONOMIC AND ECOLOGICAL ADVANTAGES

RHA fillers claim two distinct advantages over other commercial fillers. First, the economic factor; being the unwanted by-product of rice mills, RHA is readily available at extremely low cost. Of course other costs, such as grinding to size, may be involved for further processing but that is not high. RHA obtained from burners can be ground directly while the semi burnt RHA collected from open field burning sites may require additional burning (for WRHA production).

Secondly, with the world's ever-increasing sensitivity to the environmental issue, disposal of rice husks waste poses a growing problem, thanks to the low bulk density of the husks. At present, the common method of disposal is dumping the RHA on waste land, consequently creating environmental pollution and land dereliction problems. Finding useful applications for RHA will certainly help to alleviate problems related to the disposal of the waste husks. Even if the rice husk ashes do not perform superbly as reinforcing fillers, the economic and ecological benefits plus their ability to perform as low cost extenders (without significantly affecting the properties of interest in the composites) speaks volumes in advocating their application as fillers in polypropylene.

REFERENCES

1. Fuad, M.Y.A., Ismail, Z., Ishak, Z.A.M. and Omar, A.K.M. (1993) *Intern. J. Polymeric Mater.*, **19**, 75.
2. Fuad, M.Y.A., Ismail, Z., Ishak, Z.A.M. and Omar, A.K.M. (1995) *Eur. Polym. J.*, **31**, 885.
3. United States Patent, 4,420,582 (1983).
4. United States Patent, 3,574,816 (1971).
5. Bigg, D.M. (1987) Mechanical properties of particulate filled polymers, *Polym. Composites*, **8**, 115.

Keywords: filler, rice husk ash, mechanical properties, polypropylene, silane coupling agent.

See also: Fillers;
Fillers: their effect on the failure mode of plastics.

Scorch inhibitors for flexible polyurethanes

Robert L. Gray and Robert E. Lee

INTRODUCTION

Polyurethane foams are produced through the polymerization of aromatic isocyanates and aliphatic polyols. The process involves a simultaneous polymerization and expansion by blowing agents such as chlorofluorocarbon (CFC) or internally generated carbon dioxide. (See the separate entry on 'Blowing agents' in this book.) As both the polymerization and decarboxylation reactions are highly exothermic, thermo-oxidation of the polyurethane foam can occur. Discoloration resulting from the oxidation is commonly known as scorch.

Recent environmental concerns have led to significant changes in the flexible slabstock foam industry. Replacement of environmentally damaging CFCs by carbon dioxide produced by the reaction between isocyanate and water has placed new demands on the stabilization packages used in this application. The use of higher levels of water to induce carbon dioxide formation and release result in increased curing temperatures. In inadequately stabilized foams, this can lead to scorch and potentially self-ignition.

The standard approach to reducing scorch involves the addition of one or more antioxidants. Butylated hydroxytoluene (BHT, see Appendix 1) has been the traditional foundation of the stabilization package. Phenolic antioxidants are used to protect the polyol from oxidation during manufacture, minimize hydroperoxide formation during storage, and reduce scorch during PUR (polyurethane) foam manufacture. The relatively high volatility of BHT can be a concern as it is easily volatilized out of

Plastics Additives: An A–Z Reference
Edited by G. Pritchard
Published in 1998 by Chapman & Hall, London. ISBN 0 412 72720 X

the foam bun during processing and BHT build-up problems can occur. Additionally, BHT is known to cause staining. These effects have caused the polyol/polyurethane industry to look for effective non-staining, low volatility antioxidants.

Further studies have focused on novel ways to combat foam degradation. Co-additives can have both positive and negative influences on scorch development and discoloration. Secondary antioxidants such as phosphites and thioesters contribute to maintenance of polymer integrity and color stability. An understanding of co-additive mechanisms and interactions aids the successful development of stabilization packages specifically formulated for defined applications. The impact of both primary and secondary antioxidants on scorch inhibition will be reported in this work.

MECHANISMS

PUR degradation processes

Polyurethane foams typically have a segmented structure of hard and soft segments. The hard segments consist of diisocyanate and chain extender units. The soft segments are hydroxy-terminated polyether or polyester-based polyols.

The hard segments are relatively stable towards thermal oxidation. Degradation of the urethane segment typically does not occur below its thermolytic decomposition temperature. Aromatic isocyanates are generally less susceptible to thermal degradation than aliphatic isocyanates [1].

The potential for thermo-oxidative degradation is greater in the soft segments of the PUR, particularly if the polyol is polyether based. Polyester diols are relatively resistant to oxidative degradation. This difference in stability can be attributed to the high affinity for oxidation of the methylene α to the ether oxygen [2].

As outlined in a simplified mechanism in Figure 1, PUR and polyol degradation proceeds through a radical chain mechanism [3,4]. Initiation typically occurs through exposure to heat generated during the foam production. Trace metal impurities such as copper or iron accelerate radical formation. Reactive hydroperoxides are formed after reaction of the carbon-centered radical with oxygen. Thermally induced homolytic cleavage of hydroperoxides lead to additional reactive radical formation and subsequent polymer chain scission. Uncontrolled degradation of this type results in a phenomenon known as scorch.

Phenolic antioxidants

An effective method of thermal stabilization is through the use of a radical terminating antioxidant. The most common class of antioxidant for

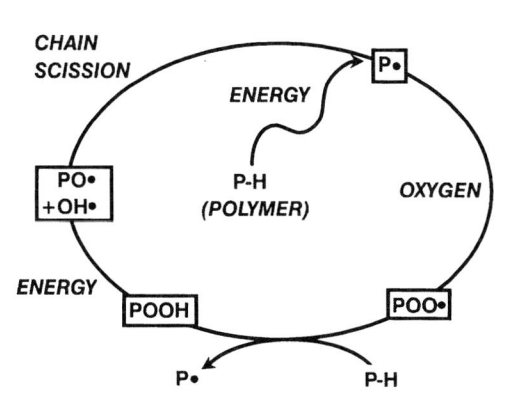

Figure 1 Auto-oxidation of PUR.

radical termination is a hindered phenol [5]. Phenolic antioxidants are highly effective at relatively low concentrations (i.e. <0.5%) in inhibiting scorch formation. The mechanism involves a chain-breaking donation of a hydrogen atom from the antioxidant to the reactive peroxy-radical (Figure 2). This produces a less reactive, resonance stabilized phenolic radical. Peroxycyclohexadienones can then be formed after reaction with a second peroxy-radical [6]. Each phenolic moiety is capable of trapping a total of two radicals before it is completely consumed.

Phosphite and thioester and antioxidants

Preventative or secondary antioxidants act at the initiation stage of the radical chain mechanism to prevent the formation of radical products. Their primary mechanism involves the decomposition of hydroperoxides

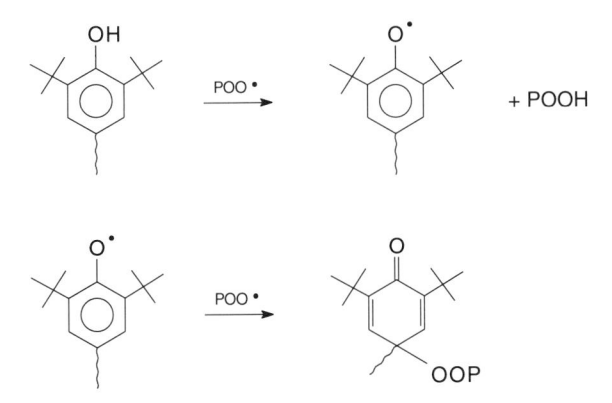

Figure 2 Radical trapping mechanism of phenolic antioxidants.

to form stable non-radical products. In the absence of peroxide scavengers, hydroperoxides thermally or photolytically decompose to radical products. The most common secondary antioxidants are sulfur or phosphorus based.

Thioesters have been shown to decompose several moles of hydroperoxide per mole of stabilizer [7]. The hydroperoxide is typically reduced to an alcohol and the thioester is transformed into a variety of oxidized sulfur products including sulfenic and sulfonic acids. Synergistic combinations with phenolic antioxidants are often used to enhanced thermal stability in polyolefins at elevated temperatures (>100°C).

Phosphites are also commonly used in combination with phenolic antioxidants to inhibit polymer degradation and improve color. As in the case of thioesters, phosphites reduce hydroperoxides to the corresponding alcohols and are transformed into phosphates.

Other antioxidants

Alkylated diphenylamines and phenothiazine are also used alone or in combination with phenolic antioxidants as antiscorch agents. As a class of antioxidant, alkylated diphenylamines are quite cost effective but often have a propensity to discolor. Phenothiazine is also an effective antiscorch agent when used in combination with phenolic antioxidants.

ANTIOXIDANT VOLATILITY

The issue of BHT volatility is becoming increasingly important in the flexible polyurethane bunstock industry. This is particularly true for manufacturers of low density high water foam or manufacturers using air cooling technology. Low volatility components can be forced from flexible polyurethane bun stock by high exotherms. This is especially true as blowing agents, with their evaporative cooling effect, are being eliminated. Material forced from the bun cools and is deposited as a residue. These residues must be routinely removed and disposed of at considerable cost to the manufacturer.

Volatility and migration are also two issues affecting performance. While some degree of volatility is generally considered desirable, excessive migration and volatility can have a deleterious impact on scorch inhibition. It has been demonstrated that large scale migration of volatile BHT can occur during normal foam production. This migration results in a dramatic decrease in antioxidant concentration at the center of the bun where the highest exotherms are encountered. Higher molecular weight antioxidants such as the materials evaluated in this review would be expected to have a greatly reduced rate of migration and loss.

Performance data

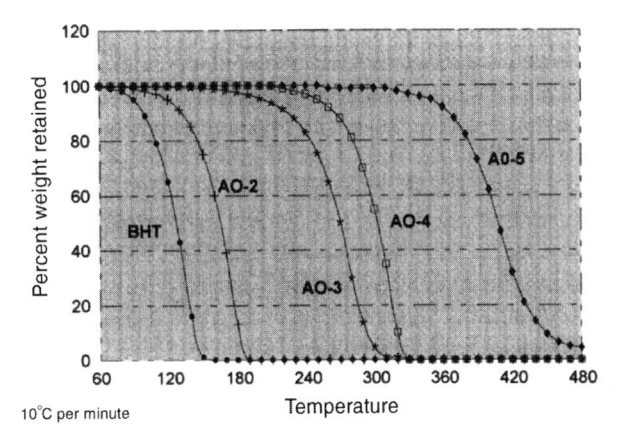

Figure 3 Thermogravimetric analysis of phenolic antioxidants.

Figure 3 shows thermogravimetric analysis (TGA) data for selected phenolic antioxidants. BHT is the most voltatile of the group, with a 5% weight loss at about 90°C and a 90% loss at about 142°C. AO-2 has a range for the 5% and 90% weight losses of 120°C to 182°C, respectively. AO-3, AO-4 and AO-5 have TGA volatility values that are predominantly above the peak exotherm temperature of flexible slabstock polyurethane foam. They would be expected to volatilize significantly less during foam cure if incorporated.

The decreased volatility is achieved by increasing the molecular weight of the antioxidant. This is accomplished through the addition of long hydrocarbon chains in the case of AO-3, AO-4, and AO-2. (See Appendix 1 for the chemical structures.) AO-5 achieves a high molecular weight without dilution of the active content by coupling four phenolic moieties. AO-2 and AO-3 are not only higher in molecular weight than BHT but are also liquids. Liquid antioxidants are often easier to handle and meter in urethane production than solids like BHT.

TGA analysis of two secondary antioxidants evaluated in this study also indicated that loss of stabilizer would not be expected under typical curing conditions. The phosphite, PHOS-1, has a 5% weight loss at 189°C and a 90% weight loss at 266°C. The thioester, DTDTDP, has a 5% weight loss at 227°C and a 90% weight loss at 328°C. Some antioxidant structures are given in Appendix 1.

PERFORMANCE DATA

A variety of structurally dissimilar phenolic antioxidants were evaluated for performance as scorch inhibitors. The results of screening work are shown in Figure 4. In general, all of the antioxidants evaluated provided

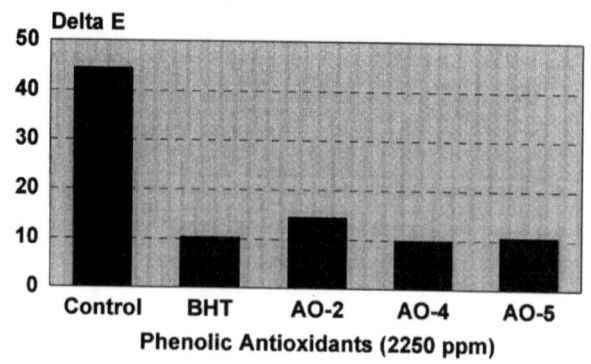

Figure 4 Anti-scorch performance of selected antioxidants.

a high level of scorch protection as measured by delta E. The reactive performance is as follows:

$$AO\text{-}4 = AO\text{-}3 \text{ (not shown)} > BHT > AO\text{-}5 > AO\text{-}2$$

The traditional antioxidant, BHT, was among the best performers. Foams prepared with AO-2 showed a slightly higher level of discoloration as compared with BHT. This small difference in performance may be structurally related. AO-2 is similar to BHT, differing only in the substitution of the para methyl group with a nonyl hydrocarbon chain. This substitution results in over a 50% increase in molecular weight with no increase in active substituents. The dilution effect of the alkyl chain may contribute to the observed marginal decrease in performance.

In this study, color measurement (delta E) has been chosen as the method to evaluate extent of scorch. This method of evaluation should be carefully considered as color development can result from several very different sources. Actual degradation of the PUR foam is the primary concern and probably contributes the majority of measured discoloration. Discoloration can also result from the highly colored transformation products of phenolic antioxidants. Discoloration from BHT transformation to quinones has been widely studied.

All phenolic antioxidants can be expected to undergo transformations similar to BHT. Discoloration of phenolics in other substrates varies widely. These differences can be attributed to the dramatic variation in the extinction coefficient of the various transformation products [8]. Extinction coefficients for transformation products of BHT and AO-4 are shown in Table 1.

One would expect the transformation products of AO-2 and BHT to have similar structures and extinction coefficients. The most significant difference in transformation products may be in molecular weight. In

Table 1 Extinction coefficients of phenolic transformation products

Antioxidant	Transformation product	$\varepsilon\ (l\,mol^{-1}\,cm^{-1})$
BHT	Stilbenquinone	106 000
	Diphenoquinone	72 700
AO-4	Conjugated bisquinonemethide	34 800
	Unconjugated bisquinonemethide	11
	Biscinnamate	106

this case, the relatively high volatility of the highly colored BHT transformation products may prove to be an advantage.

As AO-3, AO-4 and AO-5 are all derived from the same base active moiety, their transformation products have similar low extinction coefficients, relative to BHT type chemistry. This minimal color development upon consumption of these antioxidants may be key in the high level of performance observed. For this class of hindered phenolic antioxidant, dilution of the active functional group appears to have less of a negative influence on performance than in the case of AO-2.

Discoloration during warehouse storage is also an important concern. This phenomenon, commonly known as gas fade, is a result of interaction between phenolic antioxidants and oxides of nitrogen generated by combustion of natural gas by warehouse equipment (e.g. forklifts, heaters etc.). The color bodies associated with gas fade are often related to the same general class of quinones formed during foam production.

The efficacy of peroxide scavengers in combustion with a phenolic antioxidant (BHT) was also evaluated. Figure 5 shows the extent of scorch (as measured by delta E) in each formulation. The performance profile of

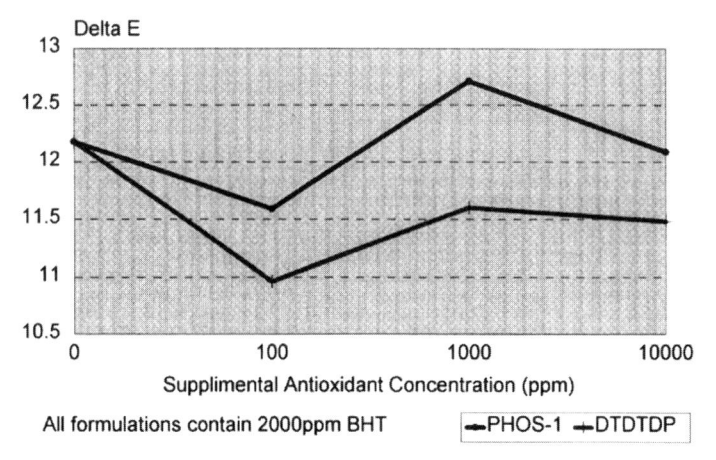

Figure 5 Anti-scorch performance of secondary antioxidants.

both types of secondary antioxidant was similar. At low concentrations (100 ppm), a decrease in color is observed. The benefit of both the thioester (DTDTDP) and phosphite (PHOS-1) is clearly diminished as concentrations approach and exceed 1000 ppm. The performance advantage of DTDTDP over PHOS-1 is most probably related to temperature effects. Thioesters function at temperatures as low as 100°C. The maximum bun temperature of 160–180°C is near the lower limit of the phosphite performance range.

CONCLUSIONS

The requirements of flexible polyurethane slabstock foam manufacturers vary due to sensitivity to antioxidant volatility, preference in physical form (liquid or solid), and degree of anti-scorch protection. As a result of this broad variation in antioxidant performance demands, several alternative antioxidants to BHT are available. AO-2 and AO-3 combine relatively low volatility and good anti-scorch performance with the advantage of a liquid product form. AO-4 and AO-5 have significantly lower volatility than BHT and improved performance. Addition of peroxide scavengers such as DTDTDP and PHOS-1 can further enhance the scorch resistance of the polyurethane foam.

It is important to bear in mind that anti-scorch evaluations are often developed in a laboratory setting where the actual conditions of industrial scale production may not be accurately represented. In industrial foam production where high internal temperature may be maintained for times in excess of six hours, the effect of volatility may be much more dramatic. Therefore, one might expect the performance results with the lower volatility antioxidants presented in this review to be significantly improved over that reported in these studies.

APPENDIX 1   ANTIOXIDANT STRUCTURES

BHT

$HO-C_6H_2(t\text{-}Bu)_2-CH_2CH_2CO(CH_2)_nCH_3$

$n = 14\text{–}15$

$n = 14\text{–}15$ – AO-3: Anox BF
$n = 17$ – AO-4: Anox PP 18, Irganox 1076

AO-2: Lowinox DBNP

AO-5: Anox 20, Irganox 1010

PHOS-1: Lowinox OS 330

DTDTDP

REFERENCES

1. Antipova, V.F., Marei, A.I., Apukhtina, N.P., Mozzhukina, L.V., Melamed, V.I. (1970) *Polym. Sci. USSR*, **12**, 2542.
2. Sing, A., Weissbein, L., Mollica, J.C. (1966) *Rubber Age*, **98**, 77.
3. Bolland, J.L., Gee, G. (1946) *Trans. Faraday Soc.*, **42**, 236.
4. Bolland, J.L. (1948) *Trans. Faraday Sci.*, **44**, 669.
5. Das, P.K., Encinas, M.V., Scaiano, J.C., Steenken, S. (1981) *J. Am. Chem. Soc.*, **103**, 4162.
6. Pospisil, J. (1980) *Adv. in Polym. Sci.*, **36**, 69.
7. Armstrong, C., Husbands, M.J, Scott, G. (1979) *Eur. Polym. J.*, **15**, 241.
8. Klemchuk, P.P., Horng, P.L. (1991) *Polym. Degradation and Stability*, **34**, 333.

Keywords: foam, BHT, antioxidant, phenolic, discoloration, phosphite, degradation.

See also: Antioxidants: an overview;
           Antioxidants: hindered phenols.

Smoke suppressants

Stewart White

INTRODUCTION

Fewer people are dying in fires in the UK and USA than ever before. In the US fire deaths fell by 7.8% to 4275 between 1993 and 1994. This compares with more than 5000 in 1990, more than 6000 in 1988 and more than 7000 in 1979 [1]. Since 1977, fire deaths have declined by about 42%. The decline in the UK is not as dramatic as that in America, but the number of people dying in fires has fallen significantly in the last few years.

A major cause of deaths in fires (50%) is asphyxiation. In real fires so-called 'secondary effects' occur as a result of the overall burning process. The most important of these secondary effects is the formation of smoke and some particularly lethal fire gases (including carbon monoxide).

SMOKE FORMATION

The amount of smoke produced in a fire depends on several factors including the source of ignition, oxygen availability, heat flux and the chemical makeup and properties of the fuel.

Smoke production in fires results from incomplete combustion. It is usually thought of as being a dispersion of solid or liquid particles in a carrier gas consisting of the combustion gases and hot air. The liquid particulates are tar-like droplets or mists composed of the liquid products arising from pyrolysis, or their partially oxidized derivatives and, of course, water. The solid component of smoke often contains carbon flakes, soot, ash and sublimed pyrolysis products.

Polymers containing purely aliphatic structural units of polyethylene (PE) and polypropylene (PP) produce relatively little smoke when they

Plastics Additives: An A–Z Reference
Edited by G. Pritchard
Published in 1998 by Chapman & Hall, London. ISBN 0 412 72720 X

burn in air. As the carbon chain length of the structural repeat units in these polyolefins increases (e.g. polyethylene to polypropylene to poly-4-methylpent-1-ene) smoke formation tends to increase, because more carbon is available.

Polymers with aliphatic chains having pendant aromatic sub-groups have a greater tendency to generate smoke when burning, e.g. polystyrene (PS) and acrylonitrile-butadiene-styrene (ABS). On the other hand polymers with aromatic groups in the main polymer chain produce intermediate amounts of smoke, probably because on thermal decomposition they produce large proportions of char.

The chemical action of some flame retardants may increase the formation of smoke during burning of the polymer due to there being incomplete oxidation of the volatile decomposition products from the polymer. Those flame retardants which act mainly in the gas phase are generally the ones which react most readily with the active flame propagating species such as hydrogen and hydroxyl radicals; these radicals are thus no longer available for the chain reaction which promotes further burning. Cracking of some polymer breakdown products therefore tends to occur at the expense of oxidation and this leads eventually to the formation of larger amounts of carbon rich solids which constitute the major part of smoke.

SMOKE SUPPRESSANTS

The mechanism of smoke suppressants has not been as extensively studied as flame retardance mechanisms. Certain compounds are believed to perform their smoke suppressant function in a number of ways including (1) promoting char formation, (2) diluting the polymer content, (3) dissipating heat and (4) altering the chemical reactions in the condensed phase. These processes can be identified by an increase in the char residue. Other compounds are believed to act in the gas phase by promoting oxidation of the carbon or soot in the smoke and modifying the chemical reactions in the gas phase.

Compounds which have shown some activity as smoke suppressants include oxides of some of the heavy metal elements, such as vanadium in Group VB, chromium and molybdenum in Group VIB, manganese in Group VIIB, iron, cobalt, and nickel in Group VIII, copper in Group IB, zinc and cadmium in Group IIB, aluminium in Group IIIA, tin and lead in Group IVA, and possibly antimony in Group VA. Compounds containing non-metallic elements such as boron and phosphorus have also shown some limited smoke suppressant activity. However, many of these potential smoke suppressants are not commercially viable, mainly due to price. Of the more commonly used additives, molybdenum compounds perform most of their smoke suppressant action in the condensed phase. Chemical analysis has established that over 90% of the molybdenum remains in a char. Molybdenum trioxide (MoO_3) probably acts as a crosslinking agent by promoting the formation of stable char. Commercially available molybdenum compounds include molybdenum trioxide, ammonium octamolybdate (AOM), calcium and zinc molybdates. Iron is known to act as a smoke suppressant by promoting the formation of stable char and by acting as an oxidation catalyst, converting polymer carbon into oxides of carbon. Blending polyvinyl chloride (PVC) with ABS, a high smoke producing polymer, i.e. 70 ABS/30 PVC gives only a marginal reduction in smoke formation (8%) despite the fact that rigid PVC has a smoke value of about half that of ABS when measured using an NBS smoke density chamber, in the flaming mode.

However, adding 5% hydrated iron(III) oxide to the above formulation increases char formation from 6% to 20%, at 650°C, and reduces smoke emission by over 50% [2]. The most effective iron compounds investigated to date appear to be the oxides, although some organometallic compounds of iron such as ferrocene are active in PVC. (See also the entry entitled 'Flame retardants: iron compounds, their effect on fire and smoke in halogenated polymers' for more detail.) See also Figure 1.

Zinc compounds also act as smoke suppressants, mainly in the condensed phase. Zinc oxide is an effective smoke suppressant for PVC, but when used alone it can cause thermal breakdown of PVC and early release of hydrogen chloride, a phenomenon known as 'zinc failure'.

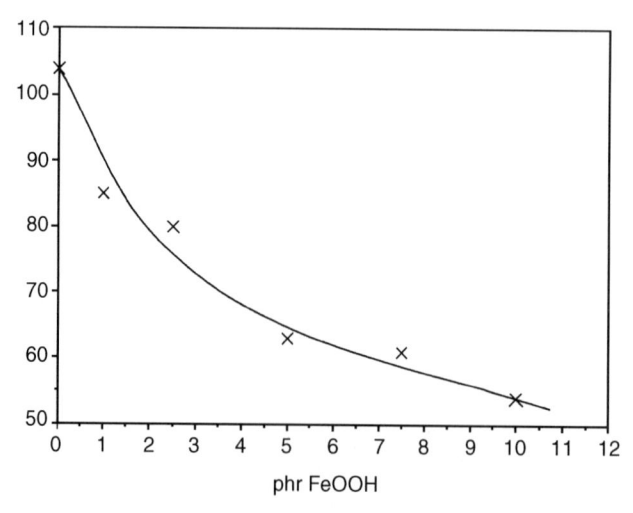

Figure 1 Smoke density as $D_{max}\,g^{-1}$ against phr FeOOH (0–10) in a 70 ABS/30 PVC blend.

Certain compounds, e.g. Ongard 2, a zinc/magnesium complex (Anzon Ltd) can reduce this tendency by 'masking' the zinc oxide in a magnesium oxide lattice. Hydrated zinc borates are multi-functional in flammability terms. Apart from acting as general flame retardants, they suppress smoke and afterglow, are char promoters and they also act as anti-arcing agents, thereby improving the tracking index. The major use is in plasticized PVC. Zinc borates can be prepared with a wide range of stoichiometries, with varying water levels of hydration. However, only a selected range are commercially acceptable, due to their thermal stability characteristics. For example: ZB 223 has the composition $2ZnO \cdot 2B_2O_3 \cdot 2H_2O$ and ZB 467 is $4ZnO \cdot 6B_2O_3 \cdot 7H_2O$, with ZB 467 having a superior thermal stability towards PVC, due to its lower ZnO content.

During heating, e.g. in a fire, the boric oxide content can fuse and form a glassy layer, protecting the underlying polymer from further fire chemical decomposition, whilst the zinc oxide content causes crosslinking of the PVC, thereby reducing smoke emission. The release of crystallized water during polymer combustion also reduces flame propagation by cooling and diluting the flame.

Zinc in combination with tin as zinc stannate $ZnSnO_3$ and zinc hydroxystannate $ZnSnOH_6$ (Flamtard S and H) are also used as flame retardants/smoke suppressants in polymers.

Zinc compounds are also particularly effective when used as flame retardants in unplasticized PVC. Typical loadings for the additives so far described would be between 2 and 5%.

FILLERS

Chemically inactive smoke suppressants in the form of fillers are also available, the most widely used being alumina trihydrate, known as ATH ($Al_2O_3 \cdot 3H_2O$) and magnesium hydroxide ($MH(OH)_2$). These fillers are used at loadings ranging from 30 to 150 phr, or perhaps even higher if the formulation can tolerate a high filler level.

ATH is the largest tonnage flame retardant/smoke suppressant used today in plastics. ATH is widely used in the formulation of flame retardant carpets and in unsaturated polyester resins. On heating to above 200°C ATH decomposes to give aluminium oxide and water vapour:

$$2Al(OH)_3 \rightarrow Al_2O_3 + 2H_2O$$

This reaction is strongly endothermic, absorbing 1.97 kJ of heat per gramme of ATH, thus removing heat from the reaction zone of the polymer [3]. On decomposition ATH also releases 34.6% of its weight as water vapour. Most of the water vapour is evolved over the temperature range 200–400°C, which coincides with the decomposition temperature of many polymers.

The disadvantage of ATH is its unsuitability for polymers which are processed at temperatures above 200°C. ATH is therefore little used in acrylonitrile-butadiene-styrene (ABS), polycarbonate (PC), polyamides (PA) and polypropylene (PP), and it is also ineffective at attainable filler loadings in polymethyl methacrylate (PMMA) and polystyrene (PS). ATH is successfully used in PVC, polyethylene (PE), ethyl vinyl acetate (EVA), ethylene propylene diene monomer (EPDM) and polyester, epoxy, phenolic, methacrylic and urethane thermosets.

The smoke suppressant property of ATH is not yet fully understood, but it is believed that heat dissipation in the burning polymer favours crosslinking reactions over pyrolysis; this leads to the formation of a char in preference to soot particles.

For polymers processed above 200°C magnesium hydroxide ($Mg(OH)_2$) can be used. Magnesium hydroxide begins to decompose at 330–340°C and releases 31% of its mass as water in an endothermic reaction. This absorbs heat from the system and dilutes flammable fire gases. The magnesium oxide formed acts as an insulating protective barrier [4]. Magnesium hydroxide is used as a smoke suppressant in high impact polystyrene (HIPS) and PP.

Surface coated grades of ATH and $Mg(OH)_2$ are also available. The special coating appears to reduce the adverse effect of the high loading on physical properties and polymer appearance.

TEST METHODS

The importance of smoke and toxic combustion products as factors leading to deaths in fires, and the possibility that burning some plastics may contribute to the hazard, are well recognized. Of these, smoke is probably the most problematical because it restricts visibility, disorientates potential victims and panic may result. A conservative estimate of the proportion of fatal casualties due to these hazards in fire situations lies around 60%.

At present there are more than fifteen widely used different test methods to evaluate smoke, each employing its own unique set of heating conditions, sample size and orientation, gas flow and means of smoke measurement. The most frequently used tests are those based on optical methods, i.e. attentuation of a light beam due to the sample formulations burning. There are also mechanical methods, those based on separation of liquid and solid aerosol particles from the smoke gases, the Arapahoe method and electrical methods (generation of electrical charges in an ionization chamber).

Optical methods of measuring smoke density can be carried out under static conditions (e.g. NBS smoke density chamber ASTM E-662-83, see Figure 2), or dynamic conditions, an open system where smoke is

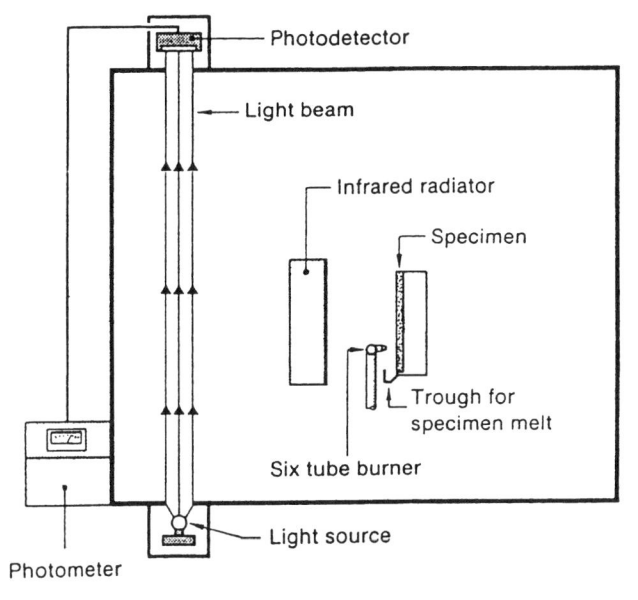

Figure 2 Principle of the NBS smoke chamber.

measured as it escapes from the apparatus (e.g. Ohio State University O.S.U. heat and visible smoke release apparatus ASTM E-906-83) [5].

When light penetrates a smoke filled or partially filled space, its intensity is reduced because of absorption and scattering by smoke particles. The level of attenuation depends on particle size and shape, refractive index, wavelength and angle of incidence of the light.

With the static NBS smoke density test, the apparatus is designed so that all the smoke generated is contained in a closed chamber. The chamber volume is $0.51\,m^3$ and the test specimens are heated by means of an electrically heated radiant heat source, calibrated to $2.5\,W/cm^2$. Smoke tests using the radiant heat source only are said to be done in 'smouldering mode'. Optionally, a multi-angled burner may be positioned so that the flamelets impinge on the sample surface to simulate 'flaming mode'. Generally, most polymers produce more smoke in flaming than in smouldering mode.

The specimen holders are designed to expose an average of 65 mm × 65 mm of specimen to the heat source. The photometric system consists of a vertically mounted light source and a photocell. The photocell consists of a photomultiplier tube with a spectral response which is considered to be close to that of the human eye. In line with the photomultiplier tube is connected a sensitive amplifier and chart recorder or PC. The amplifier has provision for accurately measuring light transmittance down to 0.001% and it can be extended down to 0.00001% by a

Table 1 Smoke density values for some polymers tested in flaming mode

Polymer	Smoke density, $D_{max}\,g^{-1}$
ABS	113
PVC	58
PP	14
CPVC (65% Cl)	11
HDPE	7

special filter located in the photometer. Smoke density values can be presented as specific optical density, as a function of time, or as maximum specific optical density (D_{max}). In Table 1 the smoke values, quoted as $D_{max}\,g^{-1}$, are given for some polymers tested in flaming mode, using specimens ~1 mm thick.

With the dynamic O.S.U. apparatus, a photometer measures the percentage of light transmitted through the gases leaving the apparatus. The O.S.U. method yields dynamic records of smoke (and heat) released during combustion. The test also provides for radiant thermal exposure of a specimen both with and without a pilot source. Chamber dimensions are 890 mm × 410 mm × 200 mm with a pyramidal top section 395 mm high connected to the outlet. Specimens are typically 150 mm × 150 mm × 2 mm and are exposed in the vertical orientation.

Recently, smoke production for burning plastics has also been reported using the Cone Calorimeter (ISO 5660). This apparatus employs an electrical heater in the form of a truncated cone, hence its name.

The smoke and combustion gases are drawn to a sampling point, where the smoke measurement is made with a low intensity helium–neon laser beam projected across the diameter of the duct. The smoke data are reported as the specific extinction area. This is defined as the area (m^2) of the smoke generated per mass (kg) of specimen decomposed; thus the units are m^2/kg. Specimens used as 100 mm × 100 mm and up to 50 mm thick. The heat flux can be varied from 1 to 100 kW/m^2, with horizontal or vertical specimen orientation.

Most flammability (including smoke) tests are relatively small-scale, because experimental materials are often only available in relatively small quantities and it is prohibitively costly to burn complete structures. Small-scale tests are also more easily and more economically replicated than large-scale fires. Many tests have an inherent deficiency in that they fail to reproduce the massive effect of heat produced in a large-scale fire, and therefore may give results that can be misleading if applied in the wrong context.

REFERENCES

1. Fire Prevention LPC publications (Oct. 1995).
2. Carty, P. and White, S. (1995) 'Flammability studies – plasticised and non-plasticised PVC/ABS blends', Paper accepted for publication, *Polymer Networks and Blends*.
3. Brown, S.C. and Herbert, M.J. (1992) *Flame Retardants '92*, Elsevier Applied Science.
4. Rothan, R. (1990) *Flame Retardants '90*, Elsevier Applied Science.
5. Troitzsch, J. (1983) *International Plastics Flammability Handbook*, Hanser Publications, Munich.

Keywords: combustion, smoke, flame, char, heavy metal, aluminium hydroxide, molybdenum, zinc, fillers, cone calorimeter.

See also: Series of entries 'Flame retardants:...';
Flame retardants: iron compounds, their effect on fire and smoke in halogenated polymers.

Surface-modified rubber particles for polyurethanes

Bernard D. Bauman

INTRODUCTION

A new type of material, i.e. surface-modified rubber particles, is a novel reinforcing, elastomeric filler. The combination of these particles with polyurethane represents a very important development for the polyurethane industry. Rubber and polyurethane are both recognized as families of materials having desirable and unique physical properties. For many applications, such as industrial wheels, the two materials compete with each other. Surface modification of the rubber particles enables one to effectively combine polyurethane and rubber. This results in the formation of composites, in which rubber particles are dispersed in a continuous phase of polyurethane. There is a huge number of possible rubber particle/polyurethane composites that can be formed. Variations include different types of rubber particles, different sizes of rubber particles, different proportions of rubber and polyurethane, and different types of polyurethanes. The resulting new class of polyurethane/rubber composites can have lower costs than pure polyurethane formulations, and can have performance enhancements for specific applications. Surface-modified rubber particles hold promise to expand significantly the markets for polyurethane, by altering its performance/cost ratio and by making it more competitive with other materials.

Almost all types of products currently made with polyurethanes can be made in rubber particle/polyurethane composites. Examples include molded wheels and rollers, adhesives, sealants, coatings and flexible foam.

Plastics Additives: An A–Z Reference
Edited by G. Pritchard
Published in 1998 by Chapman & Hall, London. ISBN 0 412 72720 X

Surface-modified rubber particles

This entry describes the technology of surface-modification of rubber particles, characterizes the properties of some rubber/polyurethane composites, discusses processing considerations, and reviews applications that have been developed to date.

SURFACE-MODIFIED RUBBER PARTICLES

Rubber and polyurethane are generally incompatible and cannot be simply combined to give homogeneous materials with good physical properties. With rubber particles being non-polar and polyurethane precursors being fairly polar, trying to combine the two is like mixing oil and water. Even if one does force them together, there is generally very little bonding at the interface between the two materials. Hence, the combination of untreated rubber particles with polyurethane generally gives materials with poor physical properties. Such materials have relatively little commercial usage, and are primarily used in applications having low performance requirements.

With a patented [1], proprietary surface modification, rubber particles are rendered compatible with polyurethanes. One surface modification is a controlled oxidation of the outermost molecules on each rubber particle. The treatment chemistry involves the reaction of halogen- and oxygen-containing gases with the rubber backbone to form pendant polar functional groups, Figure 1. In fact, the rubber surface becomes so hydrophilic that the treated particles are readily wetted by water.

Surface modification also enables excellent interfacial bonding between the rubber particles and the polyurethane. The effectiveness of this surface modification in facilitating adhesion is demonstrated by comparing the bond strength of strips of rubber with polyurethane cast on them. In T-peel tests, it was found that where polyurethane was cast onto untreated strips of rubber, the bond strength was 5 N/cm. In analogous

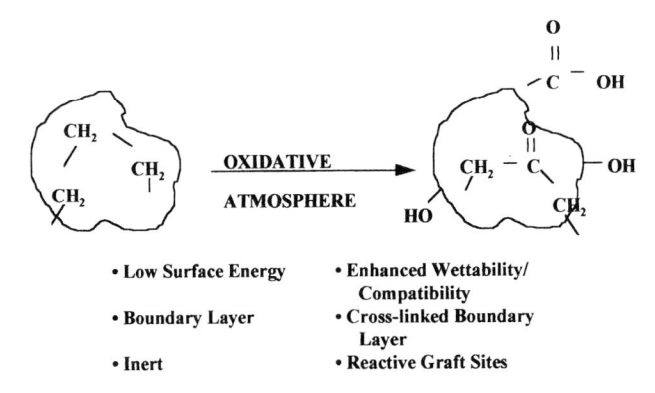

Figure 1 Surface modification imparts polar functional groups.

tests with strips of surface-treated rubber, the bond strength exceeded 250 N/cm, and the rubber tore before the adhesion bond failed. The modification is permanent. Particles treated ten years ago still exhibit excellent adhesion.

Surface-modified rubber particles are commercially available [2]. At the time of writing, the rubber is derived from scrap automotive tyres. Rubber particles consisting of other types of rubber, scrap or virgin in origin, can also be surface-modified. Currently, grades of rubber particles are available in sizes ranging from diameters of 1.5×10^{-3} m (10 mesh) down to less than 7.5×10^{-5} m (200 mesh).

Surface-modified rubber particles are quite hygroscopic. The polar surface readily absorbs moisture from the atmosphere. Laboratory studies have shown that treated rubber particles can absorb as much as 1.4% moisture by weight, and that the particles can absorb moisture at a rate greater than 0.6%/hour when exposed to the atmosphere. As many polyurethane formulations are sensitive to moisture, the treated rubber particles must be dried before use in many systems.

PROPERTIES OF RUBBER/POLYURETHANE COMPOSITES

The physical properties of composites made with surface-modified rubber particles in polyurethanes can be custom tailored over a very broad range. When rubber particles are combined with a polyurethane that has a similar hardness, the end properties are comparable. This is exemplified in Table 1, where the properties of a composite consisting of 15% treated rubber particles and 85% polyurethane are nearly indistinguishable from those of the base polyurethane. Composites comprising as much as 70% treated rubber particles have been shown to have good engineering properties.

Table 1 Physical property comparison (high performance polyurethane)

Property	Unfilled polyurethane*	15% Surface-modified rubber particles[†]/ 85% polyurethane*
Tensile strength (MPa)	26	24
% elongation	278	275
Tear resistance (Die C; kN/m)	104	92
Tear resistance (Trouser; kN/m)	20	18
% rebound	49	49
Hardness (Shore D)	50	50

* Polyurethane is AIRTHANE® PET 95A, from Air Products & Chemicals, Inc., cured with ETHACURE® 300 from Albemarle.
† Surface-modified rubber particles are VISTAMER® R4060, from Composite Particles, Inc.

Treated rubber is the only known reinforcing, elastomeric filler. The use of surface-modified rubber, in general, does not make the formulation harder and less flexible. In contrast, most other fillers are inorganic minerals, which reduce the elastomeric properties of a polyurethane formulation. Oils are used as fillers for some polyurethanes and to increase their flexibility. However, oils are certainly not reinforcing fillers, and they significantly reduce overall properties.

The dynamic properties of polyurethane composites made with treated rubber particles can also be very similar to those of the unfilled polyurethane. Comparison of dynamic mechanical analysis (DMA) curves for composites composed of surface-modified rubber particles in polyurethane and corresponding curves for the unfilled polyurethane show nearly identical hysteretic heating. End products, such as wheels and rollers, made in either unfilled polyurethane or in composites comprising treated rubber particles/polyurethane have essentially identical heat build-up in dynamic applications.

ENHANCED PROPERTIES

The incorporation of treated rubber particles in polyurethane formulations can give beneficial property enhancements. Examples include a higher coefficient of friction (especially on wet surfaces), greater thermal stability, higher modulus, reduced tendency for glazing, and reduced moisture absorption. Benefits to the coefficient of friction are illustrated in Figure 2.

PROCESSING CONSIDERATIONS

The incorporation of up to about 15% surface-modified rubber particles in polyurethane formulations is generally straightforward and requires little

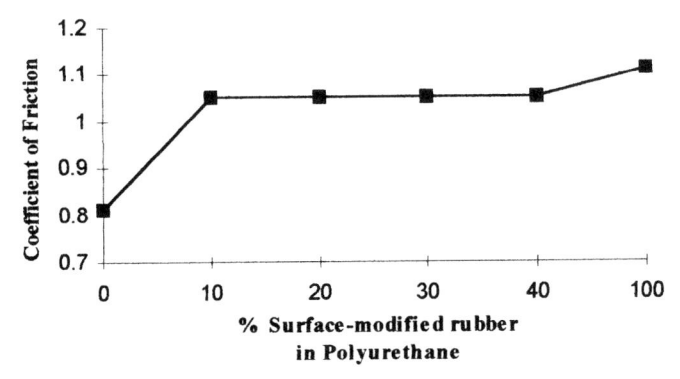

Figure 2 Inclusion of surface-modified rubber particles in polyurethane formulation increases the coefficient of friction.

or no modifications to molding equipment. When making cast items using polyurethane prepolymers, the rubber particles are combined with the prepolymer prior to combination with a cross-linker. For polyurethane RIM (reaction injection molding), the rubber particles are combined with the polyol and subsequently fed to the mix head. For systems that polymerize more slowly, it is possible to add the particles following the combination of the polyurethane precursors. Incorporation of treated rubber particles in polyurethane foam is accomplished by mixing the particles in the polyol prior to the mix head.

Manufacture of rubber/polyurethane composites with higher levels of rubber particles generally requires more drastic modifications to the molding process and equipment. This is because incorporation of rubber particles in polyol or prepolymer causes an increase in viscosity. These thicker slurries are more difficult to pump and to deaerate.

END PRODUCTS MOLDED IN TREATED RUBBER/POLYURETHANE

A number of different types of end-products, molded in treated rubber/polyurethane, have been developed and commercialized. These include rollers, wheels, flexible foams, molded goods, coatings, adhesives and sealants. It is obvious that this is just the beginning and that a large number of additional uses will be developed in the future.

CONCLUSIONS

Surface-modified rubber particulates constitute the only known reinforcing, elastomeric filler. These rubber particles can be readily combined with polyurethane to manufacture end-products that have reduced raw material costs and, in some cases, better physical properties. This new class of materials promises to open significant new markets for polyurethanes by altering their performance/price ratio and making them more competitive with other materials.

NOTES

1. U.S. Patents 5,382,635 and 5,506,283.
2. Sold under the tradename VISTAMER® rubber particles by Composite Particles, Inc., Allentown, Pa.

BIBLIOGRAPHY

Bauman, B.D. (1995) High value engineering materials from scrap rubber, *Rubber World*, **212**(2), 30–33.

Bauman, B.D. (1994) A new alchemists' dream? Scrap rubber becomes high value products, *Urethane Technology*, **28**, 8 pp.

Bauman, B.D. (1994) *Proc. Polyurethanes '94 Conference (SPI)*, October, pp. 675–679.

Keywords: surface-modification, polyurethane, rubber particles, composites, adhesion, filler, compatibility, coefficient of friction, peel tests, tear resistance.

Surface treatments for particulate fillers in plastics

Marianne Gilbert

INTRODUCTION

Most of the commonly used particulate fillers have highly polar hydrophilic surfaces, whereas many of the polymers in which they are used are non-polar and hydrophobic (e.g. polyolefins). Coatings are used to modify the character of the filler surface, and therefore the degree of interfacial adhesion at the boundary between the filler and polymer matrix. The surface energy of fillers tends to be much higher than that of organic polymers. For example, when calcium carbonate in the form of limestone is ground, broken bonds on the freshly exposed surfaces (i.e. metal ions which are not fully saturated) react with water in the air to produce hydroxyl groups, which cause particle agglomeration and the surface energy is greater than that of water, i.e. $72\,\text{mJ}\,\text{m}^{-2}$, while polymers typically have surface energies of about $35\,\text{mJ}\,\text{m}^{-2}$. In the absence of surface water the surface energies of most inorganic fillers are even higher, typically $100\,\text{mJ}\,\text{m}^{-2}$.

From surface energy theory, the work of cohesion in a liquid is equal to twice the surface energy. For a polymer this would be approximately $70\,\text{mJ}\,\text{m}^{-2}$. It has been shown that for cases when either or both the filler and polymer are non-polar, W_{PF}, the reversible work necessary to separate unit area of filler–polymer interface is given by:

$$W_{PF} = 2(\gamma_P^d \gamma_F^d)^{1/2}$$

where γ_P^d and γ_F^d are the dispersion contributions to the specific surface excess free energies of the polymer and filler respectively. From this

Plastics Additives: An A–Z Reference
Edited by G. Pritchard
Published in 1998 by Chapman & Hall, London. ISBN 0 412 72720 X

equation, for uncoated fillers, W_{PF} is approximately equal to 100 mJ m^{-2}. It has been suggested that under these conditions, the polymer molecules adsorb and link adjacent filler particles, restricting dispersion. Surface coating reduces the surface energy of fillers, thus reducing polymer/filler interaction and assisting dispersion. Hornsby and Watson have considered the effect of the strength of a filler polymer interface on failure behaviour. When the filler–polymer interface is very strong, exceeding the strength of the matrix, the failure pathway is through the matrix. This situation could apply when uncoated fillers are used. When the interface is very weak, the filler and matrix separate, and the filler serves only to dilute the polymer. When, however, the interface has intermediate strength, similar to that of the matrix, a propagating crack will follow the filler/matrix boundary. This will occur when a suitable coating is used.

PURPOSE OF FILLER COATING

The addition of fillers to plastics offers a number of benefits, with possibly the most important ones being the enhancement of stiffness or rigidity, the production of compounds with reduced flammability, and the reduction of cost. However, problems also arise when fillers are introduced, notable among these being reduction in toughness, increased water absorption and increased melt viscosity. Deterioration of colour and abrasion of processing plant are secondary issues. Suitable filler coatings can be used to ameliorate several of these disadvantages. Mechanical properties, particularly impact strength, can be improved, often through improved filler dispersion, coatings can act as lubricating agents, thus reducing the melt viscosity of the compound, and discoloration, which often arises from the abrasion of processing plant, can be reduced. Because coatings generally produce a hydrophobic surface on the filler particle, water absorption is reduced. The presence of coatings also enables higher filler loadings to be achieved.

FILLER COATING INTERACTIONS

When coating of fillers is considered, fillers can be divided into three types. A number of the commonest materials contain high concentrations of hydroxyl (OH) groups, e.g. aluminium hydroxide (more commonly known as aluminium trihydrate or ATH) and magnesium hydroxide. These hydroxyl groups react readily with many of the potential coating materials. Carbonates such as calcium carbonate and huntite react most easily with acidic coatings. Other minerals, particularly silicates, contain localized hydroxyl groups, which react best with coatings such as silane compounds. These fillers include talc and mica. Talc has platy particles with inert surface layers, and OH groups only present at the broken edges. Mica also has a platy structure based on linked SiO$_4$ tetrahedra; the overall negative charge is overcompensated by metal cations, and the excess cation charge is balanced by negatively charged OH ions, and possibly fluoride ions. As with talc, the OH concentration is therefore low.

Many of the available coatings react with the filler particles, but do not react with the polymer matrix. Most are monofunctional. A typical reaction for an organic acid and magnesium hydroxide is shown here:

$$\text{OH} + \text{CH}_3(\text{CH}_2)_{16}\text{COOH} = \text{CH}_3(\text{CH}_2)_{16}\text{COO} + \text{H}_2\text{O}$$
$$\underset{\text{Mg}}{|} \qquad\qquad\qquad\qquad \underset{\text{Mg}}{|}$$

Other difunctional coatings, which are strictly coupling agents, react with both filler and matrix polymer. Organosilanes are the most common type in this category.

TYPES OF COATING

Fatty acids and their salts

Fatty acids and their salts are the most common type of coating used for particulate mineral fillers. By far the most common of these is stearic acid. It is inexpensive and can give a number of favourable effects, in particular easier processing. The commercially available material is usually a complex blend containing both saturated and unsaturated acids, with the stearic acid content often less than 60%. Rosin acid has also been used. The use of other acids has also been considered, but they are more expensive. Figure 1 compares the effect on processability when magnesium hydroxide coated with decanoic (C$_{10}$), behenic (C$_{22}$) and stearic (C$_{18}$) acids is compounded with polyethylene in an APV twin-screw extruder. Increase in coating concentration or chain length both reduce machine torque. The use of the unsaturated oleic acid has also been reported, and it has been suggested that while stearic acid molecules tend to lie approximately normal to the filler surface, unsaturated molecules such as oleic acid have a different conformation, and tend to lie flat.

Stearic acid reacts to form stearate as shown above, and such coatings can also be applied as stearate salts. The use of salts has been shown to have favourable effects on compound properties. For example, when magnesium stearate is used as an alternative to stearic acid for coating magnesium hydroxide, tensile and impact properties (measured using an instrumented drop weight impact tester) are both improved as shown in Table 1.

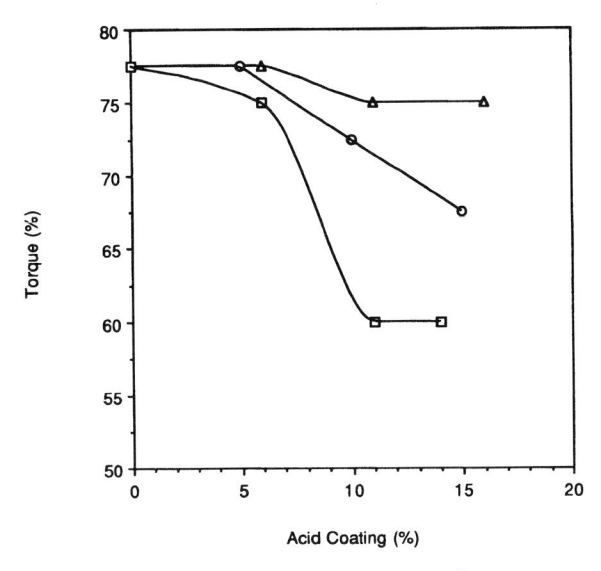

Figure 1 Effect of type and level of coating on twin-screw extruder torque: △ decanoic acid; ○ stearic acid; □ behenic acid. Polymer: Polyethylene.

Stearic acid coated calcium carbonates are commercially available, and are widely used. Stearic acid coating of magnesium hydroxide has been studied in some detail, and fatty acids could be used for coating other hydroxide and carbonate containing fillers.

Silane compounds

Silane compounds are also widely used both as coatings and coupling agents. They have a general formula R'Si(OR), where OR is a hydrolysable

Table 1 Properties of polyethylene compounds containing 20% magnesium hydroxide

Filler used	Ultimate tensile strength (MN/m²)	Elastic modulus (MN/m³)	Impact peak force (N)	Impact failure energy (J)
Uncoated Mg(OH)$_2$	23.8	847	494	2.22
Mg(OH)$_2$ + 10% stearic acid	24.5	716	550	11.3
Mg(OH)$_2$ + 7.85% magnesium stearate	30.0	954	690	20.8

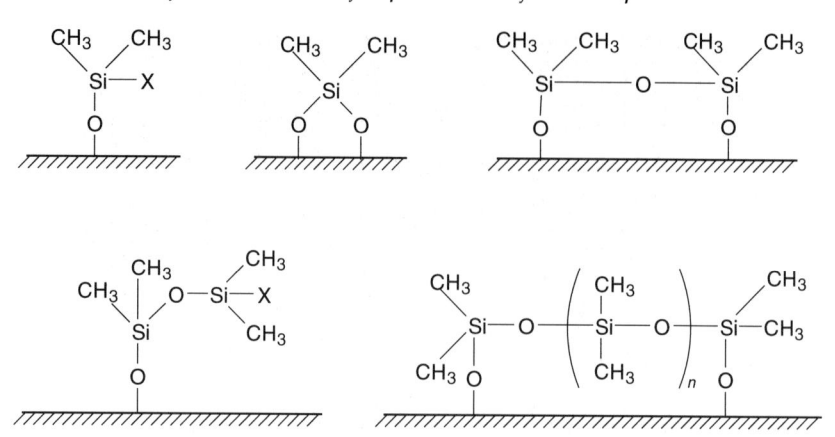

Figure 2 Some structures possible from the reaction of dimethyldihalogenosilane with a silica surface.

alkoxy group which can react with filler particles. The nature of R′ determines whether the silane functions as a coating or a coupling agent. If R′ is an alkyl group (methyl, ethyl etc.) it will not react with the polymer matrix, but if it incorporates vinyl, thiol or amine groups, for example, reaction with the polymer may be possible. Many of the silane suppliers provide a wide range of these materials for different purposes, together with literature which assists in their selection.

Silanes are particularly useful for coating silica, silicates such as calcined clay and mica, oxides and hydroxides. They are not reactive with carbonates, so they cannot readily be used with these materials.

Reactions with silanes are complex, and a range of possible reaction products which could be obtained from the reaction of dimethyldihalogenosilane with a silica surface is illustrated in Figure 2.

Silane coupling agents are more commonly used than silane coatings, but generally in reactive systems, when the reactive group in the silane can partake in any polymerization or crosslinking reactions which are occurring. These systems include thermosets, elastomers, and also crosslinkable thermoplastics. An example is the use of silanes for the improvement of filler/matrix adhesion of silicas or silicates in unsaturated polyesters, acrylics or epoxies. Certain thermoplastics can react; for example, amine groups in polyamides can undergo condensation reactions with certain reactive silanes. The improved adhesion which occurs when glass spheres, used as a filler in polyamide 6, are treated with an aminosilane is shown in Figure 3.

Many silanes are supplied in methanol, which is a flammable solvent, so suppliers' instructions for handling these materials should be strictly followed.

Types of coating

(a)

(b)

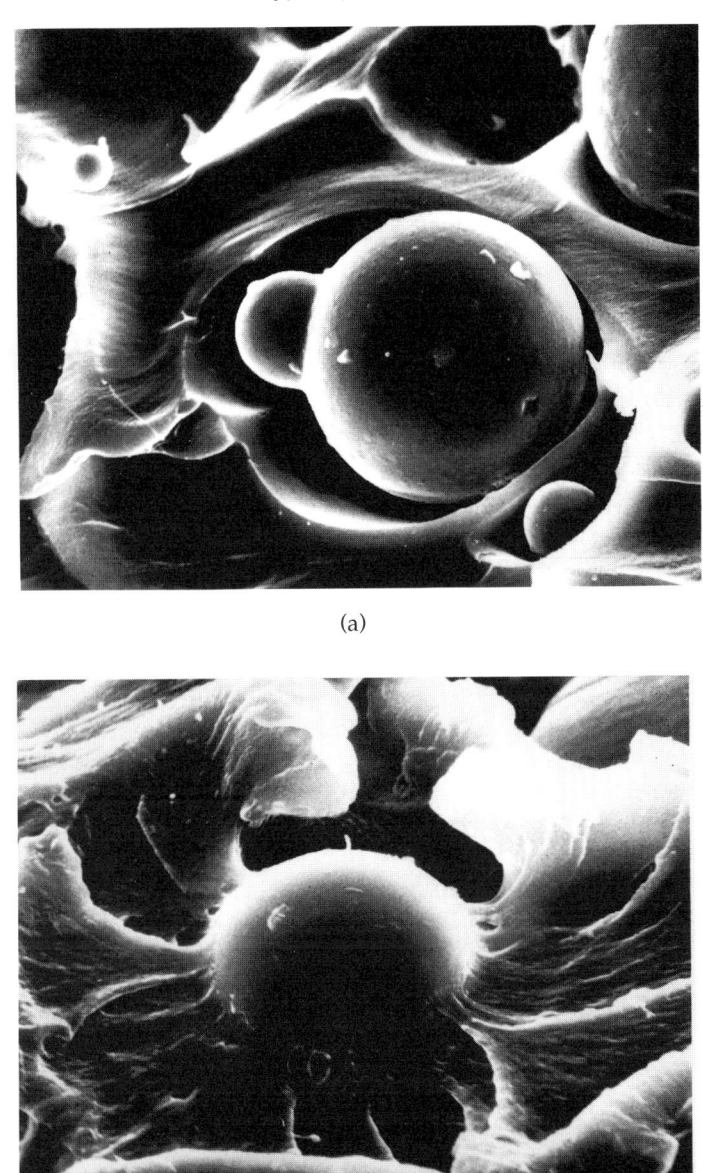

Figure 3 Effects of an aminosilane on filler/matrix adhesion in a glass sphere/polyamide 6 composite: (a) untreated; (b) with aminosilicate treatment (reprinted by permission of OSi Specialities).

Titanates

As with silanes, a wide variety of titanates are available for use as coatings or coupling agents. They have a general formula $R'_x Ti(OR)_y$. The OR groups are hydrolysed to OH, and reaction with OH groups on filler surfaces is thought to follow. As Ti–C bonds are unstable, R' groups actually link to Ti through an –O– linkage, so that R' can represent, for example, an ester, phosphate or sulphonate group. Also, in a given compound, R' may represent more than one type of group. Like silanes, titanates may be reactive or unreactive with respect to the polymer matrix. As with silanes, for unreactive non-polar polymer systems, i.e. when there are no polymerization reactions expected to occur, non-reactive titanates, e.g. isopropyl triisostearyl titanate, illustrated below, has been used with a wide variety of filler types.

$$(CH_3)_2 CH-O-Ti(-OCOC_{17}H_{35})_3$$

A wide range of different titanates have been used for different purposes in systems which undergo curing (i.e. elastomers and thermosets). In the former case their addition has enabled filler loadings to be increased, or has resulted in improved mechanical properties; for thermosets reduction in viscosity is an important objective in many cases.

The wide range of filler types amenable to surface treatment by titanates is of interest. Reaction with fillers containing hydroxyl groups would be anticipated, but their effective application to calcium carbonate is more surprising. It has been suggested that free acid, present in the titanates as impurity, or produced by hydrolysis, may contribute to the reaction.

Some titanates have low flash points, and products should be handled as instructed by the suppliers. In the presence of phenolic compounds such as antioxidants and UV stabilizers, titanates can cause discoloration.

The remaining coatings described below are used less frequently, but brief comments will be included for completeness.

Zirconates

Zirconates are analogous to titanates in their function. They are more expensive, but do not cause the discoloration experienced with titanates, and have less effect on cure rate for systems crosslinked via peroxide crosslinking reactions. Zircoaluminate surface modifiers have been considered as an alternative to silanes for improving the dispersion of a variety of particulate fillers in thermosets. Two different types of organic groups are bound to a low molecular weight zircoaluminate backbone, to produce a complex compound that functions in several ways. Aluminium acylates will bond firmly to filler surfaces, but are hydrolytically unstable and therefore unable to behave as coupling agents; the larger zirconium atom participates in stable organofunctional complexes which can interact with polymers. Zircoaluminates react rapidly at room temperature to form hydrogen/covalent bonds to fillers that are not destroyed by moisture, thus having permanent effects, for example in reducing resin viscosity. They are effective with all filler types.

Phosphate based compounds

A variety of phosphate based compounds are available for use as dispersing agents for fillers in plastics. These compounds are usually phosphate esters, with the general formula ($RO-PO_3H_2$), although compounds available tend to be mixtures rather than pure materials. Reactive and unreactive versions are again available. Phosphates have mainly been used for calcium carbonate, when evidence of surface reaction has been obtained. They are claimed to offer benefits with a range of basic fillers in both thermoplastics and thermosets, and the reactive compounds increase filler/matrix adhesion in peroxide cured elastomers. At present these coatings are not widely used.

Polymeric additives

A number of polymeric additives have been used to coat fillers. In general terms, these all consist of an anchor which will react in some way with the filler particle, and a polymer chain which will react with/entangle with the polymer matrix. The latter becomes significant when polymer coatings are used. Chain length is important: too short chains do not have the ability to form entanglements (e.g. fatty acid chains are too short), while longer chains can fold back on the filler particle, causing agglomeration.

Available compounds of this type include polyolefin backbones with carboxylic acid, polyacrylic acid end groups and maleic anhydride, polybutadiene, also functionalized with maleic anhydride, and unsaturated polymeric silane compounds. As with other coatings, reactions depend on the filler/coating/matrix chemistry. Carboxylic acid groups will react as for their short chain counterparts, unsaturated polymer chains will enter into crosslinking processes, for example during the vulcanization of rubber, and maleic anhydride groups lead to reaction with the filler surface.

METHODS OF COATING

Quite a number of coated fillers are available commercially, including stearate, stearic acid and rosin coated carbonate, and silane coated clay. Some manufacturers actually incorporate coatings when the filler is ground or milled. Alternatively the filler may be wet or dry coated. Most suppliers recommend methods of application. Wet coating is carried out by dissolving the coating in water, then adding the filler to form a slurry, and heating to enable reaction to occur. The coated filler then has to be washed and dried. When silane coatings are used, water is adjusted to a pH of 4–5 by the addition of acetic acid, to ensure that alkoxy groups are hydrolysed. Solvent slurries have also been used, but this method would involve solvent removal, which is better avoided. Dry coating can be carried out in a high speed mixer of the type used to prepare PVC compounds. The coating is added to the filler either dry, or as a spray in solvent. Coatings such as stearic acid need to be headed above their melting temperature during mixing. A fourth alternative involves addition of the coating when the filler and polymer are compounded. If the coating is not too compatible with the polymer, it will migrate to the filler surface.

COATING LEVELS

Resulting compound properties are very dependent on coating concentration. There is a widespread assumption that optimum properties are obtained with monolayer coverage, i.e. when a single layer of coating molecules surrounds the filler. However, this presents problems. First, how can it be measured? Methods vary and can be theoretical or practical. The theoretical route involves calculating a 'footprint' for the coating molecule from a knowledge of the way it interacts with the filler surface, atomic sizes and bond lengths. This area is then related to the surface area of the filler. Alternatively the number of reactive sites on the surface, and the amount of coating which would react with these, can be calculated. Practical methods require rather sophisticated techniques such as quantitative DRIFT (Diffuse Reflectance Fourier Transform Infra-Red Spectroscopy) and XPS (X-ray Photoelectron Spectroscopy). Unfortunately these methods often give different results because most filler particles are irregular in shape, not all reactive sites are necessarily accessible and coating molecules can lie at various orientations with respect to the filler surface. Also coatings can be physisorbed as well as chemisorbed on to the filler.

Manufacturers' literature offers recommendations on estimating optimum coating levels, and in practice it is usually best to test several coating levels so that the required properties can be optimized. Even the concept of optimum properties is not straightforward, as illustrated in Figure 4 for a compound containing stearic acid coated calcium carbonate in polyethylene. While yield strength is at a maximum

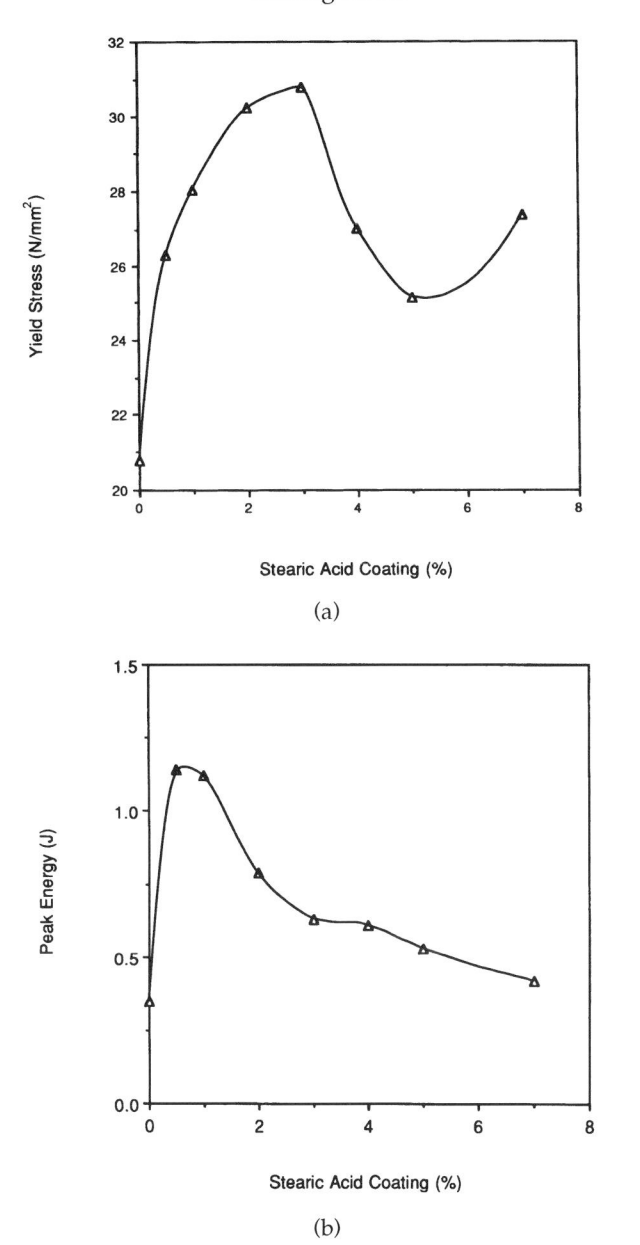

Figure 4 Mechanical properties of compounds containing stearic acid coated calcium carbonate in polyethylene: (a) tensile yield stress; (b) impact energy using instrumented falling weight impact tester.

with about 2% stearic acid, maximum impact strength (measured using an instrumented drop weight impact testing machine) occurs with lower coating levels. This figure also shows the effectiveness of filler coating.

EFFECTS ON PROPERTIES OF FILLED COMPOUNDS

There is extensive literature demonstrating the effects of filler coatings on properties. As indicated earlier, primary reasons for coating fillers are to reduce viscosity (a method frequently used for thermosets), or to improve mechanical properties. Reduction of viscosity can offer other benefits such as increased filler loading, which may in turn reduce cost or provide further benefits such as reduced flammability. Reactive coatings can enhance performance of elastomers because the fillers partake in the curing process with resultant property improvement.

A few examples will be given here to typify some of the mechanical property changes which can be achieved.

A fourfold increase in notched Izod impact strength has been reported with 1.5% titanate used as a coating for calcium carbonate in linear low density polyethylene, as shown in Figure 5, which illustrates the effect of two different proprietary titanate coatings.

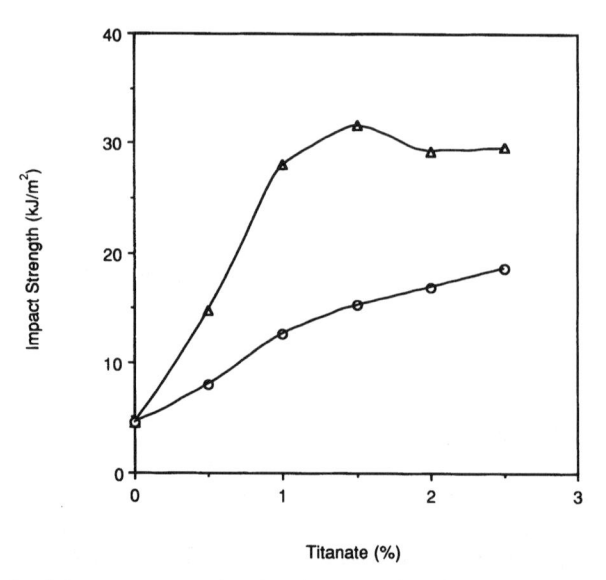

Figure 5 Izod impact strength of calcium carbonate filled linear low density polyethylene: △ Tilcom CA10; ○ Tilcom CA35 (reprinted by permission of Tioxide Chemicals).

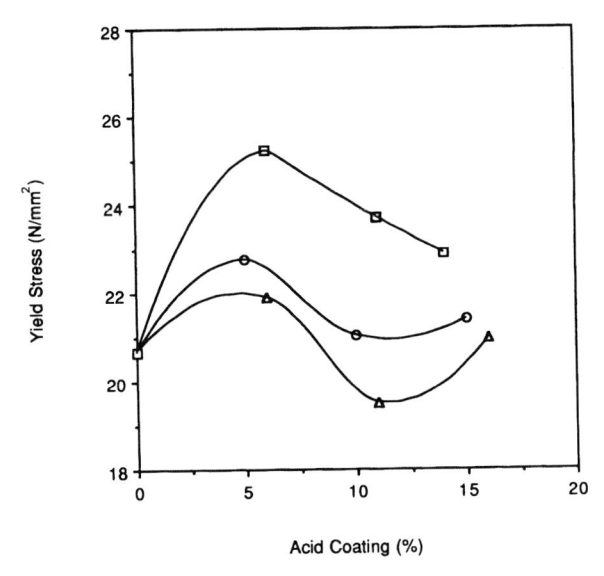

Figure 6 Effect of fatty acid chain length on yield stress of magnesium hydroxide filled polypropylene: △ decanoic acid; ○ stearic acid; □ behenic acid.

Maleinized polybutadiene has demonstrated particularly favourable effects for filled elastomers, improving the tensile strength, tear strength and modulus of a sulphur cured EPDM (ethylene propylene diene monomer rubber) containing 100 phr filler, while a stearic acid coating actually caused all these properties to deteriorate significantly.

The effect of acid chain length for a range of fatty acids on magnesium hydroxide in polyethylene is shown in Figure 6. It is seen that 6% of behenic acid increases the yield strength to above that of unfilled PE. (Unfortunately the impact strength of this compound is lower than that containing decanoic acid, but similar to that containing stearic acid.)

WHY FILLER COATINGS AFFECT PROPERTIES

As explained in the introduction, coatings reduce filler/matrix interaction, which will account for observed reductions in viscosity. Any excess coating may also act as a lubricant. Changes in rheology can modify orientation in injection mouldings, for example, and this will also change properties. A related effect is a change in filler alignment, which can be very dramatic for fillers with high aspect ratio. The most significant effect of coatings on properties is probably the result of improved filler dispersion. Filler coatings also modify packing fraction.

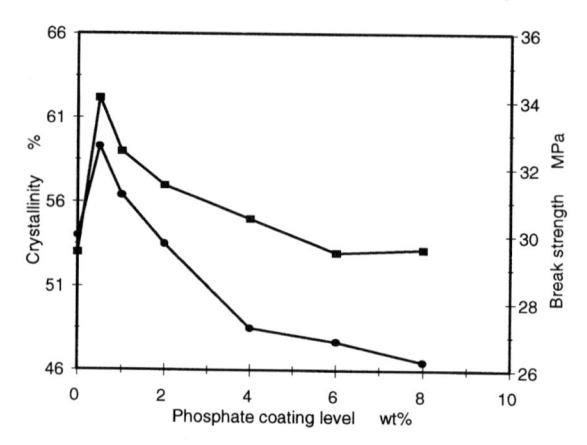

Figure 7 Effect of phosphate coating level on failure strength and crystallinity for talc filled polypropylene: ■ failure strength; ● crystallinity.

For the compounds illustrated in Figure 6, maximum packing fraction increases with coating level, which is another reason why filler loading can be increased when coated fillers are used.

For crystalline thermoplastics, fillers often act as nucleating agents, and modify crystallization behaviour. Coating can cause further changes, which may or may not affect properties. For the fatty acid coated series described above, behenic acid coatings produced the lowest polyethylene crystallinities, which could not account for the highest yield strengths. However, for phosphate coated talc in polypropylene, there was a clear relationship between tensile strength and crystallinity, measured by thermal analysis, as shown in Figure 7.

In summary, coating/structure/property relationships are complex, and few generalizations can be made. Filler coating will obviously increase compound cost, but there are many potential benefits, and many systems available, so the possibilities offered are worth serious consideration.

BIBLIOGRAPHY

Rothon, R.N. (ed.) (1995) *Particulate-filled Polymer Composites*, Longman Scientific and Technical, Harlow.
Wickson, E.J. (ed.) (1993) *Handbook of PVC Formulating*, John Wiley & Sons, New York.
Hornsby, P.R. and Watson, C.L. (1995) Interfacial modification of polypropylene composites filled with magnesium hydroxide. *Journal of Materials Science*, **30**, 5347–5355.
Eurofillers 95 (September 1995) Proceedings of Eurofillers 95 Conference, Mulhouse, France.

Monte, S.J. and Sugarman, G. (1985) Titanate and zirconate coupling agent applications in polymer composites. *Developments in Polymer Technology – 2* (eds A. Whelan and J.L. Craft), Elsevier Applied Science Publishers, Barking, UK.

Keywords: fillers, coatings, silanes, titanates, zirconates, zircoaluminates, phosphates.

See also: Fillers;
           Coupling agents.

Surfactants: applications in plastics

J.H. Clint

NATURE OF SURFACTANTS

This article should be read in conjunction with 'Surfactants: the principles'.

The word surfactant is a shortened form of the expression 'surface active agent'. As the name implies, such materials are active (adsorb) at surfaces. More generally, they adsorb at interfaces and in doing so they modify the property of that interface. Surfactants work because they are amphiphilic, which means that different parts of the molecule have different affinities for the phase which is acting as solvent. For example, a molecule of a conventional surfactant, of the type used for detergency in water, has a hydrophilic portion (called the head group) which has a strong preference for remaining in the water, and a hydrophobic portion (the tail group) which does not. The hydrophobic portion would prefer to be in an oil phase, or even projecting into air. By adsorbing at an oil/water or air/water interface, surfactant molecules can best satisfy this difference in properties of the two parts of the molecule. Figure 1 shows the simple way in which surfactant molecules are usually depicted.

SURFACTANTS IN POLYMERS

When considering the roles that surfactants play in polymer systems, similar considerations apply and molecules are active at a variety of interfaces. It is convenient to place additives in categories depending on the interface at which their effects are intended. Examples are shown in Table 1.

Plastics Additives: An A–Z Reference
Edited by G. Pritchard
Published in 1998 by Chapman & Hall, London. ISBN 0 412 72720 X

Tail Group
Soluble in non-polar media

Head Group
Soluble in polar media

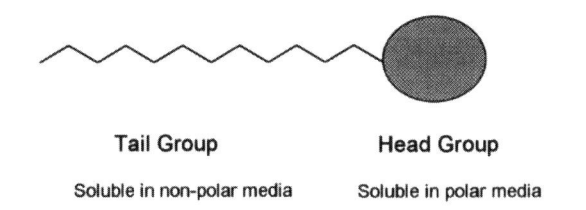

Figure 1 Simple diagrammatic representation of a surfactant molecule.

SURFACTANTS ACTIVE AT THE POLYMER/AIR INTERFACE

The surface activity of an additive is dependent on two major factors, the compatibility of the additive with the host polymer, and the surface free energy of the additive relative to that of the polymer. The former property is a measure of the tendency of the additive to migrate to the interface, and the second indicates the 'strength' of the adsorption at the interface. Complete compatibility (miscibility at the molecular level) allows mobility of the additive molecules within the polymer matrix but leads to low surface activity. Total incompatibility would cause immobility of additive molecules and a phase separation into macroscopic domains.

The best situation for high surface activity is partial compatibility where the additive is able to mix with the polymer matrix but prefers to accumulate at the polymer surface. For example, many additives are used at concentrations above their intrinsic solubility in the polymer but can be held in a supersaturated solution because their diffusion coefficient is low.

Antistatic agents

Polymers generally tend to be good electrical insulators and therefore any electrostatic charges developed on their surfaces are very slow to

Table 1 Classification of surfactant additives according to the interface at which they are active

Interface	Functional surfactant
Polymer/air	Antistatic agents
	Lubricants
	Slip additives
	Foam control agents
	Wetting agents
Polymer/polymer	Compatibilizers
Polymer/solid filler	Solids dispersants

Table 2 Surfactants used as anti-static agents

Surfactant type	Example
Quaternary ammonium compounds	alkyl-N$^+$(R)-Cl$^-$
Alkyl amines and amides	alkyl-C(=O)-NH$_2$
Alkyl aryl sulphonates	alkyl-C$_6$H$_4$-SO$_3$Na
Phosphate esters	(alkyl-O)$_2$P(=O)OH
Alkyl ethers of polyethylene glycol	$C_{16}H_{33}(OCH_2CH_2)_6OH$

dissipate. As a result, electrostatic effects can be a nuisance. Surfactants used as anti-static additives are usually hygroscopic. They migrate to the polymer surface and absorb moisture from the surrounding air. This provides a conducting layer which dissipates the charge and is even more effective if the surfactant is ionic in nature, thus providing additional conductivity to the surface layer. Major classes of anti-static agents used in plastics are shown in Table 2.

Surfactants for use as anti-static agents with plastics can be applied as a surface finish by spraying or dipping. Such external agents provide an immediate anti-static effect but one which has very little durability. More usually they are compounded into the plastic during processing and provide a long term effect by slow migration to the surface. The performance of these internal anti-static agents usually represents a compromise between compatibility and diffusibility. Increased surface conduction can be achieved by rapid diffusion of the additive to the surface, but the greatest durability of anti-static effect depends on the rate of loss of additive to the surface being slow.

Lubricants

Surfactants used as lubricants are added to polymer resins to improve the flow characteristics of the plastic during processing. In a similar way to the anti-static agents, additives can be internal or external lubricants but the distinction is between the mechanism of action, not the method of addition. Lubricants that reduce molecular friction, thus lowering the polymer's melt viscosity and improving its flow, are referred to as internal lubricants. Substances that promote resin flow by reducing friction between the melt and solid surfaces used to confine the plastic during processing, as classed as external lubricants [1].

Although many 'oily' substances can act as hydrodynamic lubricants when present as a comparatively thick layer between the contacting surfaces, surfactants whose polar groups have a strong affinity for the solid surface can be effective at much lower concentration. A single layer of oriented surfactant molecules acts as a boundary lubricant. Effective lubricants for metal components are long-chain fatty acids, metal soaps and alcohols. Because of the high stress involved when such a thin layer is sheared, it is important for the stability of the boundary layer that the polar groups of the surfactant are strongly anchored to the solid surface. As a result, the best boundary lubricants are those in which the polar group reacts chemically with the solid. For example, a fatty acid such as lauric acid will form strong chemical bonds with the surfaces of metals such as copper and zinc. The probable structure of the resulting metal/polymer interface is depicted in Figure 2.

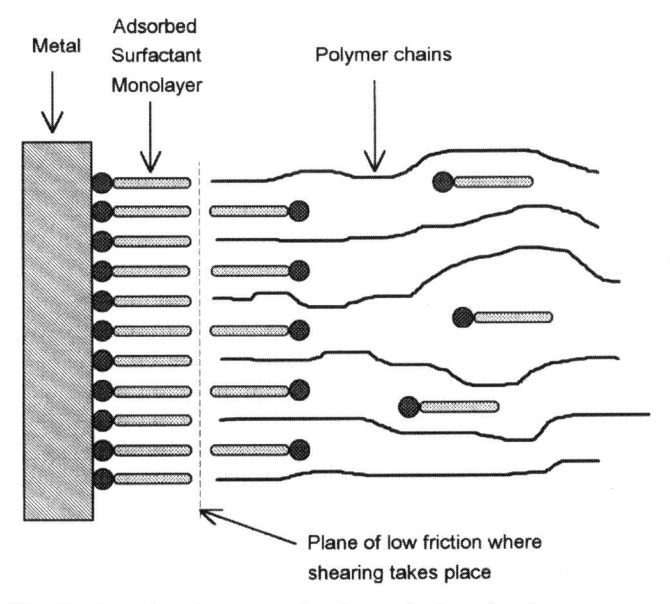

Figure 2 Distribution of surfactant molecules at the interface between a metal surface and molten polymer during processing.

Internal lubricants are chemically similar to external lubricants but their greater compatibility reduces the tendency to migrate to the polymer surface. The rules used to select lubricants can be summarized as follows [2].

1. Metal soaps, mainly stearates, possess low compatibility with all polymers and therefore act primarily as external lubricants.
2. Long chain fatty acids, alcohols and amides function as internal lubricants for polar polymers, e.g. PVC, polyamides, etc., but have relatively low compatibility with non-polar polymers, e.g. polyolefins.
3. Long-chain di-alkyl esters have medium compatibility with most polymers and can act both internally and externally, hence they are often used to obtain a balanced lubrication.
4. High molecular weight paraffin waxes have very low compatibility with polar polymers and act as external lubricants for polymers such as PVC.

'Slip additives' are used with packaging plastics such as polyolefins, polystyrene and PVC to impart lubrication, prevent films (such as cling-film) from sticking together and to reduce static charges. They are usually surfactants such as fatty acid amides which have the desired rate of 'blooming' to the polymer surface where their action is needed.

Foam control agents

In the production of plastic foams, especially those used for upholstery and packaging, it is important to control the relative abundance of closed and open cells in the foam. In this way the compressibility of the foam can be matched to the specific application. Surfactants are added during the foaming stage in order to influence the rheology of the polymer/gas interface, which in turn controls the stability of the thin polymer film between adjacent gas bubbles in the foam. In the same way the surfactant controls the size and uniformity of the gas cells. Surfactants based on silicones are most commonly used for this purpose, especially for polyurethane foams. They are usually comb polymers with a silicone backbone and a variety of side chains consisting of alkyl groups and copolymer groups of ethylene oxide and propylene oxide. Their general structure is shown in Figure 3.

Structural variations, via changes in m, n, p, x, y and z, which alter the length, abundance and distribution of the various side chains are used to produce the desired cell structure in the foam.

Wetting agents

Most plastics materials are not wetted by water. A droplet of water will adhere to the polymer/air interface but will form a finite contact angle

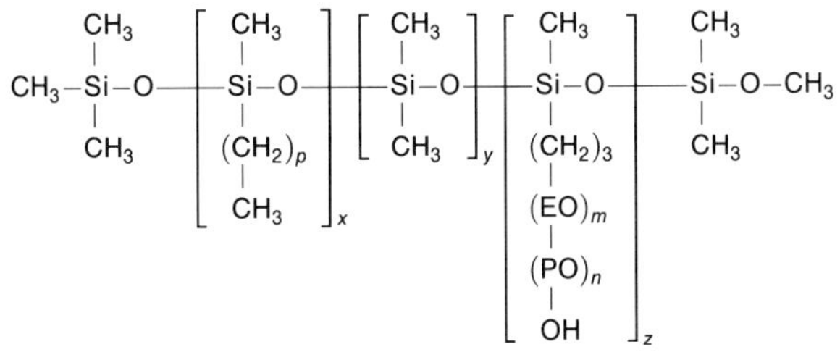

$$CH_3-\underset{\underset{CH_3}{|}}{\overset{\overset{CH_3}{|}}{Si}}-O\left[\underset{\underset{\underset{\underset{CH_3}{|}}{(CH_2)_p}}{|}}{\overset{\overset{CH_3}{|}}{Si}}-O\right]_x\left[\underset{\underset{CH_3}{|}}{\overset{\overset{CH_3}{|}}{Si}}-O\right]_y\left[\underset{\underset{\underset{\underset{\underset{OH}{|}}{(PO)_n}}{|}}{\underset{(EO)_m}{|}}}{\overset{\overset{CH_3}{|}}{Si}}-O\right]_z\underset{\underset{CH_3}{|}}{\overset{\overset{CH_3}{|}}{Si}}-O-CH_3$$

Fig. 3 Silicone based surfactants used in polymer foam control.

rather than spread over the surface. For some applications this can produce undesirable effects. For example, in plastic film used for food wrapping, the condensation of moisture arising from the food produces a misted appearance by forming droplets on the inside of the film. Surfactants in the form of wetting agents are applied to the polymer surface in order to promote complete spreading of the water as a continuous layer, thereby maintaining the clarity of the wrapping film. Food grade surfactants are best used for this application. Good examples are mono- and di-glycerides.

SURFACTANTS ACTIVE AT THE POLYMER/POLYMER INTERFACE

Compatibilizers

The blending of polymers is being increasingly used to improve the performance of plastics materials, to reduce costs and to enable recycled plastics waste to be put to use. For example, polystyrene becomes a tougher polymer with greater impact strength through the incorporation of a rubbery polymer such as polybutadiene. However, most polymer blends tend to be immiscible because of the very low entropy of mixing which prevails when molecular weights are as great as those used in plastics. The result is the phase separation of the components into discrete domains, which produces inferior mechanical properties because of the lack of penetration of polymer chains from one phase to the other. Compatibilizers are used to provide compatibility between otherwise immiscible polymers but they do not produce miscibility on the molecular scale. Their action can be thought of as reducing the interfacial tension between the components by adsorbing at the interface surrounding the domains. This process is often referred to as strengthening the interface [3]. Block copolymers are especially useful surfactants for this purpose because the two blocks can be made up from molecules of the individual polymers which it is desired to mix.

Usually the addition of only a few percent of the copolymer produces a dramatic improvement in the mechanical properties of the blend, in two main ways.

1. There is a large reduction in the mean size of the domains. The corresponding large increase in interfacial area becomes possible because of the reduction in interfacial tension at the domain boundaries. In this respect the compatibilizers act in an analogous way to emulsifiers in oil/water systems. Indeed, the reduction in domain size levels off with increasing additive concentration, possibly indicating saturation of the interface.
2. The domains tend to be more uniformly distributed throughout the polymer blend.

SURFACTANTS ACTIVE AT THE POLYMER/SOLID INTERFACE

Many of the solids used in a dispersed state in polymers have high energy, hydrophilic surfaces. Examples include titanium dioxide, coloured pigments, mica and even metal particles. Such powdered solids can be made more compatible with polymers by coating their surfaces with an adsorbed layer of surfactant in the form of a dispersant. Sodium di-alkyl sulphosuccinates, more commonly used as wetting agents, have been employed for this purpose.

STABILIZERS

Some additives used in plastics are surfactants but are not used primarily for their surface or interfacial activity. Rather they contain functional groups that, because of their chemical nature, are polar but they also contain non-polar groups, such as hydrocarbon chains, in order to provide compatibility with the polymer. Good examples are the heat stabilizers used in plastics such as PVC to prevent thermal degradation. Typical additives used for this purpose are soaps (metal salts of alkyl carboxylates) or metal salts of other organic acids such as phenols. The metal cations used most commonly are tin, calcium, barium, zinc and cadmium. Lead salts are also used in electrical cables, pipes and window frames. Tin salts are especially useful for stabilizing clear, rigid PVC bottles.

Although such protection is usually essential during processing (when Zn/Ca stearate would be commonly used), stabilizers based on phenol can also extend the useful life of the polymer when used under hot conditions. When PVC is heated, scission of C–Cl bonds takes place in the weakest points of the polymer chains, e.g. allylic or tertiary positions which occur at branching sites in the polymer or at sites adjacent to unsaturated terminal bonds. The Cl radicals abstract hydrogen from adjacent CH groups and create another weak allylic C–Cl bond which becomes susceptible to scission, thus giving rise to HCl unzipping reactions. Basic metal soaps are commonly used to prevent such degradation since they have a variety of actions. They neutralize HCl, which stops autocatalytic chain reactions. They preferentially displace labile chlorine from the polymer chains and replace it with an alkyl ester group. They prevent free-radical processes, such as oxidation reactions, and also they disrupt conjugation in the polymer chains from which HCl has been removed, thus inhibiting discoloration of the plastic.

An important factor is the homogeneity of distribution of the stabilizer in the polymer matrix. A homogeneous distribution on a molecular scale would obviously lead to the most efficient stabilizer but is unlikely to be achieved in view of the chemical nature of the additives. Metal soaps in a polymer matrix will behave like surfactants in a non-aqueous solvent and either form inverse micelles or crystals, depending on the purity, especially uniformity of chain length, of the soap. Inverse micelles are small (usually spherical) aggregates of surfactants in which the carboxylate head groups associate to form a polar inner core and the hydrocarbon chains form a shell whose lack of polarity improves the compatibility of the soap in the polymer. Aggregates of this nature would have a diameter approximately equal to twice the fully stretched out length of the soap molecules, i.e. about 5 nm. For a stabilizer concentration of 2%, such aggregates would be spaced out at intervals of a few tens of nanometres. Aggregates of the type described would not be fixed entities, for this would severely restrict the ability of individual molecules to partake in stabilizing reactions. Instead they will be in dynamic equilibrium with single molecules at a concentration equal to the 'critical micelle concentration'. For Zn/Ca stearate in PVC, the existence of such aggregates in equilibrium with unassociated molecules has been demonstrated by solid state nuclear magnetic resonance studies [4].

Metal soaps are not surface active at the polymer/air interface and as a result they show an insignificant amount of migration to the surface of plastics. Because of the high polarity of the carboxylate group, they have a finite tendency to adsorb at the polymer/water interface. However, they show little tendency to cause tainting of foods wrapped in polymers containing such stabilizers. Nevertheless, heavy metal soaps are being phased out for environmental reasons.

REFERENCES

1. Ainsworth, S.J. (1992) Plastics additives. *Chemical & Engineering News*, **70**(35), 34–55.
2. Mascia, L. (1974) *The Role of Additives in Plastics*, Edward Arnold, London.

3. Kramer, E.J., Norton, L.J., Dai, C.-A., Sha, Y. and Hui, C.-Y. (1994) Strengthening polymer interfaces. *Faraday Discuss.*, **98**, 31–46.
4. Barendswaard, W., Moonen, J. and Neilsen, M. (1993) Analysis of polymer stabilisers by means of solid state NMR: some case studies. *Analytica Chimica Acta*, **283**, 1007–1024.
5. Porter, M.R. (1994) *Handbook of Surfactants*, 2nd edn, Chapman & Hall, London.

Keywords: surfactant, interface, surface, adsorption, monolayer, miscibility, compatibility, anti-static agent, diffusion, lubricant, foam, foam control agent, wetting agent, compatibilizer, dispersant, stabilizer, surfactant aggregates.

See also: Surfactants: the principles;
Antistatic agents;
Compatibilizers for recycled polyethylene.

Surfactants: the principles

Gregory G. Warr

SURFACTANTS

Surfactants are compounds that vary widely in structural detail, but which have one characteristic structural feature in common. All surfactants consist of hydrophobic and hydrophilic groups covalently bonded together. The generic structure is shown schematically in Figure 1, together with some typical surfactant structures.

The hydrophobic 'tail' of a surfactant is usually a hydrocarbon chain containing 8–16 methylenes. This may be either a straight or branched chain. Industrially a mixture of isomers is also common. Although single tailed surfactants are by far the most common, a number of double tailed compounds are also widely used [1].

The hydrophilic head group may either be a charged or an uncharged polar group. The first case denotes an ionic surfactant which, like all salts, must contain a counterion in order to be electrically neutral. Cationic surfactants are most commonly based around quaternary nitrogens, and are widely commercially available as chloride or bromide salts (see Figure 1(b)). Among anionic surfactants, alkylsulfates and alkylbenzenesulfonates typify the common structures. Two examples are shown in Figures 1(c) and (d).

Non-ionic surfactants contain uncharged polar groups, and by far the most common of these is an oligomeric ethylene oxide chain (Figure 1(e)). The number of ethylene oxide units in such a non-ionic surfactant may be as few as two, or as many as 100 or more. Industrial non-ionic surfactants of this type are usually made by a condensation reaction of ethylene oxide with an alcohol or alkylphenol, and hence contain a statistical distribution of ethylene oxide chains [2]. Other types of non-ionic

Plastics Additives: An A–Z Reference
Edited by G. Pritchard
Published in 1998 by Chapman & Hall, London. ISBN 0 412 72720 X

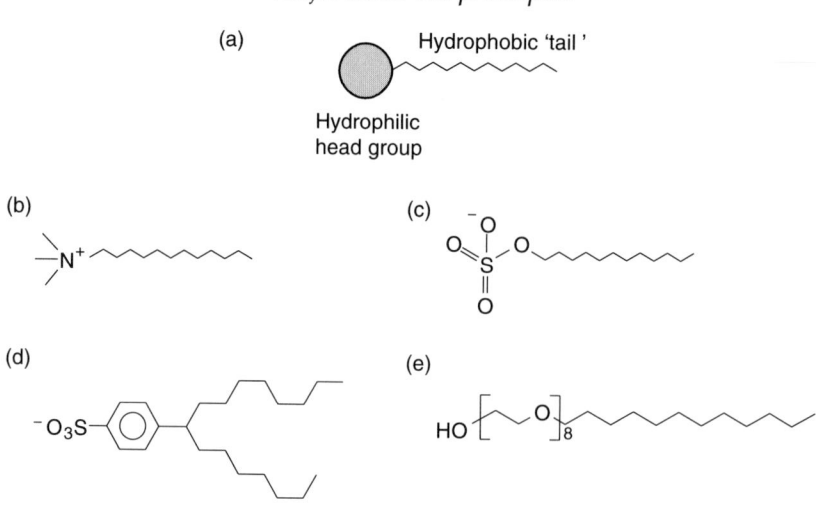

Figure 1 Surfactant structures showing (a) generic surfactant structural features, together with (b) cationic dodecyltrimethylammonium, (c) anionic dodecylsulfate, (d) anionic hexadecylbenzenesulfonate, and (e) non-ionic octa(ethylene glycol)-mono-*n*-dodecyl ether.

surfactants include those employing sugar residues as their hydrophilic segment, and are principally of interest for their biodegradability.

The term surfactant is a contraction of surface active agent, which concisely describes the action of these compounds. They are also known as amphiphiles, which refers to the competition or antagonism between their hydrophobic and hydrophilic components.

ADSORPTION

The singular property of surfactants or amphiphiles is their propensity to adsorb at an interface and there to form an oriented monolayer. Adsorption of surfactants occurs due to the unfavourable short-range interactions between the hydrophobic chain of the surfactant and water molecules. Contact between the hydrocarbon and water is minimized by the accumulation of these chains at the interface between water and a less-polar material. The attractive interactions between the polar moiety and water prevent the surfactant from reaching a conventional solubility limit and forming a separate phase, as do simple hydrocarbons. Maintaining contact between hydrophilic groups and water induces the surfactants to form an oriented monolayer in which the hydrophobic parts of the molecules are separated from water by a thin layer of hydrophilic groups to which they are attached. The situation is shown schematically in Figure 2.

Adsorption

Air or Oil

hydrophobic tails

hydrophilic head groups

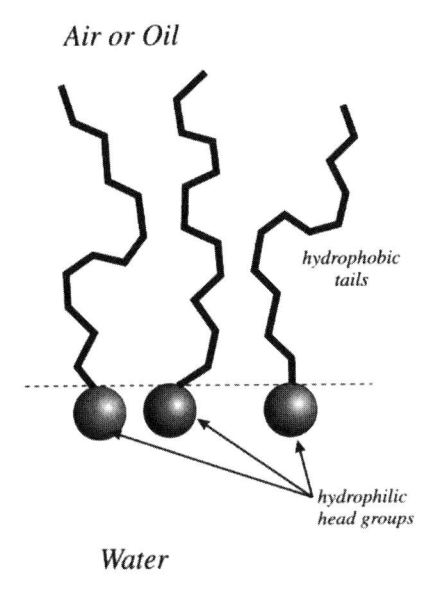

Water

Figure 2 Conformation of surfactants adsorbed at polar/apolar interface such as air/solution or oil/water.

This depicts the conventional view of the arrangement of surfactant molecules at an interface between water and a less polar material, such as air, oil, or a hydrophobic solid (e.g. paraffin wax, graphite, polythene). Thus bubbles or oil droplets dispersed in a surfactant solution will gather an equilibrium coating of surfactant up to a maximum density corresponding to a close-packed monolayer. It is this surface coating which imparts stability to foams (air/water) and to emulsions (oil/water) by inhibiting bubble or droplet coalescence.

The surface concentration of a surfactant, referred to as adsorption density or surface excess (Γ mol m^{-2}), typically increases linearly with increasing surfactant concentration, eventually 'rolling over' and asymptoting to close-packed monolayer coverage. This is illustrated in Figure 3 for the adsorption of tetradecyltrimethylammonium bromide at the air/water interface. As a surfactant becomes more hydrophobic, say by increasing the length of the hydrocarbon chain, it adsorbs more strongly at the interface. That is to say, adsorption occurs at lower concentration. Maximum coverage, however, is determined by the closest-packing density of the surfactant, and typically does not depend on the hydrophobic chain. In order to change the maximum coverage, the size or behaviour of the head-group must be changed.

Even at the surface of many mildly hydrophilic solids it has long been though that this general picture is correct. The driving force for surfactant

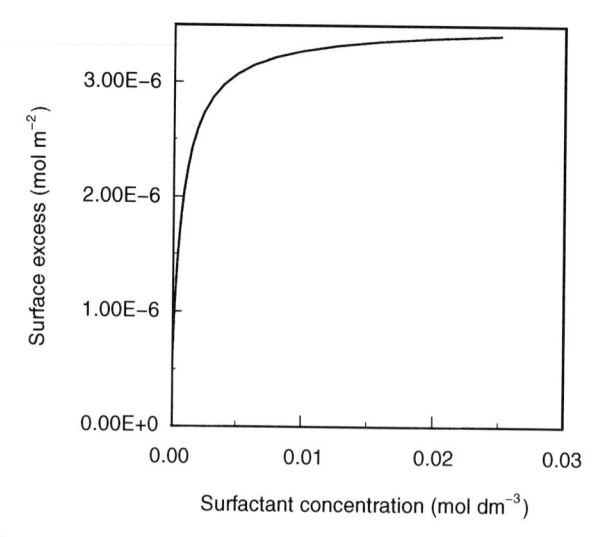

Figure 3 Adsorption isotherm for tetradecyltrimethylammonium bromide at the air/water interface, showing saturation adsorption corresponding to the formation of a close-packed monomolecular film.

adsorption is the expulsion of hydrocarbon chains from the water hydrogen-bond network. As this driving force is so large, and the hydrocarbon molecules are relatively indifferent to their surroundings, only surfaces which are highly polar (i.e. which effectively participate in hydrogen bonding to water molecules) would not accumulate an oriented monolayer of surfactant.

The hydrophobic interaction also occurs in other solvents. 'Solvophobic' interactions have been exploited in nonaqueous media using surfactants based around silicone oils or fluorocarbons. The expulsion of both of these molecular groups by bulk liquids leads to adsorption at air–solution interfaces and to the formation of stable, nonaqueous foams, that are used in the production of both rigid and flexible isocyanate-based polymeric foams for insulation and cushioning applications [3].

Missing from the picture developed so far is charge. As the electrostatic interaction is both strong and long-ranged, it plays an important role in the adsorption behaviour of ionic surfactants onto charged surfaces. Many solids generate charge either through surface dissociation reactions or, in the case of electrodes, by direct application [4]. The range and strength of the electrostatic interaction between a charged surfactant and a charged interface can overwhelm hydrophobic effects. This leads to two extreme situations: surfactants adsorb only weakly onto surfaces of like charge, but onto surfaces of opposite charge, they can adsorb with their charged, hydrophilic head-groups oriented towards the

SELF-ASSEMBLY INTO MICELLES

Solutions of surfactants are also affected by the amphiphilic character of these unusual solutes. Above a critical concentration, surfactant molecules aggregate together to form nanoscopic droplets known as micelles. This process, micellization, is the first step in a hierarchy of self-assembly patterns which surfactants exhibit. Typically, micelles form in solution at concentrations slightly above those where available surfaces have been completely covered by surfactant. Micelle formation is an abrupt occurrence, and this is perhaps best illustrated by surface tension measurements on surfactant solutions.

When surfactant is added to water, it forms an equilibrium film at the air–solution interface. This spontaneous adsorption lowers the surface tension (γ), at first a little and then – as concentration is increased and the surfactant film covers more of the interface – a lot. A typical surface tension curve for the surfactant tetradecyltrimethylammonium bromide ($C_{14}TAB$) is shown in Figure 4. The steepest part of the curve occurs as the surfactant film approaches close packing, and in fact the gradient, $\partial \gamma / \partial \ln c$, is proportional to the adsorption density.

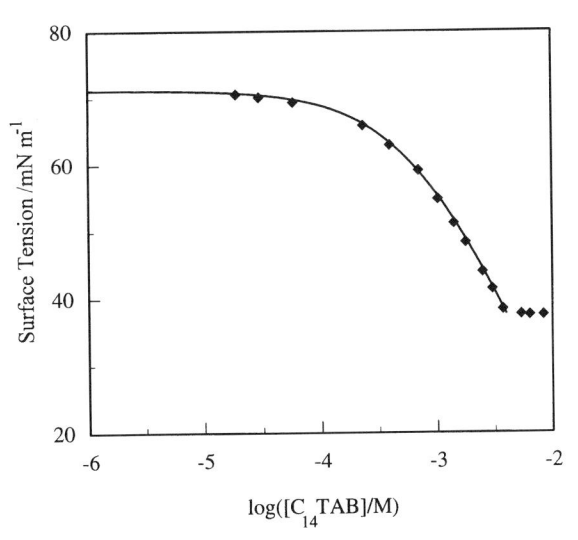

Figure 4 Surface tension versus concentration for tetradecyltrimethylammonium bromide. The steep decrease in surface tension corresponds to the surfactant-saturated air/water interface, and the break in the curve is the critical micelle concentration.

$r = 15\text{-}25\text{Å}$

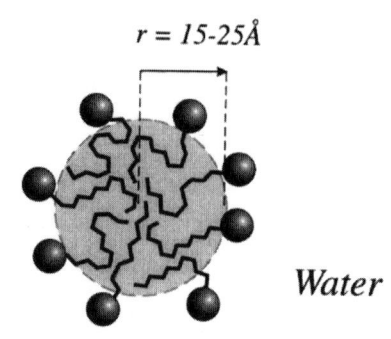

Water

Figure 5 Schematic diagram of a spherical micelle, showing hydrophobic core of radius ≤ length of a single surfactant chain.

However, at a particular concentration, the critical micelle concentration (c.m.c.), the surface tension curve abruptly becomes flat, and no amount of additional surfactant will reduce the surface tension further. The reason for this change in solution behaviour, which can also be detected by light scattering, osmotic pressure, conductivity, or from diffusion coefficient measurements, is the formation of micelles.

Micelles are formed by the assembly of tens, hundreds, or even thousands of individual surfactant molecules into a single aggregate. The mechanism for micellization is essentially the same as that of adsorption. The driving force is the expulsion of apolar hydrocarbon chains from the polar water environment, balanced by the maintenance of contact between hydrophilic groups and water. A stable micelle is formed when the hydrocarbon chains aggregate together to form a nanodroplet of oil, stabilized by a surface coated with hydrophilic groups. A typical micelle is shown schematically in Figure 5.

The essential features of micelles are as follows. (1) No point inside the micelle can be more than the length of one hydrocarbon chain away from water. As most surfactants have chains of between 12 and 25 Å (1.2 to 2.5 nm) in length, this makes micelles much too small to be seen. (2) The micelle's core is like an oil droplet, with the hydrocarbon tails in liquid-like disorder, and packed at liquid hydrocarbon density. It contains N surfactant tails, where N is the aggregation number of the micelle. (3) The surface is covered by N surfactant head-groups, and these are packed at approximately the same close-packing as occurs at macroscopic (oil–water) interface.

In addition, if the surfactant used is ionic, the micelle will carry a net charge equal to $N\times$(valence of surfactant ion), and will accumulate around it a balancing diffuse layer of charge, due to its counterions, in the same manner as any charged colloidal particle [4].

Table 1 Critical micelle concentrations of some typical surfactants (mol/L)

Head group: Counterion:	Anionic $-OSO_4^-$ Sodium	Cationic $-N(CH_3)_3^+$ Bromide	Non-ionic $-(OCH_2CH_2)_6OH$ –
Hydrocarbon chain length, n			
8	0.133	–	9.9×10^{-3}
10	3.3×10^{-2}	6.5×10^{-2}	9.0×10^{-4}
12	8.3×10^{-3}	1.56×10^{-2}	8.7×10^{-5}
14	2.1×10^{-3}	3.6×10^{-3}	–
16	$5 \times 10^{-4}*$	9.2×10^{-4}	–

* At 40°C.

Depending on the close-packed area of the surfactant head-group, and on the hydrocarbon chain length and volume, micelles may form spheres, rods or even disks (although evidence of the latter is scant). A very wide range of possible architectures arises from the packing of hydrocarbon chains and head-groups, of which spherical micelles are both the most common and the simplest. Even rod-like micelles lead to interesting behaviour: as rods grow in length they become flexible and entangled, leading to rheological properties similar to solutions of entangled polymers.

Micelles form in dilute solution. Sodium dodecyl sulfate, a widely used anionic surfactant, has a critical micelle concentration (c.m.c.) of 8.3×10^{-3} M (approx. 2.3 g/L) at room temperature. As with adsorption, increasing the hydrophobicity of a surfactant lowers its c.m.c. Table 1 lists c.m.c.s for typical cationic, anionic and non-ionic surfactants. As can easily be seen, non-ionic surfactants have much lower c.m.c.s than their ionic counterparts, indicating the lower hydrophilicity of the non-ionic head group. As hydrocarbon chain length increases, the c.m.c.s of all three types of surfactants decreases markedly. As a rule of thumb, the c.m.c. of ionic surfactants decreases by about a factor of four for every two CH_2 groups added. For non-ionic surfactants, the decrease is much greater, being over 10 for addition of two CH_2 groups. This difference is due to the role of the counterions in ionic surfactants.

Some surfactants, including many with two hydrocarbon chains, cannot efficiently pack themselves into either spherical or cylindrical micelles. While disks seem to be the solution, disk edges result in a large amount of hydrocarbon being exposed to water, and are energetically unfavourable. These surfactants, which are similar in structure to biological lipids, form vesicles or liposomes instead of micelles. The 'edge problem' is solved by forming a nearly flat bilayer of surfactant molecules, which gently curves back on itself, encapsulating water inside a spherical cavity. As with micelles, no point within the bilayer is more than the length of one hydrocarbon tail away from water, but the overall size of these vesicles is much larger than that of micelles. Vesicles are usually very polydisperse, with radii that can range from a few hundred Ångströms (tens of nm) to more than 1 µm.

FORMATION OF LYOTROPIC LIQUID CRYSTALS

At much higher concentrations, surfactant aggregates undergo a higher degree of self-organization to form a range of liquid crystal phases. The best recognized of these are the hexagonal phase, also known as the middle phase, and the lamellar phase, also known as the neat phase. These are made up of micelle-like aggregates which order themselves as follows.

Hexagonal phase

A hexagonal phase is formed by the parallel packing of rod-like micelles onto a hexagonal lattice, as shown in Figure 6(a). The constituent micelles in a hexagonal phase thus display both alignment and long range positional order: a characteristic diffraction pattern may be measured. There is some evidence for a nematic phase in surfactant systems, where the micelles are aligned but not ordered on lattice points, but this is less well known.

Lamellar phase

A lamellar phase consists of planar bilayers, separated by a constant, repeat distance, as shown in Figure 6(b). As with hexagonal phases, lamellar phases have a characteristic diffraction pattern indicating the high degree of order in the system.

Both phases are optically anisotropic, and may easily be seen through crossed polarizers. They arise due to interactions between surfactant aggregates, principally excluded volume interactions, and therefore are usually observed at concentrations well above the c.m.c.

Cubic phases

Cubic phases are the third concentrated self-assembly state commonly formed by surfactants. They include a variety of structures, the simplest being analogous to atomic crystals but in which spherical micelles take the place of individual atoms on the lattice points. Other more complex structures are also known or postulated, including ordered, interpenetrating networks of branched cylinders and deformed bilayers which form

Formation of lyotropic liquid crystals 621

(a)

(b)

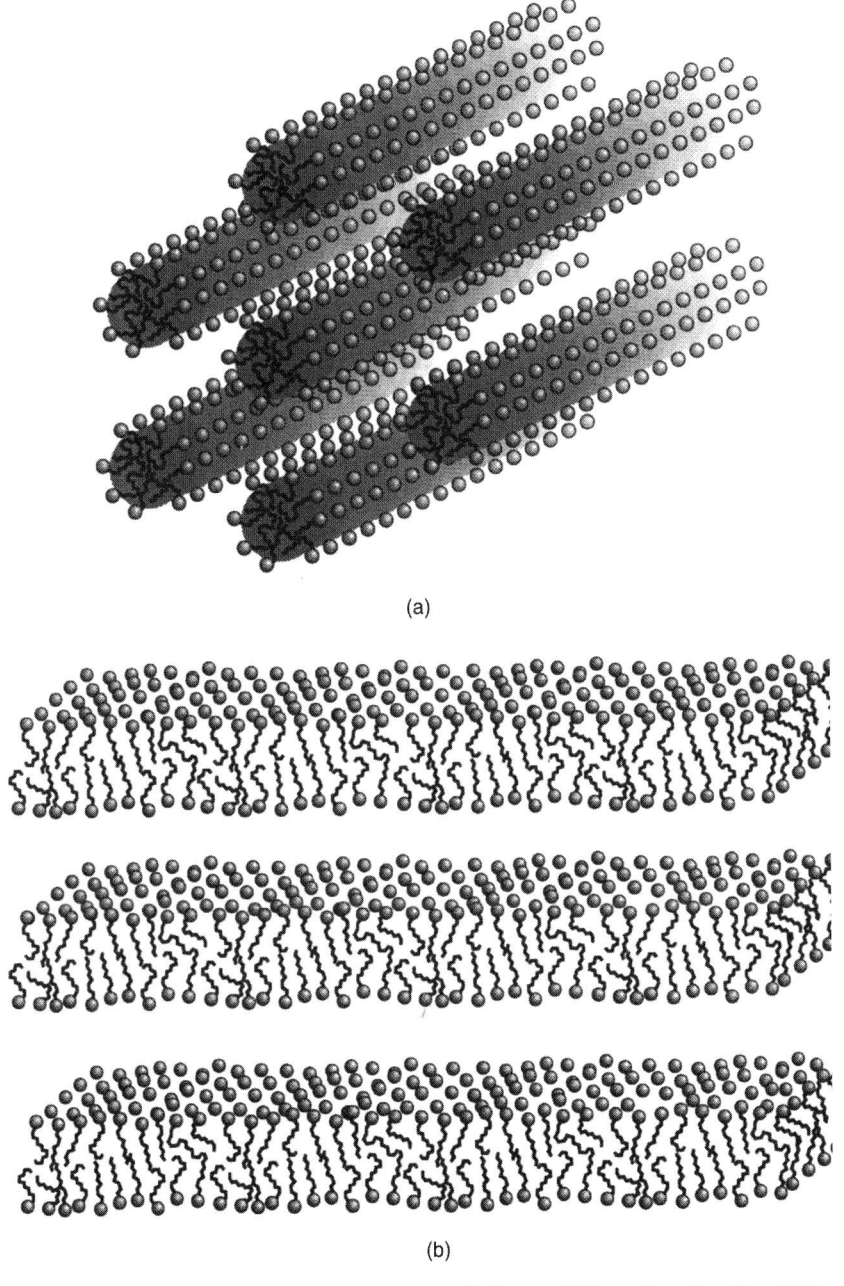

Figure 6 Schematic diagram of (a) hexagonal phase of aligned, ordered rod-like micelles, and (b) lamellar phase of aligned, ordered surfactant bilayers.

triply-periodic, interconnected networks of surfaces which are everywhere saddle-points. Unlike lamellar and hexagonal phases, cubic phases are optically isotropic, exhibiting equivalent order in three dimensions. These structures are very difficult to identify unambiguously, and experimental evidence relies heavily on low-angle X-ray crystallography, together with some novel electron microscope studies.

ADDITION OF OILS: SOLUBILIZATION, MICROEMULSIONS AND REVERSE MICELLES

Solutions of micelles or vesicles are microheterogeneous. Each micelle or vesicle contains a hydrophobic microenvironment in which water-insoluble or sparingly water-soluble molecules may reside. The enhanced solubility of such materials in surfactant solutions is known as solubilization. Aromatic molecules, arenes, may have their solubilities in water enhanced by several orders of magnitude in a micellar solution. This principle is widely employed in preparing water-based aerosols.

At high loadings of oil these materials may, in turn, swell the micellar droplet or vesicle bilayer. Such swollen micelles, or oil-in-water (o/w) microemulsions, remain too small to be seen and are thermodynamically stable. Unlike emulsions, which are kinetically stable, they do not separate into layers upon standing. Chemical reactions between water-soluble and water-insoluble reactants have also been extensively investigated in micellar solutions and microemulsions. Here solubilization can lead to dramatic enhancements of reactions rates.

In emulsion and microemulsion polymerization [5] the surfactant stabilizes the reactive dispersion of monomer against coagulation, as well as providing a high surface area for exchange of initiator and short polymer chains with the aqueous environment. Stabilization may be electrostatic if ionic surfactants are used, or by provision of a physical or steric barrier using non-ionic surfactants. The surfactant must be present in sufficiently high concentration to coat the emulsion droplets and prevent oil/water contact. Hence it is also present in significant concentrations in the final latex.

Monomer may also be incorporated into micelles or vesicles, and preparation of polymer latices by micellar or microemulsion polymerization is well known. Of particular interest in vesicles is the preparation of hollow polymer particles for controlled release applications.

In addition to the swelling of micelles by oil to form microemulsions, oil-soluble surfactants can form inverted aggregates which solubilize water. These reverse micelles have a hydrophilic core, often containing trace amounts of water together with the surfactant head-groups, surrounded by the hydrophilic tails of the surfactant. Many reverse micelles can be swollen by addition of water to form sub-micron sized water droplets, or water-in-oil (w/o) microemulsion. There has been much interest in the use of such microemulsions to extract proteins selectively from solution by enclosing them within the water droplets.

The final class of behaviour in oil + water + surfactant systems is the formation of bicontinuous liquids. In an oil-swollen micelle or o/w microemulsion, the oil is clearly inside the surfactant aggregate, and when it is swollen beyond its limit, bulk excess oil will be observed. Likewise in w/o emulsions, the system can take up only so much water within the surfactant-coated droplets. But there is also a third class of microemulsions in which neither oil nor water can be regarded as inside or outside of the aggregates. These are referred to as bicontinuous (or more properly tricontinuous) liquids, in which the surfactant forms a flexible monolayer within the liquid separating oil from water microdomains, but without enclosing either. Thus both liquids, as well as the surfactant film, are continuous throughout the liquid, as depicted schematically in Figure 7.

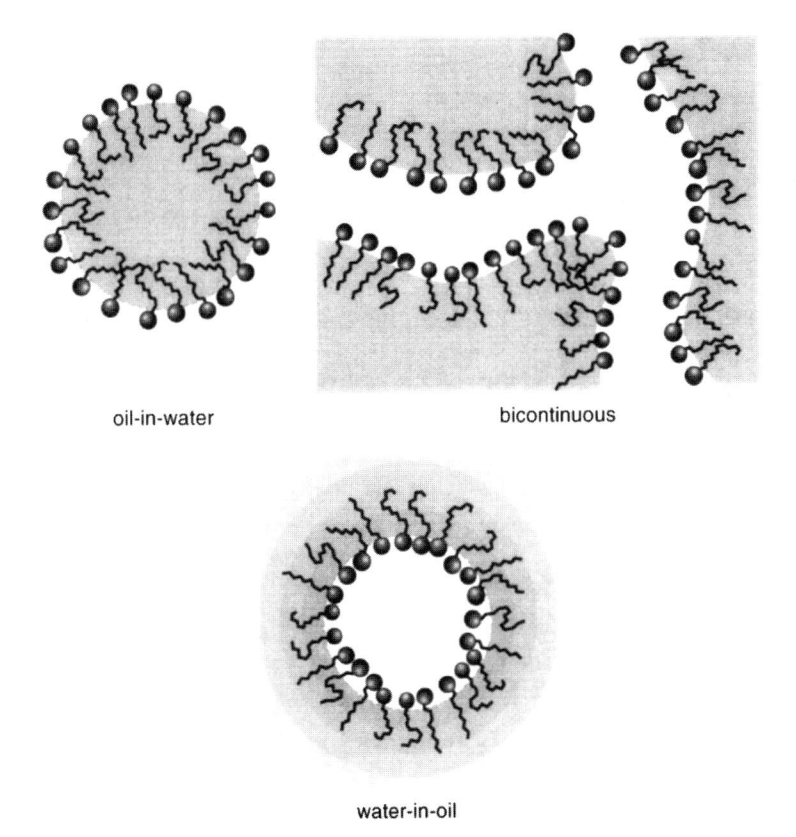

Figure 7 Schematic diagram of the various structures of microemulsions.

Bicontinuous structures are not as uncommon as one might think: a common sponge is an interconnected network into which water is absorbed to fill all the air space, and a bicontinuous microemulsion may be visualized by regarding the sponge material itself as the oil. The surfactant monolayer resides at the interface, lowering the surface tension and preventing direct contact between oil and water. Such microemulsions may have very high oil contents, but remain electrically conducting through the interconnected water pathways. Sponge phases are closely related to bicontinuous microemulsions, but are distinguished by the fact that they contain two discrete, interpenetrating water networks separated by a bilayer of surfactant swollen with oil. As with vesicles, many of these highly organized self-assembly structures have been examined using polymerizable oils. Bicontinuous structures, including cubic phases, microemulsions and sponge phases are regarded as candidates for the manufacture of novel gels and membranes with widely varied and controllable pore structures and porosities.

REFERENCES

1. Patrick, H.N. and Warr, G.G. (1996) *Self-Assembly Patterns in Double and Triple Chained Ionic Surfactants*, Chapter 2 in *Specialist Surfactants*, I.D. Robb (Ed.), Blackie Academic and Professional, Glasgow.
2. Schick, M.J. (Ed.) (1967) *Nonionic Surfactants*, Surfactant Science Series, Vol. 1, Marcel Dekker, New York.
3. Schmidt, D.L. (1995) *Nonaqueous Foams*, Chapter 7 in *Foams: Theory, Measurements, and Applications*, R.K. Prud'homme and S.A. Khan (Eds), Surfactant Science Series, Vol. 57, Marcel Dekker, New York.
4. Hunter, R.J. (1989) *Foundations of Colloid Science, Vols I and II*, Oxford University Press, Oxford.
5. Gilbert, R.J. (1995) *Emulsion Polymerization. A Mechanistic Approach*, Academic Press, New York.

Keywords: hydrophilic, hydrophobic, cationic, anionic, non-ionic, amphiphiles, adsorption, micelles, aggregates, vesicles, liquid crystals, microemulsions.

See also: Surfactants: applications in plastics.

Index

Accelerators 11, 198
Acid scavengers
   colour stabilization, their role in 47
   and crystallization 44
   DHT as 46, 48
   and film production 44
   and HALS stabilizers 44, 47
   lactates as 45
   metal stearates as 45, 48
   synthetic hydrocalcites as 44
   and Ziegler–Natta catalysts 44, 47
   zinc oxide as 46
Agricultural applications 8, 67
Algicides 125
Alkylamines, ethoxylated 112
Alumina trihydrate (ATH), *see* Aluminium hydroxide
Aluminium fillers 172, 174, 176–7, 250
Aluminium hydroxide
   effect on viscosity 288–9
   as flame retardant/smoke suppressant 287–90, 579–80
Amides as lubricants 453
Amines in epoxy resin cure 204–6
Anaerobic biodegradation 32
Analysis of additives 26, 80
Anatase, *see* Titanium dioxide
Anhydrides in epoxy resin cure 204
Antacids, *see* Acid scavengers
Anthraquinone dyes 219–20
Antiblocking agents 11, 49
   factors affecting effectiveness of 52–3
   incorporation of 53
   particle size of 53
   silica-based 52
   suppliers of 54
Antifogging agents 11
Antifouling additives 119
Antimicrobial agents, *see* Biocides
Antimony compounds 292–3, 327–9
   compared with tin compounds, for smoke emission 344
Antioxidants
   analysis of 80
   in biodegradable plastics formulations 138
   biological 70
   blooming of 69
   chain breaking variety (primary) 57–9, 65
   characterization 81
   classification 56
   colour stability 68, 70
   evaluation of 105–6
   extraction from polymers 81–3
   function 11, 56, 95
   function during processing 62
   leaching, prevention 69
   mechanisms of 57
   melt processing variety 62
   phenolic type 58, 62–3, 70, 73–8
   photo-antioxidants 64–7
      see also Light stabilizers
   preventive variety (secondary) 59, 61
   primary antioxidants 57–9, 96
   reactive antioxidants 69

Index

Antioxidants *contd*
  as scorch inhibitors 568–74
  secondary antioxidants 61, 96
  solubility in polymers 69
  synergistic effects 70–1
  thermooxidative antioxidants 67–8
  volatility 570–1
Antistatic agents 11
  anionic 113
  cationic 113
  effectiveness, measuring 113
  internal 109–10
  non-ionic 112
  polymeric 110
  suppliers of 113
Aramid fibres 556–7
Asbestos 550–1
Aspergillus niger, growth rates 129, 133
Autoxidation 55, 138, 140
Azine dyes 224
Azo compounds
  as blowing agents 145
  as dyes 218–19
  as pigments 493–5

Bactericides 125
Benzodifuranone dyes 224
Biocides 11
  a classification 118, 125
  measuring the effectiveness of 128–33
  requirements of 123
  toxicity of 119, 123–5
Biodegradation
  aerobic 32
  measures of 34
  molecular weight and 33
  monitoring 37
  of packaging film 136–7
  promotion of 135
  susceptibility of polymers to 35
  tests for completion of 36
Blends
  characterization 516
  miscibility 513
  preparation methods 515–16
  thermodynamics of formation 514

Block copolymers as compatibilizers 166
Blocking
  countering, *see* Antiblocking agents
  film blocking 49, 51
  measurement of 49–50
Blooming, *see* Migration
Blowing agents
  chemical 145–7
  function 12
  in intumescent systems 305
  suppliers of 147
Boron compounds
  examples listed 270
  mechanism of action of flame retardants 274
  suppliers of 273
  thermogravimetric analysis of 272
Bromine compounds, *see* Halogen compounds
Butadiene polymers as impact modifiers/toughening agents 379–81, 400–1

Calcium carbonate
  applications in non-PVC polymers 151–2
  applications in PVC 150–1
  natural forms 148
  synthetically precipitated 149
Carbodiimides 101
Carbon, varieties of 153
Carbon black, structure 154–5
Carbon fibres 553–4
Catalyst residues, and acidity 44
Cellulose, its role in char formation 300–2
Ceramic fibres 552
Chain-breaking donors 429
Char, *see under* Flame retardants
Chlorine compounds, *see* Halogen compounds
Chlorofluorocarbons 144, 147
Chromatography, column, for additives analysis 29, 83–4
Cling, in polymer film 108
Colour
  and carbon black 154, 156

Index

of epoxy resins, their diluents and modifiers 212
index 217
Compatibilizers 12, 609–10
   blends, for use with 166
   commercial examples of 540
   copolymers as 164–5
   functional polymer as 165
   mechanisms of 163
   polyethylene blends, for use with 166–7
   principle 536–7
   selection of 168
Compounding, see Mixing
Conducting polymers as additives 181–4
Conductivity, increased by carbon black 156–7
Cone calorimeter 322–4, 342, 582
Copper filler 171
Copper phthalocyanines 495–6
Coupling agents 12, 229–32
   see also Fibres; Silanes
Curing agents 12
   for bismaleimides 208–9
   for epoxies 203–7
   for phenolics 208
   for polyesters 199–202
   suppliers of 210
   for vinyl ester resins 202–3

DABCO as curing additive for bismaleimides 209
Defoaming agents 12
Density of fillers 245–6
Dichloromethane, see Methylene chloride
Diels–Alder adducts as flame retardants 327–38
Differential scanning calorimetry in analysis of additives 86
Diluents 12, 212
Dispersion/distribution of additives during mixing of carbon black 158–61
Doping agents 181
Dyes
   anthraquinone type 219–20
   azine 224
   azo type 218–19
   benzodifuranones 224
   methine and polymethine 223
   perinones 221–2
   photochromic 224–5
   quinopthalones 220–1
   sulfur type 223
   vat type 222–3
   see also under Pigments

Elastomers as impact modifiers 379–84, 400, 408–10
Electrical conductivity 171–2, 181, 187
   and shielding effectiveness 172
Electromagnetic shielding 170, 172–8, 185–6
   and antistatic agents 111
   costs of materials for 177
   health and safety considerations 177
   solder-like alloys 174–5
   suppliers of flakes and fibres for 179
Energy saving 8
Equivalent spherical diameter of filler particles 243–4
Esters as lubricants 454–5
Ethylene polymers as impact modifiers/toughening agents 381–3, 394–5
Excited-state quenchers 429
Exotherm modifiers 12
Extenders 214–15

Fatty acids
   as coatings for fillers 592–3
   as lubricants 455
Fibres
   coupling agents for 189–94
   critical length 549
   effect on mechanical properties 233–40
   electromagnetic shielding applications 174, 176
   length 234
   length reduction during mixing 228, 233
   mixing with plastics 226–7, 230–3
   in reinforcing plastics 545–50

Fibres *contd*
  suppliers of metallic fibres 179
  surface treatment 229–32
  see also Coupling agents *and under specific reinforcements*
Fillers 12
  coating 591–600
  density 245–6
  embrittlement by 257–9
  and hardness 245
  health and safety 251
  in intumescent systems 283
  metallic, *see under* Metal; Flakes; Fibres
  and modulus 246–7
  particle characteristics 242–5
  and refractive index 245
  rubber particles as 585–6
  surface energy 590
  surface treatment 195, 591–600
  and tensile deformation 253–7
  used in recycling 538–9
  see also Particle shape, Particle size *and under specific fillers*
Film blocking, *see* Blocking
Fire retardants, *see* Flame retardants
Flakes
  for electromagnetic shielding 175–6
  suppliers of 179
Flame retardants
  alumina trihydrate based, *see* Aluminium hydroxide
  antimony based, *see* Antimony compounds
  borate-based, *see* Boron compounds *and under* Zinc borates
  brominated systems, *see* Halogenated systems
  and char formation 265, 277–8, 297, 300–2, 309, 311–16, 325, 348–50
  chlorinated systems, *see* Halogenated systems
  and dilution of volatile products 265
  halogenated systems 327–38
  halogen-free systems 277–86
  and heat release rate 322–3
  and intumescent coatings 266, 298–300
  intumescent systems (not coatings) 300, 303
  iron compounds, *see under* Iron
  magnesium hydroxide based, *see* Magnesium oxide and hydroxide
  mode of action of 263–6
  performance of, assessing 261–2, 278–9, 341–2
  phosphorus type, *see under* Phosphorus compounds
  polyvinyl alcohol 315–19
  selection of 267
  silicon compounds 319–24
  stannates, *see* Zinc stannates
  synergistic systems 327–30, 333–8
  zinc compounds 270, 272–4, 294–5
Fluoropolymers 519–20
  effect on finished properties 523
  see also Processing aids
Foam catalysts 12
Foam control agents 608
Fragrance modifiers, *see* Odour modifiers
Fumigants 122
Fungicides 125–7

Glass fibres 226–30, 551–2
Glass-forming compounds in intumescent systems 305–6
Glycidyl ethers as reactive diluents for epoxies 213
Graphite fibres, *see* Carbon fibres

Halogen compounds
  in synergistic flame retardants 327–38
  and tin compounds 343
HALS
  early development 353
  interaction with hydrogen halides 366, 367
  interaction with phenolic antioxidants 364
  interaction with pigments 366
  interaction with thioethers 366
  mechanisms 354–6
  monomeric and oligomeric 361

performance improvement by acid
    scavengers 47–8
  as photoantioxidants 66
  polymer bound 359
  selection 357
  structures 355, 369–70
  suppliers 370
  UV stabilization 361–2
Heat stabilizers 13
Heterocyclic compounds as
    biocides 118
Hexamethylene tetramine in phenolic
    resin cure 208
Hindered amines
  hindered piperidine as light
      stabilizers 435–8
  as optical brighteners 472–3
  see also under HALS
Hindered phenol antioxidants 63, 70,
    73ff, 98–9
  structures 74–8
Hollow microspheres 13
  applications 374
  and mechanical properties 373
  production 373
Hydrocarbons as lubricants 453–4
Hydrochlorofluorocarbons 144, 147
Hydrogen chloride scavengers 59–60
Hydroperoxides
  and crosslinking of polyesters 200
  and crosslinking of vinyl ester
      resins 203
  and polymer oxidation 55–6, 64, 95
Hydroxybenzoate stabilizers 431

Impact modifiers 13, 379–84
  core/shell systems 391, 410
  effect on the processing of
      thermosets 420–1
  effect on other properties of
      thermosets 423
  for engineering plastics 393
  measurement of effectiveness 387–9,
      399–400, 406–8, 421–2
  morphological features 390
  in recycling 539–40
  for SMC 424–5
  suppliers 393

see also under Toughening *for
    mechanism, etc. and under specific
    impact modifier type*
Impact strength
  effect of filler surface treatment
      600–1
  of engineering plastics 396
  of recycled plastics 163
Infrared spectroscopy, *see* Spectroscopy
Inhibitors of free radical cure
    reactions 198
Interphase thickness, effect on
    properties of glass fibre
    composites 238–40
Intumescent systems 281–3, 300–6
Iron oxides 295–6, 307–8, 310, 330

Kaolin 248
'Kicker' chemicals in blowing agent
    compositions 145

Lactates as acid scavengers 45–6
Latent acid curing agents 208
Lewis acids as epoxy curing
    additives 204, 207
Light stabilizers 13
  a classification 427
  pigment type 428
  selection of 440
  synergism and antagonism 438–40
  see also under HALS; Antioxidants
Limiting oxygen index 325, 331–7, 341
  of borate-based PVC
      compositions 273
  of phosphate-based
      polypropylene 282
Liptov's model for interphase design in
    glass fibre composites 238
Low profile additives 13, 442
  applications 443
  formulations 444
  mechanism of shrinkage control 447
  suppliers 448
Lubricants 13, 450, 606–8
  classification 453–6
  effect on processing 457–8
  fluoropolymers as 520–22
  testing 457

Magnesium oxide and hydroxide 250, 290–1, 330
Maleic anhydride in polypropylene–glass coupling 231
Masterbatch, manufacture of 159
Mechanical properties
  of filled plastics 252–9
  of glass fibre reinforced plastics 233–40
Melamine 285
Melt flow index, changes during recycling 537–8
Melt stabilization, by acid scavengers 47
Metal complexes as stabilizers 432–5
Metal deactivators 59, 68
Metal filaments, fibres 553
Metal flakes, *see under specific metals*
Metal hydroxides as flame retardants 287–92
Metal ions, their role in polymer oxidation 97
Metal stearates
  as acid scavengers 45–8
  as lubricants 455–6
Metals, finely divided
  geometry effects 170
  health and safety 177
Metals, sources of: flake, powder, filaments 179
Metallic soaps as lubricants 453
Methine dyes 223
Methylene chloride, as blowing agent 144
Mica
  applications 462–3
  composition 459
  mineral deposits 460
  surface treatment 460–1
Microorganisms in plastics 115
  effects of 117
  polymers and additives with susceptibility to them 116
Microspheres, *see* Hollow microspheres
Migration
  of antioxidants 69

of antistatic agents 110
Mixing additives into polymers
  antiblocking agents 53
  carbon black 159–60
  compatibilizers 168
  by extruder 21–5
  fibres 226–7
  by internal mixer 19–21
  properties, effect on 162, 169
  by two-roll mill 16–19
  using pellets 158
Molybdenum trioxide 295

Natural fibres 554
Nickel complexes as antioxidants 60–1, 65
Nickel fillers 176–7
Nitroxyl radicals, their stabilizing role 66
Nucleating agents 13, 466
  role of acid scavengers 44
Nucleation 464–71
  of polymer blends 470–1
Nylon fibres, *see* Polyamide fibres

Odour modifiers 13
Optical brighteners 13, 472–3
Organometallic biocides 118
Organophosphorus compounds, *see under* Phosphorus compounds
Organosilanes, *see* Silanes
Organosulfur biocides 118
Ortho-hydroxy aromatic stabilizers 429
Outdoor applications, protection for 7, 65
Oxidation reactions in polymers, catalysis by metal ions 56
Oxidative induction time (OIT) 86–7
Oxygen index, *see* Limiting oxygen index

Paper
  laminating process 476–8
  for resin bonded laminates 474, 479–83
Parahydroxybenzoate stabilizers 431
Particle shape 242–4

Particle size
  antiblocking agents 53
  carbon black 154
  distribution 244
  fillers 242
  surfaces 244–5
Pentane blowing agents 144
Percolation threshold for
    conductivity 171, 184–5
Perinone dyes 221–2
Peroxide-decomposing
    antioxidants 59
Peroxides 14
  activities of 201
  as radical cure initiators 199–200
PET, antioxidants for 97
Phenolic antioxidants, in relation to
    scorch inhibitors 568–9
Phenolic biocides 118
Phosphate based dispersing agents for
    fillers 597, 602
Phosphorus compounds
  in intumescent systems 281–3
  organophosphorus compounds as
      flame retardants 283–4
  phosphites as antioxidants 61, 70,
      99–101
  red phosphorus 279–80
Photo-antioxidants 60, 64
  testing of 66
  see also under HALS
Photochromic dyes 224–5
Photodegradable plastics 137
Phthalate esters as plasticizers 501–2
Pigments 5, 217
  black 491
  coloured inorganic 491–3
  dispersion 488
  fastness 487
  fluorescent 497
  inorganic 489
  metallic 497–8
  migration 487–8
  organic 493–7
  pearlescent 497
  requirements 486
  toxicological and environmental
      considerations 489
  white 490–1
Piperidines as antioxidants 66
Plasticizers 14, 499
  compatibility 215
  for epoxies 214–15
  LD_{50} values of 505
  liver effects 506
  mode of action 500–1
  for PVC 501–2
  reproductive effects 507
  requirements of 502
Polyacrylates as impact modifiers/
    toughening agents 384–5
Polyamide fibres 556
  see also Aramid fibres
Polyaniline as conducting additive 186
Polyester fibres 556
Polyethylene fibres 555
Polymer-bound additives 9
Polypropylene fibres 554
Polypyrrole as conducting
    additive 185–6
Polysiloxanes as impact modifiers/
    toughening agents 384
Polyurethanes as impact modifiers/
    toughening agents 383–4
Polyvinyl alcohol 315–19, 560
Processing aids 4, 14, 519, 526
  effect on melt homogeneity 529–31
  effect on melt strength and
      elasticity 532
  effect on PVC fusion 528
  for vinyl foam processes 527
  see also Lubricants

Quartz 249
Quaternary ammonium compounds as
    biocides 118
Quinacridone pigments 496
Quinones 57
Quinopthalone dyes 220–1

Radioisotopes in biocide
    evaluation 128
Reactive diluents 213–14
Recycling
  basic concepts of 536–7
  effect on impact strength 162

Release agents 14, 559–60
  external variety 559–60
  internal variety 559
Rice husk ash 561
  economic and ecological
    advantages 565
  effect on mechanical properties 562–5
Rubber particles as fillers 585–6
  effect on polyurethane
    processing 587–8
  effect on polyurethane
    properties 586–7
Rubbers as impact modifiers, see under
  Elastomers
Rutile, see Titanium dioxide

Scorch inhibitors
  concept 567
  mechanisms 568
  phenolic antioxidants, in relation
    to 568–9
Separation of additives for
  analysis 28–9
Shielding, EM, see Electromagnetic
  shielding
Silanes, as coupling agents 191–4,
  593–5
Silica filler 248–9
  synthetic, for antiblocking 52
Silicates 247–8
Silicon compounds as flame
  retardants 319–24
Slip agents 14
Smoke
  generation 308–11, 576–7
  NBS smoke box test 342, 581
  suppressants 14, 577–9
  test methods 580–2
Solvent extraction of additives 28–9,
  81–5
Spectroscopy
  GC-MS, in analysis of additives 31,
    83–4
  infrared, in analysis of additives
    88–92
  near infrared 89–90
  in the study of filler coatings (DRIFT
    and XPS) 598

X-ray fluorescence (XRF) 92–4
Stabilizers 610–11
Starch, in biodegradable plastics 139
Static electricity 108–9, 180–1
Stearates as lubricants 455–6
Structure (of carbon black) 154–5
Sulfides in antioxidant systems 61
Sulfur dyes 223
Supercritical fluids in analysis of
  additives 84–5
Surface area of carbon black 154
Surface energy of fillers 590
Surfactants 14, 604
  amphiphilic character 617
  as antistatic agents 111, 605–6
  as compatibilizers 609–10
  conformation of 615
  critical micelle concentration 618–19
  description of 613–14
  electrostatic interaction 616
  in emulsion polymerization 622–4
  liquid crystal formation 620
  as lubricants 606–8
  as stabilizers 610–11
  as wetting agents 608–9
Synthetic hydrotalcites, see under
  Acid scavengers

Talc 330–3
Thermal analysis of antioxidants 86
Thermal degradation of PET 103–5
Thermal separation of additives for
  analysis 30
Thermogravimetric analysis of
  borates 272
Thermo-oxidative degradation (of
  PET) 102–3
Thermoplastics as impact
  modifiers 401–2, 404, 410–12
  in combination with rubbers 402–3
  see also under Ethylene polymers, etc.
Thickening agents 14
Thioesters as antioxidants 61
Thiolates as antioxidants 65
Tin compounds
  as antifouling additives 119
  compared with antimony trioxide,
    for smoke emission 344

as flame retardants 339–41
and halogens 343
mechanism of flame retardancy 347
synergistic systems 342–6
Tin (IV) oxide as flame retardant 340
Titanates 596
Titanium dioxide 249–50
Toughening
  additives 379
  criteria for 389
  measuring extent of, *see under* Impact modifiers
  mechanisms 376, 389–90, 412–15
  microstructural features of 417–18
  see also under Impact modifiers
Toxicity
  of biocides 119, 123–5
  of phthalates 506–10
  of pigments 218, 489
Transition metals
  in biodegradable plastics systems 138, 140
  complexes, photostability of 65
Trimellitate ester plasticizers 503

Ultraviolet quenchers 429

Ultraviolet stabilizers 15, 60, 65, 157–8, 427
  mechanisms 430
  see also Light stabilizers

Vat dyes 223
Viscosity modifiers 212, 214
Vitamin E 70
Volume resistivity 171–2, 186

Waxes 560
Weathering and antioxidants 65–7
Wetting agents 15, 608–9
Wire and cable, flame retardant 330–2
Wollastonite 247–8

X-ray fluorescence (XRF) 92–4

Zinc borates as flame retardants 294–5, 328–9
Zinc pyrithione based biocides 118
Zinc stannates, hydroxystannates as flame retardants 294, 329–30, 340–1, 350–1
Zirconates 596–7

Property of
American Plastics Council
Automotive Learning Center